Compound Poisson Approximation

Compound Poisson approximation appears naturally in situations where one deals with a large number of rare events. It has important applications in insurance, extreme value theory, reliability theory, mathematical biology, etc. *Compound Poisson Approximation* synthesizes the most important recent research in the field in a single volume. With an extensive list of references, open problems, and exercises, it will become the standard reference book on the topic.

Features

- Provides a comprehensive overview of this rapidly expanding field
- Synthesizes the most important research results of recent years
- Presents an array of special topics
- Provides the reader with a set of tools needed for research and education

The book is of interest to researchers and postgraduate students from probability, statistics, and mathematics.

V. Čekanavičius is a professor of mathematics at the Institute of Applied Mathematics, Faculty of Mathematics and Informatics, Vilnius University. He is the author of *Approximation Methods in Probability Theory*, Springer, 2016.

S. Y. Novak works at Middlesex University London. He is the author of *Extreme Value Methods with Applications to Finance*, Chapman & Hall/CRC Press, 2011.

Compound Poisson Approximation

V. Čekanavičius and S. Y. Novak

CRC Press
Taylor & Francis Group
Boca Raton London New York

CRC Press is an imprint of the
Taylor & Francis Group, an **informa** business
A CHAPMAN & HALL BOOK

First edition published 2025
by CRC Press
2385 Executive Center Drive, Suite 320, Boca Raton, FL 33431, U.S.A.

and by CRC Press
4 Park Square, Milton Park, Abingdon, Oxon, OX14 4RN

CRC Press is an imprint of Taylor & Francis Group, LLC

Library of Congress Cataloging-in-Publication Data
Names: Čekanavičius, Vydas, author. | Novak, Serguei Y., author.
Title: Compound poisson approximation / V. Čekanavičius & S.Y. Novak.
Description: First edition. | Boca Raton, FL, U.S.A. : CRC Press, 2024. | Includes bibliographical references and index. | Summary: "Compound Poisson Approximation appears naturally in situations where one deals with a large number of rare events. It has important applications in insurance, extreme value theory, reliability theory, mathematical biology, etc. This book synthesizes the most important recent research in the field in a single volume. With an extensive list of references, open problems and exercises, it will become the standard reference book on the topic. Features: Provides a comprehensive overview of this rapidly expanding field Synthesizes the most important research results of recent years Presents an array of special topics Provides the reader with a set of tools needed for research and education The book is of interest to researchers and postgraduate students from probability, statistics, and mathematics"-- Provided by publisher.
Identifiers: LCCN 2023057329 (print) | LCCN 2023057330 (ebook) | ISBN 9781032762562 (hbk) | ISBN 9781032763521 (pbk) | ISBN 9781003478164 (ebk)
Subjects: LCSH: Poisson distribution. | Approximation theory.
Classification: LCC QA273.6 .C424 2024 (print) | LCC QA273.6 (ebook) | DDC 519.2/4--dc23/eng/20240326
LC record available at https://lccn.loc.gov/2023057329
LC ebook record available at https://lccn.loc.gov/2023057330

ISBN: 978-1-032-76256-2 (hbk)
ISBN: 978-1-032-76352-1 (pbk)
ISBN: 978-1-003-47816-4 (ebk)

DOI: 10.1201/9781003478164

Typeset in Nimbus Roman
by KnowledgeWorks Global Ltd.

Publisher's note: This book has been prepared from camera-ready copy provided by the authors.

To my wife, Virgilija,
always supportive and
frequently the driving
force in my achievements

V. Čekanavičius

Ksenia Ivanovna

To my great
grandmother,
Ksenia Ivanovna

S.Y. Novak

Contents

Preface

Compound Poisson approximation appears naturally in situations where one deals with a large number of rare events. It has important applications in insurance, extreme value theory, reliability theory, mathematical biology, and more. It is a backbone of the vast and important topic of infinitely divisible approximation to the distribution of a sum $S_n = X_1 + \cdots + X_n$ of independent random variables $X_1, ..., X_n$ (Kolmogorov's problem).

Yet the merits of compound Poisson approximation appear barely recognisable.

Consider, for instance, an example where S_n has Binomial distribution $\mathbf{B}(n,p)$. Recall that $\mathbf{B}(n,p)$ often plays the role of a "testing stone" in Probability Theory: the central limit theorem and the Poisson approximation were first derived in the case of a Binomial distribution; Esseen [116] used it in order to establish a lower bound to the constant in the Berry–Esseen inequality; and it was involved in establishing a particular property of the distribution of Student's statistic (see [221], Example 12.3), etc.

The accuracy of normal approximation to $\mathbf{B}(n,p)$, according to the Berry–Esseen inequality, is

$$c/\sqrt{np}\,,$$

where c is an absolute constant. Note that bound $c/\sqrt{np}$ looses sense when p is "small".

The accuracy of Poisson approximation to $\mathbf{B}(n,p)$ is

$$\sup_{0\le p\le 1/2}\ \sup_A |\mathbb{P}(S_n \in A) - \mathbb{P}([np^2]+\pi_{n,p} \in A)| \le c/\sqrt{n}$$

(cf. Theorem 3.2), where $\pi_{n,p}$ is a Poisson random variable with parameter $np(1{-}p){+}\{np^2\}$ and c is an absolute constant. Clearly, bound $c/\sqrt{n}$ is preferable to $c/\sqrt{np}$.

Even sharper is the accuracy of compound Poisson approximation to $\mathbf{B}(n,p)$: according to Theorem 7.1, there exists an absolute constant c and a compound Poisson random variable $\zeta_{n,p}$ such that

$$\sup_{0\le p\le 1/2}\ \sup_A |\mathbb{P}(S_n \in A) - \mathbb{P}(\zeta_{n,p} \in A)| \le cn^{-2/3}\,.$$

This example illustrates the advantages of compound Poisson approximation. However, the fact that compound Poisson approximation can be preferable to normal approximation remains virtually unknown to researchers. Textbooks mention normal and Poisson approximation but not the fact that Poisson approximation may have advantages over the normal approximation; advantages of compound Poisson approximation are not mentioned at all.

We would like to attract attention of researchers to the merits of compound Poisson approximation.

The book aims to bring a reader to the edge of research on the topic of compound Poisson approximation. It provides a comprehensive overview of the vast and actively growing field of Probability Theory – the theory of compound Poisson approximation. It equips a reader with relevant facts and tools and encourages further research.

We believe the popularity of compound Poisson approximation will eventually overcome that of normal approximation. We wish our readers all the best in establishing new results concerning the accuracy of compound Poisson approximation.

Authors

V. Čekanavičius is a professor of mathematics at the Institute of Applied Mathematics, Faculty of Mathematics and Informatics, Vilnius University. He is the author of *Approximation Methods in Probability Theory*, Springer, 2016.

S.Y. Novak works at Middlesex University London. He is the author of *Extreme Value Methods with Applications to Finance*, Chapman & Hall/CRC Press, 2011.

Notation and abbreviations

$\alpha(\cdot)$, $\beta(\cdot)$, $\varphi(\cdot)$	mixing coefficients
$b_n = O(a_n)$	$\limsup_{n\to\infty} b_n/a_n < \infty$
$b_n = o(a_n)$	$\lim_{n\to\infty} b_n/a_n = 0$
ch.f.	characteristic function
CP	compound Poisson
C	absolute constant
$\mathbb{CP}$	class of (shifted) compound Poisson distributions
Δg	$\Delta g(\cdot) = g(\cdot+1) - g(\cdot)$
$\Delta^2 g$	$\Delta(\Delta g)$
$d_{TV}(X;Y)$	total variation distance
$d_K(X;Y)$	uniform (Kolmogorov's) distance
$d_G(X;Y)$	Gini-Kantorovich distance
$\mathcal{D}$	class of infinitely divisible distributions
$\exp(M)$	exponent of measure M
F^{*n}	n-fold convolution of F
$\mathcal{F}$	class of all distributions
$\mathcal{F}_s$	class of symmetric distributions
$\mathcal{F}_\alpha \subset \mathcal{F}_s$	class of distributions with ch.f. φ obeying $\varphi(t) \geq -1+\alpha$ $(\forall t)$
$\mathcal{F}_+$	the class of distributions with non-negative characteristic functions
$\mathcal{F}_{0,\beta}$	the class of zero-mean distributions such that $\mathbb{E}\|X\|^\beta < \infty$
i.i.d.	independent and identically distributed
i	imaginary unit: $\mathrm{i}^2 = -1$
I_a	distribution concentrated at a
I	I_0
$\mathcal{L}(X)$	distribution of r.v. X
LHS, RHS	the left-hand side, the right-hand side
$M * V$	convolution of measures M and V
$\widehat{M}$	Fourier transform of measure M
MB	Markov Binomial distribution
$\\|M\\|_{TV}$	the total variation of measure M
$\mathbb{N}$	the set of natural numbers
π_λ	Poisson random variable with parameter λ
$\mathbf{\Pi}(\lambda)$	Poisson distribution
$\mathbf{\Pi}(\lambda, X)$	compound Poisson distribution
$Q_F(\cdot)$	concentration function of distribution F
$\mathbb{R}$	the set of real numbers
r.v.	random variable
r.v.s	random variables

S_n	$X_1 + X_2 + \cdots + X_n$
$\tilde{S}_n$	sum of accompanying random variables
SCP	signed compound Poisson
Θ	any measure satisfying $\|\Theta\|_{TV} \leq 1$
$\lfloor x \rfloor$	integer part of x
w. p. 1	with probability 1
w.l.o.g.	without loss of generality
$\mathbb{Z}$	the set of integer numbers
$\mathbb{Z}_+$	$\mathbb{N} \cup \{0\}$
$\square$	end of proof

1

Preliminaries

1.1 Notation

Let $\mathcal{L}(X)$ denote the distribution of a random variable (r.v.) X. By an integer-valued r.v. we usually mean a r.v. taking values in $\mathbb{Z}$ with a maximal span equal to 1.

Given random variables (r.v.s) $X_1, ..., X_n$, we denote

$$S_n = X_1 + \cdots + X_n .$$

Distributions of r.v.s $X_1, ..., X_n$ may depend on n, i.e., $\{X_1, \ldots, X_n\}_{n\geq 1}$ may be a triangular array of r.v.s.

Measure M on $\mathbb{R}$ (equipped with a particular sigma-algebra $\mathcal{A}$) is called *finite* if

$$-\infty < \inf_{A\in\mathcal{A}} M(A) \leq \sup_{A\in\mathcal{A}} M(A) < \infty.$$

Measure Q on $\mathbb{R}$ is called a *signed measure* if there exists a measurable set A such that $Q(A)<0$.

Signed measure Q on $\mathbb{R}$ is called a *unit measure* if $Q(\mathbb{R})=1$.

If M is a finite measure, let

$$\widehat{M}(t) = \int_{-\infty}^{\infty} \mathrm{e}^{\mathrm{i}tx} M(\mathrm{d}x) \qquad (t\in\mathbb{R})$$

denote its *Fourier transform.*

If F_X is a d.f. of $\mathcal{L}(X)$, then $\widehat{F}_X$ denotes the characteristic function (ch.f.) of F_X.

Since $\widehat{F}_X$ is continuous and $\widehat{F}_X(0)=1$, a logarithm of $\widehat{F}_X$ can be defined in a neighbourhood of 0. If the kth moment of $\mathcal{L}(X)$ exists, then kth *cumulant* is

$$\kappa_k = \mathrm{i}^{-k}\frac{\mathrm{d}^k}{\mathrm{d}t^k}\ln \widehat{F}_X(t)\Big|_{t=0}.$$

One can check that

$$\kappa_1 = \mathbb{E}X, \quad \kappa_2 = \mathrm{var}\,X, \quad \kappa_3 = \mathbb{E}(X-\mathbb{E}X)^3, \quad \kappa_4 = \mathbb{E}(X-\mathbb{E}X)^4 - 3\mathrm{var}\,X.$$

For any $x\in\mathbb{R}, k\in\mathbb{N}$,

$$x^{(k)} = x(x-1)\cdots(x-k+1)$$

is called the *kth* factorial of x. We set $x^{(0)} = 1$.

If r.v. X takes values in $\mathbb{Z}_+$, then $\mathbb{E}X^{(k)}$ is called the *kth* factorial moment of X. Factorial moments appear in Taylor's expansion of the factorial moment generating function

$$\mathbb{E}(1+t)^X = 1 + t\mathbb{E}X + \frac{t^2}{2!}\mathbb{E}X^{(2)} + \frac{t^3}{3!}\mathbb{E}X^{(3)} + \cdots$$

DOI: 10.1201/9781003478164-1

Factorial cumulants $\varkappa_k$ come from Taylor's expansion of the logarithm of the moment generating function:

$$g(t) \equiv \ln \mathbb{E}(1+t)^X = \varkappa_1 t + \frac{t^2}{2!}\varkappa_2 + \frac{t^3}{3!}\varkappa_3 + \cdots \qquad (t \geq 0).$$

For example,

$$\varkappa_1 = \mathbb{E}X, \quad \varkappa_2 = \operatorname{var} X - \mathbb{E}X, \quad \varkappa_3 = \mathbb{E}X^{(3)} - 3\mathbb{E}X\mathbb{E}X^{(2)} + 2\mathbb{E}^3 X.$$

If $\mathcal{L}(X) = \mathbf{\Pi}(\lambda)$, then $g(t) = \lambda t$, hence $\varkappa_1 = \lambda$, $\varkappa_k = 0$ $(\forall k > 1)$.

Let $\mathbf{\Pi}(\lambda)$ denote a Poisson distribution with parameter λ. We usually denote by π_λ a Poisson $\mathbf{\Pi}(\lambda)$ random variable (r.v.): $\mathbb{P}(\pi_\lambda = k) = \mathrm{e}^{-\lambda}\lambda^k/k!$ $(k \in \mathbb{Z}_+)$.

Random variable Y has a *compound Poisson distribution* $\mathbf{\Pi}(\lambda, X) \equiv \mathbf{\Pi}(\lambda, \mathcal{L}(X))$ if

$$Y \stackrel{d}{=} X_0 + \ldots + X_{\pi_\lambda}, \tag{1.1}$$

where Poisson $\mathbf{\Pi}(\lambda)$ random variable π_λ is independent of $\{X_i\}_{i \geq 1}$, $X_0 = 0$, random variables $X, X_1, X_2, \ldots$ are independent, $X_i \stackrel{d}{=} X$ $(i \geq 1)$.

Compound Poisson distribution dates back to Lundberg [200] and Khintchine [175], Chapter 2.

The characteristic function of $\mathbf{\Pi}(\lambda, X)$ is

$$\widehat{\mathbf{\Pi}}(\lambda, X) = \exp(\lambda(\varphi_X - 1)),$$

where φ_X is a ch.f. of $\mathcal{L}(X)$. We call $\mathcal{L}(X)$ a compounding or multiplicity distribution.

A compound Poisson distribution with a geometric multiplicity distribution is called sometimes a Pólya-Aeppli distribution, cf. [167], p. 410.

Given a set of non-negative numbers $\{\lambda_j\}_{j \geq 1}$ such that $\lambda := \sum_{j \geq 1} \lambda_j < \infty$, denote

$$Z = \sum_{j=1}^{\infty} j \pi_{\lambda_j}, \tag{1.2}$$

where $\{\pi_{\lambda_j}\}$ are independent Poisson $\mathbf{\Pi}(\lambda_j)$ random variables ($\pi_{\lambda_j} \equiv 0$ if $\lambda_j = 0$). Then Z is a compound Poisson random variable with characteristic function

$$\mathbb{E}\exp(\mathrm{i}tZ) = \exp\left(\sum_{j=1}^{\infty} \lambda_j (\mathrm{e}^{\mathrm{i}tj} - 1)\right). \tag{1.3}$$

In other words, $Z \stackrel{d}{=} X_0 + \ldots + X_{\pi_\lambda}$, where $\mathbb{P}(X = j) = \lambda_j/\lambda$.

Negative Binomial distribution $\mathbf{NB}(r,p)$. Random variable $S_{r,p}$ has distribution $\mathbf{NB}(r,p)$ with parameters $p \in (0,1)$ and $r > 0$ if

$$\mathbb{P}(S_{r,p} = j) = \frac{\Gamma(r+j)}{\Gamma(r)\, j!}(1-p)^r p^j \qquad (j \geq 0), \tag{1.4}$$

where

$$\Gamma(y) = \int_0^\infty x^{y-1}\mathrm{e}^{-x}\,\mathrm{d}x.$$

The characteristic function of $\mathbf{NB}(r,p)$ is $(1-p)^r/(1-p\mathrm{e}^{\mathrm{i}t})^r$. It is known that the Negative Binomial distribution is a particular compound Poisson distribution.

If $n \in \mathbb{N}$, then $S_{n,p} \stackrel{d}{=} \xi_1 + ... + \xi_n$, where $\xi_1, ..., \xi_n$ are independent r.v.s with geometric $\mathbf{\Gamma}_0(p)$ distribution,

$$\mathbb{E}S_{n,p} = np/(1-p), \ \ \mathrm{var}\, S_{n,p} = np/(1-p)^2 .$$

Convolution $P * G$ *of measures* P *and* G. We denote by $P{*}G$ the convolution of finite variation measures P and G: for an arbitrary Borel set A

$$(P{*}G)(A) = \int P(A-x)G(\mathrm{d}x).$$

It is known that $\widehat{P{*}G} = \widehat{P}\widehat{G}$. If P and G are concentrated on integers, then

$$(P{*}G)(k) = \sum_{j=-\infty}^{\infty} P(k-j)G(j).$$

For any independent r.v.s X and Y, the distribution of the sum $\mathcal{L}(X+Y)$ is the convolution of the distributions: $\mathcal{L}(X+Y) = \mathcal{L}(X) * \mathcal{L}(Y)$.

Exponent of a measure. Let $I \equiv I_{\mathrm{o}}$ denote the distribution concentrated at 0, i.e.,

$$I(\{0\})=1, \ \ I(\mathbb{R}\backslash\{0\})=0.$$

In the multivariate case I denotes the distribution concentrated at $\bar{0} = (0, \ldots, 0)$. Let I_a be the distribution concentrated at a.

Given a finite measure P, we denote

$$\exp(P) := \sum_{j\geq 0} P^{*j}/j! = I + \sum_{j\geq 1} P^{*j}/j!$$

Here powers $P^j \equiv P^{*j}$ are understood in the convolution sense, $P^0 \equiv P^{*0} = I$.

Note that for any $a \in \mathbb{R}$

$$\exp(aI) = \mathrm{e}^a I, \ \ \exp(P - aI) = \mathrm{e}^{-a} \exp(P).$$

Poisson distribution $\mathbf{\Pi}(\lambda)$ can be presented as

$$\mathbf{\Pi}(\lambda) = \exp\left(\lambda(I_1 - I)\right) = \mathrm{e}^{-\lambda} \exp(\lambda I_1).$$

Compound Poisson distribution $\mathbf{\Pi}(\lambda, X)$ with $P_X = \mathcal{L}(X)$ can be presented as

$$\mathbf{\Pi}(\lambda, X) = \exp\left(\lambda(P_X - I)\right) = \mathrm{e}^{-\lambda} \exp(\lambda P_X).$$

If P and Q are two finite measures, then

$$\mathrm{e}^P * \mathrm{e}^Q = \mathrm{e}^{P+Q}, \ \ \ \mathrm{e}^P = I + P * \int_0^1 \mathrm{e}^{tP} \mathrm{d}t. \tag{1.5}$$

More generally, for any $k \in \mathbb{Z}_+$

$$\mathrm{e}^P = I + P + P^{*2}/2 + \ldots + \frac{P^{*k}}{k!} + \frac{P^{*(k+1)}}{k!} * \int_0^1 \mathrm{e}^{tP}(1-t)^k \mathrm{d}t. \tag{1.6}$$

One can verify (1.6) using the definition of e^P. Indeed, (1.5) means (1.6) is valid if $k=0$. We can use induction. If (1.6) is valid for a particular $k \geq 0$, then substituting equation $\mathrm{e}^P = I + P * \int_0^1 \mathrm{e}^{tP} \mathrm{d}t$ into (1.6) shows (1.6) is valid with $k+1$ instead of k.

In particular (Θ denotes a measure such that $\|\Theta\|_{TV} < 1$),

$$\exp((P-I)) - P = (P-I)^{*2} * \int_0^1 \mathrm{e}^{t(P-I)}(1-t)\mathrm{d}t = \tfrac{1}{2}(P-I)^{*2} * \Theta. \tag{1.6*}$$

If $\|F - I\|_{TV} < 1$, then the logarithm of F can be defined as

$$\ln F := \sum_{j=1}^{\infty}(-1)^{j+1}(F-I)^{*j}/j.$$

Example 1.1 Let $\mathbb{P}$ be a unit measure $\mathbb{P}(\{0\}) = 1+p$, $\mathbb{P}(\{1\}) = -p$, where $p \in (0;1)$. One can say that there is a "generalised random variable" τ with "probabilities"

$$\mathbb{P}(\tau=0) = 1+p, \quad \mathbb{P}(\tau=1) = -p \qquad (p \in (0;1)).$$

The Fourier transform of this unit measure is

$$\varphi(t) = 1+p-p\mathrm{e}^{\mathrm{i}t}.$$

Note that $\varphi(\pi) = 1+2p > 1$, $\mathrm{var}\,\tau = -p(1+p)$. □

Example 1.2 We denote by $\mathbf{\Pi}(-\lambda, X)$ a signed compound Poisson (SCP) unit measure:

$$\mathbf{\Pi}(-\lambda, X) \equiv \mathbf{\Pi}(-\lambda, \mathcal{L}(X)) = \exp(-\lambda(P_X - I)) = \mathrm{e}^{\lambda}\exp(-\lambda P_X),$$

where $P_X = \mathcal{L}(X)$, $\lambda > 0$. In particular, $\mathbf{\Pi}(-\lambda) = \exp(-\lambda(I_1 - I))$ is a signed Poisson unit measure:

$$\mathbf{\Pi}(-\lambda)(\{k\}) = \mathrm{e}^{\lambda}(-\lambda)^k/k! \ (k \in \mathbb{Z}_+).$$

The Fourier transform of $\mathbf{\Pi}(-\lambda, X)$ is

$$\widehat{\mathbf{\Pi}}(-\lambda, X) = \exp(-\lambda(\mathrm{e}^{\mathrm{i}t} - 1)).$$

With some abuse of notation we denote by $\pi_{-\lambda}$ a "generalised Poisson random variable" meaning we deal with Poisson unit measure $\mathbf{\Pi}(-\lambda)$. Saying that π_λ, π_μ are "independent" Poisson "r.v.s" means we deal with a convolution of Poisson unit measures. In particular, $\pi_2 + \pi_{-1}$ is a Poisson $\mathbf{\Pi}(1)$ r.v., though π_{-1} is not (π_{-1} is a "generalised r.v."). □

Accompanying distribution. Given a r.v. X, let $\pi_1, X_1, X_2, \ldots$ be independent r.v.s, where π_1 is a Poisson $\mathbf{\Pi}(1)$ r.v., $X_i \stackrel{d}{=} X$ $(\forall i)$. Set

$$\tilde{X} = \sum_{j=1}^{\pi_1} X_j. \tag{1.7}$$

Then $\mathcal{L}(\tilde{X})$ is called an "accompanying distribution", $\tilde{X}$ is called an accompanying X r.v. (cf. Gnedenko-Kolmogorov [135]). The definition is valid for random elements taking values in a general measurable space as well. Clearly,

$$\mathcal{L}(\tilde{X}) = \mathbf{\Pi}(1, X) = \exp(\mathcal{L}(X) - I).$$

If $\mathbb{E}|X| < \infty$, then $\mathbb{E}\tilde{X} = \mathbb{E}X$. If $\mathbb{E}X^2 < \infty$ and $\mathbb{E}X = 0$, then $\mathrm{var}\,\tilde{X} = \mathrm{var}\,X$.

If

$$X \stackrel{d}{=} \tau X',$$

where τ is independent of X', $\mathcal{L}(\tau)=\mathbf{B}(p)$, then (cf. (6.26) in [221])

$$\mathcal{L}(\tilde{X})=\mathbf{\Pi}(p,X')=\exp\left(p(\mathcal{L}(X')-I)\right). \tag{1.8}$$

Given a sequence $\{X_1,...,X_n\}$ or a triangular array $\{X_1\equiv X_{n,1},...,X_n\equiv X_{n,n}\}_{n\geq 1}$ of random variables, we denote by

$$\tilde{S}_n=\tilde{X}_1+\cdots+\tilde{X}_n \tag{1.9}$$

the sum of independent accompanying random variables. Clearly,

$$\mathcal{L}(\tilde{S}_n)=\exp\left(\sum_{i=1}^{n}(\mathcal{L}(X_i)-I)\right). \tag{1.9*}$$

If $X,X_1,...,X_n$ are identically distributed r.v.s, then $\mathcal{L}(\tilde{S}_n)=\mathbf{\Pi}(n,X)$.

Random variables $\{X_{n,1},...,X_{n,k_n}\}_{n\geq 1}$ are called *infinitesimal* if

$$\lim_{n\to\infty}\max_{1\leq k\leq k_n}\mathbb{P}(|X_{n,k}|>\varepsilon)=0 \qquad (\forall\varepsilon>0). \tag{1.10}$$

A distribution $P=\mathcal{L}(X)$ is said to be *infinitely divisible* if for each positive integer number n there exists a distribution G_n such that $P=G_n^{*n}$ (that is, $\widehat{F}=\widehat{G}_n^n$). In other words, $X\stackrel{d}{=}Y_{1,n}+\ldots+Y_{n,n}$, where r.v.s $Y_{1,n},...,Y_{n,n}$ are independent and identically distributed. In particular, any compound Poisson r.v. is infinitely divisible.

The notion of the infinitely divisible distribution was introduced by de Finetti in 1925. A well-known result due to Khintchine [177] states that the class $\mathcal{D}$ of infinitely divisible distributions coincides with the class of weak limits of compound Poisson distributions. Thus, the topics of compound Poisson and infinitely divisible approximations are closely related.

A sequence of random variables $\{X_i\}_{i\geq 1}$ is called m-dependent if random vectors $X_1,\ldots,X_k$ and $X_l,X_{l+1},\ldots,X_n$ are independent for all k,l such that $1\leq k<l<\infty$, $l-k>m$.

Observe that by grouping consecutive summands one can present the sum of m-dependent random variables as a sum of 1-dependent ones.

A sequence of random variables $X_1,X_2,\ldots X_n$ is *strictly stationary* if for arbitrary integer numbers $r,k,i_1<i_2<\ldots<i_r$, the distribution of $X_{k+i_1},\ldots,X_{k+i_r}$ does not depend on k. In particular, r.v.s $X_1,\ldots X_n$ are identically distributed.

Concentration function. We denote by

$$Q_Y(h)\equiv Q_{\mathcal{L}(Y)}(h)=\sup_x\mathbb{P}(x\leq Y\leq x+h)$$

the concentration function of $\mathcal{L}(Y)$.

Sometimes we may use the following variants of the concentration function:

$$Q_Y^+(h)\equiv Q_{\mathcal{L}(Y)}^+(h)=\sup_x\mathbb{P}(x<Y\leq x+h),$$

$$Q_Y^-(h)\equiv Q_{\mathcal{L}(Y)}^-(h)=\sup_x\mathbb{P}(x\leq Y<x+h).$$

Metrics. The accuracy of Poisson or compound Poisson approximation is usually evaluated in terms of the total variation, uniform, the Gini-Kantorovich, and point metrics. The total variation distance between the distributions of r.v.s X and Y is

$$d_{TV}(X;Y) \equiv d_{TV}(\mathcal{L}(X);\mathcal{L}(Y)) = \sup_{A\in\mathcal{A}} |\mathbb{P}(X\in A) - \mathbb{P}(Y\in A)|\,,$$

where $\mathcal{A}$ is a Borel σ-field.

The uniform (Kolmogorov's) distance $d_K(X;Y)$ between the distributions of random variables X and Y with distribution functions (d.f.s) F_X and F_Y is defined as

$$d_K(X;Y) \equiv d_K(\mathcal{L}(X);\mathcal{L}(Y)) = \sup_x |F_X(x) - F_Y(x)|$$

(in the multidimensional case F_X and F_Y denote multivariate distribution functions). Evidently, $d_K(X;Y) \le d_{TV}(X;Y)$.

The Gini-Kantorovich distance $d_G(X;Y)$ between the distributions of r.v.s X and Y with finite first moments (known also as the Kantorovich-Wasserstein distance) is given by

$$d_G(X;Y) \equiv d_G(\mathcal{L}(X);\mathcal{L}(Y)) = \sup_{g\in\mathcal{L}_1} |\mathbb{E}g(X) - \mathbb{E}g(Y)|\,, \tag{1.11}$$

where $\mathcal{L}_1 = \{g\colon |g(x)-g(y)| \le |x-y|\}$ is the set of Lipschitz functions.

If r.v.s X and Y take values in $\mathbb{Z}$, then the *point metric* is defined as

$$d_o(X;Y) \equiv d_o(\mathcal{L}(X);\mathcal{L}(Y)) = \sup_{k\in\mathbb{Z}} |\mathbb{P}(X=k) - \mathbb{P}(Y=k)|\,.$$

Given distributions P_1 and P_2, set

$$\chi^2(P_1;P_2) = \int_{\mathrm{supp}P_2} (\mathrm{d}P_1/\mathrm{d}P_2 - 1)^2\,\mathrm{d}P_2\,.$$

Then $\chi(P_1;P_2)$ is the χ^2-distance.

If distributions P_1 and P_2 have densities f_1 and f_2 with respect to a measure μ, set

$$d_H^2(P_1;P_2) = \frac{1}{2}\int \left(f_1^{1/2} - f_2^{1/2}\right)^2 \mathrm{d}\mu = 1 - \int \sqrt{f_1 f_2}\,\mathrm{d}\mu.$$

Then d_H denotes the *Hellinger* distance.

Let

$$d_{KL}^2(P_1;P_2) = \int_{\mathrm{supp}P_2} \ln(\mathrm{d}P_1/\mathrm{d}P_2)\,\mathrm{d}P_1$$

denote the *Kullback–Leibler* divergence.

Given $\varepsilon \ge 0$, the *Dudley* divergence is defined as

$$\rho_\varepsilon(P_1;P_2) = \inf_{X,Y} \mathbb{P}(|X-Y|>\varepsilon) \qquad (\varepsilon\ge 0),$$

where the infimum is taken over all random pairs (X,Y) such that $\mathcal{L}(X) = P_1$, $\mathcal{L}(Y) = P_2$. The Dudley divergence is a generalisation of the total variation distance: $d_{TV}(P_1;P_2) = \rho_0(P_1;P_2)$.

Concerning properties of metrics and norms, see Section A and references therein.

We may use the same symbol C to denote different absolute constants (with or without indexes). Symbols $C(F)$, C_F, C_X denote constants that depend on a distribution $\mathcal{L}(X)$ with d.f. F.

If function $h:\mathbb{Z}\to\mathbb{R}$, set

$$\Delta h(\cdot)=h(\cdot+1)-h(\cdot),\quad \Delta^k h(\cdot)=\Delta(\Delta^{k-1}h(\cdot)),\quad \|h\|=\sup_{j\in\mathbb{Z}}|h(j)|.$$

As usual, $a_n\sim b_n$ means that $\lim_{n\to\infty}a_n/b_n=1$. We write $f(n)=O(g(n))$ if $f(n)/g(n)\leq C<\infty$ for all large enough n.

For any $x\in\mathbb{R}$, let $\lfloor x\rfloor$ and $\{x\}$ denote the integer and the fractional parts of x.

Below symbol Θ may denote any measure with total variation $\|\Theta\|_{TV}\leq 1$; multiplication is superior to division.

1.2 Basic properties of a compound Poisson distribution

Recall that *compound Poisson* (CP) distribution $\mathbf{\Pi}(\lambda,\zeta)\equiv\mathbf{\Pi}(\lambda,\mathcal{L}(\zeta))$ is the distribution of a random variable

$$\sum_{i=0}^{\pi_\lambda}\zeta_i, \tag{1.1$^+$}$$

where $\zeta_0\equiv 0,\pi_\lambda,\zeta,\zeta_1,\zeta_2,...$ are independent r.v.s, $\zeta_i\stackrel{d}{=}\zeta$ $(i\geq 1)$, $\mathcal{L}(\pi_\lambda)=\mathbf{\Pi}(\lambda)$.

Typically $\zeta\neq 0$ *w.p.* 1. The requirement $\zeta\neq 0$ *w.p.* 1 may be omitted. Indeed, denote $p=\mathbb{P}(\zeta\neq 0)$. By Khintchine's formula ([175], Chapter 2) any random variable ζ obeys

$$\zeta\stackrel{d}{=}\tau_p\zeta', \tag{1.12}$$

where τ_p and ζ' are independent r.v.s, $\mathcal{L}(\zeta')=\mathcal{L}(\zeta|\zeta\neq 0)$, $\mathcal{L}(\tau_p)=\mathbf{B}(p)$. Note that

$$\mathbf{\Pi}(\lambda,\tau_p\zeta')=\mathbf{\Pi}(\lambda p,\zeta'), \tag{1.8*}$$

i.e., (1.1$^+$) can be rewritten as

$$\sum_{i=0}^{\pi_\lambda}\zeta_i\stackrel{d}{=}\sum_{i=0}^{\pi_{\lambda p}}\zeta_i'$$

(cf. (6.26) in [221]). Therefore, one usually deals with $\mathbf{\Pi}(t,X)$, where $\mathbb{P}(X\neq 0)=1$.

Denote $P_\zeta=\mathcal{L}(\zeta)$. Recall that $\mathbf{\Pi}(\lambda,\zeta)=\mathbf{\Pi}(\lambda,P_\zeta)$ can be presented as

$$\exp(\lambda(P_\zeta-I))=\mathrm{e}^{-\lambda}\exp(\lambda P_\zeta)=\mathrm{e}^{-\lambda}\sum_{k=0}^{\infty}\lambda^k P_\zeta^{*k}/k!$$

If X is a compound Poisson r.v. and c is a constant, then cX is obviously a compound Poisson r.v. as well. A sum of two independent compound Poisson random variables is a compound Poisson random variable. Indeed,

$$\exp\Big(\lambda(\widehat{F}(t)-1)\Big)\exp\Big(\mu(\widehat{G}(t)-1)\Big)=\exp((\lambda+\mu)\,(\varphi(t)-1)),$$

where $\varphi(t)=\frac{\lambda}{\lambda+\mu}\widehat{F}(t)+\frac{\mu}{\lambda+\gamma}\widehat{G}(t)$ is a characteristic function.

Compound Poisson distribution is infinitely divisible.

A random variable X with support on $[0;\infty]$ and $\mathbb{P}(X=0)>0$ is infinitely divisible if and only if it is compound Poisson (van Harn [147]).

$\mathbf{\Pi}(\lambda,\zeta)$ is not absolutely continuous since it has an atom at zero.

Though the sum $\pi_a+\pi_b$ of two independent Poisson r.v.s π_a and π_b is a Poisson r.v. π_{a+b}, the difference $\pi_a-\pi_b$ is a compound Poisson r.v.:

$$\mathcal{L}(\pi_a-\pi_b) = \exp(a(I_1-I) + b(I_{-1}-I)) .$$

$\mathcal{L}(\pi_a-\pi_b)$ is sometimes called the *Skellam* distribution, see [283, 124].

Raikov [249] has shown that if the sum $X+Y$ of two independent r.v.s is Poisson, then both X and Y are Poisson.

Supplements.

- Expectations of functions of a sum of r.v.s can be bounded in terms of expectations of an "accompanying" r.v. $\tilde{S}_n = \tilde{X}_1 + \cdots + \tilde{X}_n$, where $\{\tilde{X}_i\}$ are defined by (1.9).

  Suppose that function f is twice differentiable, f'' is convex, and there exists constant $c_f \in (0;\infty)$ such that

  $$|f(x+y)| \leq c_f(1+|f(y)|(1+|f(y)|) \qquad (\forall\, x,y).$$

  Let $X_1,...,X_n$ be independent r.v.s. Assume that $\sum_{i=1}^n \mathbb{E}f(X_i) < \infty$. Then ([300], Theorem 4)

  $$\mathbb{E}f(S_n) \leq \mathbb{E}f(\tilde{S}_n).$$

- Prokhorov [243] gives upper and lower bounds to $\mathbb{P}(\pi'_{\lambda/2} - \pi''_{\lambda/2} > k)$, where $\lambda > 0$, $\pi'_{\lambda/2}$, $\pi''_{\lambda/2}$ are independent Poisson $\mathbf{\Pi}(\lambda/2)$ r.v.s, $\lambda \geq 0$. In particular, he shows that

  $$\mathbb{P}(\pi'_{\lambda/2}-\pi''_{\lambda/2}>k) \leq \exp\left(-\frac{k}{2}\mathrm{arsh}(k/2\lambda)\right) \qquad (k \in \mathbb{Z}_+).$$

- There is a close relation between the Skellam distribution and Bessel functions. Denote

  $$J_\alpha(x) = \sum_{k\geq 0}(-1)^k (x/2)^{2k+\alpha}/k!\Gamma(k+1+\alpha) \qquad (x\geq 0,\ \alpha\geq 0).$$

  J_α is called a Bessel function. A modified Bessel function is

  $$I_\alpha(x) = \sum_{k\geq 0}(x/2)^{2k+\alpha}/k!\Gamma(k+1+\alpha) \qquad (x\geq 0).$$

  Let $\pi'_{x/2}$, $\pi''_{x/2}$ be independent Poisson $\mathbf{\Pi}(x/2)$ r.v.s. Then

  $$\mathbb{P}(\pi'_{x/2}-\pi''_{x/2}=m) = e^{-x} I_{|m|}(x) \qquad (m \in \mathbb{Z}). \tag{1.13}$$

  If $\pi'_{x/2}$, $\pi''_{-x/2}$ are independent Poisson "r.v.s" (recall that symbol $\pi_{-x/2}$ means we deal with a convolution of Poisson unit measures), then

  $$J_m(x) = \mathbb{P}(\pi'_{x/2}-\pi''_{-x/2}=m) \qquad (m \in \mathbb{Z}_+).$$

Bessel functions appear in mathematical description of many physical phenomena (flow of heat, propagation of electromagnetic waves, motions of fluids, vibration of a membrane, etc.).

Exercises

1.1. Let $\mathbf{\Pi}(-\lambda)$ be a Poisson unit measure with parameter $-\lambda$, where $\lambda>0$. Prove that

$$1 \le \|\mathbf{\Pi}(-\lambda)\|_{TV} \le e^{2\lambda}.$$

Recall that $\|\cdot\|_{TV}$ denotes the total variation norm, see Section A.

1.2 Note that measures $\mathbf{\Pi}(\lambda, X)$ and $\mathbf{\Pi}(-\lambda, X)$ are finite, and

$$\widehat{\mathbf{\Pi}}(-\lambda, \zeta) = 1/\widehat{\mathbf{\Pi}}(\lambda, \zeta) \qquad (\lambda>0).$$

Let F be an absolutely continuous distribution with bounded density and finite variance σ^2. Prove that if $1/\widehat{F}$ is a Fourier transform of a unit measure G, then $\|G\|_{TV}=\infty$.

Hint. Use estimate (D.8).

1.3. Let X be a r.v. taking values in on $\mathbb{Z}_+$, and let $\mathbb{E}X<\infty$. Denote by φ_X the ch.f. of $\mathcal{L}(X)$. Prove that

$$|\varphi_X(t)-1| \le 2\min\{\mathbb{P}(X\neq 0); \mathbb{E}X|\sin(t/2)|\} \qquad (t\in\mathbb{R}). \tag{1.14}$$

Hint. Check that $|e^{itk}-1| \le k|e^{it}-1|$.

1.4 Prove (1.5).

1.5 Let X, Y be independent integer-valued r.v.s. Check that

$$d_{TV}(X,Y) = 1 - \sum_{k=-\infty}^{\infty} \min\{\mathbb{P}(X=k); \mathbb{P}(Y=k)\}.$$

2

Poisson limit theorem

Two students prepare for an exam.
– *This Poisson approximation looks so complicated!* – said one student.
– *Not complicated but compounded,* – replied the other student.
– *Compounded Poisson approximation?? Sounds even more complicated!*

Student folklore

2.1 Independent random variables

Let $\{X_{n,1},...,X_{n,k_n}\}_{n\geq 1}$, where $\{k_n\}$ is a non-decreasing sequence of natural numbers, be a triangle array of independent random variables (r.v.s).

Denote $S_n = X_{n,1} + ... + X_{n,k_n}$, and let

$$B_\varepsilon = (-\varepsilon;\varepsilon)\cup(1-\varepsilon;1+\varepsilon).$$

The following Poisson limit theorem is due to Gnedenko [133] and Marcinkiewicz [203].

Theorem 2.1 [133, 203] *If* $\{X_{n,k}\}$ *are infinitesimal r.v.s, then*

$$\mathcal{L}(S_n)\Rightarrow \mathbf{\Pi}(\lambda) \tag{2.1}$$

as $n\to\infty$ *if and only if for any* $\varepsilon\in(0;1)$, *as* $n\to\infty$,

$$\sum_k \mathbb{P}(|X_{n,k}-1|<\varepsilon)\to\lambda, \tag{2.2}$$

$$\sum_k \mathbb{P}(X_{n,k}\notin B_\varepsilon)\to 0,\quad \sum_k \mathbb{E}X_{n,k}\mathbb{1}\{|X_{n,k}|<\varepsilon\}\to 0, \tag{2.3}$$

$$\sum_k \left(\mathbb{E}X_{n,k}^2\mathbb{1}\{|X_{n,k}|<\varepsilon\} - \mathbb{E}^2X_{n,k}\mathbb{1}\{|X_{n,k}|<\varepsilon\}\right)\to 0. \tag{2.4}$$

Let $\{X_{n,1},...,X_{n,k_n}\}$ be independent infinitesimal r.v.s. Gnedenko [134] has shown that condition

$$\lim_{n\to\infty}\sum_{k=1}^{k_n}\mathbb{E}(X_{n,k}-\mathbb{E}X_{n,k})^2\mathbb{1}\{|X_{n,k}-\mathbb{E}X_{n,k}-1|\geq\varepsilon\}=0,$$

$$\lim_{n\to\infty}\sum_{k=1}^{k_n}\mathbb{E}(X_{n,k}-\mathbb{E}X_{n,k})^2\mathbb{1}\{|X_{n,k}-\mathbb{E}X_{n,k}-1|<\varepsilon\}=\lambda\qquad(\forall\varepsilon>0)$$

DOI: 10.1201/9781003478164-2

is necessary and sufficient for (2.1) to hold simultaneously with the convergence $\operatorname{var} S_n \to \lambda$ as $n \to \infty$ (see also Gnedenko & Kolmogorov [135], Section 4.21).

The following corollary presents necessary and sufficient conditions for the weak convergence of a sum of independent and identically distributed (i.i.d.) non-negative integer-valued r.v.s to a Poisson random variable.

Corollary 2.2 *If* $\{X_{n,1}, ..., X_{n,k_n}\}_{n\geq 1}$ *is a triangle array of independent random variables taking values in* $\mathbb{Z}_+$ *such that* $X_{n,i} \stackrel{d}{=} X_{n,1}$ $(1 \leq i \leq k_n)$, *then (2.1) holds if and only if*

$$k_n \mathbb{P}(X_{1,n}=1) \to \lambda \quad \text{and} \quad \mathbb{P}(X_{n,1}\geq 2)/\mathbb{P}(X_{n,1}\geq 1) \to 0. \tag{2.5}$$

Note that (2.5) yields $\mathbb{P}(X_{n,1}\geq 1) \sim \mathbb{P}(X_{n,1}=1)$ as $n \to \infty$.

The second relation in (2.5) means

$$X_n' \underset{p}{\to} 1 \qquad (n \to \infty),$$

where r.v. X_n' has the distribution

$$\mathcal{L}(X_n') = \mathcal{L}(X_{n,1} | X_{n,1} \neq 0).$$

Example 2.1 Let $\{X_{n,1}, ..., X_{n,n}\}$ be i.i.d. random variables with the distribution

$$\mathbb{P}(X=0) = 1-\lambda/n-1/n^{1.5}, \ \mathbb{P}(X=1) = \lambda/n, \ \mathbb{P}(X=n) = 1/n^{1.5} \qquad (\lambda>0).$$

Then (1.10) and (2.5) hold, hence $\mathcal{L}(S_n) \Rightarrow \mathbf{\Pi}(\lambda)$. Note that $\mathbb{E} S_n \not\to \lambda$. □

In the case of Bernoulli $\mathbf{B}(p_{n,k})$ random variables relations (2.3) and (2.4) trivially hold, (2.2) means

$$\sum_k p_{n,k} \to \lambda \qquad (n \to \infty), \tag{2.2*}$$

while (1.10) states that $p_n^* = \max_k p_{n,k} \to 0$ as $n \to \infty$. The latter together with (2.2*) yields

$$\sum_k p_{n,k}^2 \to 0 \qquad (n \to \infty). \tag{1.10*}$$

Thus, conditions (1.10*) and (2.2*) are necessary and sufficient for the weak convergence (2.1) of a sum of independent infinitesimal 0-1 r.v.s.

Let function $f: \mathbb{N} \to \mathbb{R}$ obey $\mathbb{E}|f(\pi_\lambda)| < \infty$. If $\lambda := \sum_k p_{n,k}$ does not depend on n and condition (1.10*) holds, then

$$\sum_{k=0}^{\infty} f(k) \Big| p_{n,k} - \lambda^k e^{-\lambda}/k! \Big| \to 0 \qquad (n \to \infty) \tag{2.6}$$

(Simons & Johnson [282]). Generalisations of (2.6) are given in [23, 57, 102, 270], see also (3.28).

Sufficient conditions for the convergence to a Poisson law in various metrics follow from the estimates of the accuracy of approximation. For example, let $X_1 = X_{n,1}, ..., X_n = X_{n,n}$ be a triangular array of non-negative integer-valued r.v.s having finite variances. If

$$\lim_{n\to\infty} \sum_{j=1}^n \mathbb{E} X_j = \lambda > 0, \quad \lim_{n\to\infty} \sum_{j=1}^n \mathbb{E} X_j(X_j-1) = 0 \quad \lim_{n\to\infty} \max_{1\leq j\leq n} \mathbb{E} X_j = 0, \tag{2.7}$$

then $d_K(\mathcal{L}(S_n); \mathbf{\Pi}(\lambda)) \to 0$ as $n \to \infty$ (Franken [120], Satz 9).

2.2 Dependent Bernoulli random variables

Recall that α denotes the mixing coefficient (see Appendix).

Condition (Δ) is said to hold if $\alpha(l_n)\to 0$ for a sequence $\{l_n\}$ of natural numbers such that $0\le l_n \ll n$. Therefore, we sometimes refer to this condition as $(\Delta(\{l_n\}))$.

Let $\{X_{n,1},...,X_{n,n}\}_{n\ge 1}$ be a triangle array of 0-1 random variables such that sequence $X_{n,1},...,X_{n,n}$ is strictly stationary for each $n\in\mathbb{N}$ and obeys condition (Δ).

For instance, in extreme value theory one often has

$$X_{n,k} = \mathbb{I}\{Y_k > u_n\},$$

where $\{Y_i\}_{i\ge 1}$ is a stationary sequence of random variables and $\{u_n\}$ is a sequence of "high" levels (a special case where $\{Y_i, i\ge 1\}$ is a moving average is related to the topic of the Erdös–Rényi partial sums, cf. [221], Chapter 2).

Note that there exists a sequence $\{r_n\}$ of natural numbers such that

$$n \gg r_n \gg l_n \gg 1,\ n r_n^{-1}\alpha^{2/3}(l_n) \to 0 \qquad (n\to\infty) \tag{2.8}$$

(for instance, one can take $r_n = \left[\sqrt{n\max\{l_n; n\alpha(l_n)\}}\,\right]$). We denote the class of such sequences $\{r_n\}$ by $\mathcal{R}$.

Set

$$S_n = X_{n,1} + ... + X_{n,n}, \quad \lambda_n = \mathbb{E}S_n,$$

and let $\zeta_{r,n}$ be a r.v. with the distribution

$$\mathcal{L}(\zeta_{r,n}) = \mathcal{L}(S_r \,|\, S_r > 0). \tag{2.9}$$

In extreme value theory $\mathcal{L}(\zeta_{r,n})$ is known as the cluster size distribution.

Theorem 2.3 *If condition* (Δ) *holds and*

$$S_n \Rightarrow \pi_\lambda \qquad (\exists\lambda > 0), \tag{2.10}$$

as $n\to\infty$, *then*

$$\zeta_{r,n} \xrightarrow[p]{} 1 \qquad (n\to\infty) \tag{2.11}$$

for any sequence $\{r = r_n\}\in\mathcal{R}$.

If there exists the limit

$$\lim_{n\to\infty} \mathbb{P}(S_n = 0) = \mathrm{e}^{-\lambda} \qquad (\exists\lambda > 0) \tag{2.12}$$

and (2.11) holds for some $\{r = r_n\}\in\mathcal{R}$, *then* $S_n \Rightarrow \pi_\lambda$ *as* $n\to\infty$.

Theorem 2.3 generalises Corollary 2.2 to the case of dependent α-mixing r.v.s.

Condition (2.12) is an analogue of (2.2); it means that $\mathbb{P}(X_{n,i}\neq 0)$ are "properly small".

Condition (2.11) prohibits asymptotic clustering of rare events. In the case of independent r.v.s taking values in $\mathbb{Z}_+$ assumption (2.11) means $X'_{n,1} \xrightarrow[p]{} 1$ as $n\to\infty$, where r.v. $X'_{n,1}$ has the distribution $\mathcal{L}(X'_{n,1}) = \mathcal{L}(X_{n,1}|X_{n,1}\neq 0)$.

Remark 2.1. The following condition (D') has been widely used in extreme value theory (cf. [189, 221]):

$$\lim_{n\to\infty} n \sum_{i=1}^{r-1} \mathbb{P}\left(X_{n,i+1}\neq 0, X_{n,1}\neq 0\right) = 0 \tag{D'}$$

for any sequence $\{r = r_n\}$ such that $n \gg r_n \gg 1$. Condition (D') means that there is no asymptotic clustering of extremes. It was introduced by Loynes [199].

Closely related is the following condition

$$\lim_{n\to\infty} \sum_{i=1}^{r-1} \mathbb{P}\left(X_{n,i+1} \neq 0 | X_{n,1} \neq 0\right) = 0. \tag{$\tilde{D}'$}$$

If (2.12) hold, then (D') is equivalent to $(\tilde{D}')$.

Indeed, one can check that (2.12) and (2.20) yield

$$\mathbb{P}(S_r > 0) \sim \lambda r/n \qquad (n \to \infty). \tag{2.13}$$

Denote $p = \mathbb{P}(X_{n,1} \neq 0)$. Then

$$\lambda r/n \sim \mathbb{P}(S_r > 0) \leq rp.$$

Hence $np \geq \lambda + o(1)$, and $(D') \Rightarrow (\tilde{D}')$.

In order to show that $(\tilde{D}') \Rightarrow (D')$ we notice that

$$\begin{aligned}
\lambda r/n \;\sim\; \mathbb{P}(S_r > 0) &\geq r\mathbb{P}(X_{n,1} \neq 0) - \mathbb{P}\Big(\cup_{1\leq i<j\leq r} \{X_{n,i} \neq 0, X_{n,j} \neq 0\}\Big) \\
&\geq rp - rp \sum_{i=1}^{r-1} \mathbb{P}\left(X_{n,i+1} \neq 0 | X_{n,1} \neq 0\right)
\end{aligned}$$

by Bonferroni's inequality. Therefore,

$$1 \;\geq\; \mathbb{P}(S_r > 0)/rp \geq 1 - \sum_{i=1}^{r-1} \mathbb{P}\left(X_{n,i+1} \neq 0 | X_{n,1} \neq 0\right), \tag{2.14}$$

$$\lambda + o(1) \;\geq\; \Big(1 - \sum_{i=1}^{r-1} \mathbb{P}\left(X_{n,i+1} \neq 0 | X_{n,1} \neq 0\right)\Big) np. \tag{2.15}$$

Thus, $np \sim \lambda$, and (D') is equivalent to $(\tilde{D}')$.

Remark 2.2. Condition (2.11) is weaker than (D'): if (2.12) holds, then (D') entails (2.11). Indeed, $\zeta_{r,n} \geq 1$ by construction. Note that

$$\begin{aligned}
\mathbb{P}(S_r > 1) &= \mathbb{P}\Big(\cup_{1\leq i<j\leq r} \{X_{n,i} \neq 0, X_{n,j} \neq 0\}\Big) \\
&\leq r \sum_{i=1}^{r-1} \mathbb{P}\left(X_{n,i+1} \neq 0, X_{n,1} \neq 0\right).
\end{aligned}$$

Thus, $\mathbb{P}(S_r > 1) = o(r/n)$ if (D') holds. In view of (2.13), $\mathbb{P}(\zeta_{r,n} > 1) \to 0$ as $n \to \infty$, i.e., (2.11) holds.

Remark 2.3. If (D') holds, then (2.12) is equivalent to

$$\lim_{n\to\infty} n\mathbb{P}(X_{n,1} \neq 0) = \lambda. \tag{2.2$'$}$$

Indeed, this follows from (2.13), (2.14), and (2.20) (see also [189], Theorem 3.4.1).

A generalisation of Corollary 2.2 to the case of stationary φ-mixing r.v.s has been given by Utev [301], Theorem 10.1, who has shown that conditions (2.2′) and (D') are necessary and sufficient for (2.10). Sufficient conditions for Poisson convergence without assuming stationarity have been provided by Sevastyanov [277]. A Poisson limit theorem in the case of a two-dimensional random field $\{X_{i,j}\}$ has been given by Banys [16].

Theorem 2.4 [157] *Let* $\{X, X_1, ...\}$ *be a strictly stationary sequence of* m*-dependent Bernoulli random variables. Weak convergence (2.1) holds if and only if*

$$n\mathbb{P}(X=1) \to \lambda, \ n\,\mathrm{cov}(X_1; X_j) \to 0 \qquad (1<j\leq m+1).$$

A generalisation of Theorem 2.4 to the multivariate case has been given by Kobus [178].

Example 2.2 Let $\xi_1, \ldots, \xi_{n+1}$ be independent identically distributed Bernoulli $\mathbf{B}(p)$ r.v.s. Set $X_i = \xi_i \xi_{i+1}$. Then $S_n = X_1 + \cdots + X_n$ is the so-called 2-runs statistic. Obviously, $\{X_i\}$ are 1-dependent r.v.s, $n\mathbb{P}(X_1 = 1) = np^2$, $n\,\mathrm{cov}(X_1, X_2) = np^3(1-p)$. Condition $\lim_{n\to\infty} np^2 = \lambda$ is necessary and sufficient for the weak convergence of $\mathcal{L}(S_n)$ to the Poisson law. □

It is possible to formulate a limit theorem for dependent Bernoulli random variables in terms of the so-called *Binomial moments.*

Let $X_1, X_2, \ldots, X_n$ be Bernoulli random variables (no assumption is made on the structure of dependence of $\{X_j\}$). A Binomial moment $p_n(k)$ is defined as $p_n(0) = 1$,

$$p_n(k) = \sum_{i_1<i_2<\cdots<i_k} \mathbb{P}(X_{i_1} = X_{i_2} = \cdots = X_{i_k} = 1) \Big/ \binom{n}{k} \qquad (k\geq 1).$$

Theorem 2.5 [172, 290] *If* $\lim_{n\to\infty} n^k p_n(k) = \lambda^k$ $(\forall k\geq 1)$, *then* $\mathcal{L}(S_n) \Rightarrow \mathbf{\Pi}(\lambda)$ *as* $n\to\infty$. *The converse is true if* $\max_n n^k p_n(k)$ *is bounded for each* $k\geq 1$.

Sufficient conditions for Poisson convergence in the case of a stationary Markov chain have been given by Gudynas [145].

2.3 Proofs

The proof of (2.7) follows from estimates (3.55) and (C.12). Indeed, denote $\mu = \mathbb{E}S_n$. Then

$$d_{TV}(S_n; \pi_\lambda) \leq d_{TV}(S_n; \pi_\mu) + d_{TV}(\pi_\mu; \pi_\lambda) \leq \sum_{i=1}^{n} (\mathbb{E}^2 X_i + \mathbb{E}X_i(X_i-1)) + |\mu-\lambda|.$$

Observe that

$$\sum_{i=1}^{n} \mathbb{E}^2 X_i \leq \left(\max_j \mathbb{E}X_j\right) \sum_{i=1}^{n} \mathbb{E}X_i\,,$$

so that $d_{TV}(S_n; \pi_\lambda) \to 0$ as $n \to \infty$, and (2.7) follows. □

The proof of Theorem 2.1 can be found in Gnedenko & Kolmogorov [135].

If condition $(\Delta(\{l_n\}))$ holds, then there exists a sequence $\{r_n\}$ of natural numbers such that

$$n \gg r_n \gg l_n \geq 0, \quad nr_n^{-1}\alpha^{2/3}(l_n) \to 0, \quad r_n l_n^{-1}\alpha(l_n) \to 0 \qquad (n \to \infty) \tag{2.16}$$

(for instance, one can take $r_n = \left[\max\{\sqrt{nl_n}\,; n\sqrt{\alpha(l_n)}\,; 1\} \wedge (l_n+1)/\sqrt{\alpha(l_n)}\,\right]$).

We denote by $\mathcal{R}$ the class of such sequences $\{r_n\}$.

The proof of Theorem 2.3 uses the following Propositions 2.6 and 2.7 from the addendum to [224].

Proposition 2.6 *If condition* (Δ) *holds, then there exists a sequence* $\{l_n\}$ *such that* $l_n \in \mathbb{Z}_+$, $0 \leq l_n \ll n$ *and*

$$\alpha(l_n)n/(l_n+1) \to 0 \qquad (n \to \infty). \tag{2.17}$$

In the case of independent r.v.s (2.17) holds with $l_n = 0$. If $\{X_{n,i}\}$ are m-dependent r.v.s, them (2.17) holds with $l_n = m$.

We denote by $\mathcal{M}$ the class of sequences $\{l_n\}$ such that $l_n \in \mathbb{Z}_+$, $0 \leq l_n \ll n$ and (2.17) holds. Obviously, if $\{l_n\} \in \mathcal{M}$, then condition $(\Delta(\{l_n\}))$ holds.

Proposition 2.7 *If* $\{l_n\} \in \mathcal{M}$, $\{r_n\} \in \mathcal{R}$ *and*

$$\lim_{n\to\infty} \mathbb{P}(S_n = 0) = \mathrm{e}^{-\lambda} \qquad (\exists \lambda \geq 0), \tag{2.18}$$

then, as $n \to \infty$,

$$\frac{n}{l_n+1}\mathbb{P}(S_{l_n} > 0) \leq \lambda + o(1), \quad \frac{n}{r_n+1}\mathbb{P}(S_{r_n} > 0) \leq \lambda + o(1).$$

Proof of Theorem 2.3. Let $\{r = r_n\}$ be an arbitrary sequence from $\mathcal{R}$. Condition Δ and the standard argument (cf. Theorem 5.2 in [221]) imply that for any $t \in \mathbb{R}$

$$\mathbb{E}\exp(\mathrm{i}tS_n) = \exp\left(\frac{n}{r}\mathbb{P}(S_r > 0)\mathbb{E}\left\{\mathrm{e}^{\mathrm{i}tS_r} - 1 | S_r > 0\right\}\right) + o(1), \tag{2.19}$$

$$\mathbb{P}(S_n = 0) = \mathbb{P}^{n/r}(S_r = 0) + o(1) = \exp\left(-\frac{n}{r}\mathbb{P}(S_r > 0)\right) + o(1) \tag{2.20}$$

as $n \to \infty$.

Indeed, let $\{l_n\} \in \mathcal{M}$, $\{r_n\} \in \mathcal{R}$. Using Bernstein's blocks approach (splitting the sample into blocks of length $r = r_n$ separated by sub-blocks of length $l = l_n$, cf. [189, 221]), we get

$$|\mathbb{P}(S_n = 0) - \mathbb{P}^{\lfloor n/r \rfloor}(S_r = 0)| \leq (\mathbb{P}(S_l > 0) + \alpha_n(l))\, n/r,$$

$$\left|\mathbb{E}\exp(\mathrm{i}tS_n) - \left(1 + \mathbb{P}(S_r > 0)\mathbb{E}\left\{\mathrm{e}^{\mathrm{i}tS_r} - 1 | S_r > 0\right\}\right)^{\lfloor n/r \rfloor}\right|$$

$$\leq 2\,(\mathbb{P}(S_l > 0) + 8\alpha_n(l))\, n/r$$

(we have applied the Volkonskiy-Rozanov inequality, see [310] or (14.66) in [221]).

If (2.10) holds, then so does (2.18): $\mathbb{P}(S_n = 0) \to \mathrm{e}^{-\lambda}$ as $n \to \infty$. Note that (2.18) and (2.20) yield (2.13). Since

$$\mathbb{E}\mathrm{e}^{\mathrm{i}tS_n} \to \exp\bigl(\lambda(\mathrm{e}^{\mathrm{i}t} - 1)\bigr) \qquad (\forall t \in \mathbb{R}) \tag{2.10*}$$

by the assumption, (2.19) and (2.13) entail $\mathbb{E}\mathrm{e}^{\mathrm{i}t\zeta_{r,n}} \to \mathrm{e}^{\mathrm{i}t}$, i.e., (2.11) holds.

On the other hand, if (2.11) and (2.18) hold for some $\{r = r_n\} \in \mathcal{R}$, then (2.13) is valid. Relations (2.13) and (2.19) yield (2.10*). □

Exercises

2.1. Let $G, F_1, F_2, \dots$ be a sequence of integer-valued distributions such that

$$\sup_{t\in[-\pi,\pi]} |\widehat{F}_n(t) - \widehat{G}(t)| \to 0 \qquad (n\to\infty).$$

Prove that $d_o(F_n; G)\to 0$ as $n\to\infty$.

2.2. Let $X, X_1, X_1, \dots$ be a sequence of integer-valued r.v.s such that $\mathbb{P}(X_n=k) \to \mathbb{P}(X=k)$ as $n\to\infty$ $(\forall k\in\mathbb{Z})$. Prove that $d_{TV}(X_n; X)\to 0$, i.e., point-wise convergence entails convergence in total variation.

2.3. Let $\{\lambda_n\}$ be a sequence of positive numbers, $\lambda_n \to \lambda > 0$ as $n \to \infty$. Prove that $d_G(\pi_{\lambda_n}; \pi_\lambda)\to 0$.

2.4. Let $\mathcal{L}(X_n) = \mathbf{NB}(n; p)$, where $p\equiv p(n)\in(0;1)$, see (1.4). Prove that $d_{TV}(X_n; \pi_\lambda)\to 0$ as $n\to\infty$ if $np\to\lambda>0$.

Notes

Poisson approximation was apparently introduced by de Moivre [213], problems 5–7. Theorem 2.3 is from [224], see also Theorem 5.1 in [221].

3

Accuracy of Poisson approximation

The problem of evaluating the accuracy of Poisson approximation to $\mathcal{L}(S_n)$ has attracted a lot of attention among researchers (cf. [21, 223] and references wherein). This chapter presents results on the accuracy of Poisson approximation.

Recall that $S_n = X_1 + \ldots + X_n$. In this chapter, we set $\lambda = \mathbb{E}S_n$.

3.1 Independent Bernoulli r.v.s

First, we consider the problem of evaluating the accuracy of Poisson approximation to the Binomial distribution $\mathbf{B}(n,p)$ with parameters n and p; $\lambda = \sum_{j=1}^{n} p_j$.

It seems natural to approximate $\mathbf{B}(n,p)$ by the Poisson distribution. For instance, in the case of identically distributed Bernoulli $\mathbf{B}(p)$ r.v.s $\{X_i\}$ one has

$$\mathbb{P}(S_n = k) \equiv \mathbb{P}(S_n = k | N_n = n) = \mathbb{P}(\pi_n(p) = k | \pi_n(1) = n), \tag{3.1}$$

where $N_n \equiv n$ is the total number of 0's and 1's among $X_1, X_2, \ldots, X_n$, and $\{\pi_n(t), t \in [0;1]\}$ is a Poisson jump process on $[0;1]$ with intensity rate n. Thus,

$$\mathbf{B}(n,p) = \mathcal{L}(\pi_n(p) | \pi_n(1) = n). \tag{3.1*}$$

Tsaregradskii [291] has shown that

$$d_K(\mathbf{B}(n,p); \mathbf{\Pi}(np)) \le p\pi^2 e^{2p(2-p)}/16(1-p) \qquad (p \in (0;1/2]). \tag{3.2}$$

Note that $\pi^2/16 \approx 0.617$. Inequality (3.2) seems to be the first estimate of the accuracy of Poisson approximation with an explicit constant.

If $np \to \infty$ as $n \to \infty$, then $\mathbf{B}(n,p)$ can be approximated by the normal distribution, and the rate of convergence is $O(1/\sqrt{np})$. Both normal and Poisson approximations are applicable if $np \to \infty$, $p \to 0$ as $n \to \infty$. Poisson approximation is more accurate than the normal one if $p \ll n^{-1/3}$ (cf. [221], p.91).

In the case of non-identically distributed Bernoulli $\mathbf{B}(p_i)$ random variables Franken [120] has proved that

$$d_K(S_n; \pi_\lambda) \le 0.6 p_n^*$$

if $p_n^* := \max_{i \le n} p_i \le 1/4$.

Denote

$$\theta = \sum_{i=1}^{n} p_i^2/\lambda, \quad \lambda_2 = \sum_{i=1}^{n} p_i^2, \quad \lambda_k = \sum_{i=1}^{n} p_i^k \quad (k \ge 3).$$

Shorgin [279] has proved that

$$d_K(S_n; \pi_\lambda) \le c_1\theta/(1-\sqrt{\theta}) \qquad (\theta < 1),$$

DOI: 10.1201/9781003478164-3

where $c_1 = (1+\sqrt{\pi/2})/2 < 1.13$. According to Daley (see (5.5) in Serfling [276]),

$$d_K(S_n;\pi_\lambda) \le \sum_{i=1}^n p_i^2/2.$$

Daley & Vere-Jones [96] have shown that

$$d_K(S_n;\pi_\lambda) \le 0.36\theta.$$

Roos [260] has shown that

$$\begin{aligned} d_K(S_n;\pi_\lambda) &\le \Big(1/2\mathrm{e} + 1.2\sqrt{\theta}/(1-\sqrt{\theta})\Big)\theta, \\ d_o(S_n;\pi_\lambda) &\le \left(\frac{1}{2}\left(\frac{3}{2\,\mathrm{e}}\right)^{3/2} + \frac{\sqrt{\theta}(6-4\sqrt{\theta})}{3(1-\sqrt{\theta})^2}\right)\frac{\theta}{\lambda}. \end{aligned}$$

Note that $1/2\mathrm{e} \le 0.184$, $\frac{1}{2}(1.5/\mathrm{e})^{1.5} \le 0.205$.

For $0 \le p_i \le 1$ Kruopis [185] has proved that

$$d_o(S_n;\pi_\lambda) \le \lambda_2 \min\{\sqrt{\mathrm{e}/\pi}\,\sigma^{-3};1\}, \tag{3.3}$$

where $\sigma^2 = \sum_{j=1}^n p_j(1-p_j)$.

According to Nagaev [217],

$$\begin{aligned} d_K(\mathbf{B}(n,p);\mathbf{\Pi}(np)) &= \mathrm{e}^{-np} - (1-p)^n \quad \text{if } np \le 2-\sqrt{2}, \\ d_K(\mathbf{B}(n,p);\mathbf{\Pi}(np)) &\le 0.44np^2 \qquad\quad \text{if } np > 2-\sqrt{2}. \end{aligned}$$

Kontoyiannis et al. [181] have shown that

$$d_H^2(S_n;\pi_\lambda) \le \lambda^{-1}\sum_{i=1}^n p_i^3/(1-p_i),$$

while Zacharovas & Hwang [321] present the estimate

$$\chi^2(S_n;\pi_\lambda) \le 2(\sqrt{\mathrm{e}}-1)^2\theta^2/(1-\theta)^3.$$

Note that $2(\sqrt{\mathrm{e}}-1)^2 \le 0.8417$; $\theta^2 \le \lambda^{-1}\sum_{i=1}^n p_i^3$.

Borisov & Vorozheikin [49] have obtained sharp lower and upper bounds to $\chi^2(\mathbf{B}(n,p);\mathbf{\Pi}(np))$:

$$0 \le \chi^2(\mathbf{B}(n,p);\mathbf{\Pi}(np)) - p^2/2 - 2p^3/3n \le p^4/(1-p) + p^8(23-20p)/(1-p)^2.$$

Harremoës & Ruzankin [148] present lower and upper bounds to $d_{KL}^2(\mathbf{B}(n,p);\mathbf{\Pi}(np))$. In particular, they have shown that

$$2d_{KL}^2(\mathbf{B}(n,p);\mathbf{\Pi}(np)) = (-p - \ln(1-p))(1+O(1/n)).$$

According to Bobkov et al. [45], if $\lambda_2 \le \lambda/2$, then there exists an absolute constant $c \ge 2$ ($c \le 15$ if $\lambda \le 1/2$) such that

$$(\lambda_2/2\lambda)^2 \le d_{KL}^2(S_n;\pi_\lambda) \le \chi^2(S_n;\pi_\lambda) \le c(\lambda_2/\lambda)^2.$$

Borisov [46] has shown that for any $\varepsilon \ge 0$

$$\rho_\varepsilon(\mathbf{B}(n,p);\mathbf{\Pi}(np)) \le C\mathrm{e}^{-c\varepsilon}\Big((np^2)^{[\varepsilon]+1}\mathbb{I}\{np \ge 1\} + np^{[\varepsilon]+2}\mathbb{I}\{np < 1\}\Big),$$

where C is an absolute positive constant.

Ruzankin [268] presents the following bounds $(p \le 1/2,\ k \in \mathbb{Z}_+)$:

$$\rho_k(\mathbf{B}(n,p);\mathbf{\Pi}(np)) < \tfrac{7}{3}e^{-k^2/27np^3}\,\mathbb{I}\{k<np^2\} + e^{-np/15}\,\mathbb{I}\{k\ge np^2\}, \tag{3.4}$$

$$\rho_k(\mathbf{B}(n,p);\mathbf{\Pi}(np)) < \exp\left(-\tfrac{1}{4}\sqrt{nk\ln^3(k/np^2e^2)}\right) \qquad (np^2e^4 \le k \le n(1-p)/\ln(1/p)),$$

see also Ruzankin & Borisov [271] where these bounds have been extended to a non-i.i.d. case. Estimate (3.4) yields

$$\rho_{\lceil np^2\rceil}(\mathbf{B}(n,p);\mathbf{\Pi}(np)) \le e^{-np/15}.$$

Ruzankin & Borisov [271] provide a lower bound to $\rho_k(\mathbf{B}(n,p);\mathbf{\Pi}(np))$.

Let $X_1,...,X_n$ be i.i.d. Bernoulli $\mathbf{B}(p)$ r.v.s, and let $Y_1,...,Y_n$ be i.i.d. Poisson $\mathbf{\Pi}(p)$ r.v.s. Set $S_k^X = X_1+...+X_k$, $S_k^Y = Y_1+...+Y_k$ $(k\in\mathbb{N})$. Denote $S^X = (S_1^X,...,S_n^X)$, $S^Y = (S_1^Y,...,S_n^Y)$,

$$\rho_k(S^X;S^Y) \equiv \rho_k(\mathcal{L}(S^X);\mathcal{L}(S^Y)) = \inf \mathbb{P}\left(\max_{1\le k\le n}|S_k^X - S_k^Y| > k\right),$$

where the infimum is taken over all random vectors $X_1,...,X_n,Y_1,...,Y_n$ defined on a common probability space with $\mathcal{L}(X_i)=\mathbf{B}(p)$, $\mathcal{L}(Y_i)=\mathbf{\Pi}(p)$ $(\forall i)$. Ruzankin & Borisov [271] show that

$$\rho_k(S^X;S^Y) \le 14np^{2+k}e^{-k/3} \qquad \text{if } np\le 1/2, \tag{3.5}$$

$$\rho_k(S^X;S^Y) \le 5(np^2)^{1+k/2}e^{-k/8} \qquad \text{if } np^2\le 1/3. \tag{3.6}$$

Liao et al. [193] present an estimate of the accuracy of Poisson approximation in terms of d_2.

3.1.1 Estimates in terms of the total variation distance

Note that

$$d_K(X;Y) \le d_{TV}(X;Y) \le d_G(X;Y)$$

if X and Y are integer-valued r.v.s.

Many authors worked on the problem of evaluating the total variation distance $d_{TV}(S_n;\pi_\lambda)$ and the Gini-Kantorovich distance $d_G(S_n;\pi_\lambda)$ (cf. [21, 225] and references wherein).

Let $\{X_i\}$ be independent Bernoulli $\mathbf{B}(p_i)$ r.v.s. Prokhorov [240] has established the existence of an absolute constant C such that

$$d_{TV}(\mathbf{B}(n,p);\mathbf{\Pi}(np)) \le Cp. \tag{3.7}$$

According to Kolmogorov ([179], Lemma 5), there exists an absolute constant C such that

$$d_{TV}(S_n;\pi_\lambda) \le C\sum_{i=1}^n p_i^2. \tag{3.8}$$

Thus,

$$d_{TV}(\mathbf{B}(n,p);\mathbf{\Pi}(np)) \le C\min\{p;np^2\}. \tag{3.9}$$

There are two remarkable aspects of estimate (3.7). First, estimate (3.7) is more accurate than (3.8) if $\sum_{i=1}^n p_i$ is "large". Since p differs from np^2 by the factor $1/np$, the latter is sometimes called "the magic factor", cf. [21]. Second, an important fact following from (3.7) and (2.2*) is that Poisson approximation can hold even if $\sum_{i=1}^n p_i$ is "large".

Prokhorov [240] presents also a discrete version of normal approximation to $\mathcal{L}(S_n)$. He points out that Poisson approximation is more accurate than the normal one if $p \equiv p(n) = o(n^{-1/3})$. In other words, Poisson approximation can be more accurate than the normal one in situations where both Poisson and normal approximations hold.

Le Cam [190] attributes inequality

$$d_{TV}(S_n;\pi_\lambda) \le \sum_{i=1}^n p_i^2 \tag{3.10}$$

to Khintchine [175]. Bound (3.10) is sharp: according to (2.10) in Deheuvels & Pfeifer [104],

$$d_{TV}(S_n;\pi_{np}) \ge np^2(1+O(p)) \tag{3.10°}$$

in the case of i.i.d. Bernoulli $\mathbf{B}(p)$ r.v.s such that $p \equiv p(n) = o(1/n)$ as $n \to \infty$.

Note that (3.10) is a consequence of the property of d_{TV} and the following fact:

$$d_{TV}(\mathbf{B}(p);\mathbf{\Pi}(p)) = (1-\mathrm{e}^{-p})p \le p^2\,. \tag{3.11}$$

Indeed, denote $\bar{X} = (X_1,...,X_n)$, $\bar{\pi} = (\pi_{p_1},...,\pi_{p_n})$, where $\{\pi_{p_i}\}$ are independent Poisson $\mathbf{\Pi}(p_i)$ r.v.s. Then

$$d_{TV}(\bar{X};\bar{\pi}) \le \sum_{i=1}^n d_{TV}(X_i;\pi_{p_i}) \le \sum_{i=1}^n p_i^2. \tag{3.10′}$$

According to Kennedy & Quine [173], if $0<np<2-\sqrt{2}$, then

$$d_{TV}(\mathbf{B}(n,p);\mathbf{\Pi}(np)) = \mathbb{P}(S_n=1) - \mathbb{P}(\pi_{np}=1).$$

Kerstan [174] has shown that

$$d_{TV}(S_n;\pi_\lambda) \le 1.05\theta. \tag{3.12}$$

Romanowska [254] has noticed that

$$d_{TV}(\mathbf{B}(n,p);\mathbf{\Pi}(np)) \le p/2\sqrt{1-p}\,. \tag{3.13}$$

Presman [238] has established an estimate of $d_{TV}(S_n;\pi_\lambda)$ with the constant 0.83 at the leading term. In the case of i.i.d. Bernoulli $\mathbf{B}(p)$ r.v.s Presman's bound becomes

$$d_{TV}(\mathbf{B}(n,p);\mathbf{\Pi}(np)) \le 0.83p/(1-p)(1-1/n). \tag{3.14}$$

Xia [317] has derived an estimate with the constant 0.6844 at the leading term. The popular estimate

$$d_{TV}(S_n;\pi_\lambda) \le \lambda^{-1}(1-\mathrm{e}^{-\lambda})\sum_{i=1}^n p_i^2 \tag{3.15}$$

is effectively due to Barbour & Eagleson ([17], Lemma 4). Note that $\lambda^{-1}(1-\mathrm{e}^{-\lambda}) \le \min\{1;1/\lambda\}$.

Roos [260] (see also Čekanavičius & Roos [83]) has obtained a bound with a correct constant $3/4\mathrm{e} \approx 0.276$ at the leading term: if $\theta<1$, then

$$d_{TV}(S_n;\pi_\lambda) \le 3\theta/4\mathrm{e}(1-\sqrt{\theta}\,)^{3/2}\,. \tag{3.16}$$

Note that $\theta/(1-\sqrt{\theta})^{3/2} \geq \theta(1+1.5\sqrt{\theta}+3.75\theta)$. Roos [260] has shown also that

$$d_{TV}(S_n;\pi_\lambda) \sim 3\theta/4\mathrm{e}$$

if $\theta \to 0$ and $\lambda \to 1$ as $n \to \infty$. Thus, constant $3/4\mathrm{e}$ cannot be improved.

Denote

$$p_n^* = \max_{i \leq n} p_i, \qquad \varepsilon = \min\left\{1; (2\pi\lfloor\lambda - p_n^*\rfloor)^{-1/2} + 2\delta/(1-p_n^*/\lambda)\right\},$$

$$\delta = (1-\mathrm{e}^{-\lambda})\sum_{i=1}^n p_i^2/\lambda, \qquad \delta^* = (1-\mathrm{e}^{-\lambda})\sum_{i=1}^n p_i^3/\lambda.$$

Clearly, $\delta^2 \leq \delta^*$. The following inequality from [221], Theorem 4.12, sharpens the second-order term of the right-hand side (RHS) of estimate (3.16):

$$d_{TV}(S_n;\pi_\lambda) \leq 3\theta/4\mathrm{e} + 2\delta^*\varepsilon + 2\delta^2. \tag{3.17}$$

In the case of $\mathcal{L}(S_n) = \mathbf{B}(n,p)$ estimate (3.17) becomes

$$d_{TV}(S_n;\pi_{np}) \leq 3p/4\mathrm{e} + 2(1-\mathrm{e}^{-np})p^2\varepsilon_n + 2(1-\mathrm{e}^{-np})^2p^2, \tag{3.17*}$$

where $\varepsilon_n = \min\{1; (2\pi\lfloor(n-1)p\rfloor)^{-1/2} + 2(1-\mathrm{e}^{-np})p/(1-1/n)\}$. The second-order term in (3.17*) is of order $p^2 \wedge np^3$.

In applications one often has $\lambda \equiv \lambda(n) \to \infty$ as $n \to \infty$. Hence estimates with the "magic factor" $(1-\mathrm{e}^{-\lambda})/\lambda$ attract special interest.

A possibility of the "super–magic" factor $\mathrm{e}^{-\lambda}$ when one approximates $\mathbb{P}(S_n \in A)$ for a bounded A has been discussed in Novak [221], Section 4.5 (such approximations are of interest in extreme value theory). For instance, if $\{X_i\}$ are independent Bernoulli $\mathbf{B}(p_i)$ r.v.s and $A = \{0\}$, then

$$0 \leq \mathbb{P}(\pi_\lambda = 0) - \mathbb{P}(S_n = 0) \leq \mathrm{e}^{-\lambda + p_n^*}\sum_{i=1}^n p_i^2/2. \tag{3.18}$$

Indeed, set $c_i = \prod_{j=1}^{i-1}\mathrm{e}^{-p_j}\prod_{j=i+1}^n(1-p_j)$. Since $\mathrm{e}^{-p_i}-1+p_i \leq p_i^2/2$ by Taylor's formula,

$$\mathbb{P}(\pi_\lambda = 0) - \mathbb{P}(S_n = 0) = \sum_{i=1}^n(\mathrm{e}^{-p_i} - 1 + p_i)c_i \leq \mathrm{e}^{-\lambda + p_n^*}\sum_{i=1}^n p_i^2/2.$$

In the case of Binomial $\mathbf{B}(n,p)$ distribution (3.18) becomes

$$0 \leq \mathbb{P}(\pi_{np} = 0) - \mathbb{P}(S_n = 0) \leq \tfrac{1}{2}np^2\mathrm{e}^{-(n-1)p} \leq \frac{2n}{\mathrm{e}^2(n-1)^2}. \tag{3.18*}$$

Note that $2/\mathrm{e}^2 \approx 0.2707$.

A trivial lower bound is

$$d_{TV}(S_n;\pi_\lambda) \geq \mathbb{P}(\pi_\lambda = 0) - \mathbb{P}(S_n = 0) = \frac{1}{2}\sum_{i=1}^n p_i^2 c_i \mathrm{e}^{-\hat{\tau}_i p_i} \geq \frac{1}{2}\sum_{i=1}^n p_i^2\prod_{j=1}^n(1-p_j), \tag{3.18°}$$

where $\{\hat{\tau}_i\} \in [0;1]$. If $p \equiv p(n) = o(1/n)$ as $n \to \infty$, then (3.18°) yields

$$d_{TV}(\mathbf{B}(n,p); \mathbf{\Pi}(n,p)) \geq np^2/2(1+o(1)).$$

According to Ruzankin & Borisov [271],

$$d_{TV}(S_n;\pi_\lambda) \geq 1 - \exp\left(-\sum_{i=1}^n p_i^2(1-p_i)/2\right),$$

cf. (3.10°).

Bound (3.17) is a consequence of inequality

$$|\mathbb{P}(S_n \in A) - \mathbb{P}(\pi_\lambda \in A)| \leq |\mathbb{P}(\pi_\lambda + 1 \in A) - \mathbb{P}(\pi_\lambda^\star \in A)|\theta/2 + 2\delta^*\varepsilon + 2\delta^2. \qquad (3.17^\star)$$

The first term on the RHS of (3.17*) has the "super–magic" factor if A is finite.

The standard choice of a parameter of the approximating Poisson distribution is $\lambda = \mathbb{E}S_n$. However, a different choice of a parameter may sometimes be preferable. Set

$$\tilde{p}_i = -\ln(1-p_i) \quad (1 \leq i \leq n), \quad \mu = \sum_{i=1}^n \tilde{p}_i\,.$$

Then

$$d_{TV}(S_n;\pi_\mu) \leq \sum_{i=1}^n \tilde{p}_i^2(1+\tilde{p}_i)/2$$

(Borowkow [50], Chapter 5). According to Serfling [276],

$$d_{TV}(S_n;\pi_\mu) \leq \sum_{i=1}^n \tilde{p}_i^2/2. \qquad (3.19)$$

Novak [221], Proposition 4.4, has shown that

$$d_{TV}(S_n;\pi_\mu) \leq \frac{1}{2}\sum_{i=1}^n \tilde{p}_i(1-e^{-\tilde{p}_i}) \leq \frac{1}{2}\left(1-e^{-\theta_+}\right)\sum_{i=1}^n \tilde{p}_i\,, \qquad (3.20)$$

where $\theta_+ = \sum_{i=1}^n \tilde{p}_i^2 / \sum_{i=1}^n \tilde{p}_i$. Note that (3.20) entails (3.19).

Lower bounds.

A lower bound to $d_{TV}(S_n;\pi_\lambda)$ has been given by Barbour & Hall [18]:

$$d_{TV}(S_n;\pi_\lambda) \geq \min\left\{1/\lambda; 1\right\}\sum_{i=1}^n p_i^2/32. \qquad (3.21)$$

Presman ([8], Lemma 8.4.1) has shown that

$$d_K(\mathbf{B}(n,p);\mathbf{\Pi}(n,p)) \geq np^2/8e^2 \qquad (np \leq 2)\,.$$

A lower bound to $d_K(S_n;\pi_\lambda)$ with an explicit constant has been derived by Kruopis [185].

Bound (3.21) has been improved by Sason [273]:

$$d_{TV}(S_n;\pi_\lambda) \geq \frac{e}{2\lambda}\cdot\frac{1-\tilde{\omega}^{-1}(3+7\lambda^{-1})}{\tilde{\omega}+2+e^{-1/2}}\sum_{i=1}^n p_i^2\,, \qquad (3.22)$$

where $\tilde{\omega} = 3 + 7\lambda^{-1} + \lambda^{-1}\sqrt{(3\lambda+7)((3+2e^{-1/2})\lambda+7)}$.

Karymov [171] has shown that

$$d_o(S_n;\pi_\lambda) \geq \frac{\lambda_2}{2}\prod_{i=1}^n(1-p_i).$$

3.1.2 Estimates in terms of the Gini-Kantorovich distance

According to Roos [260],

$$d_G(S_n;\pi_\lambda) \le \sum_{i=1}^{n} p_i^2 .$$

Witte [316] has shown that

$$d_G(S_n;\pi_\lambda) \le \frac{-\sqrt{e\lambda}}{2\sqrt{2\pi}} \ln(1-2\theta e^{2p_n^*}). \tag{3.23}$$

According to Barbour & Xia [33],

$$d_G(S_n;\pi_\lambda) \le \tfrac{4}{3}\sqrt{2/e\lambda}\sum_{i=1}^{n} p_i^2 . \tag{3.24}$$

Note that $\frac{4}{3}\sqrt{2/e} \approx 1.1438$.

Recall that $\mu = -\sum_{i=1}^{n} \ln(1-p_i)$. If $p_n^* \le 1/2$, then

$$d_G(S_n;\pi_\mu) \le \sum_{i=1}^{n} p_i^2/2(1-p_i) \tag{3.25}$$

(Deheuvels et al. [106]).

Roos [260] has shown that

$$d_G(S_n;\pi_\lambda) \le \Big(1/\sqrt{2e} + 1.6\sqrt{\theta}(2-\theta)/(1-\sqrt{\theta})^2\Big)\,\theta\sqrt{\lambda} . \tag{3.26}$$

Constant $1/\sqrt{2e} \approx 0.429$, in general, cannot be improved.

If $\mathcal{L}(X_i) = \mathbf{B}(p)$ $(\forall i)$, then (3.70) yields

$$d_G(S_n;\pi_{np}) \le np^2 \min\Big\{1; 1/\sqrt{2\pi\lfloor np\rfloor}\,; 1/\sqrt{2enp}\Big\} + \hat{\gamma}_n , \tag{3.27}$$

where $\hat{\gamma}_n = 2p^2\big((1\wedge\frac{2}{3}np) + (np\wedge c_\star\sqrt{np})\big)$, $c_\star = \frac{4}{3}\sqrt{2/e}$.

If $np \ge 1/(1-e/\pi)$, then $\pi\lfloor np\rfloor \ge enp$, and (3.27) becomes

$$d_G(S_n;\pi_{np}) \le np^2/\sqrt{2\pi\lfloor np\rfloor} + \hat{\gamma}_n . \tag{3.27*}$$

Estimate (3.27*) has a sharper constant at the leading term than (3.26).

Ibragimov & Presman [163] have introduced distance

$$d_{t,a}(\mathcal{L}(X);\mathcal{L}(Y)) = \sum_k |k-a|^t |\mathbb{P}(X=k) - \mathbb{P}(Y=k)| \qquad (-1/2<t\le 0, a\ge 0)$$

for distributions concentrated on $\mathbb{Z}_+$. They have shown that for any $t \in (-1/2;0]$, there exists constant $C_t > 0$ such that

$$d_{t,0}(\mathbf{B}(n,p);\mathbf{\Pi}(np)) \le C_t p(np)^t \tag{3.28}$$

(cf. (3.7)). They state that Prokhorov's [240] argument yields

$$d_{1,np}(\mathbf{B}(n,p);\mathbf{\Pi}(np)) \le Cp\sqrt{np} \qquad (\exists C>0).$$

Large deviations.

The following large deviations inequality is due to Bernstein [43], p. 168:

$$\mathbb{P}((S_n - np)/\sqrt{npq} > t\,(1+\gamma_{p,n})) < \exp(-t^2/2) \qquad (t>0),$$

where $q = 1-p$, $\gamma_{p,n} = t_{p,n}(q-p)/6 + t_{p,n}^2(p^3+q^3)/12$, $t_{p,n} = t/\sqrt{npq}$.

In the case of the Binomial distribution $\mathcal{L}(S_n) = \mathbf{B}(n,p)$ the following result has been proved by Ivchenko [165]. Suppose that $p \to 0$, $np \to \infty$ as $n \to \infty$, where $\lambda = np$. Then uniformly in $k \in \{k\colon \sqrt{\lambda} \ll |k-\lambda| \ll \lambda\}$ as $\lambda \to \infty$,

$$\frac{\mathbb{P}(S_n \le k)}{\mathbb{P}(\pi_\lambda \le k)} = \exp\left(-nq\sum_{r=2}^{\infty}\frac{(-1)^r}{r(r-1)}\left(\frac{\lambda-k}{nq}\right)^r\right)(1+o(1)) \quad (k<\lambda),$$

$$\frac{\mathbb{P}(S_n > k)}{\mathbb{P}(\pi_\lambda > k)} = \exp\left(-nq\sum_{r=2}^{\infty}\frac{1}{r(r-1)}\left(\frac{k-\lambda}{nq}\right)^r\right)(1+o(1)) \quad (k>\lambda).$$

Okamoto [227] has shown that $(y>0)$

$$\mathbb{P}(S_n - np \le y) \le e^{-y^2/2npq} \text{ if } p \le 1/2, \quad \mathbb{P}(S_n - np \ge y) \le e^{-y^2/2npq} \text{ if } p \ge 1/2.$$

Besides ([221], formula (14.50)),

$$\begin{aligned}\mathbb{P}((S_n - np)/\sqrt{npq} \ge y) &\le \mathbb{E}\exp\left(-\frac{y^2}{2}\left(1+\frac{\tau y\sqrt{q}}{\sqrt{np}}\right)^{-1}\left(1-\frac{\tau y\sqrt{p}}{\sqrt{nq}}\right)^{-1}\right)\\ &\le \exp\left(-\frac{y^2}{2}\left(1+\frac{y(1-2p)}{\sqrt{npq}}\right)^{-1}\right),\end{aligned} \tag{3.29}$$

where $\mathcal{L}(\tau) = \mathbf{U}(0;1)$, $q = 1-p$. In particular,

$$\mathbb{P}\left((S_n - np)/\sqrt{npq} \ge y\right) \le \exp\left(-y^2/2\right) \qquad \text{if } p \ge 1/2. \tag{3.29*}$$

If $0<p<x<1$ and $c_1\sqrt{nx^3}<1$, then

$$\mathbb{P}(S_n \ge nx)/\mathbb{P}(\pi_{np} \ge nx) = e^{-n(x-p)^2/2}(1+(5.74\sqrt{nx^3}+p/x)\gamma),$$

where $|\gamma| \le 1$, $c_1 = 2e^{1/12}\sqrt{2\pi}$ (Nagaev & Chebotarev [218]). Kruglov [184] presents bounds to $\mathbb{P}(\sup_{k\ge n} S_k/k \ge p+x)$ if $p \in [1/2;1]$ and to $\mathbb{P}(\inf_{k \ge n} S_k/k \le p-x)$ if $p \in [0;1/2]$.

Probabilities of large deviations have been studied by Aleškevičienė & Statulevičius [3]. Probabilities of moderate deviations have been presented by Liu & Xia [197]. Results concerning probabilities of large and moderate deviations for the distribution of a sum of independent 0-1 r.v.s can be found in Barbour et al. [23], Corollary 4.3, Roos [256].

Some other results.

Let $\{X_i\}$ be i.i.d. Bernoulli $\mathbf{B}(p)$ r.v.s, $S_n = X_1 + \ldots + X_n$. Pinelis [233] has shown that

$$\begin{aligned}\mathbb{P}(S_n \ge m) &< \mathbb{P}(\pi_{np} \ge m) \qquad (n \ge m \ge 1+np),\\ \mathbb{P}(S_n < m) &< \mathbb{P}(\pi_{np} < m) \qquad (np \ge m).\end{aligned}$$

A non-uniform bound to $|\mathbb{P}(S_n = k) - \mathbb{P}(\pi_\lambda = k)|$ has been given by Arak [5]:

$$|\mathbb{P}(S_n = k) - \mathbb{P}(\pi_{np} = k)| \le \tfrac{4}{n}\left(k + (k-np)^2\right)\left(\mathbb{P}(S_n = k) + \mathbb{P}(\pi_{np} = k)\right).$$

Let $\{X_i\}$ be independent Bernoulli $\mathbf{B}(p_i)$ r.v.s, $\lambda = \mathbb{E}S_n$. A non-uniform bound to $|\mathbb{P}(S_n=k) - \mathbb{P}(\pi_\lambda=k)|$ has been given by Karymov [171]:

$$|\mathbb{P}(S_n=k+2) - \mathbb{P}(\pi_\lambda=k+2)| \le \frac{\lambda_2}{2}(1+\lambda^2/k^2)\mathrm{e}^{-\lambda}(\lambda\mathrm{e}/k)^k \qquad (k\ge\lambda),$$

$$|\mathbb{P}(S_n\le k+1) - \mathbb{P}(\pi_\lambda\le k+1)| \le \frac{\lambda_2}{2}(1+\lambda^2/k^2)\mathrm{e}^{-\lambda}(\lambda\mathrm{e}/k)^k\frac{k}{k-\lambda\mathrm{e}} \qquad (k>\lambda\mathrm{e}).$$

According to Zacharovas & Hwang [321], if $k\ge0$, then

$$|\mathbb{P}(S_n \le k) - \mathbb{P}(\pi_\lambda \le k)| \le \frac{\sqrt{2}\,(\sqrt{\mathrm{e}}-1)\theta}{(1-\theta)^{3/2}}\mathrm{e}^{-(k-\lambda)^2/4(k+\lambda)},$$

$$|\mathbb{P}(S_n = k) - \mathbb{P}(\pi_\lambda = k)| \le \frac{\sqrt{6}\,(\sqrt{\mathrm{e}}-1)\theta}{(1-\theta)^2\sqrt{\lambda}}\mathrm{e}^{-(k-\lambda)^2/4(k+\lambda)}.$$

Zacharovas & Hwang ([160], Theorem 6) present asymptotics of $d_K(\mathbf{B}(n,p);\mathbf{\Pi}(np))$ and $d_o(\mathbf{B}(n,p);\mathbf{\Pi}(np))$.

Borisov & Ruzankin ([47], Lemma 2) have showed that

$$\sup_k \mathbb{P}(S_n=k)/\mathbb{P}(\pi_\lambda=k) \le (1-p_n^*)^2.$$

Sharp non-Poisson approximation to the Binomial $\mathbf{B}(n,p)$ distribution function has been given by Zubkov & Serov [335]. Denote by Φ the standard normal d.f.. Let

$$\Lambda(x) = x\ln(x/p) + (1-x)\ln((1-x)/(1-p)) \qquad (0<x<1)$$

denote the rate function of the Bernoulli distribution $\mathbf{B}(p)$ (cf. [221], p. 322). Set

$$Z_{n,p}(k) = \Phi\Big(\mathrm{sgn}(k/n-p)\sqrt{2n\Lambda(k/n)}\,\Big).$$

Then

$$Z_{n,p}(k) \le \mathbb{P}(S_n\le k) \le Z_{n,p}(k+1).$$

3.1.3 Asymptotics of $d_{TV}(S_n;\pi_\lambda)$

In the case of identically distributed Bernoulli $\mathbf{B}(p)$ r.v.s the asymptotics of $d_{TV}(S_n;\pi_\lambda)$ has been established by Prokhorov [240]: if $p\equiv p(n)\to0$, $np\to\infty$ as $n\to\infty$, then

$$d_{TV}(\mathbf{B}(n,p);\mathbf{\Pi}(np)) = p/\sqrt{2\pi\mathrm{e}}\Big(1+O\big(p+1/\sqrt{np}\,\big)\Big). \tag{3.30}$$

Kerstan [174], Deheuvels & Pfeifer [103, 105], Deheuvels et al. [106], and Roos [256] have generalised (3.30) to the case of non-identically distributed 0-1 r.v.s. Hwang & Zacharovas [160] have improved (3.30):

$$d_{TV}(\mathbf{B}(n,p);\mathbf{\Pi}(np)) = \frac{p}{\sqrt{2\pi\mathrm{e}}}\Big(1+p/2+O\big(p^2+1/np\big)\Big) \tag{3.31}$$

if $p\to0$, $np\to\infty$.

Given a non-negative integer-valued random variable Y, denote by $Y^\star$ a random variable with the distribution

$$\mathbb{P}(Y^\star=k) = \mathbb{P}(Y=k)(k-\lambda)^2/\lambda \qquad (k\in\mathbb{Z}_+). \tag{3.32}$$

The next bound is a consequence of Theorem 3.16.

Theorem 3.1 *If $X_1, ..., X_n$ are independent Bernoulli $\mathbf{B}(p_i)$ r.v.s, then*

$$|d_{TV}(S_n;\pi_\lambda) - \theta d_{TV}(\pi_\lambda^\star;\pi_\lambda{+}1)/2| \le 2\delta^*\varepsilon + 2\delta^2 .$$

One can check that

$$d_{TV}(\pi_\lambda^\star;\pi_\lambda{+}1) = \sqrt{2/\pi \mathrm{e}} + O\big(1/\sqrt{\lambda}\big) \tag{3.33}$$

as $\lambda \to \infty$ (cf. (4.59) in [221]). Thus,

$$d_{TV}(S_n;\pi_\lambda) = \theta/\sqrt{2\pi\mathrm{e}}\left(1{+}O\big(\theta{+}1/\sqrt{\lambda}\big)\right) \tag{3.34}$$

if $\lambda \to \infty$ and $\theta \to 0$ as $n{\to}\infty$.

Deheuvels & Pfeifer [103] and Hwang & Zacharovas [160] present the asymptotics of $d_{TV}(S_n;\pi_\lambda)$ in the case where $\lambda{\equiv}\lambda(n){\to}\text{const}$ as $n{\to}\infty$.

Example 3.1 Let $p_i = 1/i,\ i \in \mathbb{N}$. Then $\lambda = \lambda(n) \to \infty,\ \theta \to 0$ as $n{\to}\infty$, and (3.34) entails

$$d_{TV}(S_n;\pi_\lambda) \sim \theta/\sqrt{2\pi\mathrm{e}} .$$

Note that $p_n^* = 1$. □

Deheuvels et al. ([106], Corollary 2.2) have shown that

$$\left|d_G(S_n;\pi_\lambda) - \lambda_2\mathrm{e}^{-\lambda}\lambda^{\lfloor\lambda\rfloor}/\lfloor\lambda\rfloor!\right| \le 2(2\theta)^{3/2}\sqrt{\lambda}/(1{-}\sqrt{2\theta}) \qquad (\theta{<}1/2).$$

3.1.4 Shifted (translated) Poisson approximation

The mean of a Poisson random variable is equal to its variance. Therefore, it is natural to expect that pure Poisson approximation is effective in situations where $\mathbb{E}S_n$ is "close" to $\operatorname{var} S_n$.

If $\mathbb{E}S_n$ is not "close" to $\operatorname{var} S_n$, one may consider shifted Poisson approximation

$$S_n \approx a + \pi_b ,$$

thus adding an additional parameter. If one wants to retain the lattice structure (which is important for the total variation distance and the point distance), then parameter a must be an integer number (i.e., only approximate matching of the first two moments can be ensured).

Denote $\lfloor x\rfloor = \max\{k{\in}\mathbb{Z}\colon k{\le}x\}$, $\{x\} = x{-}\lfloor x\rfloor$. The choice of parameters a, b is based on matching the first two moments of S_n: one would like

$$\mathbb{E}(a{+}\pi_b) = a + b = \lambda, \quad \operatorname{var}(a{+}\pi_b) \equiv b = \sigma^2 .$$

The exact match can be achieved if $b{=}\sigma^2$, $a = \lambda{-}\sigma^2 \in \mathbb{Z}$.

Example 3.2 Let X be a Bernoulli $\mathbf{B}(p)$ r.v. Set $Y = p^2{+}\pi_{pq}$, where $q{=}1{-}p$. Then

$$\mathbb{E}X = p = \mathbb{E}Y,\ \ \mathbb{E}X^2 = p^2 = \mathbb{E}Y^2 .$$

Thus, first two moments of $\mathcal{L}(X)$ and $\mathcal{L}(Y)$ coincide. This suggests that $\mathcal{L}(\lfloor np^2\rfloor{+}\pi_{npq})$ may approximate Binomial distribution $\mathbf{B}(n,p)$ better than $\mathcal{L}(\pi_{np})$ (cf. Theorem 3.11). □

In the case of integer-valued r.v.s one would take

$$a = \lfloor \lambda - \sigma^2 \rfloor, \quad b = \sigma^2 + \{\lambda - \sigma^2\} = \lambda - \lfloor \lambda - \sigma^2 \rfloor,$$

i.e., the approximating r.v. is

$$Y = a + \pi_b \,.$$

Note that $0 \leq \mathrm{var}\, Y - \mathrm{var}\, S_n < 1$. Since

$$d_{TV}(S_n; Y) = d_{TV}(S_n - a; \pi_b),$$

one can view shifted Poisson approximation as a pure Poisson approximation to $S_n - a$.

Let $X_1, ..., X_n$ be independent 0-1 r.v.s. Set $p_i = \mathbb{P}(X_i = 1)$, $q_i = 1 - p_i$,

$$\lambda = \mathbb{E} S_n, \quad \sigma^2 = \mathrm{var}\, S_n \quad \lambda_2 = \lambda - \sigma^2 \,.$$

Čekanavičius & Vaitkus [80] have shown that

$$d_{TV}(S_n; Y) \leq \frac{\{\lambda_2\}}{\sigma^2 + \{\lambda_2\}} + \frac{0.93\, \lambda_2}{\sigma^3} + \mathrm{e}^{-\sigma^2/4} \,. \tag{3.35}$$

if $\sigma^2 \geq 4$. A similar result has been given by Röllin [253]:

$$d_{TV}(S_n; Y) \leq 2\sigma^{-2} + \Big(\sum_{i=1}^{n} p_i^3 (1 - p_i)\Big)^{1/2} \sigma^{-2} \,. \tag{3.36}$$

If $\mathcal{L}(S_n) = \mathbf{B}(n, p)$ and $q = 1 - p$, then (3.36) becomes

$$d_{TV}(S_n; Y) \leq 2/npq + \sqrt{p/nq} \,.$$

Kruopis [185] used a slightly different choice of parameters. Let

$$\tilde{Y} = \lfloor \lambda_2 + 0.5 \rfloor + \pi_{\lambda - \lfloor \lambda_2 + 0.5 \rfloor} \,.$$

Then $\mathbb{E}\tilde{Y} = \lambda$, $\mathrm{var}\, \tilde{Y} = \lambda - \lfloor \lambda_2 + 0.5 \rfloor$, and

$$d_{\mathrm{o}}(S_n; \tilde{Y}) \leq \sum_{j=1}^{n} p_j^2 \min\left(1; \sqrt{\frac{\mathrm{e}}{\pi}} \Big(\sum_{j=1}^{n} p_j (1 - p_j)\Big)^{-3/2}\right).$$

Kruopis [185] has also established a similar to (3.36) bound to $d_K(S_n; \tilde{Y})$.

If $\{X_j\}$ are i.i.d. Bernoulli $\mathbf{B}(p)$ r.v.s, set $q = 1 - p$,

$$\mu = npq + \{np^2\}, \quad \kappa_{S_n} = np^2 \,.$$

Theorem 3.11 entails

$$d_{TV}\Big(S_n; \lfloor np^2 \rfloor + \pi_{npq + \{np^2\}}\Big) \leq \{np^2\} \bar{\varepsilon} + 2p\varepsilon_n^{\mathrm{o}} \,, \tag{3.37}$$

where $\varepsilon_\mu^* = 1 \wedge 1/\sqrt{2\pi \lfloor \mu \rfloor}$, $\varepsilon_n' = \max_k \mathbb{P}(S_{n-1} = k)$,

$$\begin{aligned} \varepsilon_n^{\mathrm{o}} &= \min\Big\{ \frac{\sqrt{2/\pi}}{\sqrt{1/4 + (n-1)p}} ; \frac{1}{\sqrt{2\pi[(n-1)p]}} + 2p \frac{1 - \mathrm{e}^{-np}}{1 - 1/n} ; \varepsilon_n' \Big\}, \\ \bar{\varepsilon} &\leq 2\varepsilon_\mu^* \sqrt{2/\mathrm{e}\mu} + 2\mu^{-1}(1 - \mathrm{e}^{-\mu}) \left(\{np^2\}/np + 2p\varepsilon_n^{\mathrm{o}}\right). \end{aligned}$$

If $np\to\infty$, $p\equiv p(n)\ll n^{-1/3}$ as $n\to\infty$, then $\varepsilon_n^o\sim 1/\sqrt{2\pi np}$, $\bar\varepsilon\sim 2\varepsilon_\mu^*\sqrt{2/enp}$, and the RHS of (3.37) is $\sim 2\{np^2\}/\sqrt{\pi e np}+\sqrt{2p/\pi n}$.

Shifted Poisson approximation can be compared to pure Poisson approximation.

Consider Binomial distribution $\mathbf{B}(n,p)$, where $p\le 1/2$. If $np^2\ge 1$, then the RHS of (3.35), (3.36), (3.37) are

$$O\Big(\sqrt{p/n}+1/np\Big), \tag{3.38}$$

while the RHS of (3.15) is $p\ge 1/np$; $p\ge\sqrt{p/n}$ if $np\ge 1$.

One may use (3.10) or (3.19) if $p<1/n$, (3.16) or (3.17) if $1/n\le p<1/\sqrt{n}$, a (3.38)-type estimate if $p\ge 1/\sqrt{n}$.

Estimates (3.35), (3.36), (3.75*) have advantages over the Berry–Esseen inequality if p is "small" and np is "large". Some applications can be found in [29, 33, 35, 223, 253].

Shifted Poisson approximation in terms of d_G.

An estimate of the accuracy of shifted Poisson approximation to the distribution of a sum S_n of Bernoulli $\mathbf{B}(p_i)$ r.v.s in terms of the Gini-Kantorovich distance has been given by Barbour & Xia [33]:

$$d_G(S_n;\lfloor\lambda_2\rfloor+\pi_{\lambda-\lfloor\lambda_2\rfloor})\le 4\lambda_2/(\lambda-\lambda_2)+\tfrac{4}{3}\sqrt{2/e}\,\{\lambda_2\}/\sqrt{\lambda-\lambda_2}\,. \tag{3.39}$$

If $\{X_i\}$ are independent Bernoulli $\mathbf{B}(p)$ r.v.s, set $q=1-p$. Then (3.39) becomes

$$d_G(S_n;\lfloor\lambda_2\rfloor+\pi_{\lambda-\lfloor\lambda_2\rfloor})\le 4p/q+\tfrac{4}{3}\sqrt{2/e}\,\{np^2\}/\sqrt{npq}\,.$$

Let $\{X_i\}$ be independent Bernoulli $\mathbf{B}(p)$ r.v.s. Theorem 3.12 yields

$$d_G(S_n;\lfloor np^2\rfloor+\pi_{npq+\{np^2\}})\le 2p(1\wedge c_\mu\varepsilon_n^o)+\{np^2\}\min\Big\{1;c_\star/\sqrt{npq}\,;2\varepsilon_n^o\Big\}, \tag{3.40}$$

where

$$c_\star=\tfrac{4}{3}\sqrt{2/e},\quad c_\mu=\mu\wedge c_\star\sqrt{\mu},\quad \mu=npq+\{np^2\}\,.$$

Thus,

$$d_G(S_n;\lfloor np^2\rfloor+\pi_{npq+\{np^2\}})=O\Big(p+1/\sqrt{np}\Big). \tag{3.38*}$$

The accuracy of shifted Poisson approximation in terms of the Gini-Kantorovich distance differs from (3.38) by a factor of $\sqrt{np}$.

If $np\to\infty$, $p\equiv p(n)\ll n^{-1/3}$ as $n\to\infty$, then $\varepsilon_n^o\sim 1/\sqrt{2\pi np}$, and the RHS of (3.40) is $\sim pc_\star\sqrt{2/\pi}+\{np^2\}\sqrt{2/\pi np}$.

3.1.5 Uniform in p Poisson approximation

Let $X_1,...,X_n$ be i.i.d. Bernoulli $\mathbf{B}(p)$ r.v.s, i.e., $\mathcal{L}(S_n)=\mathbf{B}(n,p)$. Then

$$Y\equiv Y_{n,p}=\lfloor np^2\rfloor+\pi_{npq+\{np^2\}},\quad \mathbb{E}Y=np,\quad \operatorname{var}Y=npq+\{np^2\}.$$

The following theorem presents a uniform in $p\in[0;1/2]$ bound to $d_{TV}(S_n;Y)$. Note that one only needs to consider $p\le 1/2$ since $\mathcal{L}(n-S_n)=\mathbf{B}(n,1-p)$.

Theorem 3.2 [223] *If $n>4$, then*

$$\sup_{0\le p\le 1/2} d_{TV}(S_n;Y)\le\frac{2/\sqrt{\pi e}}{\sqrt{n}-2}+\frac{0.9n^{1/4}}{n-1}+\frac{2+1.8/n^{1/4}}{(\sqrt{n}-1)^2}\,. \tag{3.41}$$

Note that $2/\sqrt{\pi e} \approx 0.6844$. The rate $n^{-1/2}$ in (3.41) is correct: according to (3.30) and (3.31), $d_{TV}(\mathbf{B}(n,p);\mathbf{\Pi}(np)) \sim c/\sqrt{2\pi e n}$ as $n\to\infty$ if $p = c/\sqrt{n}$, $c \in (0;1)$.

Inequality (3.41) has advantages over (3.15), (3.30), (3.35), (3.36) etc.: consider, e.g., "small" or "large" values of p. Bound (3.41) is preferable to (3.15), (3.16), (3.17*) if $p \geq 8\sqrt{e}/3\sqrt{\pi n}\,(1+o(1))$. For all small enough p the right-hand side of estimate (3.41) is sharper than that of the Berry–Esseen bound (concerning the Berry–Esseen bounds, see, e.g., [278]). Estimate (3.41) has another advantage over the Berry–Esseen inequality: a uniform in $p\in(0;1/2]$ Berry–Esseen bound is infinite.

Comparing (3.41) to the corresponding Meshalkin's [206] and Pressman's [237] results, note that the constants in (3.41) are explicit (which matters in applications); besides, the structure of the approximating distribution $\mathcal{L}(Y)$ is simpler and does not assign mass to negative numbers.

A bound $\sup_{0<p\leq 1/2} d_o(S_n;Y) \leq Cn^{-3/4}$ in terms of the point metric, where C is an absolute constant, is available in [226].

The following theorem presents a uniform in $p \in [0;1/2]$ bound in terms of the Gini-Kantorovich distance.

Theorem 3.3 [225] *If* $n\geq 10$, *then*

$$\sup_{0<p\leq 1/2} d_G(S_n; \lfloor np^2\rfloor + \pi_\mu)/\sqrt{np} \leq \frac{2c_\star}{\sqrt{\pi}}\Big(1/\sqrt{n-\sqrt{n}} + \sqrt{2\pi}/n^{3/4}\Big). \tag{3.42}$$

Note that $2c_\star/\sqrt{\pi} \approx 1.29$.

If $p \equiv p(n) \gg 1/\sqrt{n}$, then approximation (3.42) is sharper than that of (3.26).

Bound (3.42) can be compared with the estimate of the accuracy of normal approximation to $\mathcal{L}(S_n)$ in terms of d_G. In the case of the Binomial distribution one has

$$d_G(S_n; np+\xi\sqrt{npq}\,)/\sqrt{np} \leq (q^2+p^2)/\sqrt{np}, \tag{3.43}$$

where ξ is a standard normal r.v. (Goldstein [132] and Tyurin [294]). For all small enough p the right-hand side of estimate (3.42) is sharper than that of (3.43). Moreover, a uniform in $p\in(0;1/2]$ bound to the RHS of (3.43) would be infinite.

Note also that (3.42) allows for approximating $\mathbb{E}h(S_n)$ for a class of unbounded functions beyond the class $\mathcal{L}_1$ of Lipschitz functions.

Open problem

3.1. Generalise Theorems 3.2 and 3.3 to the case of m-dependent r.v.s.

3.2 Poisson approximation to the multinomial distribution

Results on the accuracy of Poisson approximation to the distribution of a sum of Bernoulli r.v.s can be generalised to the case of a multinomial distribution.

Let $\bar{S}_n$ be a random vector with multinomial distribution $\mathbf{B}(n, p_1, ..., p_m)$:

$$\mathbb{P}(\bar{S}_n = \bar{l}\,) = \frac{n!}{l_1!...l_m!(n-l)!} p_1^{l_1}...p_m^{l_m}(1-p)^{n-l}, \tag{3.44}$$

where $l_i \in \mathbb{Z}_+$ $(\forall i)$, $\bar{l} = (l_1, .., l_m)$, $l = l_1 + ... + l_m \leq n$, $p = p_1 + ... + p_m$.

Formula (3.44) describes, in particular, the joint distribution of the increments of the empirical d.f.

Note that

$$\bar S_n \stackrel{d}{=} \bar\xi_1 + ... + \bar\xi_n\,,$$

where $\bar\xi, \bar\xi_1, ..., \bar\xi_n$ are i.i.d. random vectors with the distribution

$$\mathbb{P}(\bar\xi = \bar 0) = 1-p,\ \mathbb{P}(\bar\xi = \bar e_j) = p_j \qquad (1 \le j \le m),$$

vector $\bar 0$ has all coordinates equal to 0, vector $\bar e_j$ has the jth coordinate equal to 1 and the other coordinates equal to 0.

Let $\pi_n(\cdot)$ denote a Poisson jump process on $[0;1]$ with intensity rate n. Denote by

$$\bar\pi = \left(\pi_{(1)}, \dots, \pi_{(m)}\right)$$

a vector of independent Poisson r.v.s with parameters $np_1, ..., np_m$. Then $\bar\pi$ is a vector of increments of Poisson process $\pi_n(\cdot)$:

$$\pi_{(1)} \stackrel{d}{=} \pi_n(p_1),\ \dots,\ \pi_{(m)} \stackrel{d}{=} \pi_n(p) - \pi_n(p-p_m).$$

Note that (cf. (3.1))

$$\mathbb{P}(\bar S_n = \bar l\,) = \mathbb{P}\left(\pi_n(p_1) = l_1, ..., \pi_n(p) - \pi_n(p-p_m) = l_n \,|\, \pi_n(1) = 1\right). \tag{3.44*}$$

Arenbaev ([9], formulas (5)–(9′)) has shown that

$$d_{TV}(\bar S_n; \bar\pi) = d_{TV}(\mathbf{B}(n,p); \mathbf{\Pi}(np)). \tag{3.45}$$

Using (3.45), he has derived the asymptotics of $d_{TV}(\bar S_n; \bar\pi)$:

$$d_{TV}(\bar S_n; \bar\pi) = p/\sqrt{2\pi \mathrm{e}}\,(1 + O(1 \wedge 1/\sqrt{np}\,)) \tag{3.46}$$

(the term $1/\sqrt{np}$ in (3.46) apparently needs to be replaced with $p + 1/\sqrt{np}$, cf. (3.30)).

Using (3.45) and (3.17), we deduce

$$d_{TV}(\bar S_n; \bar\pi) \le 3p/4\mathrm{e} + 4(1 - \mathrm{e}^{-np})p^2 \tag{3.47}$$

(cf. (6.32*) in [221]). According to Deheuvels & Pfeifer [104],

$$|d_{TV}(\bar S_n; \bar\pi) - K_{n,\lambda}| \le \max\{16p^2; 5np^3\}, \tag{3.48}$$

where $K_{n,\lambda} = np^2 \mathrm{e}^{-np}\left((np)^{\alpha - np}(\alpha - np)/\alpha! - (np)^{\beta - np}(\beta - np)/\beta!\right)/2$,

$$\alpha = np + 1/2 + \sqrt{np + 1/4}\,,\quad \beta = np + 1/2 - \sqrt{np + 1/4}\,.$$

The case of non-identically distributed random vectors $\bar\xi_1, ..., \bar\xi_n$ has been treated by Roos [266]. A generalisation of (3.47) to the case of a stationary sequence of *dependent* r.v.s has been given in [221], Theorem 6.8. Estimates of the accuracy of multivariate compound Poisson approximation to the distribution of a sum of 1-*dependent* 0-1 random vectors have been given by Čekanavičius & Vellaisamy [92].

3.3 Dependent Bernoulli r.v.s

Let $X_1, ..., X_n$ be (possibly dependent) Bernoulli r.v.s. We present below generalisations of (3.10) and (3.15) to the case of dependent Bernoulli r.v.s.

Set $p_i = \mathbb{P}(X_i = 1|X_1, ..., X_{i-1})$. A generalisation of (3.10) has been given by Serfling [275]:

$$d_{TV}(S_n; \pi_\lambda) \le \sum_{i=1}^{n} (\mathbb{E}p_i)^2 + \sum_{i=1}^{n} \mathbb{E}|p_i - \mathbb{E}p_i|, \tag{3.10*}$$

$$d_K(S_n; \pi_\lambda) \le \frac{2}{\pi} \sum_{i=1}^{n} (\mathbb{E}p_i)^2 + \sum_{i=1}^{n} \mathbb{E}|p_i - \mathbb{E}p_i|. \tag{3.49}$$

Chen [56] pioneered the use of Stein's method in deriving estimates of the accuracy of Poisson approximation, and obtained an estimate of the accuracy of Poisson approximation to the distribution of a sum of φ-mixing r.v.s.

Locally dependent random variables.

Let $\{X_a, a \in J\}$ be a family of (possibly dependent) Bernoulli $\mathbf{B}(p_a)$ random variables. Assign to each $a \in J$ a "neighbourhood" $B_a \subset J$ such that $\{X_b, b \in J \setminus B_a\}$ are "almost independent" of X_a (for instance, if $\{X_b\}$ are m–dependent r.v.s and $J = \{1, ..., n\}$, then one would set $B_a = [a-m; a+m] \cap J$).

The idea of splitting the sample into "strongly dependent" and "almost independent" parts goes back to Bernstein [42].

Denote

$$\begin{aligned} S &= \sum_{a \in J} X_a, \quad \lambda = \mathbb{E}S, \\ \delta_1 &= \sum_{a \in J} \sum_{b \in B_a} \mathbb{E}X_a \mathbb{E}X_b\,, \quad \delta_2 = \sum_{a \in J} \sum_{b \in B_a \setminus \{a\}} \mathbb{E}X_a X_b, \\ \delta_3 &= \sum_{a \in J} \mathbb{E}\left|\mathbb{E}X_a - \mathbb{E}\left\{X_a \middle| \textstyle\sum_{b \in J \setminus B_a} X_b\right\}\right|. \end{aligned}$$

Theorem 3.4 *There holds*

$$d_{TV}(S; \pi_\lambda) \le \frac{1 - e^{-\lambda}}{\lambda}\Big(\delta_1 + \delta_2\Big) + \min\{1; 1.4/\sqrt{\lambda}\}\delta_3\,, \tag{3.50}$$

$$d_G(S; \pi_\lambda) \le \Big(1 \wedge \tfrac{4}{3}\sqrt{2/e\lambda}\Big)\,(\delta_1 + \delta_2) + \delta_3\,. \tag{3.51}$$

According to Remark 10.2.4 in [21], the term $1.4/\sqrt{\lambda}$ in (3.50) can be replaced with $\sqrt{2/e\lambda}$.

In the case of independent random variables one can choose $B_a = \{a\}$, then (3.50) coincides with (3.15). In the case of m-dependent Bernoulli $\mathbf{B}(p)$ r.v.s one has $\delta_1 \le (2m+1)\lambda p$, $\delta_2 \le 2m\lambda p$, $\delta_3 = 0$, and (3.50), (3.51) become

$$d_{TV}(S; \pi_\lambda) \le (1 - e^{-\lambda})(4m+1)p, \quad d_G(S; \pi_\lambda) \le \Big(1 \wedge \tfrac{4}{3}\sqrt{2/e\lambda}\Big)(4m+1)\lambda p.$$

Theorem 3.4 has applications to the problem of Poisson approximation to the distribution of the number of long head runs in a sequence of Bernoulli r.v.s, and to the problem of Poisson approximation to the distribution of the number of long match patterns in two sequences (e.g., DNA sequences, see [21, 224] and references therein).

The topic of Poisson approximation to $\mathcal{L}(S_n)$ in the case of stationary dependent r.v.s has applications in extreme value theory [189, 221]. The case where the sequence $X_1, ..., X_n$ is a moving average is related to the topic concerning the so-called Erdös–Rényi maximum of partial sums (cf. [221], Chapter 2).

Estimates of the accuracy of Poisson approximation for special types of dependence among $\{X_a, a \in J\}$ can be found in Barbour et al. [21]. Poisson and shifted Poisson approximations have been applied to sums of negatively dependent Bernoulli random variables by Daly et al. [98].

An estimate of the accuracy of shifted Poisson approximation to the distribution of a sum of dependent Bernoulli $\mathbf{B}(p_i)$ r.v.s in terms of the total variation distance has been given by Čekanavičius & Vaitkus [80]. Poisson and shifted Poisson approximation in the case of the Markov Binomial distribution has been investigated by Čekanavičius & Roos [85].

A generalisation of (3.50) to the case of compound Poisson approximation has been given by M.Roos [255].

Open problem

3.2. Improve the constants in (3.50), (3.51).

3.4 Independent integer-valued r.v.s

The topic of Poisson approximation to the distribution of a sum of integer-valued r.v.s has applications in extreme value theory, insurance, reliability theory, etc. (cf. [15, 21, 189, 221]). For instance, in insurance applications the sum $S_n = \sum_{i=1}^{n} Y_i \mathbb{I}\{Y_i > y_i\}$ of integer-valued r.v.s allows to account for the total loss from the claims exceeding excesses $\{y_i\}$. Since $\{Y_i > y_i\}$ are rare events, Poisson or compound Poisson approximation to $\mathcal{L}(S_n)$ may be applicable.

In extreme value theory one often deals with the number of extreme (rare) events represented by a sum $S_n = X_1 + ... + X_n$ of 0-1 r.v.s (indicators of rare events). Random variables $X_1, ..., X_n$ can be dependent. One way to cope with dependence is to split the sample into blocks, which can be considered almost independent (the so-called Bernstein's blocks approach [42]). The sum of r.v.s in a block is an integer-valued r.v.; thus, the number of rare events is a sum of almost independent integer-valued r.v.s.

In all such situations one deals with a sum of non-negative integer-valued r.v.s that are non-zero with small probabilities, and Poisson approximation to $\mathcal{L}(S_n)$ appears plausible. An estimate of the accuracy of Poisson approximation to the distribution of S_n can indicate whether Poisson approximation is applicable.

A simple way to derive an estimate of the accuracy of Poisson approximation to the distribution of a sum of integer-valued r.v.s is by using truncation. Denote

$$\lambda = \mathbb{E} S_n\,, \quad \lambda^* = \sum_{i=1}^{n} \mathbb{P}(X_i = 1), \quad \lambda_2^* = \sum_{i=1}^{n} \mathbb{P}(X_i = 1)^2 .$$

Then (cf. [174, 120, 18])

$$d_{TV}(S_n;\pi_{\lambda^*}) \leq \sum_{i=1}^{n} \mathbb{P}(X_i \geq 2) + (1-e^{-\lambda^*})\lambda_2^*/\lambda^* . \tag{3.52}$$

The following result is due to Franken [120]. Let $X_1, ..., X_n$ be independent non-negative integer-valued r.v.s. Set

$$\tilde{\lambda} = \sum_{j=1}^{n} (\mathbb{E}X_j - \mathbb{E}^2 X_j - \mathbb{E}X_j(X_j-1)).$$

If Franken's condition

$$2\mathbb{E}X_j > \mathbb{E}^2 X_j + \mathbb{E}X_j^2 \qquad (j=1,\ldots,n) \tag{3.53}$$

holds, then

$$d_K(S_n;\pi_\lambda) \leq \frac{1}{\pi} \min\left(2; \sqrt{e}/\tilde{\lambda}\right) \sum_{i=1}^{n} (\mathbb{E}^2 X_i + \mathbb{E}X_i(X_i-1)). \tag{3.54}$$

Franken's condition is fulfilled in the case of Bernoulli $\mathbf{B}(p_i)$ r.v.s such that $0<p_i<1$.

Condition (3.53) is restrictive: it does not hold if $\mathcal{L}(X_i) = \mathbf{\Pi}(\lambda_i)$, where $\lambda_i \geq 1$. Using Stein's method, Barbour [19] has shown that

$$d_{TV}(S_n;\pi_\lambda) \leq \min\{1;\lambda^{-1}\} \sum_{i=1}^{n} (\mathbb{E}^2 X_i + \mathbb{E}X_i(X_i-1)) \tag{3.55}$$

without assuming Franken's condition.

Franken [120] used the method of characteristic functions. The method of generating functions has been applied by Hwang [159].

Applications of Poisson approximation to statistics of empty boxes have been given by Vatutin & Mikhailov [304]. Poisson approximation to hyper-geometric distribution has been studied by Kruopis [185]. An estimate of the accuracy of Poisson approximation to the distribution of the number of fixed size cycles in generalised random graphs has been given by Bobkov et al. [44].

Given a random variable Y that takes values in $\mathbb{Z}_+$ and has a finite first moment, let Y^* denote a random variable with the distribution

$$\mathbb{P}(Y^* = m) = (m+1)\mathbb{P}(Y = m+1)/\mathbb{E}Y \qquad (m \geq 0). \tag{3.56}$$

Distribution (3.56) differs by a shift from the distribution introduced by Stein [288], p. 171. Note that $Y^* \stackrel{d}{=} Y$ if and only if $\mathcal{L}(Y)$ is Poisson.

Observe that (3.56) leads to

$$\mathbb{E}Xf(X) = \mathbb{E}X\mathbb{E}f(X^*+1) \tag{3.57}$$

for any function f such that $\mathbb{E}|Xf(X)| < \infty$. In particular, (3.57) yields

$$\mathbb{E}X^* = \mathbb{E}X(X-1)/\mathbb{E}X \tag{3.58}$$

if $\mathbb{E}X^2 < \infty$.

The next theorem presents a generalisation of (3.15) to the case of independent integer-valued r.v.s (see [221], Section 4.4).

Theorem 3.5 *As* $n \geq 1$,

$$d_{TV}(S_n;\pi_\lambda) \leq \lambda^{-1}(1-e^{-\lambda})\sum_{i=1}^{n} d_G(X_i;X_i^*)\mathbb{E}X_i\,, \tag{3.59}$$

$$d_G(S_n;\pi_\lambda) \leq \min\Big\{1;\tfrac{4}{3}\sqrt{2/e\lambda}\Big\}\sum_{i=1}^{n} d_G(X_i;X_i^*)\mathbb{E}X_i\,. \tag{3.60}$$

In the case of Bernoulli $\mathbf{B}(p_i)$ r.v.s one has $X_i^* \equiv 0$, and (3.59) coincides with (3.15). In the case of i.i.d.r.v.s (3.59) becomes

$$d_{TV}(S_n;\pi_\lambda) \leq (1-e^{-\lambda})\mathbb{E}|X-X^*|\,.$$

Here X^* may be chosen independent of X, although one would prefer to define X and X^* on a common probability space in order to make $\mathbb{E}|X-X^*|$ smaller.

3.4.1 Estimates in terms of d_{TV}

A generalisation of (3.17) to the case of independent integer-valued r.v.s has been given by Novak [224].

Let $X_1,...,X_n$ be independent non-negative integer-valued r.v.s. Given a random variable X and a random pair (ξ,η) with finite second moments, denote

$$\kappa_X = \mathbb{E}X - \mathrm{var}X,\ \ \gamma_{\xi,\eta} = \mathbb{E}|\xi(\xi-1) - \eta(\eta-1)|.$$

Taking into account (3.58), one can notice that

$$\kappa_X = \mathbb{E}X\mathbb{E}(X-X^*). \tag{3.61}$$

We set $X_0 = 0$, $\varepsilon_\lambda^* = 1 \wedge 1/\sqrt{2\pi\lfloor\lambda\rfloor}$,

$$\lambda_{(i)} = \lambda - \mathbb{E}X_i\,,\ \ u_i = 1 - d_{TV}(X_i;X_i+1)\ \ (i\geq 0),\ \ U = \sum_{i=1}^{n} u_i\,,$$

$$\varepsilon_1 = \lambda^{-1}(1-e^{-\lambda})\sum_{i=1}^{n}\min\{2\mathbb{E}|X_i-X_i^*|;\gamma_{X_i,X_i^*}\varepsilon_{i,n}\}\mathbb{E}X_i\,,$$

$$\varepsilon_2 = 2\lambda^{-1}(1-e^{-\lambda})\sum_{i=1}^{n}\mathbb{E}X_i|\kappa_{X_i}|\varepsilon_{i,n}\,,\ \ \varepsilon_3 = 2\lambda^{-1}(1-e^{-\lambda})|\kappa_{S_n}|\varepsilon_{0,n}^+\,,$$

$$\varepsilon_{i,n} = 1 \wedge \sqrt{2/\pi}\Big/\Big(1/4+U-u_i\Big)^{1/2} \wedge \Big(\varepsilon_{\lambda_{(i)}}^* + 2\varepsilon_{i,n}^+\Big),$$

$$\varepsilon_{i,n}^+ = \frac{1-e^{-\lambda}}{\lambda_{(i)}}\sum_{j=1}^{n} d_G(X_j;X_j^*)\mathbb{E}X_j\,.$$

Theorem 3.6 *If* $X_1,...,X_n$ *are independent non-negative integer-valued random variables with finite second moments, then*

$$d_{TV}(S_n;\pi_\lambda) \leq 3\theta^*/4e + \varepsilon_1 + \varepsilon_2 + \varepsilon_3\,, \tag{3.62}$$

where $\theta^* = \Big|\sum_{i=1}^{n}\kappa_{X_i}\Big|/\lambda$.

In the case of independent and identically distributed r.v.s (3.62) becomes

$$d_{TV}(S_n;\pi_\lambda) \le \tfrac{3}{4\mathrm{e}}|\kappa_X|/\mathbb{E}X + (1-\mathrm{e}^{-\lambda})\varepsilon_X\,, \qquad (3.62^*)$$

where

$$\varepsilon_X = \min\{2\mathbb{E}|X-X^*|;\varepsilon_{1,n}\gamma_{X,X^*}\} + 2|\kappa_X|\varepsilon_{1,n} + 2(1-\mathrm{e}^{-\lambda})|\kappa_X|d_G(X;X^*)/\mathbb{E}X.$$

The moment assumption can be relaxed at a cost of adding an extra term if one uses truncation at some levels $\{K_i\}$ (i.e., switches from $\{X_i\}$ to $\{X_i'\}$, where $X_i' = X_i\mathbb{I}\{X_i \le K_i\}$) since $d_{TV}((X_1,...,X_n);(X_1',...,X_n')) \le \sum_{i=1}^n \mathbb{P}(X_i > K_i)$.

Note that the terms ε_2 and ε_3 in (3.62) vanish if $\kappa_{X_i} = 0$ (equivalently, if $\mathbb{E}X_i = \mathbb{E}X_i^*$) for all i.

Example 3.3 Let $X, X_1, X_2, ...$ be independent geometric $\mathbf{\Gamma_0}(p)$ r.v.s. Then S_n is a negative Binomial $\mathbf{NB}(n,p)$ r.v., $\mathbb{E}S_n = np/(1-p)$.

Set $r = p/(1-p)$. It is easy to see that $\mathbb{P}(X_i^* = m) = (m+1)p^m(1-p)^2$. Hence

$$X^* \stackrel{d}{=} X + X, \quad \mathbb{E}|X-X^*| = r.$$

Note that $d_G(X;X^*) = \mathbb{E}X = r$. Estimate (3.59) entails

$$d_{TV}(S_n;\pi_\lambda) \le \left(1-\mathrm{e}^{-np/(1-p)}\right) r, \qquad (3.63)$$

which is due to Barbour [19], while (3.60) yields

$$d_G(S_n;\pi_\lambda) \le \tfrac{4}{3}\sqrt{2/\mathrm{e}}\; r\sqrt{nr}\,.$$

Vervaat [309] has shown that $d_{TV}(S_n;\pi_\lambda) \le r$, while Romanowska [254] has noticed that $d_{TV}(S_n;\pi_\lambda) \le r/\sqrt{2}$. Roos [262] has shown that

$$d_{TV}(\mathbf{NB}(n,p);\mathbf{\Pi}(nr)) \le \min\{3r/4\mathrm{e}; nr^2\}, \qquad (3.64)$$

$$d_G(\mathbf{NB}(n,p);\mathbf{\Pi}(nr)) \le nr^2\,. \qquad (3.65)$$

Estimate (3.64) has correct constant $3/4\mathrm{e}$ at the leading term. □

Example 3.4 Let $X_1, X_2, ..., X_n$ be independent r.v.s with the distributions

$$\mathbb{P}(X_i=0)=1-p_i+p_i^2/2,\ \mathbb{P}(X_i=1)=p_i-p_i^2,\ \mathbb{P}(X_i=2)=p_i^2/2 \qquad (0\le p_i\le 1).$$

Note that $\mathbb{E}X_i = \mathrm{var}X_i = p_i$. One can check that $d_{TV}(X_i;\pi_{p_i}) \le \frac{2}{3}p_i^3$. Therefore,

$$d_{TV}(S_n;\pi_\lambda) \le \sum_{i=0}^n d_{TV}(X_i;\pi_{p_i}) \le \tfrac{2}{3}\sum_{i=1}^n p_i^3\,. \qquad (3.66)$$

This inequality is due to Deheuvels & Pfeifer [103]. A bound of order $p^{3/2}n^{-1/2} \wedge np^3$ in terms of d_K for a similar distribution has been given by Kruopis [186].

By (3.56), X_i^* is a Bernoulli $\mathbf{B}(p_i)$ random variable. Note that $\mathbb{E}X_i = \mathbb{E}X_i^*$, $\kappa_{X_i} = 0$ $(\forall i)$, $\varepsilon_2 = \varepsilon_3 = 0$. One can check that

$$\varepsilon_1 = \lambda^{-1}(1-\mathrm{e}^{-\lambda})\sum_{i=1}^n p_i^3\varepsilon_{i,n},\ d_{TV}(X_i;X_i^*) = d_G(X_i;X_i^*) = \gamma_{X_i,X_i^*} = p_i^2\,.$$

Theorem 3.6 yields

$$d_{TV}(S_n;\pi_\lambda) \le \frac{1-\mathrm{e}^{-\lambda}}{\lambda}\sum_{i=1}^n p_i^3 \min\left\{1;\left(\frac{1}{\sqrt{2\pi\lfloor\lambda-p_i\rfloor}}+2\frac{1-\mathrm{e}^{-\lambda}}{\lambda-p_i}\sum_{j=1}^n p_j^3\right)\right\}. \tag{3.67}$$

Estimate (3.67) is sharper than (3.66) if $2\lambda > 3(1-\mathrm{e}^{-\lambda})$. □

The proof of Theorem 3.6 uses the following first-order asymptotic expansion to $\mathcal{L}(S_n)$ from [224].

Let $\pi_\lambda^\star$ denote a random variable with distribution (3.32). Let $\mathcal{S}$ denote the class of measurable functions taking values in $[0;1]$. Recall that the total variation distance between the distributions of r.v.s X and Y is

$$d_{TV}(X;Y) = \sup_{h\in\mathcal{S}}\left[\mathbb{E}h(X)-\mathbb{E}h(Y)\right].$$

Theorem 3.7 *Let $X_1,...,X_n$ be independent non-negative integer-valued random variables with finite second moments. For any function $h\in\mathcal{S}$*

$$\left|\mathbb{E}h(S_n)-\mathbb{E}h(\pi_\lambda)-(\mathbb{E}h(\pi_\lambda+1)-\mathbb{E}h(\pi_\lambda^\star))\sum_{i=1}^n \kappa_{X_i}/2\lambda\right| \le \varepsilon_1+\varepsilon_2+\varepsilon_3\,. \tag{3.68}$$

If $X, X_1,...,X_n$ are i.i.d.r.v.s, then (3.68) becomes

$$\left|\mathbb{E}h(S_n)-\mathbb{E}h(\pi_\lambda)-(\mathbb{E}h(\pi_\lambda+1)-\mathbb{E}h(\pi_\lambda^\star))\,\kappa_X/2\mathbb{E}X\right| \le \varepsilon_1+\varepsilon_2+\varepsilon_3. \tag{3.68*}$$

The right-hand side of (3.68*) is typically of order $(\mathbb{E}X)^2$.

A straightforward consequence of (3.68) is the following relation:

$$|d_{TV}(S_n;\pi_\lambda)-d_{TV}(\pi_\lambda^\star;\pi_\lambda+1)\theta^*/2| \le \varepsilon_1+\varepsilon_2+\varepsilon_3\,. \tag{3.69}$$

Using (3.33), we derive

$$d_{TV}(S_n;\pi_\lambda) = \theta^*/\sqrt{2\pi\mathrm{e}}+O\left(\theta^*/\sqrt{\lambda}+\varepsilon_1+\varepsilon_2+\varepsilon_3\right). \tag{3.30*}$$

Remark 3.1. Note that

$$\mathbb{E}h(\pi_\lambda+1)-\mathbb{E}h(\pi_\lambda^\star) = -\lambda\mathbb{E}\Delta^2 h(\pi_\lambda).$$

Thus, (3.68) can be rewritten as

$$\left|\mathbb{E}h(S_n)-\mathbb{E}h(\pi_\lambda)+\mathbb{E}\Delta^2 h(\pi_\lambda)\sum_{i=1}^n \kappa_{X_i}/2\right| \le \varepsilon_1+\varepsilon_2+\varepsilon_3 \qquad (h\in\mathcal{S}). \tag{3.68*}$$

Remark 3.2. Denote by P_n the unit measure

$$P_n(\cdot) = \mathbb{P}(\pi_\lambda=\cdot)+\left(\mathbb{P}(\pi_\lambda+1=\cdot)-\mathbb{P}(\pi_\lambda^\star=\cdot)\right)\sum_{i=1}^n \kappa_{X_i}/2\lambda.$$

Theorem 3.7 states that

$$\|\mathcal{L}(S_n)-P_n\| \le 2(\varepsilon_1+\varepsilon_2+\varepsilon_3),$$

where $\|\cdot\|$ denotes the total variation norm.

3.4.2 Estimates in terms of d_G

Let $\{X_i\}_{i\geq 1}$ be independent non-negative integer-valued r.v.s. If $\lambda=0$, then $S_n=\pi_0=0$, and hence $d_G(S_n;\pi_\lambda)=0$. Therefore, w.l.o.g. we may assume that $\lambda>0$.

If $\mathbb{E}X_i=0$ as $i\in A$ for a particular $A\subset J:=\{1,2...,n\}$, then $S_n=\sum_{j\in J\setminus A}X_j$. Thus, w.l.o.g. we may assume that $\mathbb{E}X_i>0$ $(\forall i)$.

Let

$$\begin{aligned}
c_\star &= \tfrac{4}{3}\sqrt{2/\mathrm{e}},\ c_\lambda=\lambda\wedge c_\star\sqrt{\lambda},\ c_\lambda^*=1\wedge c_\lambda,\ c_\lambda^{\mathrm{o}}=1\wedge\tfrac{2}{3}\lambda,\\
c_{i,\lambda} &= c_\lambda^{\mathrm{o}}\wedge c_\lambda\varepsilon_{i,n},\ c_{i,\lambda}^*=2c_\lambda\wedge 2\lambda\varepsilon_{i,n}\wedge\Big((c_\lambda\wedge 2\lambda\varepsilon_{\lambda_i^*}^*)+2c_\lambda\varepsilon_{i,n}^+\Big),\\
\varepsilon_{(1)} &= \lambda^{-1}\sum_{i=1}^n\min\{c_{i,\lambda}^*d_G(X_i;X_i^*);c_{i,\lambda}\gamma_{X_i,X_i^*}\}\mathbb{E}X_i,\\
\varepsilon_{(2)} &= 2\lambda^{-1}\sum_{i=1}^n c_{i,\lambda}|\kappa_{X_i}|\mathbb{E}X_i,\ \ \hat{\varepsilon}=\sum_{j=1}^n d_G(X_j;X_j^*)\mathbb{E}X_j,\\
\varepsilon_{(3)} &= 2\lambda^{-1}\theta^*c_\lambda\hat{\varepsilon}.
\end{aligned}$$

Theorem 3.8 *If $X_1,...,X_n$ are independent non-negative integer-valued random variables with finite second moments, then*

$$d_G(S_n;\pi_\lambda)\leq\Big|\sum_{i=1}^n\kappa_{X_i}\Big|\min\Big\{1;1/\sqrt{2\pi\lfloor\lambda\rfloor};1/\sqrt{2\mathrm{e}\lambda}\Big\}+\varepsilon_{(1)}+\varepsilon_{(2)}+\varepsilon_{(3)}\,. \tag{3.70}$$

If $\lambda\geq 1/(1-\mathrm{e}/\pi)$, then $\pi\lfloor\lambda\rfloor\geq\mathrm{e}\lambda$, and (3.70) becomes

$$d_G(S_n;\pi_\lambda)\leq\Big|\sum_{i=1}^n\kappa_{X_i}\Big|/\sqrt{2\pi\lfloor\lambda\rfloor}+\varepsilon_{(1)}+\varepsilon_{(2)}+\varepsilon_{(3)}\,. \tag{3.70*}$$

Estimate (3.70*) has a sharper constant at the leading term than (3.26).

The moment assumption in Theorem 3.8 can be relaxed at a cost of adding an extra term $r_n^*=|\sum_{i=1}^n\mathbb{E}(X_i-K_i)\mathbb{1}\{X_i>K_i\}|$ if one uses truncation at some levels $\{K_i\}$ (i.e., switches from $\{X_i\}$ to $\{X_i'\}$, where $X_i'=X_i\mathbb{1}\{X_i\leq K_i\}+K_i\mathbb{1}\{X_i>K_i\}$) since $d_G(\sum_{i=1}^n X_i;\sum_{i=1}^n X_i')\leq r_n^*$.

Example 3.5 If $\{X_i\}$ are independent Bernoulli $\mathbf{B}(p)$ r.v.s, then $\varepsilon_{(1)}=0$, $\kappa_X=p^2$, $\theta^*=p$, $\hat{\varepsilon}=np^2$, $\varepsilon_{(2)}\leq 2p^2(1\wedge\frac{2}{3}np\wedge c_\star\sqrt{np}\,\varepsilon_{1,n})$, $\varepsilon_{(3)}=2p^2(np\wedge c_\star\sqrt{np}\,)$, and (3.70) yields

$$d_G(S_n;\pi_{np})\leq np^2\min\Big\{1;1/\sqrt{2\pi\lfloor np\rfloor};1/\sqrt{2\mathrm{e}np}\Big\}+\gamma_n\,, \tag{3.71}$$

where $\gamma_n=2p^2\big((1\wedge\frac{2}{3}np)+(np\wedge c_\star\sqrt{np}\,)\big)$.

Estimate (3.71) has a correct constant at the leading term. □

Recall that

$$d_G(S_n;\pi_\lambda)=\sup_{h\in\mathcal{L}_1}(\mathbb{E}h(S_n)-\mathbb{E}h(\pi_\lambda)),$$

where $\mathcal{L}_1=\{g\colon |g(x)-g(y)|\leq|x-y|\}$ is the class of Lipschitz functions.

The next theorem presents a first-order asymptotic expansion to $\mathbb{E}h(S_n)-\mathbb{E}h(\pi_\lambda)$.

Theorem 3.9 *Let $X_1,...,X_n$ be independent non-negative integer-valued random variables with finite second moments. If $h\in\mathcal{L}_1$, then*

$$\left|\mathbb{E}h(S_n)-\mathbb{E}h(\pi_\lambda)-(\mathbb{E}h(\pi_\lambda+1)-\mathbb{E}h(\pi_\lambda^\star))\sum_{i=1}^n \kappa_{X_i}/2\lambda\right| \tag{3.72}$$
$$\leq \varepsilon_{(1)}+\varepsilon_{(2)}+\varepsilon_{(3)}\,.$$

Note that

$$\mathbb{E}h(\pi_\lambda+1)-\mathbb{E}h(\pi_\lambda^\star)=-\lambda\mathbb{E}\Delta^2h(\pi_\lambda).$$

Thus, the statement of Theorem 3.9 can be rewritten as

$$\left|\mathbb{E}h(S_n)-\mathbb{E}h(\pi_\lambda)+\mathbb{E}\Delta^2h(\pi_\lambda)\sum_{i=1}^n\kappa_{X_i}/2\right|\leq\varepsilon_{(1)}+\varepsilon_{(2)}+\varepsilon_{(3)}\qquad(h\in\mathcal{L}_1). \tag{3.72*}$$

Note also that (3.72*) is an estimate of the accuracy of approximation by a particular unit measure.

If $X,X_1,...,X_n$ are i.i.d.r.v.s, then (3.72) becomes

$$\left|\mathbb{E}h(S_n)-\mathbb{E}h(\pi_\lambda)-(\mathbb{E}h(\pi_\lambda+1)-\mathbb{E}h(\pi_\lambda^\star))\,\kappa_X/2\mathbb{E}X\right|\leq\varepsilon_{(1)}+\varepsilon_{(2)}+\varepsilon_{(3)}\,. \tag{3.72*}$$

If S_n is Binomial $\mathbf{B}(n,p)$ r.v., then (3.72*) yields

$$\left|\mathbb{E}h(S_n)-\mathbb{E}h(\pi_{np})-\big(\mathbb{E}h(\pi_{np}+1)-\mathbb{E}h(\pi_{np}^\star)\big)\,p/2\right|\leq 2c_{np}^{\circ}p^2+2p^2(np\wedge c_\star\sqrt{np}\,).$$

A straightforward consequence of Theorem 3.9 is the following

Corollary 3.10 *If $X_1,...,X_n$ are independent non-negative integer-valued random variables with finite second moments, then*

$$|d_G(S_n;\pi_\lambda)-d_G(\pi_\lambda^\star;\pi_\lambda+1)\theta^*/2|\leq\varepsilon_{(1)}+\varepsilon_{(2)}+\varepsilon_{(3)}\,.$$

3.4.3 Shifted Poisson approximation in terms of d_{TV}

A number of authors dealt with shifted Poisson approximation to the distribution of a sum S_n of integer-valued r.v.s, cf. [29, 223, 225] and references therein. Set $\lambda=\mathbb{E}S_n$,

$$\sigma^2=\mathrm{var}\,S_n,\ \ a=\lfloor\kappa_{S_n}\rfloor,\ \ b=\{\kappa_{S_n}\},\ \ \mu=\sigma^2+b,$$

where $\lfloor x\rfloor$, and $\{x\}=x-\lfloor x\rfloor$ denote the integer and the fractional parts of x. Recall that $\kappa_{S_n}=\mathbb{E}S_n-\mathrm{var}\,S_n$.

Barbour & Čekanavičius [29] have shown that

$$d_{TV}(S_n;a+\pi_\mu)\leq(1\wedge\sigma^{-2})\Big(b+d_n\sum_{i=1}^n\psi_i\Big)+\mathbb{P}(S_n<a), \tag{3.73}$$

where

$$\begin{aligned} d_n &= \max_{i\leq n} d_{TV}(S_{n,i};S_{n,i}+1),\ \ S_{n,i}=S_n-X_i,\ \ \sigma_i^2=\mathrm{var}X_i\,,\\ \psi_i &= \sigma_i^2\mathbb{E}X_i(X_i-1)+|\mathbb{E}X_i-\sigma_i^2|\mathbb{E}(X_i-1)(X_i-2)+\mathbb{E}|X_i(X_i-1)(X_i-2)|. \end{aligned}$$

By Chebyshev's inequality, $\mathbb{P}(S_n < a) \leq \sigma^{-2}$. Lemma C.4 yields

$$d_n \leq \sqrt{2/\pi \max\{1/4; U - 3/4\}},$$

where $U = \sum_{i=1}^n (1 - d_{TV}(X_i; X_i + 1))$.

If $\mathcal{L}(S_n) = \mathbf{B}(n,p)$, then the RHS of (3.73) is $O\big(\sqrt{p/n} + 1/np\big)$.

As estimate in terms of the point metric is given in [252]:

$$d_o(S_n; a + \pi_\mu) \leq \Big(2 + d'_n \sum_{i=1}^n \psi_i\Big)\sigma^{-2}. \tag{3.74}$$

where ψ_i are the same as in (3.73); $d'_n = \max_i d_{TV}(S_n - X_i; S_n - X_i + 1)$ can be estimated by (C.16): $d'_n \leq 17.1/U$.

Estimates in terms of the point metric have been presented by Barbour et al. [37] and Novak [226].

Example 3.6 To demonstrate the limitations of (3.73), let us consider Wilcoxon signed rank statistic $S_n = \sum_{j=1}^n X_j$, where $\{X_j\}$ are independent r.v.s, $\mathbb{P}(X_j = 0) = \mathbb{P}(X_j = j) = 1/2$ [246]. Then

$$\mathbb{E}S_n = n(n+1)/4, \quad \operatorname{var} S_n = n(n+1)(2n+1)/12,$$
$$\sigma_i^2 = j^2/4, \quad \psi_i \leq Ci^4, \quad \sum_{i=1}^n \psi_i \leq Cn^5, \quad d_n \leq C,$$

and (3.73) becomes

$$d_{TV}(S_n; a + \pi_\mu) \leq C(n^{-3} + d_n n^2).$$

Lemma C.4 does not help to avoid trivial case $d_n n^2 \to \infty$, since it does not provide a "good" estimate of d_n for Wilcoxon statistic.

Meanwhile, by the triangle inequality and normal approximation (H.1),

$$d_K(S_n; a + \pi_\mu) \leq Cn^{-1/2}.$$

The task of deriving a sharp estimate of the accuracy of compound Poisson approximation to $\mathcal{L}(S_n)$ in terms of d_{TV} remains open. □

An estimate of the accuracy of shifted Poisson approximation to the distribution of a random sum of i.i.d. integer-valued r.v.s has been given by Röllin [252]. Vaitkus & Čekanavičius [302] investigated shifted Poisson approximation under certain assumptions concerning factorial cumulants. Non-uniform estimates have been presented by Barbour & Choi [31].

Let $X, X_1, \ldots, X_n$ be i.i.d.r.v.s concentrated on integers with the greatest common divisor equal to 1. Assume that $\mathbb{E}X^2 < \infty$. Let $0 < \delta < 1$ be a fixed number. Then [72]

$$d_{TV}(S_n; a + \pi_{\sigma^2}) = O(n^{-\delta/2}) \qquad (n \to \infty)$$

if and only if

$$\mathbb{E}(X - \mathbb{E}X)^2 \mathbb{1}\{|X - \mathbb{E}X| > x\} = O(x^{-\delta}) \qquad (x \to \infty).$$

The next theorem deals with shifted Poisson approximation

$$\mathcal{L}(S_n) \approx \mathcal{L}(a + \pi_\mu),$$

where $\mu = \sigma^2 + \{\kappa_{S_n}\}$. Note that

$$a + \mathbb{E}\pi_\mu = \mathbb{E}S_n, \quad |\operatorname{var} \pi_\mu - \operatorname{var} S_n| < 1.$$

Let $\{\tilde{X}_j\}$ denote independent copies of $\{X_j\}$, $\bar{Y} := Y - \mathbb{E}Y$, $\varepsilon_n = \varepsilon_{0,n}$,

$$\bar{\varepsilon}_\mu = 2\varepsilon^*_\mu\sqrt{2/\mathrm{e}\mu} + 2\mu^{-1}(1-\mathrm{e}^{-\mu})\varepsilon^\star_\mu, \quad \varepsilon^\star_\mu = \mu^{-1}(1-\mathrm{e}^{-\mu})|\{\kappa_{S_n}\}| + \varepsilon^\#_n,$$

$$\varepsilon^\#_n = 2\mu^{-1}(1-\mathrm{e}^{-\mu})\sum_{i=1}^n \sigma_i^2\varepsilon_{i,n}\mathbb{E}X_i + \sum_{i=1}^n \delta^{(\mu)}_{X_i,\tilde{X}_i},$$

$$\delta^{(\mu)}_{X_i,\tilde{X}_i} = \mu^{-1}(1-\mathrm{e}^{-\mu})\min\Big\{2\mathbb{E}|\bar{X}_i||X_i-\tilde{X}_i|; \mathbb{E}|\bar{X}_i||X_i(X_i-1) - \tilde{X}_i(\tilde{X}_i-1)|\varepsilon_{i,n}\Big\}.$$

Theorem 3.11 *If* $X_1, ..., X_n$ *are independent non-negative integer-valued random variables with finite second moments, then*

$$d_{TV}(S_n; \lfloor\kappa_{S_n}\rfloor + \pi_\mu) \le |\{\kappa_{S_n}\}|\min\{\mu^{-1}(1-\mathrm{e}^{-\mu}); \bar{\varepsilon}_\mu\} + \varepsilon^\#_n. \tag{3.75}$$

Inequality (3.75) is sharp in the following sense: if all $\{X_j\}$ are constants, then (3.75) becomes the equality.

3.4.4 Shifted Poisson approximation in terms of d_G

Recall that $\mu = \mathrm{var}\, S_n + \{\kappa_{S_n}\}$, where $\{x\} = x - \lfloor x\rfloor$. The next theorem deals with shifted (translated) Poisson approximation $S_n \approx \lfloor\kappa_{S_n}\rfloor + \pi_\mu$ in terms of the Gini-Kantorovich distance.

Denote

$$\delta^{(\mu)}_{X_i} = \min\{c^*_{i,\mu}\mathbb{E}|\bar{X}_i|X_i\mathbb{I}\{X_i \ge 2\}; c^\circ_\mu\mathbb{E}|\bar{X}_i|X_i(X_i-1)\},$$

$$\varepsilon^\#_G = 2\mu^{-1}\sum_{i=1}^n \sigma_i^2 c_{i,\mu}\mathbb{E}X_i + \mu^{-1}\sum_{i=1}^n \delta^{(\mu)}_{X_i} \quad \text{if} \quad \mu > 0, \quad \varepsilon^\#_G = 0 \quad \text{otherwise.}$$

Theorem 3.12 [225] *If* $X_1, ..., X_n$ *are independent non-negative integer-valued random variables with finite second moments, then*

$$d_G(S_n; \lfloor\kappa_{S_n}\rfloor + \pi_\mu) \le |\{\kappa_{S_n}\}|\min\Big\{1; c_\star/\sqrt{\mu}; 2\varepsilon_{0,n}\Big\} + \varepsilon^\#_G. \tag{3.76}$$

If $X, X_1, ..., X_n$ are i.i.d. non-degenerate r.v.s, then (3.76) becomes

$$d_G(S_n; \lfloor\kappa_{S_n}\rfloor + \pi_\mu) \le |\{n\kappa_X\}|\min\Big\{1; c_\star/\sigma_X\sqrt{n}; 2\varepsilon_{0,n}\Big\} + 2(1\wedge c_\star\varepsilon_{1,n}\sqrt{\mu})\mathbb{E}X + \delta^{(\mu)}_X/\mathrm{var}\, X.$$

If $\mathcal{L}(S_n) = \mathbf{B}(n,p)$, set $q = 1-p$. Then $\mu = npq + \{np^2\}$, $\delta^{(\mu)}_{X_i} = 0$, $c^\circ_\mu = 1\wedge\frac{2}{3}\mu$, and (3.76) yields

$$d_G(S_n; \lfloor np^2\rfloor + \pi_\mu) \le |\{np^2\}|\min\Big\{1; c_\star/\sqrt{npq}; 2\varepsilon_{0,n}\Big\} + 2p(1\wedge c_\mu\varepsilon_{1,n}), \tag{3.76*}$$

where $\varepsilon_{0,n} \le \varepsilon_{1,n} \le 1\wedge\sqrt{2/\pi}\Big/(1/4+(n-1)p)^{1/2}\wedge\Big(1/\sqrt{2\pi[(n-1)p]} + 2p/(1-1/n)\Big)$. Thus,

$$d_G(S_n; \lfloor np^2\rfloor + \pi_\mu) = O\big(p + 1/\sqrt{np}\big).$$

Open problem

3.3. Improve the constants in (3.75), (3.76).

3.5 Dependent integer-valued r.v.s

Let $X_1, ..., X_n$ be (possibly dependent) non-negative integer-valued r.v.s. Set $p_i = \mathbb{P}(X_i = 1|X_1, ..., X_{i-1})$. A generalisation of (3.10*), (3.49) has been given by Serfling [275]:

$$d_{TV}(S_n; \pi_\lambda) \le \sum_{i=1}^{n} \left(\mathbb{E}^2 p_i + \mathbb{E}|p_i - \mathbb{E}p_i| + \mathbb{P}(X_i \ge 2)\right), \tag{3.10$^+$}$$

$$d_K(S_n; \pi_\lambda) \le \sum_{i=1}^{n} \left(\tfrac{2}{\pi}\mathbb{E}^2 p_i + \mathbb{E}|p_i - \mathbb{E}p_i| + \mathbb{P}(X_i \ge 2)\right). \tag{3.49$^+$}$$

Below we present a generalisation of Theorem 3.4.

Let $\{X_a, a \in J\}$ be a family of r.v.s taking values in $\mathbb{Z}_+$. Suppose one can choose the "neighbourhoods" $\{B_a\}$ so that r.v.s $\{X_b, b \in J \backslash B_a\}$ are independent of X_a. We call this assumption the "local dependence" condition.

Recall that π_λ denotes a Poisson $\mathbf{\Pi}(\lambda)$ r.v. Set

$$\delta_1^* = \sum_{a \in J} \sum_{b \in B_a \backslash \{a\}} \mathbb{E}X_a \mathbb{E}X_b, \quad \delta_4 = \sum_{a \in J} d_G(X_a; X_a^*)\mathbb{E}X_a,$$

and let $\delta_1, \delta_2, \delta_3$ be defined as in Theorem 3.4.

Theorem 3.13 *If* $\{X_b, b \in J \backslash B_a\}$ *are independent of* X_a, *then*

$$d_{TV}(S_n; \pi_\lambda) \le \frac{1 - e^{-\lambda}}{\lambda}(\delta_1^* + \delta_2 + \delta_4). \tag{3.77}$$

In Theorem 3.14 we drop the local dependence condition assumed in Theorem 3.13.

Theorem 3.14 *Denote* $\delta_5 = \sum_{a \in J} \mathbb{E}X_a(X_a - 1)\mathbb{I}\{X_a \ge 2\}$. *Then*

$$d_{TV}(S_n; \pi_\lambda) \le \frac{1 - e^{-\lambda}}{\lambda}\left(\delta_1 + \delta_2 + \delta_5\right) + \min\{1; \sqrt{2/e\lambda}\}\delta_3. \tag{3.78}$$

Theorem 3.5 can be applied to a sample of size $n = 1$ consisting of a single r.v. S_n, where S_n is a sum of (possibly dependent) r.v.s.

Corollary 3.15 *Let* S_n *be a random variable taking values in* $\mathbb{Z}_+$ *such that* $\mathbb{E}S_n < \infty$. *Then*

$$d_{TV}(S_n; \pi_\lambda) \le (1 - e^{-\lambda}) d_G(S_n; S_n^*), \tag{3.79}$$

$$d_G(S_n; \pi_\lambda) \le \min\left\{\lambda; \tfrac{4}{3}\sqrt{2\lambda/e}\right\} d_G(S_n; S_n^*). \tag{3.80}$$

Bounds (3.79), (3.80) yield estimates of $d_{TV}(S_n; \pi_\lambda)$ and $d_G(S_n; \pi_\lambda)$ if one can define S_n and S_n^* on a common probability space so that S_n^* is "close" to S_n. For instance, if $\mathcal{L}(S_n) = \mathbf{B}(n, p)$, then $S_n^* \stackrel{d}{=} S_{n-1}$ (cf. [221], p. 74, ex. 19), and (3.79), (3.80) yield bounds of correct order.

Ruzankin [270] presents an estimate of the accuracy of Poisson approximation to $\mathbb{E}h(S_n)$, where h is an unbounded function. Large deviations for sums of 1-dependent random variables satisfying Franken's condition have been proved by Čekanavičius & Vellaisamy [90].

Open problem

3.4. Improve the constants in (3.77), (3.78).

3.6 Asymptotic expansions

Uspensky [298] appears to be the first to present formal asymptotic expansions in the Poisson limit theorem. A first-order asymptotic expansion has been given by Makabe & Morimura [201]. Grigelionis [143] has suggested a first-order Poisson asymptotic expansion for sums of integer-valued random variables. Franken [120] has introduced another expansion based on regrouping of Charlier coefficients (see also discussion in [2, 150]).

Shorgin [279] extended Uspensky's result to the case of non-identically distributed Bernoulli variables. He provides full asymptotic expansions with explicit estimates of the error terms.

Kerstan [174], Kruopis [185], Deheuvels & Pfeifer [105], Novak [221] present first-order asymptotic expansions in the case of independent Bernoulli random variables.

Asymptotic expansions to $\mathbb{E}h(S_n) - \mathbb{E}h(\pi_\lambda)$ in the case of independent 0-1 r.v.s and unbounded function h have been given by Barbour et al. [23] and Borisov & Ruzankin [47].

Recall that $S_n = X_1 + \cdots + X_n$, $\lambda = \mathbb{E}S_n$,

$$\lambda_k = \sum_{i=1}^n p_i^k \quad (k \geq 2), \qquad \Delta h(\cdot) = h(\cdot + 1) - h(\cdot).$$

Let h be a polynomial of order $m \geq 2$. The following result is due to Borisov & Ruzankin [47]: if $\mathbb{E}|h(\pi_\lambda)|\pi_\lambda^4 < \infty$, then

$$\begin{aligned} &\Big|\mathbb{E}h(S_n) - \mathbb{E}h(\pi_\lambda) + \lambda_2\mathbb{E}\Delta^2 h(\pi_\lambda)/2\Big| \\ &\leq \frac{e^{p_n^*}}{(1-p_n^*)^2}\Big(\lambda_3\mathbb{E}|\Delta^3 h(\pi_\lambda)|/3 + \lambda_2^2\mathbb{E}|\Delta^4 h(\pi_\lambda)|/8\Big). \end{aligned} \tag{3.81}$$

Let $p_1 = ... = p_n = p$. Denote $\gamma_k = \mathbb{E}|\Delta^k h(\pi_{np})|$. If $p = p(n) \to 0$, $np \to \infty$, then [47]

$$\begin{aligned} &|\mathbb{E}h(S_n) - \mathbb{E}h(\pi_{np})| \sim np^2\gamma_2/2, \\ &|\mathbb{E}h(S_n) - \mathbb{E}h(\pi_{np}) + \tfrac{1}{2}np^2\mathbb{E}\Delta^2 h(\pi_{np})| \sim np^3\gamma_3/3 + n^2p^4\gamma_4/8, \\ &|\mathbb{E}h(S_n) - \mathbb{E}h(\pi_{np}) + \tfrac{1}{2}np^2\mathbb{E}\Delta^2 h(\pi_{np}) - \tfrac{1}{3}np^3\mathbb{E}\Delta^3 h(\pi_{np}) - \tfrac{1}{8}n^2p^4\mathbb{E}\Delta^4 h(\pi_{np})| \\ &\sim np^4\gamma_4/4 + n^2p^5\gamma_5/6 + n^3p^6\gamma_6/48. \end{aligned}$$

Assumption $\mathbb{E}|h(\pi_\lambda)|\pi_\lambda^k < \infty$, where $k \in \mathbb{N}$, is equivalent to $\mathbb{E}|\Delta^k h(\pi_\lambda)| < \infty$, see Proposition 1 in [47].

Below we present first-order asymptotic expansions.

Let $X_1, ..., X_n$ be independent Bernoulli $\mathbf{B}(p_i)$ r.v.s. Denote

$$\begin{aligned} Q_\lambda(A) &= \Big[\mathbb{P}(\pi_\lambda \in A) + \mathbb{P}(\pi_\lambda + 2 \in A) - 2\mathbb{P}(\pi_\lambda + 1 \in A)\Big]\Big/2, \\ \varepsilon &= \min\Big\{1; (2\pi\lfloor\lambda - p_n^*\rfloor)^{-1/2} + 2\delta/(1 - p_n^*/\lambda)\Big\}, \quad p_n^* = \max_{i \leq n} p_i. \end{aligned}$$

Let $\pi_\lambda^\star$ denote a random variable with distribution (3.32). Then

$$Q_\lambda(A) = [\mathbb{P}(\pi_\lambda^\star \in A) - \mathbb{P}(\pi_\lambda + 1 \in A)]/2\lambda$$

(see [221], Section 4.3). The following result from [221], Chapter 4, sharpens (13) in [150] and the bound of Corollary 2.4 in [19] (Corollary 9.A.1 in [21]).

Theorem 3.16 *Let $X_1, ..., X_n$ be independent Bernoulli $\mathbf{B}(p_i)$ r.v.s. Then*

$$\left| \mathbb{P}(S_n \in A) - \mathbb{P}(\pi_\lambda \in A) + Q_\lambda(A) \sum_{i=1}^{n} p_i^2 \right| \leq 2\delta^* \varepsilon + 2\delta^2 , \qquad (3.82)$$

where $\delta = \lambda^{-1}(1-\mathrm{e}^{-\lambda}) \sum_{i=1}^n p_i^2$, $\delta^* = \lambda^{-1}(1-\mathrm{e}^{-\lambda}) \sum_{i=1}^n p_i^3$.

Note that (3.82) can be expressed as

$$d_{TV}(\mathcal{L}(S_n); \mathbf{\Pi}(\lambda) - \lambda_2 Q_\lambda) \leq 2\delta^* \varepsilon + 2\delta^2. \qquad (3.82^*)$$

Integer-valued random variables.

Let $\{X_k\}$ be non-negative integer-valued random variables.

Asymptotic expansions to $\mathbb{E}h(S_n) - \mathbb{E}h(\pi_\lambda)$, where function h is either bounded or grows at a polynomial rate, are presented by Barbour [19]. In particular, a first-order asymptotic expansion is given in [19] in the assumption that random variables $\{X_k\}$ have finite third moments.

First-order asymptotic expansions in terms of d_{TV}, d_G are provided by Theorems 3.7, 3.9.

Local asymptotic expansions have been given by Barbour & Jensen [20]. First-order asymptotic expansions in the case of discrete random variables have been presented by Aleškevičienė & Statulevičius [2], see also [319].

3.7 Sum of a random number of random variables

Let $\nu, X, X_1, X_2, ...$ be independent non-negative random variables, where r.v. ν takes values in $\mathbb{Z}_+$, $X, X_1, X_2, ...$ are i.i.d. random variables. Set

$$S_\nu = X_1 + ... + X_\nu .$$

A natural task is to evaluate the accuracy of Poisson approximation to $\mathcal{L}(S_\nu)$.

We consider first the case where $X, X_1, X_2, ...$ are Bernoulli $\mathbf{B}(p)$ r.v.s.

Denote $\tilde{\nu} = \mathbb{E}\nu$. Then $\mathbb{E}S_\nu = p\tilde{\nu}$.

Let $\mathcal{F}_2$ denote the class of functions $h : \mathbb{Z}_+ \to \mathbb{R}$ such that $||\Delta^2 h|| \leq 1$, and set

$$d_2(X;Y) = \sup_{h \in \mathcal{F}_2} |\mathbb{E}h(X) - \mathbb{E}h(Y)|.$$

Logunov [198] points out that $d_{TV}(X;Y) \leq d_2(X;Y)$, and shows that

$$d_2(S_\nu; \pi_{p\tilde{\nu}}) \leq p^2 d_*(\nu; \pi_{p\tilde{\nu}}),$$

where

$$d_*(X;Y) = \sum_{k \geq 1} k(k-1)|\mathbb{P}(X=k) - \mathbb{P}(Y=k)|/2.$$

Note that

$$d_*(X;Y) \geq d_G(X;Y).$$

Yannaros [320] has shown that

$$d_{TV}(\pi_\lambda;\pi_\mu) \le \min\{|\sqrt{\lambda}-\sqrt{\mu}\,|;|\lambda-\mu|\} \qquad (\lambda>0,\mu>0). \tag{3.83}$$

The first term in (3.83) has been improved by Roos [262] and Ruzankin [269]:

$$d_{TV}(\pi_\lambda;\pi_\mu) \le \sqrt{\frac{2}{\mathrm{e}}}\left|\sqrt{\lambda}-\sqrt{\mu}\right|. \tag{3.83*}$$

The second term in the RHS of (3.83) is a consequence of the trivial inequality

$$d_{TV}(\pi_\lambda;\pi_\mu) \le 1-\exp(-|\lambda-\mu|) \le |\lambda-\mu| \tag{3.83*}$$

that follows by defining π_λ and π_μ on a common probability space (cf. (4.10) in [221]).

It is easy to see that

$$\begin{aligned} d_{TV}(S_\nu;\pi_\lambda) &\le \sum\nolimits_k \mathbb{P}(\nu=k)d_{TV}(S_k;\pi_\lambda),\\ d_{TV}(S_k;\pi_\lambda) &\le d_{TV}(S_k;\pi_{kp})+d_{TV}(\pi_{kp};\pi_\lambda). \end{aligned}$$

Using these inequalities and (3.83), Yannaros [320] has shown that

$$d_{TV}(S_\nu;\pi_{p\tilde\nu}) \le \min\Big\{\frac{p}{2\sqrt{1-p}}\,;(1-\mathbb{E}\mathrm{e}^{-p\nu})p\Big\} + \min\left\{p\mathbb{E}|\nu-\tilde\nu|;\sqrt{(p\operatorname{var}\nu)/\tilde\nu}\right\}. \tag{3.84}$$

The term $\min\{p/2\sqrt{1-p}\,;(1-\mathbb{E}\mathrm{e}^{-p\nu})p\}$ in (3.84) is inherited from (3.13) and (3.15).

The right-hand side of (3.84) can be sharpened using (3.17), (3.83*) and (3.83*):

$$d_{TV}(S_\nu;\pi_{p\tilde\nu}) \le 3p/4\mathrm{e}+2(\delta^*+\delta^2)+\min\left\{\left(1-\mathbb{E}\mathrm{e}^{-p|\nu-\tilde\nu|}\right);\mathbb{E}|\sqrt{\nu}-\sqrt{\tilde\nu}\,|\sqrt{2p/\mathrm{e}}\right\}. \tag{3.85}$$

Note that $\mathbb{E}|\sqrt{\nu}-\sqrt{\tilde\nu}\,| \le \min\{\tilde\nu^{-1/2}\sqrt{\operatorname{var}\nu}\,;\tilde\nu^{-3/2}\operatorname{var}\nu\}$.

3.7.1 Mixed Poisson distribution

A number of authors evaluated the accuracy of Poisson approximation to the *mixed Poisson distribution*, i.e., the distribution of the random variable π_ν, where $\mathcal{L}(\pi_t)=\mathbf{\Pi}(t)$, r.v. ν takes values in $[0;\infty)$:

$$\mathbb{P}(\pi_\nu=m) = \int_0^\infty \mathbb{P}(\pi_y=m)\mathbb{P}(\nu\in\mathrm{d}y) \qquad (m\ge 0).$$

Clearly, ch.f. $\mathbb{E}\mathrm{e}^{\mathrm{i}t\pi_\nu}=\mathbb{E}\mathrm{e}^{\nu(\mathrm{e}^{\mathrm{i}t}-1)}$. If $\{X_i\}$ are independent Poisson $\mathbf{\Pi}(1)$ r.v.s, then $S_\nu \stackrel{d}{=} \pi_\nu$.

For instance, the negative Binomial distribution $\mathbf{NB}(t,p)$ is a mixed Poisson distribution with

$$\mathbb{P}(\nu\in\mathrm{d}y)/\mathrm{d}y = r^t y^{t-1}\mathrm{e}^{-yr}/\Gamma(t) \qquad (y>0),$$

where $r=p/(1-p)$, $\Gamma(y)=\int_0^\infty x^{y-1}\mathrm{e}^{-x}\mathrm{d}x$.

Roos [262] presents estimates of the accuracy of Poisson approximation to the mixed Poisson distribution with a correct constant at the leading term:

$$\begin{aligned} d_{TV}(\pi_\nu;\pi) &\le \min\left\{\tfrac{3}{2\mathrm{e}}\mathbb{E}(\nu\ln(\nu/\mathbb{E}\nu));\operatorname{var}\nu\right\} && \text{if } \tilde\nu>0,\\ d_{TV}(\pi_\nu;\pi_\lambda) &\le \min\left\{\sqrt{\tfrac{2}{\mathrm{e}}}\,\mathbb{E}|\sqrt{\nu}-\sqrt{\lambda}|;\mathbb{E}|\nu-\lambda|\right\} && \text{if } \lambda>0, \end{aligned}$$

where $\tilde{\nu} = \mathbb{E}\nu$, constant $3/2e$ cannot be improved; estimates is terms of some other metrics are presented as well. Daly [97] has proved that

$$d_{TV}(\pi_\nu; [\kappa_{\pi_\nu}]+\pi_\mu) \le \frac{1.2}{\tilde{\nu}+\sigma^2}\operatorname{var}\nu + \frac{\sigma^2}{\tilde{\nu}(\tilde{\nu}+\sigma^2)}\left(\sigma+\sqrt{\tilde{\nu}+\sigma^2}\right) + \frac{2}{\tilde{\nu}+\sigma^2},$$

where $\sigma^2 = \operatorname{var}\pi_\nu$, $\mu = \sigma^2 + \{\tilde{\nu}-\sigma^2\}$.

3.7.2 Sum of 0-1 r.v.s till the stopping time

Let $X, X_1, X_2, ...$ be i.i.d. non-negative integer-valued r.v.s. Set $S_0 = 0$,

$$S_n = X_1 + ... + X_n \quad (n \ge 1),$$

and let $\mu(t)$ denote the stopping time:

$$\mu(t) = \max\{n \ge 0 : S_n \le t\}.$$

Theorems 3.17–3.18 below are cited from see [221], Chapter 3. They provide estimates of the accuracy of Poisson approximation to the distribution of the number

$$N_t(x) = \sum_{j=1}^{\mu(t)} \mathbb{I}\{X_j \ge x\} + \mathbb{I}\{t - S_{\mu(t)} \ge x\} \tag{3.86}$$

of exceedances of a "high" level $x \in [0; t]$ until S_n reaches level t.

Note that

$$\{N_t(x) = 0\} = \{M_t < x\},$$

where

$$M_t = \max\{t - S_{\mu(t)};\ \max_{1 \le i \le \mu(t)} X_i\} \tag{3.87}$$

is the largest observation among $\{X_1, ..., X_{\mu(t)}, t - S_{\mu(t)}\}$.

Let $X_{k,t}$ denote the kth largest element among $\{X_1, ..., X_{\mu(t)}, t - S_{\mu(t)}\}$. Then

$$\{X_{k,t} < x\} = \{N_t(x) < k\}.$$

The topic has applications in finance. For instance, suppose a bank has opened a credit line for a series of operations, and the total amount of credit is t units of money. The cost of the ith operation is denoted by X_i. What is the probability that the bank will ever pay x or more units of money at once? that there will be a certain number of such payments? Information on the asymptotic properties of the distribution of random variables M_t and $N_t(x)$ can help answering such questions.

Let $\{X_i^<, i \ge 1\}$, $\{X_j^>, j \ge 1\}$ be independent r.v.s with the distributions

$$\mathcal{L}(X^<) = \mathcal{L}(X | X < x), \ \mathcal{L}(X^>) = \mathcal{L}(X | X \ge x).$$

We set $p_x = \mathbb{P}(X \ge x)$,

$$S_0(k) = 0, \quad S_m(k) \ = \ \sum_{i=0}^{k} X_i^> + \sum_{i=k+1}^{m} X_i^< \quad (m \ge 1).$$

Let K_*, K^* denote the end-points of $\mathcal{L}(X)$, and set

$$\begin{aligned} \tau_k &= \tau_k' - k, \ \tau_k' = \min\{n : S_n(k) > t - x\}, \\ \lambda_k &\equiv \lambda_k(t, x, k) = p_x(t - x - k\mathbb{E}X^>)/\mathbb{E}X^<. \end{aligned}$$

In Theorems 3.17–3.18 we assume the following condition:
there exist constants $D<\infty$ *and* $D_*\in(K_*;K^*)$ *such that*

$$\int_x^\infty \mathbb{P}(X\geq y)\mathrm{d}y \leq D\mathbb{P}(X\geq x) \qquad (x\geq D_*). \tag{3.88}$$

Condition (3.88) means the tail of $\mathcal{L}(X)$ is light (cf. (3.15) in [221]). Inequality (3.88) holds if function $g(x)=\mathrm{e}^{cx}\mathbb{P}(X\geq x)$ is non-increasing as $x>1/c$ $(\exists c>0)$. The equality in (3.88) for all $x\geq 0$ may be attained only if $\mathcal{L}(X)$ is exponential with $\mathbb{E}X=D$.

Theorem 3.17 *For any* $k\in\mathbb{Z}_+$, *as* $t\to\infty$,

$$\sup_{x\in B_+(t)}\left|\mathbb{P}(N_t(x)=k)-\mathbb{P}(\pi_{\lambda_k}=k)-\sum_{r=0}^{k-1}\left(\mathbb{P}(\pi_{\lambda_k}=r)-\mathbb{P}(\pi_{\lambda_{k-1}}=r)\right)\right|=O(1/t),$$

where $B_+(t)=(K_*;K^*\wedge t/(k+2))$.

Let $\pi(t,x)$ denote a Poisson r.v. with parameter $p_x t/\mathbb{E}X$.

Theorem 3.18 *For any* $k\in\mathbb{Z}_+$, *as* $t\to\infty$,

$$\sup_{K_*<x<K^*}|\mathbb{P}(N_t(x)=k)-\mathbb{P}(\pi(t,x)=k)|=O(t^{-1}\ln t).$$

One can show that $N_t(x)$ is "small" when x is "large":

$$\sup_{x\geq\sqrt{t}}\mathbb{P}(N_t(x)\geq 1)\leq q^{\sqrt{t}} \qquad (\exists q\in(0;1)).$$

Theorem 3.7 in [221] presents asymptotic expansions to $\mathbb{P}(N_t(x)=k)$. The asymptotic expansions to $\mathcal{L}(M_t)$ are available under a weaker moment assumption (cf. [221], Chapter 3).

The number of intervals between consecutive jumps of a Poisson process.
Consider a Poisson jump process $\{\pi_\lambda(s),s\geq 0\}$ with parameter $\lambda>0$, and let η_i denote the moment of its ith jump. Set $X_i=\eta_i-\eta_{i-1}$. Then $N_t(x)$ is the number of intervals between consecutive jumps with lengths greater or equal to x.

If the points of jumps represent catastrophic/rare events, then $N_t(x)$ can be interpreted as the number of "long" intervals without catastrophes.

Let $\pi_{t,x}$ be a Poisson r.v. with parameter $t\lambda\mathrm{e}^{-\lambda x}$. Then for any $k\in\mathbb{Z}_+$, as $t\to\infty$,

$$\sup_{0<x<t}|\mathbb{P}(N_t(x)=k)-\mathbb{P}(\pi_{t,x}=k)|=O\big(t^{-1}\ln t\big)$$

(cf. (3.12) in [221]).

Open problems

3.5. Will asymptotic expansions for $\mathcal{L}(N_x(t))$ hold under a weaker moment assumption?

3.6. Generalise the results of Theorems 3.17–3.18 to the case of

$$N_t(x)=\sum_{j=1}^{\mu(t)}Y_i\mathbb{1}\{X_j\geq x\}+Y_{\mu(t)+1}\mathbb{1}\{t-S_{\mu(t)}\geq x\},$$

where $\{(X_i,Y_i)_{i\geq 1}\}$ is a sequence of i.i.d. pairs of non-negative r.v.s.

3.8 Proofs

Theorem 3.1 is a straightforward consequence of Theorem 3.16.

Proof of Theorem 3.2. Recall that $\{X_i\}$ are independent Bernoulli $\mathbf{B}(p)$ r.v.s, $\mathcal{L}(S_n) = \mathbf{B}(n,p)$,

$$\mu = npq+\{np^2\},\ \kappa_{S_n} = np^2,\ \delta^{(\mu)}_{X_i,\tilde{X}_i} = 0,$$

where $q = 1-p$. Denote

$$d^{(n)}_p = d_{TV}\Big(S_n;\, \lfloor np^2\rfloor + \pi_{npq+\{np^2\}}\Big).$$

We will apply (3.17*) and Theorem 3.11.

Estimate (3.75) entails

$$d^{(n)}_p \le \{np^2\}\bar{\varepsilon}_\mu + 2p\varepsilon_{1,n}\,,$$

where

$$\varepsilon_{1,n} \le \min\Big\{\frac{\sqrt{2/\pi}}{\sqrt{1/4+(n-1)p}}\,;\,\frac{1}{\sqrt{2\pi\lfloor(n-1)p\rfloor}} + 2p\frac{1-\mathrm{e}^{-np}}{1-1/n}\Big\}. \tag{3.89}$$

The term $2p\varepsilon_{1,n}$ appears here because of estimate (3.95).

$\mathbf{B}(n,p)$ is known to be unimodal. For the unimodal distribution we can apply (3.95′), i.e., replace $\varepsilon_{i,n}$ with $\varepsilon'_{i,n} = \max_k \mathbb{P}(S_{n,i}=k)$:

$$d^{(n)}_p \le \{np^2\}\bar{\varepsilon}_\mu + 2p\varepsilon^{\mathrm{o}}_n\,,$$

where $\varepsilon^{\mathrm{o}}_n = \min\{\varepsilon_{1,n};\varepsilon'_{1,n}\}$. Note that

$$\begin{aligned}\varepsilon^\star_\mu &\le \{np^2\}/npq + 2p\varepsilon^{\mathrm{o}}_n,\ \varepsilon^*_\mu \le 1/\sqrt{2\pi\lfloor npq\rfloor}\,,\\ \bar{\varepsilon}_\mu &\le 2/\lfloor npq\rfloor\sqrt{\mathrm{e}\pi} + 2\varepsilon^\star_\mu/npq \le 2/\lfloor npq\rfloor\sqrt{\mathrm{e}\pi} + 2\{np^2\}/(npq)^2 + 4\varepsilon^{\mathrm{o}}_n/nq.\end{aligned}$$

Thus,

$$d^{(n)}_p \le \frac{2\{np^2\}}{\sqrt{\mathrm{e}\pi}\,\lfloor npq\rfloor} + 2\frac{\{np^2\}^2}{(npq)^2} + 4\frac{\{np^2\}}{nq}\varepsilon^{\mathrm{o}}_n + 2p\varepsilon^{\mathrm{o}}_n\,. \tag{3.90}$$

Evaluating $d^{(n)}_p$ separately for $p<1/\sqrt{n}$ and $p\in[1/\sqrt{n};1/2]$ yields (3.41). Indeed,

$$d^{(n)}_p \le 3/4\mathrm{e}\sqrt{n} + 4/n$$

by (3.17*) if $p<1/\sqrt{n}$. If $1/\sqrt{n} \le p \le 1/2$, we denote by $f(p)$ the RHS of (3.90), where $\varepsilon^{\mathrm{o}}_n$ is replaced by $\varepsilon'_{1,n}$. Note that $\varepsilon'_{1,n} \le 0.45/\sqrt{(n-1)pq}$ if $(n-1)pq \ge 3.75$ (see (2.22) in [80]).

Function f on $[1/\sqrt{n}\,;1/2]$ first declines, then grows. Hence

$$\sup_{1/\sqrt{n}\le p\le 1/2} f(p) = \max\{f(1/\sqrt{n}); f(1/2)\} = f(1/\sqrt{n})\,,$$

yielding (3.41). □

Proof of Theorem 3.3. Let $\{X_i\}$ be independent Bernoulli $\mathbf{B}(p)$ r.v.s. Set $q=1-p$, and recall notation of Theorem 3.12. Then

$$\delta^{(\mu)}_{X_i} = 0,\ \ c^{\mathrm{o}}_\mu = 1\wedge 2\mu/3,$$

and (3.76) yields

$$d_G(S_n; \lfloor np^2 \rfloor + \pi_\mu) \le |\{np^2\}| \min\left\{1; c_\star/\sqrt{\mu}\,; 2\varepsilon_{0,n}\right\} + 2p(1 \wedge c_\mu \varepsilon_{1,n}),$$

where $\varepsilon_{0,n} \le \varepsilon_{1,n}$, cf. (3.89).

Put $c = 1.5$. Then (3.60) implies that

$$\sup_{0<p<c/n} d_G(S_n; \lfloor np^2 \rfloor + \pi_\mu)/\sqrt{np} \le \sup_{0<p<c/n} p\sqrt{np} = c^{1.5}/n.$$

If $n \ge 4$, then (3.27) entails

$$\begin{aligned} &\sup_{c/n \le p < 1/\sqrt{n}} d_G(S_n; \lfloor np^2 \rfloor + \pi_\mu)/\sqrt{np} \\ \le &\sup_{c/n \le p < 1/\sqrt{n}} \left(p \min\left\{ \sqrt{np} \Big/ \sqrt{2\pi \lfloor np \rfloor}\,; 1/\sqrt{2e} \right\} + 2p^2 (c_\star + 1/\sqrt{np}\,) \right) \\ \le &\min\left\{ 1 \Big/ \sqrt{2\pi(n - \sqrt{n}\,)}\,; 1/\sqrt{2en} \right\} + 2(c_\star + n^{-1/4})/n. \end{aligned}$$

Let now $p \ge 1/\sqrt{n}$. Then $npq \le \mu < np$. Using (3.76*), we derive

$$d_G(S_n; \lfloor np^2 \rfloor + \pi_\mu)/\sqrt{np} \le f(p),$$

where $f(p) = c_\star/np\sqrt{q} + 2\sqrt{2/\pi}\, c_\star p \Big/ \sqrt{1/4 + (n-1)p}$.

Function f on $[1/\sqrt{n}\,; 1/2]$ first declines, then grows. Hence

$$\sup_{1/\sqrt{n} \le p \le 1/2} f(p) = \max\{f(1/\sqrt{n}); f(1/2)\} \le \frac{2c_\star}{\sqrt{\pi}} \left(1 \Big/ \sqrt{n - \sqrt{n}} + \sqrt{2} n^{-3/4} \right),$$

and (3.42) follows. □

Proof of Theorem 3.4 (sketch) uses Stein's method (see Appendix).

In order to evaluate, say, $\mathbb{E}h(S) - \mathbb{E}h(\pi_\lambda)$, where function h belongs to a particular class of functions, one needs to bound $\|g\|$ and $\|\Delta g\|$, where $g = g(\cdot|h)$ is a solution of the "Stein equation" (F.7). Standard arguments (cf. the proof of Theorem 4.14 in [221]) yield

$$\lambda |\mathbb{E}h(S) - \mathbb{E}h(\pi_\lambda)| \le \|\Delta g\| \, (\delta_1 + \delta_2) + \|g\| \delta_3 \,,$$

where the solution $g = g(\cdot|h)$ of (F.7) is given by (F.8).

If $h_A(x) = \mathbb{I}\{x \in A\}$, $A \subset \mathbb{Z}_+$, then

$$\|g\| \le \min\{\lambda; \sqrt{2\lambda/e}\,\}, \quad \|\Delta g\| \le 1 - e^{-\lambda} \tag{3.91}$$

(see, e.g., [21], Remark 1.1.2). If $h \in \mathcal{L}_1$, where $\mathcal{L}_1 = \{h\colon |h(x) - h(y)| \le |x - y|\}$ is the set of Lipschitz functions, then

$$\|g\| \le \lambda, \quad \|\Delta g\| \le \lambda \wedge \tfrac{4}{3}\sqrt{2\lambda/e}\,, \quad \|\Delta^2 \tilde{g}\| \le \min\{4\lambda/3; 2\} \tag{3.92}$$

(see Barbour et al. [21], Remark 1.1.6, [33]). Taking into account (3.91), we get (3.50). Using (3.92), we derive (3.51). □

Proof of Theorem 3.6. If $\lambda = 0$, then $S_n = 0$, and (3.62) trivially holds. Therefore, w.l.o.g. we may assume that $\lambda > 0$.

It is known (cf. [221], Chapter 4) that for any $A \subset \mathbb{Z}_+$

$$\begin{aligned}&(\mathbb{P}(\pi^\star_\lambda \in A) - \mathbb{P}(\pi_\lambda + 1 \in A))/\lambda \\ &= \mathbb{P}(\pi_\lambda \in A) + \mathbb{P}(\pi_\lambda - 2 \in A) - 2\mathbb{P}(\pi_\lambda - 1 \in A) \\ &\equiv \Delta^2 \mathbb{P}(\pi_\lambda \in A),\end{aligned}$$

where $\Delta \mathbb{P}(\pi_\lambda \in A) = \sum_{k \in A} \Delta \mathbb{P}(\pi_\lambda = k)$.

According to (C.2),

$$\|\Delta^2 \mathbb{P}(\pi_\lambda = \cdot)\|_1 \equiv \|(I_1 - I)^{*2} * \mathbf{\Pi}(\lambda, X)\|_{TV} \le 3/\mathrm{e}\lambda.$$

Note that

$$\|\Delta^2 \mathbb{P}(\pi_\lambda = \cdot)\|_1 = 2\Delta^2 \mathbb{P}(\pi_\lambda \in B),$$

where

$$B = \{k\colon \Delta^2 \mathbb{P}(\pi_\lambda + 2 = k) > 0\} = \{k\colon k_1 \le k \le k_2\}$$

and

$$k_1 = \lambda + 1/2 - \sqrt{\lambda + 1/4}\,, \quad k_2 = \lambda + 1/2 + \sqrt{\lambda + 1/4}$$

(see Deheuvels & Pfeifer [103]). Therefore,

$$|\mathbb{P}(\pi_\lambda + 1 \in A) - \mathbb{P}(\pi^\star_\lambda \in A)| \le 3/2\mathrm{e}$$

for any $A \subset \mathbb{Z}_+$, i.e.,

$$d_{TV}(\pi_\lambda + 1; \pi^\star_\lambda) \le 3/2\mathrm{e}. \tag{3.93}$$

An application of (3.68) now leads to

$$|\mathbb{E}h(S_n) - \mathbb{E}h(\pi_\lambda)| \le \frac{3}{4\mathrm{e}\lambda}\left|\sum_{i=1}^n \kappa_{X_i}\right| + \varepsilon_1 + \varepsilon_2 + \varepsilon_3$$

for any function $h \in \mathcal{S}$, the class of measurable functions taking values in $[0;1]$. Since

$$d_{TV}(X;Y) = \sup_{h \in \mathcal{S}} |\mathbb{E}h(X) - \mathbb{E}h(Y)|,$$

the result follows. □

A useful tool when applying Stein's method is the following observation.

Given a function $f\colon \mathbb{Z}_+ \to \mathbb{R}$, we denote $(\ell \ge 0, m \ge 0, k \ge 0)$

$$\begin{aligned} R_f(m,k,\ell) &= f(m) - f(k) - (m-k)\Delta f(\ell), \\ c_1(f) &= \sup\nolimits_{i,j} |\Delta f(i) - \Delta f(j)|, \quad c_2(f) = \|\Delta^2 f\|, \\ \delta^{(\ell)}_{m,k} &= \min\{c_1(f)|m-k|; c_2(f)|(m-\ell)(m-\ell-1) - (k-\ell)(k-\ell-1)|/2\}. \end{aligned}$$

Proposition 3.19 [224] *For any function $f\colon \mathbb{Z}_+ \to \mathbb{R}$ and any $\ell \ge 0, m \ge 0, k \ge 0$,*

$$|R_f(m,k,\ell)| \le \delta^{(\ell)}_{m,k}\,. \tag{3.94}$$

Relation (3.94) is a discrete analogue of Taylor's formula.

The proof of Theorem 3.7 uses Stein's method and the following bound. Denote

$$\varepsilon'_{i,n} = \max_k \mathbb{P}(S_{n,i} = k), \quad \tilde{\varepsilon}_{i,n} = \min\{\varepsilon_{i,n}; \varepsilon^*_{\lambda_i} + 2\varepsilon^\star_{\mu_i}\}.$$

Lemma 3.20 [224] *For any bounded function* $\tilde f$

$$|\mathbb{E}\Delta\tilde f(S_{n,i})| \le \min\left\{2\|\tilde f\|\tilde\varepsilon_{i,n}\,; (\|\Delta\tilde f\|\wedge 2\|\tilde f\|\varepsilon^*_{\lambda_i}) + 2\|\Delta\tilde f\|\varepsilon^+_{i,n}\right\}. \tag{3.95}$$

If $\mathcal{L}(S_{n,i})$ *is unimodal, then* $\tilde\varepsilon_{i,n}$ *in (3.95) can be replaced with* $\varepsilon^{\circ}_{i,n} = \min\{\tilde\varepsilon_{i,n}; \varepsilon'_{i,n}\}$.

Proof of Theorem 3.7 (sketch). We use Stein's method (see Appendix).

Let $\mathcal{S}$ denote the class of measurable functions taking values in $[0;1]$. For any $h \in \mathcal{S}$ we denote by $g \equiv g(\cdot|h)$ the solution of Stein equation (F.7). Since

$$\mathbb{E}h(S_n) - \mathbb{E}h(\pi_\lambda) = \mathbb{E}g(S_n{+}1) - \lambda^{-1}\mathbb{E}S_n g(S_n),$$

one needs to evaluate $\mathbb{E}g(S_n{+}1) - \lambda^{-1}\mathbb{E}S_n g(S_n)$.

It is known that the solution of equation (F.7) is

$$g(n) = [\mathbb{E}h(\pi_\lambda)\mathbb{I}\{\pi_\lambda{<}n\} - \mathbb{E}h(\pi_\lambda)\mathbb{P}(\pi_\lambda{<}n)]/\mathbb{P}(\pi_\lambda{=}n{-}1) \qquad (n{\ge}1) \tag{F.8*}$$

(the value of $g(0)$ is irrelevant). Set

$$S_{n,i} = S_n - X_i\,,\; g_i(\cdot) = \mathbb{E}g(S_{n,i}{+}1{+}\cdot).$$

Barbour & Eagleson [17] have shown that

$$\|\Delta g\| \le 1{-}\mathrm{e}^{-\lambda} \qquad (h{\in}\mathcal{S}). \tag{3.96}$$

An application of (3.57) yields

$$\begin{aligned}
\lambda\mathbb{E}g(S_n{+}1) - \mathbb{E}S_n g(S_n) &= \lambda\mathbb{E}g(S_n{+}1) - \sum_{i=1}^n \mathbb{E}X_i g(S_{n,i}{+}X_i)\\
&= \sum_{i=1}^n \mathbb{E}X_i\,(\mathbb{E}g(S_{n,i}{+}X_i{+}1) - \mathbb{E}g(S_{n,i}{+}X_i^*{+}1))\\
&= \sum_{i=1}^n \mathbb{E}X_i\,(\mathbb{E}g_i(X_i) - \mathbb{E}g_i(X_i^*))\,.
\end{aligned}$$

Note that

$$g_i(X_i) - g_i(X_i^*) = (X_i{-}X_i^*)\Delta g(S_{n,i}{+}1) + R_{g_i}(X_i, X_i^*, 0).$$

Taking into account (3.95) and (3.96), we derive the desired result. □

Proof of Theorem 3.8 relies on Theorem 3.9 and Lemma 3.21 below.

For any function $h{\in}\mathcal{L}_1$ an application of (3.72) and Lemma 3.21 yields

$$|\mathbb{E}h(S_n) - \mathbb{E}h(\pi_\lambda)| \le \theta^*\lambda\mathbb{P}(\pi_\lambda{=}\lfloor\lambda\rfloor) + \varepsilon_{(1)} + \varepsilon_{(2)} + \varepsilon_{(3)}\,. \tag{3.97}$$

It is known that

$$\mathbb{P}(\pi_\lambda{=}\lfloor\lambda\rfloor]) \le 1/\sqrt{2\pi\lfloor\lambda\rfloor} \qquad (\lambda{>}0) \tag{3.98}$$

(cf. [221], formula (4.58)). Note that one cannot replace $\lfloor\lambda\rfloor$ in (3.98) by λ even for large λ: using inequalities

$$\begin{aligned}
&\ln(1{+}x) \ge\; x - x^2/2 &&(x{\ge}0),\\
&\ln(1{-}x) \ge -x{-}x^2/2(1{-}x)^2 &&(0{\le}x{<}1),
\end{aligned}$$

one can check that
$$\mathbb{P}(\pi_\lambda = \lfloor\lambda\rfloor) > 1/\sqrt{2\pi\lambda}$$
if $\lambda = k+1/2$ $(\forall k \in \mathbb{N})$.

It is easy to see that
$$\mathbb{P}(\pi_\lambda = \lfloor\lambda\rfloor) = \mathrm{e}^{-\lambda} \leq 1/\sqrt{2\mathrm{e}\lambda} \tag{3.99}$$
if $\lambda < 1$. Inequality (3.99) turns into equality if $\lambda = 1/2$.

We will use the fact that
$$\max_k \mathbb{P}(\pi_\lambda = k) = \mathbb{P}(\pi_\lambda = \lfloor\lambda\rfloor) \leq 1/\sqrt{2\mathrm{e}\lambda} \qquad (\lambda > 0)$$
(cf. [21], p. 262). Therefore,
$$\mathbb{P}(\pi_\lambda = \lfloor\lambda\rfloor) \leq \min\Big\{1; 1/\sqrt{2\pi\lfloor\lambda\rfloor}\,; 1/\sqrt{2\mathrm{e}\lambda}\,\Big\}.$$
This inequality and (3.97) entail (3.70). □

Lemma 3.21 [225] *For any* $\lambda \geq 0$
$$d_G(\pi_\lambda + 1; \pi^\star_\lambda) = 2\lambda\mathbb{P}(\pi_\lambda = \lfloor\lambda\rfloor).$$

Proof of Theorem 3.9 follows the lines of the proof of Theorem 3.7.

Proof of Theorem 3.11 (sketch) uses Stein's method. We need to evaluate $|\mathbb{E}h(S_n) - \mathbb{E}h(\pi_\mu + a)|$, where $h : \mathbb{Z} \to [0;1]$.

The basic equation is
$$f(k+1) - \mu^{-1}(k-a)f(k) = h_A(k) - \mathbb{E}h_A(\pi_\mu + a) \qquad (k \geq a)$$
(cf. (F.3) below). The solution $f \equiv f_{h_A}$ is
$$f(k) = 0 \quad (k \leq a), \quad f(k) = g_{\tilde h}(k-a) \quad (k \geq a),$$
where $\tilde h(m) = h_A(m+a)$ $(m \geq 0)$ and $g_{\tilde h}$ is given by (F.8) with λ replaced with μ. Hence
$$\mathbb{E}h_A(S_n) - \mathbb{E}h_A(\pi_\mu + a) = \mathbb{E}f(S_n+1) - \mu^{-1}\mathbb{E}(S_n - a)f(S_n).$$

Note that
$$\mu = \sigma^2 + b = \lambda - a \geq 0.$$
Hence
$$\begin{aligned} \mu f(S_n+1) - (S_n - a)f(S_n) &= \mu f(S_n+1) - (\bar S_n + \mu)f(S_n) \\ &= \mu\Delta f(S_n) - \bar S_n f(S_n). \end{aligned}$$
Evaluation of $|\mu\mathbb{E}f(S_n+1) - \mathbb{E}(S_n - a)f(S_n)|$ can be carried out as in the proof of Theorem 3.6. □

Proof of Theorem 3.12 follows the lines of the proof of Theorem 3.11.

Exercises

3.1. Given a r.v. X, recall that X^* denotes a r.v. with distribution defined by (3.56).

If S_n is Binomial $\mathbf{B}(n,p)$, show that $S_n^* \stackrel{d}{=} S_{n-1}$.

3.2. Let $S_{r,p}$ be a Negative Binomial $\mathbf{NB}(r,p)$ r.v. with parameters $p\in(0,1)$ and $r>0$. Show that $S_{r,p}^* \stackrel{d}{=} S_{r+1,p}$.

3.3. Prove that

$$d_G(S_n;\pi_\lambda) \le \min\left\{\frac{2}{3};\frac{1}{\lambda}\right\}\sum_{i=1}^n p_i^3 \tag{3.100}$$

in Example 3.4. If, in addition, $\max_i p_i \le 1/2$ and $\lambda \ge 10$, then

$$d_{TV}(S_n;\pi_\lambda) \le \frac{1}{\lambda\sqrt{\lambda}}\sum_{i=1}^n p_i^3. \tag{3.101}$$

If all $p_i \equiv p$, then the RHS of (3.101) is bounded by $n^{-1/2}$.

3.4. Let X be a r.v. taking values in $\mathbb{Z}$. Suppose that factorial cumulants $\{\varkappa_k\}$ of $\mathcal{L}(X)$ satisfy

$$|\varkappa_k| \le k!\mathbb{E}X/T^{k-1} \quad (\exists T\ge 7,\ k\ge 1)$$

For instance, $|\varkappa_k| = k!np^k$ in the case of the Binomial distribution $\mathbf{B}(n,p)$, and one can put $T=1/p$. Denote $\mathbb{E}X = \lambda$. Prove that

$$d_{TV}(X;\pi_\lambda) \le \frac{11}{eT}, \quad d_G(X;\pi_\lambda) \le \frac{7}{T}\sqrt{\frac{2\lambda}{5e}}.$$

Hint. Use (B.1), (C.8), (C.2) and Lemma B.1.

Notes

Theorem 3.2 is Theorem 6 in [224]. Theorem 3.3 is from [225].
Bound (3.50) in Theorem 3.4 is from Arratia et al. [10] (see also Smith [286]).
Theorem 3.5 is Proposition 4.15 in [221]. Theorems 3.6, 3.7 are from [224].
Theorems 3.8, 3.9, 3.12 and Corollary 3.10 are from [225].
Theorem 3.11 is from [223]. Theorems 3.13, 3.14, and 3.16 are from [221], Chapter 4.

4

Applications

Results concerning the accuracy of Poisson approximation to the distribution of a sum of discrete r.v.s have applications in meteorology, reliability theory, extreme value theory, mathematical biology, random graphs, urn models, etc. (cf. [15, 52, 37, 146, 189, 221, 208]).

For instance, consider a device that has a large number of components. The device is reliable: failure of a few components does not cause a catastrophe. However, there is always a threshold (say, k): if k consecutive components fail, then the device fails (cf. the Chernobyl atomic power plant).

This leads to the problem concerning the length L_n of the longest head run (LLHR) among n "0"s and "1"s and the number $N_n(k)$ of runs of length k (or length $\geq k$).

The distribution of L_n has been extensively studied. We know that $L_n \approx [\log_{1/p} n]$, where p is the probability to get "1". The distribution of the number of runs of length k (or length $\geq k$) has been evaluated as well (cf. [221, 222] and references therein).

If p has been estimated, one can evaluate how many years an atomic power plant will work before a major disaster is likely to happen.

Using results concerning the distribution of L_n, $N_n(k)$ and similar statistics one can decide if a particular generator of "random numbers" is "good". Namely, for any ε one can determine $k_*(\varepsilon), k^*(\varepsilon)$ such that

$$\mathbb{P}(k_*(\varepsilon) \leq L_n \leq k^*(\varepsilon)) \geq 1-\varepsilon. \tag{4.1}$$

One can generate a good number of sequences of 0-1 random variables and check if the frequency of sequences obeying (4.1) is close to $1-\varepsilon$.

It is interesting that human phycology rejects the possibility of "long" patterns. The following experiment has been conducted by A.Rényi in 1970s: a group of pupils were asked to write down sequences of 100 "0"s and "1"s as randomly as they could. They did – yet none of their sequences had more than 4 heads in a row. The theory suggests $L_{100} \approx [\log_2 100] = 6$. If you toss a coin 100 times, you will typically see a head-run of length 6 or 7. This experiment shows that human intuition is a poor guide when one deals with random events.

A related question is about how far a major stock index can fall once in say 40 years (approximately 10000 daily observations)? A major stock index rarely falls by more than 2% in one day; the biggest daily fall in a year may be around 5% . Yet results concerning the distribution of L_n suggests that in a long time interval there may be a combination of coincidences that may cause a considerable price movement. A famous example of an extreme market fall is the "Black Monday" crash on October 19, 1987, when S&P500 and Dow Jones fell by more than 20%.

Interestingly, there is a measure of risk that suggest that once in 40 years major stock indexes can indeed fall by a figure close to 20% (cf. Example 10.5 in [221]).

In mathematical biology one deals with sequences of discrete random variables representing a genetic code. For instance, a DNA molecular is a sequence $X_1, ..., X_n$ where each X_i takes on one of 4 possible values. A long match pattern among $X_1, ..., X_n$ may represent a fragment responsible for a particular function. However, a large number of mutations are

DOI: 10.1201/9781003478164-4

not harmful, and hence are passed to next generations. Thus, a long match pattern may happen by chance.

Using results concerning the length M_n^* of the longest match pattern (LLMP) among $X_1, ..., X_n$, one can construct a confidence interval $[k_*(\varepsilon); k^*(\varepsilon)]$ such that

$$\mathbb{P}(k_*(\varepsilon) \leq M_n^* \leq k^*(\varepsilon)) \geq 1-\varepsilon. \tag{4.2}$$

A confidence interval can be constructed for a number $N_n^*(k)$ of match patterns of length k (or length $\geq k$). Thus, the search for meaningful patterns can be reduced to those that fall into the corresponding confidence intervals.

This chapter presents results concerning Poisson approximation to the distribution of the number of long head runs and long match patterns.

4.1 Long head runs

Let $\{\xi_i, i \geq 1\}$ be a sequence of 0-1 random variables.

We say a head run (a series of 1's) starts at $i=1$ if $\xi_1=1$; a series starts at $i>1$ if $\xi_{i-1}=0$, $\xi_i=1$. If $\xi_i=0$, $\xi_{i+1}=...=\xi_{i+k}=1$, we say the head run is of length $\geq k$.

For instance, if $n=5$ and $\xi_1=\xi_2=\xi_3=1, \xi_4=0, \xi_5=1$, there is one series (head run) of length 3 and one series of length 1.

Denote

$$A_0 = \{\xi_1 = ... = \xi_k = 1\}, \;\; A_i = \{\xi_i=0, \xi_{i+1} = ... = \xi_{i+k} = 1\} \quad (i>1).$$

Then

$$S_n(k) = \sum_{i=0}^{n-k} \mathbb{1}\{A_i\} \qquad (n \geq k \geq 1)$$

is the number of head runs of length $\geq k$ among $\xi_1, ..., \xi_n$ (NLHR).

Set

$$L_n = \max\{k : \xi_{i+1} = ... = \xi_{i+k} = 1 \; (\exists i \leq n-k)\}.$$

L_n is the *length of the longest head run* (LLHR) among $X_1, ..., X_n$. Obviously,

$$\{L_n < k\} = \{S_n(k)=0\}.$$

The problem of approximating the distribution of LLHR is a topic of active research; it has applications in reliability theory and psychology (cf. [15, 221]).

Let $\{\xi_i, i \geq 1\}$ be i.i.d. Bernoulli $\mathbf{B}(p)$ r.v.s, $p \in (0;1)$, and let π_λ denote a Poisson $\mathbf{\Pi}(\lambda)$ r.v., where $\lambda = \mathbb{E}S_n = \sum_{i=0}^{n-k} \mathbb{P}(A_i)$:

$$\lambda \equiv \lambda(n,k,p) = p^k(1+(n-k)(1-p)).$$

Theorem 3.4 with $B_i = [i-k; i+k]$ yields the following

Corollary 4.1 *As* $n \geq k \geq 1$,

$$d_{TV}(S_n(k); \pi_\lambda) \;\leq\; (1-\mathrm{e}^{-\lambda})(2k+1)p^k, \tag{4.3}$$

$$d_G(S_n(k); \pi_\lambda) \;\leq\; \min\left\{\lambda; \tfrac{4}{3}\sqrt{2\lambda/\mathrm{e}}\right\}(2k+1)p^k. \tag{4.4}$$

An open question is if estimates (4.3), (4.4) can be improved. Note, for instance, that (4.3) does not yield (4.5).

There is a close relation between $S_n(k)$ and $N_t(\cdot)$, where $N_t(\cdot)$ is given by (3.86). Let $\eta_0=0$,

$$\eta_i = \min\{k>\eta_{i-1}\colon \xi_k=0\},\ X_i = \eta_i - \eta_{i-1} \quad (i\geq 1),$$

and set $\mu(t)=\max\{n\geq 0 : X_1 + ... + X_n \leq t\}$. Then

$$S_n(k) = \sum_{j=1}^{\mu(t)} \mathbb{I}\{X_j-1\geq k\} + \mathbb{I}\{n-\eta_{\mu(n)}\geq k\}.$$

Hence

$$S_{n-1}(k) = N_n(k+1).$$

Denote $\lambda_k = n(1-p)p^k$. Theorem 3.18 entails

Corollary 4.2 *For any* $j \in \mathbb{Z}_+$, *as* $n\to\infty$,

$$\max_{1\leq k\leq n} |\mathbb{P}(S_n(k) = j) - \mathbb{P}(\pi_{\lambda_k} = j)| = O\big(n^{-1}\ln n\big). \tag{4.5}$$

According to Theorem 3.13 in [221], the rate $n^{-1}\ln n$ in (4.5) cannot be improved.

Pianoforte & Schulte [232] have noticed that

$$S_n^*(k) \stackrel{d}{=} S_n(k) - Z_\nu ,$$

where operation $Y\to Y^*$ is defined by (3.56), $Z_j = \sum_{i=j-k\vee 0}^{j+k\wedge n-k} \mathbb{I}\{A_i\}$, r.v. ν is independent of all other r.v.s and has the distribution

$$\mathbb{P}(\nu=m) = \mathbb{P}(A_m)/\lambda \quad (m\leq n-k).$$

Pianoforte & Schulte [232] have shown that

$$\max_{1\leq k\leq n} |\mathbb{P}(S_n(k)\leq j) - \mathbb{P}(\pi_{\lambda_k}\leq j)| \leq 40(j+2)^2 n^{-1}\log_2 n \qquad (j\in\mathbb{Z})$$

if $p\leq 1/2$ and $n\geq 2$.

Long non-decreasing runs.

Let $\xi_i = \mathbb{I}\{Y_i \leq Y_{i+1}\}$, where $\{Y_i\}$ are i.i.d.r.v.s with a continuous d.f. Then NLHR $S_n(k)$ is the number of non-decreasing runs of length $\geq k$ (NLNR), and LLHR is the length of the longest non-decreasing run (LLNR) among $Y_1, ..., Y_{n+1}$. We denote LLNR by L_n^+ and NLNR by $S_n^+(k)$.

The topic concerning LLNR and NLNR has applications in finance. It is well known that prices of shares and financial indexes evolve in cycles of growth and decline. Knowing the asymptotics of L_n^+ and $S_n^+(k)$ can help evaluating the length of the longest period of continuous growth/decline of a particular financial instrument as well as the distribution of the number of such long periods.

Pittel [234] has proved a Poisson limit theorem for NLNR (see also Chryssaphinou et al. [93] concerning the case of a Markov chain).

We proceed with the case of i.i.d.r.v.s with a continuous d.f. Note that $\mathcal{L}(\xi_i) = \mathbf{B}(1/2)$ and $\mathbb{P}(Y_1\leq ... \leq Y_{k+1}) = 1/(k+1)!$ Set $\lambda_{n,k} = \mathbb{E}S_n^+(k)$. Then

$$\lambda_{n,k} = 1/(k+1)! + (n-k)/k!(k+2).$$

Theorem 3.4 with $B_i = [i-k-1; i+k+1]$ yields the following

Corollary 4.3 [221] *As* $n \geq k \geq 1$,

$$d_{TV}(S_n^+(k); \pi_{\lambda_{n,k}}) \leq (1 - e^{-\lambda_{n,k}})(2k+3)/(k+1)!$$

The accuracy of compound Poisson approximation to the distribution of the number of non-decreasing runs of fixed length has been evaluated by Barbour & Chryssaphinou [27], p. 982 (continuous d.f.) and Minakov [211] (discrete d.f.). Concerning the asymptotics of LLNR, see [94, 221] and references therein.

Open problems

4.1. Improve the estimates of Corollaries 4.1 and 4.3.

4.2. Derive (3.41)-type (i.e., uniform in k) estimates of the accuracy of (shifted) Poisson approximation to $\mathcal{L}(S_n(k))$ and $\mathcal{L}(S_n^+(k))$.

4.2 Long match patterns

Closely related to the number of long head runs is the number of long match patterns (NLMP) between sequences of independent r.v.s. Information on the distribution of NLMP and the length of the longest match pattern (LLMP) can help recognising "valuable" fragments of DNA sequences (see [10, 11, 210, 214, 219]).

In this section, we present results on the accuracy of Poisson approximation to the distribution of NLMP (cf. [221], Chapter 4).

Let $X, X_1, ..., X_m$, $Y, Y_1, ..., Y_n$ be independent non-degenerate random variables taking values in a discrete state space $\mathbf{A}$. Denote $(k \in \mathbb{N})$

$$\begin{aligned} T_{ij} &= 1\!\!1\{X_{i+1} = Y_{j+1}, \ldots, X_{i+k} = Y_{j+k}\}, \\ \tilde{T}_{ij} &= T_{ij}(k) 1\!\!1\{X_i \neq Y_j\}, \\ T_{ij}^* &= \tilde{T}_{ij} \;\; (i \geq 1, j \geq 1), \;\; T_{ij}^* = T_{ij} \;\; (i=0 \text{ or } j=0). \end{aligned}$$

Then

$$M_{m,n}^* = \max\left\{k \leq \min(m,n) : \max_{(i,j) \in J} T_{ij} = 1\right\}$$

is the *length of the longest match pattern* between $(X_1 \ldots X_m)$ and $(Y_1 \ldots Y_n)$.

LLMP $M_{m,n}^*$ is a 2-dimensional analogue of LLHR L_n. If $\mathbf{A} = \{0,1\}$ and $Y_1 = ... = Y_n = 1$, then $M_{n,n}^* = L_n$.

Given $m \geq k$, $n \geq k$, let

$$J \equiv J(k,m,n) = \{(i,j) : 0 \leq i \leq m-k, 0 \leq j \leq n-k\}.$$

Denote by

$$W_{m,n} \equiv W_{m,n}(k) = \sum_{(i,j) \in J} T_{ij}^*$$

the *number of long match patterns* (patterns of length $\geq k$). Then

$$\{M_{m,n}^* < k\} = \{W_{m,n} = 0\}.$$

In the rest of this section we assume that r.v.s $X, X_1, ..., X_m$, $Y, Y_1, ..., Y_n$ are identically distributed. We set

$$\lambda \equiv \lambda_{k,m,n} = \mathbb{E}W_{m,n}, \quad m' = m-k+1, \quad n' = n-k+1.$$

Then $\lambda = (m'-1)(n'-1)(1-p)p^k + (m'+n'-1)p^k$.

Denote

$$p = \mathbb{P}(X=Y), \ p_j = \mathbb{P}(X=j), \ q_k = \sum_{j \in A} p_j^{k+1}, \ q = q_2,$$

and let

$$p_* = \max_{j \in A} p_j, \ c_+ = \log(1/q) - 1, \ c_* = \log(1/p_*),$$

where log is to the base $1/p$. Note that

$$p_*^2 < p, \ p^2 \le q \le p_* . \tag{4.6}$$

Taking into account Hölder's inequality, we conclude that

$$1 \ge c_+ \ge c_* > 1/2. \tag{4.7}$$

Note that $c_+ = c_* = 1$ if $\mathcal{L}(X)$ is uniform over a finite alphabet.

Let $\pi_{m,n}$ denote a Poisson random variable with parameter $\lambda_{k,m,n}$.

The following theorem shows that the distribution of the number of long match patterns (NLMP) can be well approximated by the Poisson law.

Theorem 4.4 [221] *If* $n \ge k$ *and* $m \ge k \ge 1$, *then*

$$d_{TV}(W_{m,n}; \pi_{m,n}) \le \frac{1-e^{-\lambda}}{\lambda} m'n'(2k+1)\left(2kq_{2k} + (m'+n'-1)(p^{2k}+q^k)\right). \tag{4.8}$$

Theorem 4.4 has been derived using Theorem 3.4 and Lemma 4.8.

Denote

$$\Delta_{m,n}(k) = |\mathbb{P}(M^*_{m,n} < k) - \exp(-\lambda)\,|.$$

Corollary 4.5 *For any constant* $C \in \mathbb{R}$, *as* $m \to \infty$, $n \to \infty$,

$$\max_{k \ge C + \log mn} \Delta_{m,n}(k) = O\left((m+n)(mn)^{-c_+}(\ln mn) + (mn)^{1-2c_*}(\ln mn)^2\right). \tag{4.9}$$

If $m \to \infty$ *and* $n \to \infty$ *in such a way that* $(\ln mn)/(\min\{m,n\}) \to 0$, *then*

$$\max_{1 \le k \le m \wedge n} \Delta_{m,n}(k) = O\Big((m+n)(mn)^{-c_+}(\ln mn)^{1+c_+} + (mn)^{1-2c_*}(\ln mn)^{1+2c_*}\Big). \tag{4.10}$$

It is easy to see that the accuracy of estimate (4.9) depends on the relation between m and n. If $\mathcal{L}(X)$ is uniform over a finite alphabet and $(\ln mn)/(m \wedge n) \to 0$, then Corollary 4.5 implies that

$$\max_{1 \le k \le m \wedge n} \left|\mathbb{P}(M^*_{m,n} < k) - e^{-\lambda}\right| = O\left(n^{-1}(\ln n)^2\right). \tag{4.11}$$

If $\mathcal{L}(X)$ is uniform over a finite alphabet and

$$c \le m/n \le 1/c$$

for some constant $c > 0$, then the right-hand side of (4.9) becomes $O(n^{-1}\ln n)$.

We conject that the correct rate of convergence in (4.11) for the uniform $\mathcal{L}(X)$ is $O\left(n^{-1}\ln n\right)$.

The reason why (4.8) does not yield such a rate is the lack of factor $e^{-\lambda}$ on the right-hand side. Results obtained for LLHR by the method of recurrent inequalities do produce such a factor (cf. Theorem 3.12 in [221]).

In a more general situation one can consider NLMP with say r mismatches allowed. An estimate of the accuracy of Poisson approximation to the distribution of the number of long r-interrupted match patterns among $X_1,...,X_m,Y_1,...,Y_n$ (match patterns of length $\geq k$ with $\leq r$ "interruptions") can be found in [210, 221]. Neuhauser [219] considers a situation where $\mathcal{L}(X)$ may differ from $\mathcal{L}(Y)$ and only insertions and deletions (but no mismatches) are allowed to occur; she states that the rate of Poisson approximation to the distribution of the number of such long patterns is logarithmic.

4.3 The Zubkov–Mikhailov statistic

Let now $Y_i = X_i$ $(\forall i)$, $m = n$. Denote

$$N_n^* \equiv N_n^*(k) = \sum_{(i,j)\in A(n,k)} T_{ij}^*\,,$$

where

$$A(n,k) = \{(i,j) : 0 \leq i < j \leq n-k\} \quad (n>k).$$

N_n^* is the number of long match patterns in one and the same sequence, $X_1,...,X_n$.

Statistic N_n^* was introduced by Zubkov & Mikhailov [334] who have shown that $\mathcal{L}(N_n^*)$ is asymptotically Poisson $\mathbf{\Pi}(\mu)$ if

$$n^2p^k(1-p)/2 \to \mu>0, \quad nk^tp_*^k \to 0 \quad (\forall t>0).$$

Denote by

$$M_n^* = \max\{k \leq n: \max_{(i,j)\in A(n,k)} T_{ij} = 1\}$$

the *length of the longest match pattern* among $X_1,...,X_n$. Then $\{M_n^*<k\} = \{N_n^*=0\}$.

The next theorem evaluates the accuracy of Poisson approximation to $\mathcal{L}(N_n^*)$.

Theorem 4.6 [221] *If* $n>3k\geq 3$, *then*

$$d_{TV}(N_n^*;\pi_{n,k}^*) \leq \frac{1-e^{-\lambda^*}}{\lambda^*}\left((n^*)^3(2k+1)\left(p^{2k}+q^k\right)+2(kn^*)^2q_{2k}\right)+2kn^*p^k,$$

where $\lambda^* \equiv \lambda_{n,k}^* = (n-3k+1)\,p^k(1+(n-3k)(1-p)/2)$, $n^* = n-k$, $\mathcal{L}(\pi_{n,k}^*) = \mathbf{\Pi}(\lambda^*)$.

Theorem 4.6 has been derived using Theorem 3.4 and Lemma 4.8.

Denote

$$\Delta^*(n,k) = |\mathbb{P}(M_n^* < k) - \exp(-\lambda_{n,k}^*)\,|.$$

Corollary 4.7 *As* $n \to \infty$,

$$\max_{k\geq C+2\log n} \Delta^*(n,k) = O\left(n^{1-2c_+}\ln n + n^{2-4c_*}(\ln n)^2\right), \tag{4.12}$$

$$\max_{1\leq k<n/3} \Delta^*(n,k) = O\left(n^{1-2c_+}(\ln n)^{1+c_+} + n^{2-4c_*}(\ln n)^{1+2c_*}\right). \tag{4.13}$$

If $\mathcal{L}(X)$ is uniform over a finite alphabet, then the right-hand side of (4.12) is $O(n^{-1}\ln n)$, the right-hand side of (4.13) is $O(n^{-1}(\ln n)^2)$.

The key result behind Theorems 4.4 and 4.6 is the following

Lemma 4.8 [221] *For all natural* i, j, i', j' *such that* $(i,j) \neq (i',j')$,

$$\mathbb{P}(T^*_{ij} = T^*_{i'j'} = 1) \leq q_{2k}\,. \tag{4.14}$$

A generalisation of (4.14) to the case where r mismatches are allowed can be found in [221], formula (3.48).

Denote by

$$\tau_k = \min\{n\colon N^*_n(k) \neq 0\}$$

the first instance a match pattern of length k appears in the sequence $\{X_i, i \geq 1\}$. Then

$$\{\tau_k > n\} = \{M^*_n < k\}.$$

The results concerning the asymptotics of τ_k can be derived from the corresponding results for M^*_n.

NLMP with a small number of mismatches has been considered by several authors (see [210, 221] and references therein).

Some other applications of the results concerning the accuracy of Poisson approximation can be found in [224, 232, 44].

Open problems

4.3. Derive uniform in k estimates of (possibly shifted) Poisson approximation to $\mathcal{L}(W_{m,n})$ and $\mathcal{L}(N^*_n)$.

4.4. Find the 2nd-order asymptotic expansions for $\mathbb{P}(W_{m,n} \in \cdot)$ and $\mathbb{P}(N^*_n \in \cdot)$.

4.5. Check if the correct rate of convergence in (4.10) and (4.13) in the case of uniform $\mathcal{L}(X)$ is $O\left(n^{-1}\ln n\right)$.

4.6. Improve the estimate of the rate of convergence in the limit theorem for the length of the longest r-interrupted match pattern.

Exercises

4.1 Let S be a r.v. taking values in $\mathbb{Z}_+$. Assuming $\mathbb{E}S^2 < \infty$, show that

$$\mathbb{P}(S = 0) \leq (\operatorname{var} S)/(\mathbb{E}S)^2\,. \tag{4.15}$$

4.2 Check that (4.6) holds.

4.3 Prove (4.7).

4.4 Let $X, X_1, \ldots$ be i.i.d. Bernoulli $\mathbf{B}(p)$ r.v.s. Given $\varepsilon > 0$, find $k_*(\varepsilon)$, $k^*(\varepsilon)$ in (4.1).

4.5 Let $X, X_1, \ldots$ be i.i.d. Bernoulli $\mathbf{B}(p)$ r.v.s. Given $\varepsilon > 0$, find $k_*(\varepsilon)$, $k^*(\varepsilon)$ in (4.2).

Notes

Corollary 4.3 is cited from [221], Corollary 3.23.
Theorem 4.4 and Lemma 4.8 are from [221], Chapter 3.

5

Compound Poisson limit theorem

Compound Poisson limit theorem plays important role in the theory of sums of r.v.s. From a theoretical point of view, the interest to the topic arises in connection with Kolmogorov's problem concerning the accuracy of approximation of the distribution of a sum of independent r.v.s by infinitely divisible laws (see [8, 191, 237, 240] and references therein). Recall that the class of infinitely divisible distributions coincides with the class of weak limits of compound Poisson distributions (Khintchine [177], Theorem 26).

The topic has applications in extreme value theory, insurance, reliability theory, patterns matching, etc. (cf. [15, 21, 27, 189, 221]).

For instance, extreme value theory deals with the number of extreme (rare) events represented by a sum of 0-1 r.v.s (indicators of rare events), the largest (smallest) sample element, the kth largest sample element $X_{n:k}$, etc. Given $x \in \mathbb{R}$, let

$$N_n(x) = \sum_{i=1}^{n} \mathbb{1}\{X_i > x\}$$

denote the number of exceedances of x. Then

$$\{X_{n:k} \leq x\} = \{N_n(x) < k\}.$$

In particular,

$$\{\max_{1 \leq i \leq n} X_i \leq x\} = \{N_n(x) = 0\},$$

$\{X_{2,n} \leq x\} = \{N_n(x) \leq 1\}$, etc. Thus, results concerning the distribution of sample extremes can be derived from the corresponding results concerning $N_n(x)$. The distribution of $N_n(x)$ can often be approximated by a Poisson or a compound Poisson law.

In (re)insurance applications the sum $S_n = \sum_{i=1}^{n} Y_i \mathbb{1}\{Y_i > x_i\}$ of integer-valued r.v.s allows to account for the total loss from the claims $\{Y_i\}$ that exceed excesses $\{x_i\}$. If the probabilities $\mathbb{P}(Y_i > x_i)$ are small, $\mathcal{L}(S_n)$ can be accurately approximated by a Poisson or a compound Poisson law.

The indicators can be dependent. A well-known approach consists of grouping observations into blocks which can be considered almost independent [42]. The number of r.v.s in a block is an integer-valued r.v., hence the number of rare events is a sum of almost independent integer-valued r.v.s that are non-zero with small probabilities.

In molecular biology long match patterns between DNA sequences may indicate "valuable" fragments. A natural question is if such long patterns appear by chance. Information on the distribution of the number of long match patterns (NLMP) between sequences of r.v.s can help answering that question. The distribution of NLMP can often be approximated by a Poisson or a compound Poisson law.

More information concerning applications can be found in [15, 21, 127, 189].

In this chapter, we present compound Poisson limit theorems for $\mathcal{L}(S_n)$.

DOI: 10.1201/9781003478164-5

5.1 Compound Poisson limit theorem for independent r.v.s

Let $\{X_{n,1},...,X_{n,n}\}_{n\geq 1}$ be a triangle array of independent random variables. In the sequel we often write $X_1,...,X_n$ instead of $X_{n,1},...,X_{n,n}$. As before, $S_n = X_1 + ... + X_n$. If $P, P_1, P_2, ...$ are probability distributions, then $P_n \Rightarrow P$ denotes weak convergence of P_n to P.

Random variables that are zero with large probabilities.

Khintchine ([175], Section 2.3) was probably the first to prove a compound Poisson limit theorem.

Suppose that r.v.s $X, X_1, ..., X_n$ are independent and identically distributed (i.i.d.) in each row ($\mathcal{L}(X)$ may depend on n). Denote

$$p \equiv p(n) = \mathbb{P}(X \neq 0).$$

Theorem 5.1 [175] *If there exists* $\lambda > 0$ *and a random variable* X' *such that* $p \sim \lambda/n$ *and*

$$\mathcal{L}(X|X \neq 0) \Rightarrow \mathcal{L}(X') \tag{5.1}$$

as $n \to \infty$*, then*

$$\mathcal{L}(S_n) \Rightarrow \mathbf{\Pi}(\lambda, X') \qquad (n \to \infty). \tag{5.2}$$

To be precise, Khintchine [175] assumed $\mathcal{L}(X|X \neq 0) = \mathcal{L}(X')$ but the argument holds if merely (5.1) is assumed.

The proof of (5.2) is based on Khintchine's formula (1.12), which yields

$$S_n \stackrel{d}{=} \tau_1 X_1' + ... + \tau_n X_n', \tag{5.3}$$

where $\tau_1, X_1', ..., \tau_n, X_n'$ are independent r.v.s,

$$\mathcal{L}(X_i') = \mathcal{L}(X|X \neq 0), \;\; \mathcal{L}(\tau_i) = \mathbf{B}(p) \quad (\forall i).$$

Since $\{X_i'\}$ are i.i.d.r.v.s,

$$S_n \stackrel{d}{=} \sum_{i=1}^{\nu_n} X_i', \tag{5.4}$$

where Binomial $\mathbf{B}(n,p)$ r.v.

$$\nu_n = \tau_1 + ... + \tau_n \tag{5.5}$$

is independent of $\{X_1', ..., X_n'\}$.

Estimates of the accuracy of compound Poisson approximation follow from the estimates of the accuracy of pure Poisson approximation thanks to the following observation by Kolmogorov [179], formula (30): since $\{X_i'\}$ are identically distributed r.v.s,

$$d_{TV}\left(\sum_{i=1}^{\nu_n} X_i'; \sum_{i=1}^{\pi_\lambda} X_i'\right) \leq d_{TV}(\nu_n; \pi_\lambda). \tag{5.6}$$

In particular, Khintchine's compound Poisson limit theorem is a consequence of the Poisson limit theorem: if $p \sim \lambda/n$, where $\lambda > 0$, then weak convergence $\nu_n \Rightarrow \pi_\lambda$ together with (5.1) entails (5.2).

The following proposition states that (5.1) is necessary for (5.2) assuming $p \sim \lambda/n$. Proposition 5.2 is a consequence of Theorem 5.3.

Proposition 5.2 *Suppose that* $\exists \lim_{n\to\infty} np := \lambda$. *If there exist a r.v.* S *such that* $S_n \Rightarrow S$ *as* $n\to\infty$, *then there exist a r.v.* X' *such that* $\mathcal{L}(S) = \mathbf{\Pi}(\lambda, X')$. *If* $\lambda>0$, *then (5.1) holds.*

Example 5.1 Let $\{X, X_1, ..., X_n\}_{n\geq 1}$ be a triangular array of i.i.d.r.v.s,

$$\mathbb{P}(X=0) = 1-\lambda/n, \ \mathbb{P}(X=1) = \lambda/2n, \ \mathbb{P}(X=2) = \lambda/2n.$$

Set $S_n = X_1 + ... + X_n$. Then $X' \stackrel{d}{=} 1+\eta$ and

$$\mathcal{L}(S_n) \Rightarrow \mathbf{\Pi}(\lambda, 1+\eta) \qquad (n\to\infty),$$

where η is a Bernoulli $\mathbf{B}(1/2)$ r.v. □

The following example shows that the limiting distribution of S_n can be a mixture of compound Poisson and normal distributions.

Example 5.2 Let $X', X_1', X_2', ..., X'', X_1'', X_2'', ..., \tau, \tau_1, ...$ be independent r.v.,

$$X_i' \stackrel{d}{=} X' \ (\forall i), \ \ \mathcal{L}(\tau_i) = \mathbf{B}(\lambda/n), \ \ \mathbb{P}(X'' = -1/\sqrt{n}\,) = \mathbb{P}(X'' = 1/\sqrt{n}\,) = 1/2.$$

Denote $P' = \mathcal{L}(X'), \ P'' = \mathcal{L}(X'')$. Set

$$S_n = \sum_{i=1}^{n} (\tau_i X_i' + (1-\tau_i)X_i'').$$

Let ξ denote a standard normal r.v., $\mathcal{L}(\eta) = \mathbf{\Pi}(\lambda; F)$. Then

$$d_{_K}(S_n; \xi+\eta) \to 0 \qquad (n\to\infty). \tag{5.7}$$

Note that $\mathcal{L}(\xi+\eta)$ is an infinitely divisible absolutely continuous distribution.
Proof of (5.7) is given in Section 5.3. □

The case of i.i.d. $\{X_i'\}$**.**
We consider now a special case where $\{X_i\}$ are independent but not identically distributed random variables but $\{X_i'\}$ are i.i.d.r.v.s. Denote

$$p_i = \mathbb{P}(X_i \neq 0), \ \ \lambda = p_1 + ... + p_n\,,$$

and let

$$\tau_i = \mathbb{I}\{X_i \neq 0\} \ \ (i\geq 1), \ \ \nu_n = \tau_1 + ... + \tau_n\,.$$

According to Khintchine's formula (1.12),

$$X_i \stackrel{d}{=} \tau_i X_i',$$

where X_i' are i.i.d.r.v.s, $\mathcal{L}(X_i') = \mathcal{L}(X_i | X_i \neq 0)$, $\mathcal{L}(\tau_i) = \mathbf{B}(p_i)$, and (5.4) holds.
If $\nu_n \Rightarrow \mathbf{\Pi}(\lambda)$ as $n\to\infty$, then

$$S_n \Rightarrow \mathbf{\Pi}(\lambda, X') \qquad (n\to\infty),$$

cf. (5.6). Thus, a compound Poisson limit theorem is a consequence of a Poisson limit theorem.

The non-i.i.d. case. If $\{X_i'\}$ are independent but not identically distributed random variables, it appears natural to approximate each X_i by an "accompanying" r.v. $\tilde{X}_i$, and the sum $S_n \stackrel{d}{=} \tau_1 X_1' + ... + \tau_n X_n'$ by a compound Poisson random variable

$$\tilde{S}_n = \tilde{X}_1 + \cdots + \tilde{X}_n \tag{1.9*}$$

where $\{\tilde{X}_i\}$ are independent compound Poisson $\mathbf{\Pi}(p_i, X_i')$ r.v.s, see (6.10), (6.13).

In the assumption that $\{X_{n,1}, ..., X_{n,n}\}_{n\geq 1}$ is a triangular array of independent integer-valued random variables satisfying the infinitesimality assumption (1.10), Grigelionis [144] presents the following sufficient condition for the weak convergence (5.2): if there exists $\lambda > 0$ and a random variable X' such that

$$\lim_{n\to\infty} \sum_{j=1}^{n} p_j = \lambda, \quad \lim_{n\to\infty} \sum_{j=1}^{n} \mathbb{P}(X_{n,j} = k) = \lambda \mathbb{P}(X' = k) \quad (\forall k \neq 0),$$

then $\mathcal{L}(S_n) \Rightarrow \mathbf{\Pi}(\lambda, X')$ as $n \to \infty$.

Wang [312] presents sufficient conditions for convergence of an unbounded function of a sum S_n of non-negative integer-valued r.v.s to a corresponding function of a compound Poisson random variable (see also Chen & Roos [57]).

Example 5.3 Recall that the Skellam distribution $\mathcal{L}(\pi_a - \pi_b)$ is the distribution of the difference of two independent Poisson r.v.s., see Section 1.1. We present sufficient conditions for the convergence of a sum $\mathcal{L}(S_n)$ of independent integer-valued r.v.s to the Skellam distribution in terms of the Gini-Kantorovich distance (hence weak convergence):

$$\lim_{n\to\infty} d_G(S_n; \pi_a - \pi_b) = 0 \qquad (n \to \infty), \tag{5.8}$$

where π_a and π_b are independent Poisson r.v.s.

Let $\{X_1, \ldots, X_n\}_{n\geq 1}$ be triangular array of independent r.v.s taking values in $\mathbb{Z}$. Set

$$\begin{aligned} \kappa_{1j}^+ &= \sum_{k=1}^{\infty} k\, \mathbb{P}(X_j = k), \quad \kappa_{1j}^- = \sum_{k=1}^{\infty} k\, \mathbb{P}(X_j = -k), \\ \kappa_{2j}^+ &= \sum_{k=2}^{\infty} k(k-1) \mathbb{P}(X_j = k), \quad \kappa_{2j}^- = \sum_{k=2}^{\infty} k(k-1) \mathbb{P}(X_j = -k), \end{aligned}$$

so that $\kappa_{1j}^+ + \kappa_{1j}^- = \mathbb{E}|X_j|$. If

$$\lim_{n\to\infty} \max_{1\leq j\leq n} \mathbb{E}|X_j| = 0, \quad \lim_{n\to\infty} \sum_{j=1}^{n} \kappa_{1j}^+ = a \geq 0, \quad \lim_{n\to\infty} \sum_{j=1}^{n} \kappa_{1j}^- = b \geq 0,$$

$$\lim_{n\to\infty} \sum_{j=1}^{n} \kappa_{2j}^+ = \lim_{n\to\infty} \sum_{j=1}^{n} \kappa_{2j}^- = 0,$$

then (5.8) holds. □

5.2 Compound Poisson limit theorem for dependent r.v.s

Let $\{X, X_1, ..., X_n\}_{n\geq 1}$ be a triangle array of dependent r.v.s, strictly stationary in each row ($\mathcal{L}(X)$ may depend on n).

Recall the definitions of mixing (weak dependence) coefficients:

$$\begin{aligned}
\alpha(l) \equiv \alpha_n(l) &= \sup |\mathbb{P}(AB) - \mathbb{P}(A)\mathbb{P}(B)|, \\
\varphi(l) \equiv \varphi_n(l) &= \sup |\mathbb{P}(B|A) - \mathbb{P}(B)|, \\
\beta(l) \equiv \beta_n(l) &= \sup \mathbb{E} \sup_B |\mathbb{P}(B|\mathcal{F}_{1,m}) - \mathbb{P}(B)|,
\end{aligned}$$

where the supremum is taken over $m \geq 1$, $A \in \mathcal{F}_{1,m}$, $B \in \mathcal{F}_{m+l+1,n}$ such that $\mathbb{P}(A) > 0$, $\mathcal{F}_{l,m}$ denotes the σ-field generated by $\{X_i\}_{l\leq i\leq m}$.

Condition (Δ) is said to hold if $\alpha(l_n) \to 0$ for some sequence $\{l = l_n\}$ of natural numbers such that $0 \leq l_n \ll n$.

If condition (Δ) holds, then there exists a sequence $\{r = r_n\}$ of natural numbers obeying (2.8).

In Theorem 5.3 we assume condition (Δ).

In applications $\{X_i\}$ are typically non-negative r.v.s; they usually represent "rare" events. Therefore, in this section we assume that $X_i \geq 0$ $(\forall i)$.

Denote

$$p \equiv p(n) = \mathbb{P}(X \neq 0).$$

A common approach is to assume that there exists limit (2.12). Condition (2.12) means that r.v.s are "properly" normalised. If $\{X_i\}$ are independent, then (2.12) is equivalent to

$$np \sim \lambda \qquad (n \to \infty). \tag{5.9}$$

The same is true if $\{X_i\}$ are dependent but condition (D') holds (see, e.g., [224]).

In Theorem 5.3 below we will assume that (2.12) holds and

$$\limsup_{n\to\infty} np < \infty. \tag{5.10}$$

Note that (2.12) and Δ yield

$$\lim_{n\to\infty} p(n) = 0$$

(cf. ex. 15 in [221], p. 11).

Weaker than (5.9) is assumption (5.10). Note that (2.12) does not imply (5.10) – Denzel & O'Brien [107] present an example of an α–mixing sequence such that (2.12) holds though (5.10) does not. On the other hand, (5.10) follows from (2.12) if the sequence $\{X_1, ..., X_n\}$ is φ–mixing (cf. ex. 16 in [221], p. 11).

Theorem 5.3 *Assume conditions (2.12) and (5.10). If there exists a sequence* $\{r = r_n\} \in \mathcal{R}$ *such that*

$$\mathcal{L}(S_r | S_r \neq 0) \Rightarrow \mathcal{L}(\zeta) \qquad (n \to \infty), \tag{5.11}$$

then

$$\mathcal{L}(S_n) \Rightarrow \mathbf{\Pi}(\lambda, \zeta). \tag{5.12}$$

The limit in (5.12) does not depend on the choice of a sequence $\{r_n\} \in \mathcal{R}$.

If S_n *converges weakly to a random variable* S, *then there exists* $\lambda \geq 0$ *and a random variable* ζ *such that* $\mathcal{L}(S) = \mathbf{\Pi}(\lambda, \zeta)$, *where* $\lambda = -\ln \mathbb{P}(S=0)$. *If* $\lambda > 0$, *then (2.12) holds, and there exist a sequence* $\{r_n\} \in \mathcal{R}$ *such that (5.11) holds true.*

Theorem 5.3 states that conditions (2.12), (5.11) are necessary and sufficient for the weak convergence of $\mathcal{L}(S_n)$ to a compound Poisson law.

A random variable ζ taking values in $\mathbb{N}$ is called the *limiting cluster size* if (5.11) holds. The notion of the limiting cluster size plays important role in extreme value theory (see, e.g., [221], Chapters 5, 6).

If $\{X_i\}$ are 0-1 random variables, then the class of limiting cluster size distributions coincides with the family of all integer-valued distributions [51].

Condition (5.11) suggests the following estimator of $\mathcal{L}(\zeta)$:

$$\mathbb{P}_n(\zeta = \cdot) = \sum_{j=0}^{[n/r]} \mathbb{I}\{S_{r,j} = \cdot\} \Big/ \sum_{j=0}^{[n/r]} \mathbb{I}\{S_{r,j} \neq 0\},$$

where $S_{r,j} = \sum_{m=jr+1}^{(j+1)r \wedge n} X_m$, see Hsing [155]. Results concerning consistent estimation of $\mathcal{L}(\zeta)$ can be found, e.g., in [155, 156, 250].

Example 5.4 Let $\{\xi_i\}$ be a sequence of i.i.d.r.v.s. Suppose that $\{u_n\}$ is a sequence of threshold levels such that $\lim_{n\to\infty} \mathbb{P}\Big(\max\limits_{1\leq i\leq n} \xi_i \leq u_n\Big) = e^{-\lambda}$ $(\exists \lambda > 0)$. Set

$$X_i = \mathbb{I}\{\max\{\xi_i; \xi_{i+1}\} > u_n\}.$$

Then (5.11) holds with $\zeta \equiv 2$, $\{r\} \in \mathcal{R}$. Hence $S_n \Rightarrow 2\pi(\lambda)$, i.e., $\mathcal{L}(S_n) \Rightarrow \mathbf{\Pi}(\lambda, 2)$. □

Example 5.5 presents sufficient conditions for convergence of the Markov Binomial distribution to a compound Poisson one in terms of the Gini-Kantorovich distance.

Let 0-1 r.v.s. $X_0 = 0, X_1, X_2, \ldots$ form a Markov chain with transition probabilities

$$\mathbb{P}(\xi_i = 1 \,|\, \xi_{i-1} = 0) = \alpha, \qquad \mathbb{P}(\xi_i = 0 \,|\, \xi_{i-1} = 0) = 1 - \alpha,$$
$$\mathbb{P}(\xi_i = 1 \,|\, \xi_{i-1} = 1) = \beta, \qquad \mathbb{P}(\xi_i = 0 \,|\, \xi_{i-1} = 1) = 1 - \beta,$$

where $i \in \mathbb{N}, \alpha, \beta \in (0,1)$. $\mathcal{L}(S_n)$ is called sometimes the Markov Binomial distribution.

Let Y_b be a geometric r.v.: $\mathbb{P}(Y_b = j) = (1-b)b^{j-1}$ $(0 < b < 1,\ j = 1, \ldots)$.

If $n\alpha \to a$ and $\beta \to b < 1/3$ as $n \to \infty$, then

$$d_G(\mathcal{L}(S_n); \mathbf{\Pi}(a, Y_b)) \to 0 \qquad (n \to \infty). \tag{5.13}$$

In the case of independent r.v.s we have $\alpha = \beta$. If $n\alpha \to a$ as $n \to \infty$, then (5.13) entails $b = 0$, and $d_G(\mathcal{L}(S_n); \mathbf{\Pi}(a)) \to 0$. □

5.3 Proofs

Proof of (5.7). Let $p=\lambda/n$, $q=1-p$, and let $\Phi(\sigma^2)=\mathcal{L}(\sigma\xi)$. Then $\mathcal{L}(S_n)=(qP''+pP')^{*n}$. Hence

$$\begin{aligned}
d_K(S_n;\xi+\eta) &= \big\|(qP''+pP')^{*n}-\mathcal{N}(0;1)*\exp(\lambda(P'-I))\big\|_K\\
&\le \sum_{k=1}^{n}\binom{n}{k}q^k p^{n-k}\big\|(P''^{*k}-\Phi(k/n))*P'^{*(n-k)}\big\|_K\\
&+ \sum_{k=1}^{n}\binom{n}{k}q^k p^{n-k}\big\|(\Phi(k/n)-\Phi(q))*P'^{*(n-k)}\big\|_K\\
&+ \big\|\Phi(q)*((qI+pP')^{*n}-\exp(\lambda(P'-I)))\big\|_K\\
&+ \big\|(\Phi(q)-\Phi(1))*\exp(\lambda(P'-I))\big\|_K. \qquad (5.14)
\end{aligned}$$

By (H.2),

$$\big\|(\Phi(k/n)-\Phi(q))*F^{*(n-k)}\big\|_K \le \big\|\Phi(k/n)-\Phi(q)\big\|_K\big\|F^{*(n-k)}\big\|_{TV}\le 2|k/n-q|.$$

Applying (I.4), we prove that the second summand in the RHS of (5.14) tends to zero. Similarly,

$$\|\Phi(q)-\Phi(1)\|_K \le 2|1-q| = 2p = 2\lambda/n\to 0.$$

By (A.3),

$$\begin{aligned}
\big\|\Phi(q)*\big((qI+pP')^{*n}-\exp(\lambda(P'-I))\big)\big\|_K &\le \|((qI+pI_1)^{*n}-\exp(\lambda(I_1-I)))\|_{TV}\\
&= 2d_{TV}(\mathbf{B}(n,\lambda/n);\mathbf{\Pi}(\lambda))\to 0.
\end{aligned}$$

P''^{*k} is the distribution of $S''_k=X''_1+\cdots+X''_k$, $\operatorname{var}S''_k=k/n$, $\mathbb{E}|X''|^3=n^{-3/2}$. Therefore, $\|P''^{*k}-\Phi(k/n)\|_K\le 0.56/\sqrt{k}$ by (H.1). Since $1/\sqrt{k}\le 2/\sqrt{k+1}$ $(k\ge 1)$, it follows from (I.3) that the first summand in the RHS of estimate (5.14) also tends to zero. □

Proof of (5.8). Let $k\ge 2$. By Bergström's identity (B.9),

$$I_k = I_1^{*k} = I + k(I_1-I) + (I_1-I)^{*2} * \sum_{m=2}^{k}\binom{m-1}{2} I_1^{*(k-m)}.$$

A similar expression holds for I_{-1}. Therefore,

$$\begin{aligned}
\mathcal{L}(X_j) &= \sum_{k=-\infty}^{\infty} I_k\mathbb{P}(X_j=k) = I\mathbb{P}(X_j=0)+\sum_{k=1}^{\infty} I_k\mathbb{P}(X_j=k)+\sum_{k=1}^{\infty} I_{-k}\mathbb{P}(X_j=-k)\\
&= I+\kappa_{1j}^{+}(I_1-I)+\kappa_{1j}^{-}(I_{-1}-I)+\frac{\kappa_{2j}^{+}}{2}(I_1-I)^{*2}*\Theta+\frac{\kappa_{2j}^{-}}{2}(I_{-1}-I)^{*2}*\Theta,
\end{aligned}$$

where Θ denotes a measure satisfying $\|\Theta\|_{TV}\le 1$. For example, if $\kappa_{2j}^{+}>0$, then

$$\Theta = \frac{2}{\kappa_{2j}^{+}}\sum_{k=2}^{\infty}\mathbb{P}(X_j=k)\sum_{m=2}^{k}\binom{m-1}{2} I_{k-m}.$$

Set

$$M_j := \kappa_{1j}^{+}(I_1-I) + \kappa_{1j}^{-}(I_{-1}-I) = (I_1-I) * (\kappa_{1j}^{+}I - I_{-1}\kappa_{1j}^{-}),$$

so that $M_j^{*2} = (\kappa_{1j}^{+} + \kappa_{1j}^{-})^2(I_1-I)^{*2} * \Theta$. By (1.6),

$$\mathrm{e}^{M_j} = I + M_j + M_j^{*2} * \int_0^1 (1-\tau)\mathrm{e}^{\tau M_j}\mathrm{d}\tau = I + M_j + (\kappa_{1j}^{+} + \kappa_{1j}^{-})^2(I_1 - I)^{*2} * \Theta.$$

Note that $\exp(\tau M_j)$ is a compound Poisson distribution. Therefore, $\| \exp(\tau M_j) \|_{TV} = 1$. Since $(I_{-1} - I)^{*2} = I_{-2} * (I_1 - I)^{*2}$, we get

$$\mathcal{L}(X_j) - \mathrm{e}^{M_j} = (\kappa_{2j}^{+} + \kappa_{2j}^{-})(I_1-I)^{*2} * \Theta + (\kappa_{1j}^{+} + \kappa_{1j}^{-})^2(I_1-I)^{*2} * \Theta.$$

Consequently,

$$\begin{aligned}
\mathcal{L}(S_n) - \exp\Big(\sum_{j=1}^{n} M_j\Big) &= \sum_{j=1}^{n}(\mathcal{L}(X_j)-\exp(M_j)) * \prod_{m=1}^{*(j-1)} \mathcal{L}(X_m) * \prod_{m=j+1}^{*n} \exp(M_m) \\
&= (I_1-I)^{*2} * \left(\sum_{j=1}^{n}(\kappa_{2j}^{+} + \kappa_{2j}^{-})\Theta + \sum_{j=1}^{n}(\kappa_{1j}^{+} + \kappa_{1j}^{-})^2 * \Theta\right),
\end{aligned}$$

where we employed the fact that a convolution of distributions is a distribution as well. Set

$$a_n = \sum_{j=1}^{n} \kappa_{1j}^{+}, \quad b_n = \sum_{j=1}^{n} \kappa_{1j}^{-}, \quad Y_n = \pi_{a_n} - \pi_{b_n},$$

where Poisson random variables π_{a_n}, π_{b_n} are independent. Note that $\mathcal{L}(Y_n) = \exp\Big(\sum_{j=1}^{n} M_j\Big)$. Applying (A.9) and properties of the total variation norm, we get

$$\begin{aligned}
d_G(S_n; Y_n) &= \|\mathcal{L}(S_n)-\mathcal{L}(Y_n)\|_G \leq \|I_1 - I\|_{TV} \sum_{j=1}^{n}((\kappa_{2j}^{+} + \kappa_{2j}^{-}) + (\kappa_{1j}^{+} + \kappa_{1j}^{-})^2) \\
&\leq 2\sum_{j=1}^{n}(\kappa_{2j}^{+} + \kappa_{2j}^{-}) + 2 \max_{1\leq j\leq n} \mathbb{E}|X_j|(a_n + b_n) \to 0 \qquad (n \to \infty). \qquad (5.15)
\end{aligned}$$

Next,

$$\begin{aligned}
\mathcal{L}(Y) - \mathcal{L}(Y_n) &= \exp(a(I_1 - I)) * (\exp(b(I_{-1} - I)) - \exp(b_n(I_{-1} - I))) \\
&\quad + (\exp(a(I_1 - I)) - \exp(a_n(I_1 - I))) * \exp(b_n(I_{-1} - I)) \\
&= (b - b_n)(I_{-1} - I) * \Theta + (a - a_n)(I_1 - I) * \Theta
\end{aligned}$$

by (B.1). Applying (A.9), we get

$$d_G(Y, Y_n) \leq 2|b - b_n| + 2|a - a_n| \to 0 \qquad (n \to \infty).$$

Combining the last estimate with (5.15), we complete the proof of (5.8). □

Proof of (5.13). By the triangle inequality,

$$d_G(\mathcal{L}(S_n); \mathbf{\Pi}(a, Y_b)) \leq d_G(\mathcal{L}(S_n); \mathbf{\Pi}(n\gamma_1, Y_\beta)) + d_G(\mathbf{\Pi}(n\gamma_1, Y_\beta); \mathbf{\Pi}(a, Y_b)), \qquad (5.16)$$

where

$$\gamma_1 = \frac{\alpha(1-\beta)}{1-\beta+\alpha}, \quad \mathbb{P}(Y_\beta = j) = (1-\beta)\beta^{j-1} \qquad (j = 1, \dots)$$

By (9.5) $d_G(\mathcal{L}(S_n); \mathbf{\Pi}(n\gamma_1, Y_\beta)) \to 0$. Concerning the second summand in the RHS of (5.16),

$$I_{j+1} = (I + I_1 - I)^{*j} = I + \sum_{m=1}^{j+1} \binom{j+1}{m} (I_1 - I)^{*m} = I + (I_1 - I) * \sum_{m=1}^{j+1} I_{m-1}.$$

Therefore,

$$\mathcal{L}(Y_\beta) - I = (I_1 - I) * \sum_{j=0}^{\infty} (1-\beta)\beta^j \sum_{m=1}^{j+1} I_{m-1}.$$

Hence

$$\mathcal{L}(Y_\beta) - \mathcal{L}(Y_b) = \sum_{j=0}^{\infty} ((1-\beta)\beta^j - (1-b)b^j) \sum_{m=1}^{j+1} \binom{j+1}{m} (I_1 - I)^{*m}.$$

As before, we denote by Θ measures such that $\|\Theta\|_{TV} \le 1$.

By (B.1) and (A.9),

$$\begin{aligned}
d_G(\mathbf{\Pi}(n\gamma_1, Y_\beta); \mathbf{\Pi}(a, Y_b)) &= \| \exp(n\gamma_1(\mathcal{L}(Y_\beta) - I)) - \exp(a(\mathcal{L}(Y_b) - I)) \|_G \\
&= \|(n\gamma_1(\mathcal{L}(Y_\beta) - I) - a(\mathcal{L}(Y_b) - I)) * \Theta\|_G \\
&= \|((n\gamma_1 - a)(\mathcal{L}(Y_\beta) - I) + a(\mathcal{L}(Y_\beta) - \mathcal{L}(Y_b))) * \Theta\|_G \\
&\le |n\gamma_1 - a| \sum_{j=0}^{\infty} (1-\beta)\beta^j \Big\| \sum_{m=1}^{j+1} I_{m-1} \Big\|_{TV} + \sum_{j=0}^{\infty} \big|(1-\beta)\beta^j \\
&\quad - (1-b)b^j\big| \sum_{m=1}^{j+1} \binom{j+1}{m} \|I_1 - I\|_{TV}^m \\
&\le |n\gamma_1 - a| \sum_{j=0}^{\infty} (j+1)\beta^{j+1} + \sum_{j=0}^{\infty} \big|(1-\beta)\beta^j - (1-b)b^j\big| 3^{j+1}/2. \qquad (5.17)
\end{aligned}$$

Let $w = b/2 + 1/6$. It is easy to check that $|n\gamma_1 - a| \to 0$ if $\beta \le w < 1/3$. Hence the first summand in (5.17) tends to zero. Similarly,

$$|(1-\beta)\beta^j - (1-b)b^j| = |(b-\beta)\beta^j + (1-b)(\beta^j - b^j)| \le |b-\beta| w^j + j|\beta - b| w^{j-1}.$$

Therefore, the second summand in (5.17) is less than

$$\begin{aligned}
&3|b-\beta| \sum_{j=0}^{\infty} (3w)^j/2 + 4.5|\beta - b| \sum_{j=1}^{\infty} j(3w)^{j-1} \\
&= 1.5|\beta - b|(3(1-3w)^{-2} + (1-3w)^{-1}) \to 0 \qquad (n \to \infty),
\end{aligned}$$

and (5.13) follows. □

Exercises

5.1. Let $X, X_1, X_2, \dots, X_n$ be a sequence of independent identically distributed r.v.s, where X is a non-degenerate r.v. Denote $F = \mathcal{L}(X)$. We assume that F does not depend on n.

Let $\tau_1, \tau_2, \ldots, \tau_n$ be independent (and independent of $\{X_i\}$) identically distributed generalised r.v.s with "probabilities"

$$\mathbb{P}(\tau_i = 0) = 1+p, \quad \mathbb{P}(\tau_i = 1) = -p,$$

where $p = p(n) \sim \lambda/n$ as $n \to \infty$. Denote $P_n = \mathcal{L}(\sum_{i=1}^n \tau_i X_i)$. Prove that

$$d_{TV}(P_n; \mathbf{\Pi}(-\lambda, X)) \to 0 \qquad (n \to \infty).$$

5.2. Let $\mathbb{P}(\tau = 0) = 1+p$, $\mathbb{P}(\tau = 1) = -p$, where $p \in (0;1)$. Prove that the total variation of $\mathcal{L}(\tau)$ is greater than 1.

5.3. Let $S_n = \tau_1 X_1' + \ldots + \tau_n X_n'$, where $\mathcal{L}(\tau_i) = \mathbf{B}(p_i)$, $\{X_i'\}_{i \geq 1}$ are integer-valued r.v.s,

$$\lim_{n \to \infty} \sum_{i=1}^n p_i = \lambda > 0, \qquad \lim_{n \to \infty} \sum_{i=1}^n p_i^2 = 0.$$

Let Y be an integer-valued r.v. such that

$$\lim_{n \to \infty} \sum_{i=1}^n p_i \mathbb{P}(X_i' = k) = \lambda \mathbb{P}(Y = k) \quad (\forall k \in \mathbb{Z}).$$

Prove that $d_{TV}(\mathcal{L}(S_n); \mathbf{\Pi}(\lambda; Y)) \to 0$ as $n \to \infty$.

Hint. Use approximation by the accompanying law and the triangle inequality.

5.4. Let $X_1, \ldots, X_n$ be a triangular array of geometric $\Gamma(p_{i,n})$ r.v.s, where $0 < p_{i,n} < 1$, and let Y be a Negative Binomial $\mathbf{NB}(r, p)$ r.v. defined by (1.4), where $r \in \mathbb{N}$, $0 < p < 1$. Investigate the limiting behaviour of the sum $S_r = X_1 + \cdots + X_r$, where r is fixed. Prove that $d_{TV}(S_r; Y) \to 0$ as $n \to \infty$ if $\max_{i \leq n} |p_{i,n} - p| \to 0$.

Notes

Theorem 5.3 is essentially Theorem 5.1 from [221].

6

Accuracy of CP approximation: rare events

Compound Poisson approximation appears naturally in situations where rare events form clusters and the number of clusters is asymptotically Poisson (which is typically the case).

Sums of independent r.v.s that are non-zero with small probabilities have actuarial interpretation as aggregate claims (cf. [128, 151]). The task of evaluating the accuracy of compound Poisson approximation to the distribution of a sum of r.v.s that are non-zero with small probabilities has been approached by many authors, cf. [191, 207, 221, 238, 324].

This chapter presents results on the accuracy of compound Poisson to the distribution of a sum of r.v.s representing rare events.

6.1 Independent random variables

Let $\{X_i\}$ be independent r.v.s that are non-zero with small probabilities, i.e., $\{X_i\}$ represent rare events. Set $S_n = X_1 + ... + X_n$,

$$p_i = \mathbb{P}(X_i \neq 0) \;\; (i \geq 1), \;\; \lambda = p_1 + ... + p_n \,, \;\; \theta = \sum_{i=1}^{n} p_i^2/\lambda.$$

By Khintchine's formula (see [175] or formula (14.5) in [221]), any distribution can be presented as

$$\mathcal{L}(X) = (1-p)U + pV, \tag{6.1}$$

where $U = \mathcal{L}(X|X \in A)$ and $V = \mathcal{L}(X|X \notin A)$ for a particular measurable set A, $p = \mathbb{P}(X \notin A)$.

In particular,

$$X_i \stackrel{d}{=} \tau_i X_i' \,, \tag{1.12*}$$

where τ_i and X_i' are independent r.v.s, $\mathcal{L}(X_i') = \mathcal{L}(X_i|X_i \neq 0)$, $\mathcal{L}(\tau_i) = \mathbf{B}(p_i)$, $p_i = \mathbb{P}(X_i \neq 0)$.

Relation (1.12*) can be rewritten as

$$\mathcal{L}(X_i) = (1-p_i)I + p_i\mathcal{L}(X_i'). \tag{1.12*}$$

Hence

$$S_n \stackrel{d}{=} \tau_1 X_1' + ... + \tau_n X_n'$$

(some authors call $\mathcal{L}(S_n)$ a "compound Binomial distribution", cf. [83]), and

$$\mathcal{L}(S_n) = \prod_{i=1}^{*n} ((1-p_i)I + p_i\mathcal{L}(X_i')) = \prod_{i=1}^{*n} (I + p_i(\mathcal{L}(X_i') - I)) \,. \tag{6.2}$$

DOI: 10.1201/9781003478164-6

In (6.1) one can choose $U = \mathcal{L}(X|X \in [a;b])$ and $V = \mathcal{L}(X|X \notin [a;b])$, where $[a;b]$ is a finite interval. Since one can switch from U to $U - \mathbb{E}U$, one can assume that $\mathbb{E}U = 0$. The derivation of the estimates of the accuracy of compound Poisson and infinitely divisible approximations to $\mathcal{L}(S_n)$ is usually based on (6.1), cf. [179, 191, 241].

Concerning the sum $\tilde{S}_n$ of independent accompanying random variables (cf. (1.8), (1.9)), recall that

$$\mathcal{L}(\tilde{S}_n) = \exp\left(\sum_{i=1}^{n} p_i(\mathcal{L}(X_i') - I)\right) = \mathbf{\Pi}(\lambda, \mathcal{L}(X_\eta')), \tag{6.3}$$

where r.v. η is independent of $X_1', ..., X_n'$, $\mathbb{P}(\eta = j) = p_j/\lambda$ $(1 \le j \le n)$.

The idea behind approximation of $\mathcal{L}(S_n)$ by $\mathcal{L}(\tilde{S}_n)$ is based on the expectation that

$$\prod_{i=1}^{*n} (I + p_i(\mathcal{L}(X_i') - I)) \approx \exp\left(\sum_{i=1}^{n} p_i(\mathcal{L}(X_i') - I)\right). \tag{6.4}$$

The i.i.d. case.

Consider the situation where $X_i \stackrel{d}{=} X$ $(\forall i)$. Denote

$$\nu_n = \tau_1 + ... + \tau_n\,.$$

Then

$$S_n \stackrel{d}{=} \sum_{i=1}^{\nu_n} X_i', \quad S \stackrel{d}{=} \sum_{i=1}^{\pi_\lambda} X_i', \tag{6.5}$$

where Poisson $\mathbf{\Pi}(\lambda)$ r.v. π_λ is independent of $\{X_i'\}$.

Estimates of the accuracy of compound Poisson approximation can be derived from the estimates of the accuracy of pure Poisson approximation (cf. (6.6), (6.7) below).

Let r.v.s ν_n and ν take values in $\mathbb{Z}_+$ (ν_n, ν can be considered defined on a common probability space). In the following two lemmas r.v.s $\zeta, \zeta_1, \zeta_2, ...$ are independent of (ν_n, ν).

Lemma 6.1 *Let $\zeta_i \stackrel{d}{=} \zeta$ $(i \ge 1)$, $\zeta_0 = 0$. Then*

$$d_{TV}\left(\sum_{i=0}^{\nu_n} \zeta_i; \sum_{i=0}^{\nu} \zeta_i\right) \le d_{TV}(\nu_n; \nu). \tag{6.6}$$

Lemma 6.2 *Let $\zeta_i \stackrel{d}{=} \zeta$ $(i \ge 1)$, $\zeta_0 = 0$. If $\mathbb{E}|\zeta| < \infty$, then*

$$d_G\left(\sum_{i=0}^{\nu_n} \zeta_i; \sum_{i=0}^{\nu} \zeta_i\right) \le d_G(\nu_n; \nu)\mathbb{E}|\zeta|. \tag{6.7}$$

For instance, (6.6) and Theorem 4.12 in [221] yield

$$d_{TV}(S_n; S) \le 3\theta/4\mathrm{e} + 4\delta^*, \tag{6.8}$$

where $\theta = \sum_{i=1}^n p_i^2/\lambda$, $\delta^* = (1 - \mathrm{e}^{-\lambda}) \sum_{i=1}^n p_i^3/\lambda$.

According to (6.7),

$$d_G(S_n; S) \le d_G(\nu_n; \pi_\lambda)\mathbb{E}|X'|.$$

This and (3.24) entail

$$d_G(S_n; S) \le \left(\lambda \wedge \tfrac{4}{3}\sqrt{2\lambda/\mathrm{e}}\right)\theta\mathbb{E}|X'|. \tag{6.9}$$

The case of i.i.d. $\{X_i'\}$.

Consider the situation where

$$X_i \stackrel{d}{=} \tau_i X_i', \quad X_i' \stackrel{d}{=} X' \quad (\forall i),$$

while r.v.s $\{\tau_i\}$ are not required to be i.i.d. In such a situation (6.5) still holds:

$$S_n \stackrel{d}{=} \sum_{i=1}^{\nu_n} X_i',$$

and estimates of the accuracy of compound Poisson approximation to $\mathcal{L}(S_n)$ follows from the estimates of the accuracy of pure Poisson approximation thanks to (5.6), see [238, 207, 38, 224]. In particular, (6.8) and (6.9) remain valid. An estimate of the accuracy of "accompanying" compound Poisson approximation to $\mathcal{L}(S_n)$ in terms of the point metric is given in [226].

The task of deriving estimates of the accuracy of compound Poisson approximation by a sum of independent accompanying random variables appears non-trivial if r.v.s $\{X_i'\}$ are not identically distributed.

The non-i.i.d. case.

If r.v.s $\{X_i'\}$ are independent but not identically distributed, it appears natural to approximate each X_j by an "accompanying" r.v. $\tilde{X}_j$, so that the sum

$$S_n \stackrel{d}{=} \tau_1 X_1' + ... + \tau_n X_n'$$

is approximated by a compound Poisson random variable

$$\tilde{S}_n = \tilde{X}_1 + ... + \tilde{X}_n,$$

where $\{\tilde{X}_j\}$ are independent compound Poisson $\mathbf{\Pi}(1, X_i) = \mathbf{\Pi}(p_i, X_i')$ random variables, cf. (1.8). Recall that $\tilde{S}_n$ is a compound Poisson random variable.

A simple estimate of the accuracy of compound Poisson approximation to $\mathcal{L}(S_n)$ follows from (6.6), the property of d_{TV} and a well-known fact that $d_{TV}(\mathbf{B}(p); \mathbf{\Pi}(p)) \leq p^2$: since

$$\mathcal{L}(\tilde{X}_i) = \mathbf{\Pi}(p_i, X_i') \tag{1.7*}$$

by (1.7), one has

$$d_{TV}(S_n; \tilde{S}_n) \leq \sum_{i=1}^{n} d_{TV}(X_i; \tilde{X}_i) \leq \sum_{i=1}^{n} d_{TV}(\tau_i; \pi_{p_i}) \leq \sum_{i=1}^{n} p_i^2 . \tag{6.10}$$

The explicit proof of estimate (6.10) has been given by Le Cam [190], who attributed the idea of the proof to Khintchine [175]. Similar estimates hold in terms of some other distances.

The following result due to Roos [264] deals with situations where $\{p_i\}$ are "not too small".

Let $X_1, ..., X_n$ be independent non-negative r.v.s. Recall that $S_n \stackrel{d}{=} \tau_1 X_1' + ... + \tau_n X_n'$, where we assume that $p_i \equiv \mathbb{P}(X_i \neq 0) < 1$ $(\forall i)$ and that $\mathcal{L}(\tilde{S}_n)$ has been defined by (6.3). Denote

$$P_o = \exp\left(\frac{1}{2}\sum_{i=1}^{n} p_i(1-p_i)(\mathcal{L}(X_i') - I)\right). \tag{6.11}$$

Then [264]

$$d_K(S_n;\tilde{S}_n) \leq \frac{\pi^2}{8}\sum_{i=1}^{n}\frac{c_i p_i^2}{1-p_i}Q_{P_o}(\mathbb{E}X_i'), \tag{6.12}$$

where $c_i=1$ if $X_i'\equiv 0$, $c_i=2$ otherwise, $Q_{P_o}(\cdot)$ is a concentration function.

Zaitsev [324] has derived an estimate of the accuracy of compound Poisson approximation that can be sharper than (6.10), (6.12) if $\mathbb{E}S_n$ is "large". Denote $p_n^* := \max_{i\leq n} p_i$.

Theorem 6.3 [324] *There exists an absolute constant C such that*

$$d_K(S_n;\tilde{S}_n) \leq Cp_n^*. \tag{6.13}$$

Inequality (6.13) is instrumental in the work on Kolmogorov's problem.

We consider now a more complex sum of r.v.s. Let $S_n = \sum_{i=1}^n X_i$, where X_i are independent r.v.s. Assume that each X_i is a sum of a random number of i.i.d.r.v.s:

$$X_i = \sum_{j=1}^{\nu_i} X_{ij}',$$

where ν_i is a r.v. taking values in $\mathbb{Z}_+$, $X_i', X_{i1}', X_{i2}', \ldots$ are i.i.d.r.v.s independent of ν_i.
Denote $\tilde{\lambda} = \mathbb{E}\nu_1 + \cdots + \mathbb{E}\nu_n$. We assume that $\tilde{\lambda} \in (0;\infty)$. Set

$$\tilde{\mathbf{\Pi}} = \prod_{i=1}^{*n}\mathbf{\Pi}(\mathbb{E}\nu_i;X_i') = \exp\left(\sum_{i=1}^{n}\mathbb{E}\nu_i(\mathcal{L}(X_i')-I)\right).$$

Then $\tilde{\mathbf{\Pi}}$ is a compound Poisson distribution.
The following proposition is a generalisation of (6.10).

Proposition 6.4 *Suppose that $\mathbb{E}\nu_i^2<\infty$ $(i=1,2,\ldots,n)$. Then*

$$d_{TV}(\mathcal{L}(S_n);\tilde{\mathbf{\Pi}}) \leq \sum_{i=1}^{n}(\mathbb{E}\nu_i(\nu_i-1) + \mathbb{E}^2\nu_i). \tag{6.14}$$

Example 6.1 Let $\{\nu_i\}$ be Geometric r.v.s: $\mathbb{P}(\nu_i=k) = (1-p_i)p_i^k$ $(k\in\mathbb{Z}_+,\ 0<p_i<1)$. Then

$$d_{TV}(\mathcal{L}(S_n);\tilde{\mathbf{\Pi}}) \leq 3\sum_{i=1}^{n}\left(p_i/(1-p_i)\right)^2.$$

If $\{X_i\}$ are i.i.d.r.v.s, then (3.55) is better than (6.14). □

Open problems

6.1. Evaluate absolute constant C in Zaitsev's inequality (6.13).

6.2. Will (6.13) hold if p_n^* is replaced with θ?

6.2 Integer-valued random variables

Let $\{X_i\}$ be independent non-negative integer-valued random variables. Denote

$$\lambda_j \equiv \lambda_{n,j} = \sum_{i=1}^{n} \mathbb{P}(X_i = j) \quad (j \geq 1), \quad \lambda = \sum_{j\geq 1} \lambda_j = \sum_{i=1}^{n} \mathbb{P}(X_i \geq 1).$$

Set $Z = \sum_{j=1}^{\infty} j\pi_{\lambda_{n,j}}$ (cf. (1.2)). Clearly, Z is a compound Poisson r.v.

In fact, $Z \stackrel{d}{=} \tilde{S}_n$, where $\tilde{S}_n$ is the sum of accompanying $X_1, ..., X_n$ r.v.s. Indeed,

$$\begin{aligned} \mathcal{L}(Z) &= \exp\left(\sum_{j=0}^{\infty} \lambda_{n,j}(I_j - I)\right) = \exp\left(\sum_{i=1}^{n}\sum_{j=0}^{\infty} \mathbb{P}(X_i = j)(I_j - I)\right) \\ &= \exp\left(\sum_{i=1}^{n} (\mathcal{L}(X_i) - I)\right) = \mathcal{L}(\tilde{S}_n). \end{aligned}$$

The following result is due to Erhardsson [113]: if $\sum_{i=1}^{n} \mathbb{E}X_i < \infty$, then

$$d_{TV}(S_n; \tilde{S}_n) \leq \mathcal{M}(\lambda) \sum_{j=1}^{n} \mathbb{E}^2 X_j\,, \tag{6.15}$$

where $\mathcal{M}(\lambda) = \min\left\{1; \lambda_{n,1}^{-1}\right\} e^{\lambda}$. Due to exponential factor e^{λ} estimate (6.15) is less accurate than (6.10). However, improvements are possible for special cases.

If $i\lambda_{n,i} \geq (i+1)\lambda_{n,i+1}$ $(\forall i \geq 1)$, then $\mathcal{M}(\lambda)$ can be replaced with

$$\min\left\{1; \frac{1}{\lambda_{n,1} - 2\lambda_{n,2}} \left(\frac{1}{4(\lambda_{n,1} - 2\lambda_{n,2})} + \ln^{+}(2(\lambda_{n,1} - 2\lambda_{n,2}))\right)\right\} \tag{6.16}$$

[22], where $\ln^{+}$ denotes a positive part of the natural logarithm.

If $\theta' := \sum_{i\geq 2} i(i-1)\lambda_{n,i} / \sum_{i\geq 1} i\lambda_{n,i} < 1/2$, then [25] $\mathcal{M}(\lambda)$ can be replaced with

$$1/(1-2\theta') \sum_{i\geq 1} i\lambda_{n,i}\,. \tag{6.17}$$

For other estimates of $\mathcal{M}(\lambda)$ see [100].

Barbour & Xia [26] have proved that

$$d_K(S_n; \tilde{S}_n) \leq \min\{1/2; 1/(\lambda_{n,1} + 1)\} \sum_{j=1}^{n} \mathbb{E}^2 X_j$$

if $i\lambda_{n,i} \geq (i+1)\lambda_{n,i+1}$.

Results similar to (6.10), (6.15) have been presented in [24, 53]. In particular, let $X, X_1, ...$ be i.i.d.r.v.s such that $\mathbb{E}X^2 < \infty$, $p := \mathbb{P}(X \neq 0) < \ln 2$ and

$$\|\mathbf{\Pi}(np, X') * (I_1 - I)^{*2}\|_{TV} = O(1/np),$$

where $\mathcal{L}(X') = \mathcal{L}(X | X \neq 0)$ (recall that $\|\cdot\|_{TV}$ denotes the total variation norm). Boutsikas & Vaggelatou [53] have shown that

$$d_{TV}(S_n; \tilde{S}_n) = O(p + n^2 p^4).$$

Assume that r.v.s $X_1, \dots, X_n$ take values in a finite set $0, 1, \dots, N$ of natural numbers. A lower bound to $d_{TV}(S_n; \tilde{S}_n)$ has been given by Barbour et al. [22]:

$$d_{TV}(S_n; \tilde{S}_n) \geq \frac{1}{32N^2} \min\left(1; (\mathbb{E}S_n)^{-1}\right) \sum_{j=1}^{n} \mathbb{E}^2 X_j .$$

6.3 Asymptotic expansions

Construction of asymptotic expansions is based on (6.4).

Let f_i denote the characteristic function of $\mathcal{L}(X_i')$, and let $p_i = \mathbb{P}(X_i \neq 0)$ $(i \geq 1)$. Then

$$\begin{aligned} \mathbb{E}\exp(itS_n) &= \prod_{i=1}^{n}(1+p_i(f_i-1)) = \exp\left(\sum_{i=1}^{n}\ln(1+p_i(f_i-1))\right) \\ &= \exp\left(\sum_{i=1}^{n}\sum_{j=1}^{\infty}(-1)^{j+1}p_i^j(f_i-1)^j/j\right). \end{aligned} \tag{6.18}$$

This leads to the following asymptotic expansion:

$$\exp\left(\sum_{i=1}^{n} p_i(f_i-1)\right)\left(1-\frac{1}{2}\sum_{i=1}^{n}p_i^2(f_i-1)^2+\cdots\right).$$

One can arrive at the asymptotic expansion involving a signed compound Poisson measure:

$$\exp\left(\sum_{i=1}^{n}\sum_{j=1}^{k}\frac{(-1)^{j+1}p_i^j(f_i-1)^j}{j}\right)\left(1+\sum_{i=1}^{n}\sum_{j=k+1}^{\infty}\frac{(-1)^{j+1}p_i^j(f_i-1)^j}{j}+\cdots\right).$$

This is not the only possible asymptotic expansion. Assume that $0<p_i<1$. Then (6.18) can be rewritten as

$$\exp\left(\sum_{i=1}^{n}(\ln(1-p_i)+\ln(1+p_if_i/(1-p_i)))\right) = \exp\left(\sum_{i=1}^{n}\sum_{j=1}^{\infty}(-1)^{j+1}\left(\frac{p_i}{1-p_i}\right)^j(f_i^j-1)/j\right).$$

Truncating the number of summands in the exponent, we can get another possible SCP approximation.

Asymptotic expansions (SCP approximations) can be traced back to Uspensky [298], see also Herrmann [150].

Recall that $\tilde{S}_n$ denotes the sum of accompanying r.v.s., cf. (1.9). Set $P_i = \mathcal{L}(X_i)$,

$$\begin{aligned} G_1 &= \mathcal{L}(\tilde{S}_n) - \frac{1}{2}\sum_{i=1}^{n}p_i^2\left(\mathcal{L}(\tilde{S}_n+X_i'+X_i'') - 2\mathcal{L}(\tilde{S}_n+X_i') + \mathcal{L}(\tilde{S}_n)\right) \\ &= \prod_{i=1}^{*n}\exp(P_i-I) * \left(I-\frac{1}{2}\sum_{i=1}^{n}(P_i-I)^{*2}\right), \\ G_1' &= \prod_{i=1}^{*n}\exp(P_i-I) + \sum_{k=1}^{n}\prod_{j\neq k}^{*n}\exp(P_j-I) * (P_k - \exp(P_k-I)), \end{aligned}$$

where $X_i'' \stackrel{d}{=} X_i'$, all r.v.s are independent.

The following first-order asymptotic expansion in terms of Kolmogorov's distance is due to Čekanavičius [69]:

$$d_K(\mathcal{L}(S_n); G_1) \le C p_n^{*2}, \tag{6.19}$$

where $p_n^* = \max p_i$.

The next theorem presents estimates in terms of the total variation distance.

Theorem 6.5 *For any* $n \ge 1$

$$d_{TV}(\mathcal{L}(S_n); G_1) \le \frac{8}{3}\left(\sum_{j=1}^n p_j^2\right)^{3/2} + 2\left(\sum_{j=1}^n p_j^2\right)^2, \tag{6.20}$$

$$d_{TV}(\mathcal{L}(S_n); G_1') \le 2\left(\sum_{j=1}^n p_j^2\right)^2. \tag{6.21}$$

The rate of approximation in (6.21) is better than that in (6.20).

Next we consider SCP approximations. Given a fixed natural number s, denote

$$H_{n,s} = \exp\left(\sum_{i=1}^n \sum_{j=1}^s p_i^j (-1)^{j+1} (\mathcal{L}(X_i') - I)^{*j}/j\right), \tag{6.22}$$

$$K_{n,s} = \exp\left(\sum_{i=1}^n \sum_{j=1}^s (-1)^{j+1} \left(\frac{p_i}{1-p_i}\right)^j (\mathcal{L}(X_i')^{*j} - I)/j\right), \tag{6.23}$$

$$\tau(i,s) = (2p_i)^{s+1}/(s+1)(1-2p_i).$$

Theorem 6.6 *If* $p_n^* < 1/2$, *then*

$$d_{TV}(\mathcal{L}(S_n); H_{n,s}) \le \frac{1}{2}\left(\exp\left(\sum_{i=1}^n \tau(i,s)\right) - 1\right), \tag{6.24}$$

$$d_{TV}(\mathcal{L}(S_n); K_{n,s}) \le \frac{1}{2}\left(\exp\left(\sum_{i=1}^n \frac{2}{s+1}\left(\frac{p_i}{1-p_i}\right)^{s+1} \frac{1-p_i}{1-2p_i}\right) - 1\right). \tag{6.25}$$

Estimates (6.24) and (6.25) are of order $O(\sum_{i=1}^n p_i^{s+1})$. Therefore, (6.24), (6.25) improve (6.20) even if $s=2$. For example, if $p_i = n^{-1/2}$ for all i, then the right-hand side of (6.20) is of order $O(1)$, while RHSs of (6.24), (6.25) are of order $O(n^{-1/2})$.

Roos [264] has obtained an estimate involving a concentration function (a related result has been established in [265], see (6.12)). Set

$$\tau' = \sum_{i=1}^n (e^{\tau(i,s)-1} - 1),$$

and let P_0, c_i, $\tau(i,s)$ be defined as in (6.11).

Theorem 6.7 [264] If $\{X_i\}$ are non-negative r.v.s, $p_i < 1/2$ and $\tau' < 1$, then

$$d_K(\mathcal{L}(S_n); H_{n,s}) \le \frac{\pi^2}{4(1-\tau')} \sum_{i=1}^n c_i (e^{\tau(i,s)} - 1) Q_{P_0}(\mathbb{E}X_i'). \tag{6.26}$$

If $\mathcal{L}(X_i) = \mathbf{B}(p)$, $n(2p)^{s+1} < 1$, then the RHS of (6.26) is $O(np^{s+1}/\sqrt{np})$.

6.4 Dependent random variables

Approximation under mixing conditions. Let $\{X, X_1, ..., X_n\}$ be a stationary sequence of r.v.s that are non-zero with small probabilities, and let $S_n = X_1 + ... + X_n$. The distribution of S_n can be approximated by a compound Poisson distribution $\mathcal{L}(Y_n)$.

Given $1 \le r \le n$, set $k = [n/r]$, and let

$$p = \mathbb{P}(X \neq 0), \quad q = \mathbb{P}(S_r \neq 0).$$

Let $\nu, \pi_{n,r}, \zeta_1, \zeta_2, \dots$ be independent random variables, $\zeta_0 = 0$, $\mathcal{L}(\pi_{n,r}) = \mathbf{\Pi}(kq)$, $\mathcal{L}(\nu) = \mathbf{B}(k, q)$,

$$\mathcal{L}(\zeta_i) \;=\; \mathcal{L}(S_r | S_r \neq 0) \quad (i \ge 1).$$

Recall that α_n, β_n denote mixing coefficients (see Appendix). Set

$$Y_n = \sum_{i=0}^{\pi_{n,r}} \zeta_i.$$

Theorem 6.8 *If* $n > r > l \ge 0$, *then*

$$d_{TV}(S_n; Y_n) \;\le\; \kappa_{n,r} rp + (2kl + r')p + nr^{-1}\gamma_n(l), \tag{6.27}$$

$$d_{G}(S_n; Y_n) \;\le\; rp \min\Big\{np \,;\, \tfrac{4}{3}\sqrt{2np/\mathrm{e}}\Big\} + (2kl + r')p + n\gamma_n(l), \tag{6.28}$$

where $r' = n - rk$, $\kappa_{n,r} = \min\{1 - \mathrm{e}^{-np}\,; 3/4\mathrm{e} + (1 - \mathrm{e}^{-np})rp\}$ *and* $\gamma_n(l) = \min\{4\alpha_n(l)\sqrt{r}\,; \beta_n(l)\}$.

If $\{X_i\}$ are independent Bernoulli $\mathbf{B}(p)$ r.v.s, then (6.27), (6.28) with $r = 1$, $l = 0$ yield sharp estimates of the accuracy of pure Poisson approximation.

If random variables $\{X_i\}$ are m-dependent , then one can choose $l = m$, $r = \lceil \sqrt{mn}\, \rceil$ (the smallest integer greater than or equal to $\sqrt{mn}$) to get

$$d_{TV}(S_n; Y_n) \le 4p\lceil \sqrt{mn}\, \rceil. \tag{6.27*}$$

Locally dependent random variables.

Let $X_1, ..., X_n$ be real-valued random variables.

Suppose that there exist subsets $\mathcal{I}_{1j}$, $\mathcal{I}_{2j}$ of $\{1, 2, \dots, n\} \setminus \{j\} = \mathcal{I}_{1j} \cup \mathcal{I}_{2j}$ such that X_j and $\{X_i, i \in \mathcal{I}_{2j}\}$ are independent. According to Barbour et al. [21], Cor. 10.L.1,

$$d_{TV}(S_n; \tilde{S}_n) \le \sum_{j=1}^{n} \Big(\mathbb{P}^2(X_j \neq 0) + \sum_{i \in \mathcal{I}_{1j}} \big(\mathbb{P}(X_i \neq 0 \neq X_j) + \mathbb{P}(X_i \neq 0)\mathbb{P}(X_j \neq 0) \big) \Big), \tag{6.29}$$

where $\tilde{S}_n$ is the sum of accompanying random variables defined in (1.9).

If r.v.s $X, X_1, ..., X_n$ are identically distributed and $\mathcal{I}_{1j} = \emptyset$, then (6.29) becomes (6.10). Similar estimates have been proved in [52, 53].

Associated random variables.

Let $X_1, \dots, X_n$ be non-negative integer-valued random variables. Random variables $X_1, \dots, X_n$ are called *associated* if

$$\mathrm{cov}(f(X_1, \dots, X_n); g(X_1, \dots, X_n)) \ge 0$$

for every pair of non-decreasing functions f and g.

Random variables $X_1,\dots,X_n$ are called *negatively associated* if

$$\mathrm{cov}(f(X_i, i\in A_1); g(X_i, i\in A_2)) \le 0$$

for every pair of disjoint subsets A_1, A_2 of $\{1,2,\dots,n\}$ and non-decreasing functions f, g.

Let $\mathcal{I}(i)$ be a subset of $\{1,2,\dots,n\}\setminus\{i\}$. The choice of $\mathcal{I}(i)$ is arbitrary though it is supposed to be a neighbourhood of "strong dependence" on X_i.

Set $\lambda=\sum_{j\ge 1}\lambda_j$, where

$$\lambda_j=\sum_{i=1}^{n}\mathbb{E}X_i\mathbb{I}\{X_i+\widehat{X}_i=j\}/j, \quad \widehat{X}_i=\sum_{j\in\mathcal{I}(i)}X_j. \tag{6.30}$$

Given a set of non-negative numbers $\{\lambda_j\}_{j\ge 1}$ such that $\lambda<\infty$, we denote

$$Z_n=\sum_{j=1}^{\infty} j\pi_{\lambda_j},$$

where $\{\pi_{\lambda_j}\}$ are independent Poisson $\mathbf{\Pi}(\lambda_j)$ variables (cf. (1.2)).

Denote

$$\mathcal{M}(\lambda)=\min\left\{1;\lambda_{n,1}^{-1}\right\}\mathrm{e}^{\lambda}.$$

$\mathcal{M}(\lambda)$ is an estimate of $\|\Delta g\|$ in Stein's equation.

Theorem 6.9 [99] If $X_1,\dots,X_n$ are negatively associated r.v.s, then

$$d_{TV}(S_n;Z_n)\le\mathcal{M}(\lambda)\left(\sum_{i=1}^{n}\sum_{j\in\mathcal{I}(i)\cup\{i\}}\mathbb{E}X_iX_j-\mathrm{var}\,S_n\right).$$

If $X_1,\dots,X_n$ are associated r.v.s, then

$$d_{TV}(S_n;Z_n)\le\mathcal{M}(\lambda)\left(\mathrm{var}\,S_n-\sum_{i=1}^{n}\sum_{j\in\mathcal{I}(i)\cup\{i\}}\mathbb{E}X_iX_j+2\sum_{i=1}^{n}\sum_{j\in\mathcal{I}(i)\cup\{i\}}\mathbb{E}X_i\mathbb{E}X_j\right).$$

According to [99], $\mathcal{M}(\lambda)$ obeys (6.16)–(6.17).

Similar results for locally dependent r.v.s have been proved by Eichelsbacher & Roos [110]. Applications of Theorem 6.9 to the urn model with overflow and to k-runs can be found in [99].

If $\{X_i\}$ are independent, then one can take $\mathcal{I}(i)=\emptyset$ to arrive at (6.15).

Example 6.2 Let $\xi_1,\dots,\xi_n$ be a stationary sequence of negatively associated r.v.s. Assume that $\xi_{i+n}=\xi_i$ $(\forall i)$. Fix numbers $a_1,a_2,\dots,a_n$, and set $X_i=\mathbb{I}\{\xi_i>a_i\}$, $S_n=\sum_{i=1}^{n}X_i$. Let

$$Z_n=\sum_{j=1}^{\infty} j\pi_{\lambda_j},$$

where $\{\pi_{\lambda_j}\}$ are independent Poisson $\mathbf{\Pi}(\lambda_j)$ r.v.s. Then [99]

$$\begin{aligned} d_{TV}(S_n;Z_n) &\le \mathcal{M}(\lambda)\Bigg(\sum_{i=1}^{n}(\mathbb{P}(\xi>a_i))^2+2\sum_{i=1}^{n}\sum_{j\in\mathcal{I}(i),i<j}\mathbb{E}X_i\mathbb{E}X_j \\ &+\sum_{i=1}^{n}\sum_{j\in\mathcal{I}(i)}(\mathbb{E}X_iX_j\mathbb{I}\{j<i\}-\mathbb{E}X_iX_j\mathbb{I}\{i<j\}\Bigg). \end{aligned}$$

□

Locally dependent Bernoulli random variables.

Let $\{X_i\}_{i\in\Gamma}$ be locally dependent Bernoulli $\mathbf{B}(p_i)$ random variables, where Γ is a set of indexes. The following result is due to M.Roos [255].

Suppose that for every $i\in\Gamma$ set Γ is split into subsets $\{i\}$, Γ_i^{vs} ("very strongly" dependent on X_i), Γ_i^{vw} ("very weakly" dependent on $\{X_j\}_{j\in\Gamma_i^{vs}}$), and Γ_i^b (the rest). Denote

$$\begin{aligned} S &= \sum_{i\in\Gamma} X_i\,,\ S_i = S - X_i\,,\ \hat{X}_i = \sum_{j\in\Gamma_i^{vs}} X_j\,,\\ X_i^{(b)} &= \sum_{j\in\Gamma_i^b} X_j\,,\ Y_i = \sum_{j\in\Gamma_i^{vw}} X_j\,,\ S_{i,U} = S_i - \hat{X}_i\,,\\ D &= \max_{i\in\Gamma}|\Gamma_i^{vs}|,\ \varphi = \sum_{i\in\Gamma}(\varphi_{i,1}+\ldots+\varphi_{i,l_i}),\ l_i = 1+|\Gamma_i^{vs}|, \end{aligned}$$

where

$$\varphi_{i,j} = \mathbb{E}|\mathbb{E}X_i\mathbb{I}\{X_i+\hat{X}_i=j\} - \mathbb{E}\{X_i\mathbb{I}\{X_i+\hat{X}_i=j\}|\sigma(X_l: l\in\Gamma_i^{vw})\}|$$

and $|\Gamma_i^{vs}|$ is the number of elements in the set Γ_i^{vs}.

Let compound Poisson r.v. $S^* = \sum_{j=1}^{\infty} j\pi_{\lambda_j}$ be defined by (1.2), where

$$\lambda_j = \sum_{i=1}^{n}\mathbb{E}X_i\mathbb{I}\{X_i + \sum_{j\in\Gamma_i^{vs}} X_j = j\}/j\quad (1\le j\le D+1),\quad \lambda_j=0\ \ (j>D+1),$$

i.e., λ_j is the same as in (6.30). Set $\lambda = \lambda_1+\ldots+\lambda_{D+1}$.

Theorem 6.10 [255] *There holds*

$$d_{TV}(S;S^*) \le c_\lambda\varphi + \mathcal{M}(\lambda)\sum_{i\in\Gamma}\left(\mathbb{E}^2X_i + \mathbb{E}X_i\mathbb{E}(\hat{X}_i+X_i^b) + \mathbb{E}X_iX_i^{(b)}\right),$$

where $\mathcal{M}(\lambda)$ satisfies (6.16)–(6.17), $c_\lambda\le e^\lambda$. If $i\lambda_i \ge (i+1)\lambda_{i+1}$ $(\forall i\ge 1)$, then

$$c_\lambda \le \begin{cases} 1, & \text{if } \lambda_1-2\lambda_2\le 1,\\ (1/\sqrt{\lambda_1-2\lambda_1})(2-(1/\sqrt{\lambda_1-2\lambda_2})), & \text{if } \lambda_1-2\lambda_2>1. \end{cases}$$

Open problem

6.3. The proof of Theorem 6.8 uses Bernstein's blocks method, which is known for low accuracy of estimation (see, e.g., [221], Example 11.2). Can estimates (6.27), (6.28) be improved?

6.5 Applications

6.5.1 m-run statistic

Let $\xi_1,\xi_2,\ldots$ be independent Bernoulli $\mathbf{B}(\mathbf{p_i})$ random variables, where $0<p_i<1$. Denote

$$X_i = \xi_i\xi_{i+1}\cdots\xi_{i+m-1} = \min\{\xi_i,\ldots,\xi_{i+m-1}\},\quad S_{n,m} = X_1+\cdots+X_{n-m+1}.$$

Then $S_{n,m}$ is the number of 1-runs of length m (m-run statistic).

If $m=1$, then $X_i = \xi_i$, $S_{n,1} = X_1 + \cdots + X_n$.

Clearly, $\{X_i\}$ is a sequence of $(m-1)$-dependent Bernoulli random variables. For the sake of simplicity we will assume that $\xi_{i+n} = \xi_i$ $(1 \le i \le n)$.

Daly [99] has proved that the accuracy of compound Poisson approximation to $S_{n,m}$ in terms of the total variation distance is of the order $mp^m/(1-5p)$ if $p_i \equiv p < 1/5$ $(\forall i)$.

The accuracy of Negative Binomial approximation to $S_{n,m}$ has been evaluated by Wang & Xia [315]. Let

$$\lambda = \mathbb{E}S_{n,m}, \quad \sigma^2 = \mathrm{var}\, S_{n,m},$$

and let $Y_{n,m}$ be a Negative Binomial $\mathbf{NB}(r,p)$ r.v. defined by (1.4), where

$$r = \lambda/(\sigma^2 - \lambda), \quad p = 1 - \lambda/\sigma^2.$$

Let ϑ_k be the kth largest number among $(1-p_{i-1})^2 p_i (1-p_i) p_{i+1} \cdots p_{i+m-1}$ $(1 \le i \le n)$. Set

$$\phi = \min\left\{2;\ 4.6\left(\sum_{k=4m-1}^{n} \vartheta_k\right)^{-1/2}\right\}, \quad \hat{p}_i = \max\{p_j : i-2m+2 \le j \le i+2m-2\}.$$

Theorem 6.11 [315] If $\sigma^2 > \lambda$ and $n > 4m$, then

$$d_{TV}(S_{n,m}; Y_{n,m}) \le \frac{4.5(4m-3)(2m-1)\phi}{\lambda} \sum_{i=1}^{n} \hat{p}_i \mathbb{E}X_i. \tag{6.31}$$

If all $p_i \equiv p = o(1)$, then the RHS of (6.31) is $O(p^{2-m/2} n^{-1/2})$ as $n \to \infty$, see [55, 315]. Further reading on the topic includes [59, 60, 88, 308, 90, 91].

6.5.2 Long match patterns

Let $X, X_1, \ldots, X_m$ and $Y, Y_1, \ldots, Y_n$ be two independent sequences of independent and identically distributed random variables taking values in $\mathbb{N}$. We assume for simplicity that $X_{m+1} = X_1$, $X_{m+2} = X_2$, ..., $Y_{n+1} = Y_1$, $Y_{n+2} = Y_2$, ..., and so on. Then

$$S_{m,n,k} = \sum_{i=1}^{m} \sum_{j=1}^{n} \mathbb{1}\{(X_i, \ldots, X_{i+k-1}) = (Y_j, \ldots, Y_{j+k-1})\}$$

is the number of match patterns (NMP) of length k.

Set $R = \mathbb{P}(X = Y)$, and suppose that $R < 1/2$. Let $p_\ell = \mathbb{P}(X = \ell)$, $q_\ell = \mathbb{P}(Y = \ell)$ $(\forall \ell \in \mathbb{N})$,

$$r = \max_\ell p_\ell q_\ell, \quad S(R) = (1-R)(1-2R),$$

$$\tilde{p} = \max_\ell p_\ell \mathbb{1}\{q_\ell > 0\}, \quad \tilde{q} = \max_\ell q_\ell \mathbb{1}\{p_\ell > 0\}, \quad \lambda = mn(1-R)R^k.$$

Clearly, $\mathbb{E}S_{m,n,k} = mnR^k$.

Information on the distribution of NMP can help recognising "valuable" fragments of DNA sequences (see [210, 224, 274]) and references therein).

NMP $S_{m,n,k}$ can be approximated by a compound Poisson random variable

$$\tilde{Y} = \sum_{i=1}^{\infty} i \pi_{\theta_i},$$

where $\theta_i = \lambda(1-R)R^{i-1}$, $\{\pi_{\theta_i}\}_{i\geq 1}$ are independent Poisson $\mathbf{\Pi}(\theta_i)$ r.v.s. Note that

$$\mathbb{E}\tilde{Y} = mnR^k .$$

The following estimate of the accuracy of compound Poisson approximation to $\mathcal{L}(S_{m,n,k})$ is due to Mikhailov [209].

Theorem 6.12 [209] If $2\leq k<\min(n,m)$ and $0<R<1$, then

$$\begin{aligned} d_{TV}(S_{m,n,k};\tilde{Y}) &\leq \mathcal{M}(\lambda)\left(2k\lambda(1-R)^{-1}\left(2k(r/R)^k+m\tilde{p}^k+n\tilde{q}^k\right)\right. \\ &+ \left.\frac{2(4k-3)\lambda^2}{(1-R)^2}\left(\frac{1}{n}+\frac{1}{m}\right)\right) + \frac{4\lambda^2}{nmR(1-R)}, \end{aligned} \qquad (6.32)$$

where $\mathcal{M}(\lambda)\leq(1\wedge 1/\lambda(1-R))e^{\lambda}$. If $0<R<1/2$, then

$$\mathcal{M}(\lambda)\leq 1\wedge\left(1/4\lambda S(R)+\ln_+(2\lambda S(R))\right)/\lambda S(R).$$

In the "central zone" where $k=k(m,n)$ obeys $mnR^k\asymp const$ as $m\to\infty$, $n\to\infty$, $m\asymp n$, the RHS of (6.32) is $O(\ln(mn)(1/m+1/n))$.

Mikhailov [209] gives also an estimate of the accuracy of Poisson approximation in the case $k=1$. Note that the distribution of the number of match patterns of length $\geq k$ can be well approximated by the Poisson law (see [221, 224]).

6.5.3 Non-decreasing runs of fixed length

Let $X, X_1, \ldots, X_n$ be i.i.d. random variables with a uniform distribution

$$\mathbb{P}(X_i=k) = 1/N \qquad (k=0,1,\ldots,N-1),$$

where $N\geq 3$ is a fixed natural number.

Given $s\in\mathbb{N}$, denote $\eta_k(s)=\mathbb{I}\{X_k\leq X_{k+1}\leq\cdots\leq X_{k+s-1}\}$. Let

$$S_n(s)=\eta_1(s)+\ldots+\eta_1(s),$$

and let Z_n be defined by (1.2):

$$Z_n=\sum_{j=1}^{\infty} j\pi_{\lambda_j},$$

where $\{\pi_{\lambda_j}\}$ are independent Poisson $\mathbf{\Pi}(\lambda_j)$ variables,

$$\lambda_j=\begin{cases} N^{-s-j-1}n\varkappa_{1,s+j-1}, & j=1,\ldots,s-1,\\ j^{-1}N^{-s-j-1}n(2N\varkappa_{2,s+j-1}+(2s-2-j)\varkappa_{1,s+j-1}) & j=s,\ldots,2s-2,\\ (2s-1)^{-1}N^{-3s+2}\varkappa_{3,2s-1} & j=2s-1,\\ 0 & j=2s,2s+1,\ldots \end{cases}$$

Here

$$\varkappa_{1,k}=\binom{k+N-1}{k+2}\frac{N(k^2+k-1)-k^2-k}{N-2}, \qquad \varkappa_{2,k}=\binom{k+N}{k+1}\frac{k(N-1)}{N+k},$$

$$\varkappa_{3,k}=\binom{k+N-1}{k}.$$

Then Z_n is a compound Poisson $\mathbf{\Pi}(\lambda,\zeta)$ r.v. with $\lambda = \sum_{k=1}^{2s-1}\lambda_k$ and the multiplicity distribution $\mathcal{L}(\zeta)$ such that the ch.f. $\varphi_\zeta(t) = \sum_{j=1}^{\infty}\lambda_j e^{itj}/\lambda$.

Minakov [211], using Theorem 6.10, has shown that

$$d_{TV}(S_n(s);Z_n) \leq e^{\lambda}\frac{n(6s-5)}{(sN^{-1}+1)^2N^{2s}}\binom{s+N}{s}^2. \tag{6.33}$$

Denote $Y = \exp\Big(\tilde{\lambda}(N-1)\sum_{j=1}^{\infty}N^{-j}(I_j-I)\Big)$. If $n\to\infty$, $s\to\infty$, $s/n\to 0$ and

$$n(s+N)^{N-1}N^{-s-1}/(N-2)! \sim \tilde{\lambda} \qquad (n\to\infty),$$

then $S_n(s)$ converges weakly to the compound Poisson r.v. Y.

6.6 Proofs

Proof of Lemma 6.2. Given r.v.s $\zeta,\zeta_1,\zeta_2,...$, let ξ be a uniform $\mathbf{U}[0;1]$ r.v. independent of $\zeta,\zeta_1,\zeta_2,...$ Denote by F_{ν_n}, F_{ν_n} the d.f.s of $\mathcal{L}(\nu_n)$, $\mathcal{L}(\nu)$. Set $\nu_n' = F_{\nu_n}^{-1}(\xi)$, $\nu' = F_\nu^{-1}(\xi)$. Then

$$\nu_n' \stackrel{d}{=} \nu_n\,, \quad \nu' \stackrel{d}{=} \nu\,, \quad d_G(\nu_n;\nu) = \mathbb{E}|\nu_n'-\nu'|.$$

([242], §4). Clearly, $\mathbb{E}|\nu_n-\nu| = \mathbb{E}(\nu_n-\nu)\mathbb{I}\{\nu_n>\nu\} + \mathbb{E}(\nu-\nu_n)\mathbb{I}\{\nu>\nu_n\}$.

Denote $S_k^\zeta = \zeta_1+...+\zeta_k$ $(k\in\mathbb{Z}_+)$. Since the sequence $\zeta,\zeta_1,\zeta_2,...$ is independent of (ν_n',ν'), we have $S_{\nu_n'}^\zeta \stackrel{d}{=} S_{\nu_n}^\zeta$, $S_{\nu'}^\zeta \stackrel{d}{=} S_\nu^\zeta$.

Recall that $d_G(X;Y) = \inf_{X',Y'}\mathbb{E}|X'-Y'|$, where the infimum is taken over all random pairs (X',Y') such that $\mathcal{L}(X') = \mathcal{L}(X)$, $\mathcal{L}(Y') = \mathcal{L}(Y)$. Hence

$$d_G(S_{\nu_n}^\zeta;S_\nu^\zeta) \leq \mathbb{E}|S_{\nu_n'}^\zeta - S_{\nu'}^\zeta|.$$

Note that

$$\mathbb{E}\left|S_{\nu_n'}^\zeta - S_{\nu'}^\zeta\right| \leq \mathbb{E}\sum_{i=\nu'+1}^{\nu_n'}|\zeta_i|\mathbb{I}\{\nu_n'>\nu'\} + \mathbb{E}\sum_{i=\nu_n'+1}^{\nu'}|\zeta_i|\mathbb{I}\{\nu'>\nu_n'\} = \mathbb{E}|\zeta|\,\mathbb{E}|\nu_n'-\nu'|,$$

leading to (6.7). □

Proof of Proposition 6.4. Applying (6.6), we get

$$d_{TV}(\mathcal{L}(S_n);\tilde{\mathbf{\Pi}}) \leq \sum_{i=1}^n d_{TV}(\mathcal{L}(X_i);\mathbf{\Pi}(\mathbb{E}\nu_i;X_i')) \leq \sum_{i=1}^n d_{TV}(\nu_i;\pi_{\mathbb{E}\nu_i}).$$

Using (3.55), we derive

$$\sum_{i=1}^n d_{TV}(\nu_i;\pi_{\mathbb{E}\nu_i}) \leq \sum_{i=1}^n(\mathbb{E}\nu_i(\nu_i-1)+\mathbb{E}^2\nu_i),$$

and (6.14) follows. □

Proof of Theorem 6.5. Denote $\mathrm{e}(F_j) = \exp(F_j - I)$. Applying lemma B.3 with $s = 1$, we get

$$d_{TV}(\mathcal{L}(S_n); G_1') = \tfrac{1}{2}\|\mathcal{L}(S_n) - G_1'\|_{TV} \leq \frac{1}{2}\Big(\sum_{i=1}^{n} \|F_j - \mathrm{e}(F_j)\|_{TV}\Big)^2 . \tag{6.34}$$

Using (1.6*), we get

$$\begin{aligned}\|F_j - \mathrm{e}(F_j)\|_{TV} &\leq \|(F_j - I)^{*2}\|_{TV} \leq \|F_j - I\|_{TV}^2 \leq p_j^2\|\mathcal{L}(X_j') - I\|_{TV}^2 \\ &\leq p_j^2(\|\mathcal{L}(X_j')\|_{TV} + \|I\|_{TV})^2 \leq 4p_j^2 .\end{aligned}$$

Substituting this estimate into (6.34), we derive (6.21).

For the proof of (6.20), observe that

$$\begin{aligned}\|G_1' - G_1\|_{TV} &\leq \sum_{m=1}^{n} \Big\|(F_m - \mathrm{e}(F_m)) * (I - \mathrm{e}(F_m)) * \prod_{j\neq m}^{n} \mathrm{e}(F_j)\Big\|_{TV} \\ &\quad + \Big\|\prod_{j=1}^{*n} \mathrm{e}(F_j) * \sum_{m=1}^{n} (F_m - \mathrm{e}(F_m) + (F_m - I)^{*2}/2)\Big\|_{TV} \\ &\leq \frac{1}{2}\sum_{m=1}^{n} \|F_m - I\|_{TV}^3 + \frac{1}{6}\sum_{m=1}^{n} \|F_m - I\|_{TV}^3 \leq \frac{16}{3}\sum_{m=1}^{n} p_m^3 .\end{aligned}$$

An application of the triangle inequality completes the proof. □

Proof of Theorem 6.6. Denote $q_j = 1 - p_j$, $F_j' = \mathcal{L}(X_j')$,

$$W_{n,s} = \sum_{i=1}^{n} \sum_{j=s+1}^{\infty} (-1)^{j+1}\Big(\frac{p_i}{q_i}\Big)^j (\mathcal{L}(X_i')^{*j} - I)/j.$$

Since $\|F_j - I\|_{TV} = p_j\|F_j' - I\|_{TV} \leq 2p_j < 1$,

$$\ln F_j = \sum_{m=1}^{\infty} \frac{(-1)^{m+1}}{m!}(F_j - I)^{*m} = \sum_{m=1}^{\infty} \frac{(-1)^{m+1}p_j^m}{m!}(F_j' - I)^{*m}$$

and

$$\ln F_j = \ln(q_j(I + p_jF_j'/q_j)) = \sum_{m=1}^{\infty} \frac{(-1)^{m+1}}{m}\Big(\frac{p_i}{q_i}\Big)^m (F_j'^{*m} - I).$$

Therefore,

$$\prod_{j=1}^{*n} F_j = K_{n,s} * \exp(W_{n,s}), \quad K_{n,s} = \exp(-W_{n,s}) * \prod_{j=1}^{*n} F_j .$$

Due to (A.4), $2d_{TV}(\mathcal{L}(S_n); K_{n,s})$ equals

$$\begin{aligned}&\Big\|\prod_{j=1}^{*n} F_j - K_{n,s}\Big\|_{TV} = \Big\|\prod_{j=1}^{*n} F_j * (I - \exp(-W_{n,s}))\Big\|_{TV} \leq \|I - \exp(-W_{n,s})\|_{TV} \\ \leq\ &\Big\|\sum_{m=1}^{\infty} \frac{(-W_{n,s})^{*m}}{m!}\Big\|_{TV} \leq \sum_{m=1}^{\infty} \frac{\|W_{n,s}\|_{TV}^m}{m!} = \exp(\|W_{n,s}\|_{TV}) - 1\end{aligned}$$

and

$$\|W_{n,s}\|_{TV} \leq \sum_{i=1}^{n} \sum_{j=s+1}^{\infty} \frac{1}{j} \left(\frac{p_i}{1-p_i} \right)^j \|\mathcal{L}(X_i')^{*j} - I\|_{TV} \leq \frac{2}{s+1} \sum_{i=1}^{n} \sum_{j=s+1}^{\infty} \left(\frac{p_i}{1-p_i} \right)^j ,$$

which proves (6.25). The proof of (6.24) follows the same lines. □

Proof of Theorem 6.8. We use the blocks method, Lindeberg's device and Berbee's lemma. Denote

$$1\!\!\mathrm{I}_i = 1\!\!\mathrm{I}\{X_i > u_n\}, \ N_{r,j} = \sum_{i=jr+1}^{(j+1)r \wedge n} 1\!\!\mathrm{I}_i \qquad (0 \leq j \leq k).$$

Than $N_n = \sum_{j=0}^{k} N_{r,j}$. It is easy to see that

$$d_{TV}\left(N_n ; \sum_{j=0}^{k-1} N_{r,j} \right) \leq \mathbb{P}(N_{r,k} > 0) \leq r'p.$$

From each block $1\!\!\mathrm{I}_{jr+1}, ..., 1\!\!\mathrm{I}_{(j+1)r}$ we subtract a subblock $1\!\!\mathrm{I}_{(j+1)r-l+1}, ..., 1\!\!\mathrm{I}_{(j+1)r}$ of length l. Denote $(j < k)$

$$N_{r,j}^* = \sum_{i=jr+1}^{(j+1)r-l} 1\!\!\mathrm{I}_i , \quad N_n^* = \sum_{j=0}^{k-1} N_{r,j}^* .$$

Then $\mathbb{P}\left(\sum_{j=0}^{k-1} N_{r,j} \neq \sum_{j=0}^{k-1} N_{r,j}^* \right) \leq klp$. Therefore,

$$d_{TV}(N_n ; N_n^*) \leq r'p + klp.$$

Let $\{\hat{N}_{r,j}^*\}$ be independent copies of $N_{r,0}^*$. Put

$$S_i = \sum_{j=0}^{i-1} N_{r,j}^* + \sum_{j=i+1}^{k-1} \hat{N}_{r,j}^* \qquad (0 < i < k).$$

Notice that $S_j + \hat{N}_{r,j}^* = S_{j-1} + N_{r,j-1}^*$.

We apply Lindeberg's device in order to replace $\{N_{r,i}^*\}$ by $\{\hat{N}_{r,i}^*\}$:

$$\mathbb{P}\left(\sum_{j=0}^{k-1} N_{r,j}^* \in A \right) - \mathbb{P}\left(\sum_{j=0}^{k-1} \hat{N}_{r,j}^* \in A \right) = \sum_{j=1}^{k-1} \Big\{ \mathbb{P}(S_j + N_{r,j}^* \in A) - \mathbb{P}(S_j + \hat{N}_{r,j}^* \in A) \Big\}.$$

According to Berbee's lemma (see below), $\sum_{l=0}^{j-1} N_{r,l}^*$, $N_{r,j}^*$ and $\hat{N}_{r,j}^*$ can be defined on a CPS in such a way that $\mathbb{P}(N_{r,j}^* \neq \hat{N}_{r,j}^*) \leq \beta(l)$. Therefore,

$$\begin{aligned} & \mathbb{P}(S_j + N_{r,j}^* \in A) - \mathbb{P}(S_j + \hat{N}_{r,j}^* \in A) \\ = \ & \mathbb{E}\left[\mathbb{P}\left(\sum_{l=0}^{j-1} N_{r,l}^* + N_{r,j}^* \in A_j \right) - \mathbb{P}\left(\sum_{l=0}^{j-1} N_{r,l}^* + \hat{N}_{r,j}^* \in A_j \right) \right] \leq \beta(l), \end{aligned}$$

where $A_j = A - \sum_{i=j+1}^{k-1} \hat{N}_{r,i}^*$. Hence

$$\left| \mathbb{P}\left(\sum_{j=0}^{k-1} N_{r,j}^* \in A \right) - \mathbb{P}\left(\sum_{j=0}^{k-1} \hat{N}_{r,j}^* \in A \right) \right| \leq k\beta(l).$$

Lemma G.2 with $m=1$, $y=1$- and $K=r_+$ implies that $\beta(l)$ can be replaced by $4\alpha(l)\sqrt{r}$. Evidently, $\mathbb{P}(\sum_{j=0}^{k-1}\hat N_{r,j} \neq \sum_{j=0}^{k-1}\hat N^*_{r,j}) \le klp$. Thus,

$$\left|\mathbb{P}(N_n\in A) - \mathbb{P}(\hat N_n\in A)\right| \le k\gamma(l) + r'p + 2klp,$$

where $\hat N_n$ is a sum of k independent copies of N_r.

Denote $Z_m = \zeta_0 + ... + \zeta_m$,

$$\nu = \sum_{j=0}^{k-1} \mathbb{I}\{\hat N_{r,j} > 0\}.$$

By Khintchine's formula (5.4),

$$\hat N_n \stackrel{d}{=} Z_\nu\,, \tag{6.35}$$

where $\mathcal{L}(\nu) = \mathbf{B}(k,q)$. Using Lemma 6.1 and an estimate of the total variation distance between the Binomial and Poisson distributions, we derive (6.27).

Note that $\mathbb{E}|N_n - N^*_n| \le r'p + klp$. According to Lemmas G.1 and G.2, $\sum_{l=0}^{j-1} N^*_{r,l}$, $N^*_{r,j}$ and $\hat N^*_{r,j}$ can be defined on a CPS in such a way that

$$\mathbb{P}(N^*_{r,j} \neq \hat N^*_{r,j}) \le \min\{4\alpha(l)\sqrt{r}\,;\beta(l)\}.$$

Since

$$\mathbb{E}|h(S_j + N^*_{r,j}) - h(S_j + \hat N^*_{r,j})| \le \mathbb{E}|N^*_{r,j} - \hat N^*_{r,j}|\mathbb{I}\{N^*_{r,j} \neq \hat N^*_{r,j}\} \le r\mathbb{P}(N^*_{r,j} \neq \hat N^*_{r,j}),$$

we have $\mathbb{E}|h(N^*_n) - h(\sum_{j=0}^{k-1}\hat N^*_{r,j})| \le n\gamma_n(l)$. Therefore,

$$d_G(N_n; \hat N_n) \le r'p + 2klp + n\gamma_n(l).$$

Taking into account (6.7), (6.35), and (3.24), we derive (6.28). □

Proof of Theorem 6.9. Given a random variable X taking values in $\mathbb{Z}_+$ with a finite positive first moment, recall that X^* denotes a random variable with the distribution

$$\mathbb{P}(X^* = m) = (m+1)\mathbb{P}(X = m+1)/\mathbb{E}X \qquad (m \ge 1),$$

see (3.56). In particular,

$$\mathbb{E}Xf(X) = \mathbb{E}X\mathbb{E}f(X^*+1) \tag{6.36}$$

for any function f such that $\mathbb{E}|Xf(X)| < \infty$. If $\mathbb{E}X^2 < \infty$, then (6.36) with $f(x) = x-1$ yields

$$\mathbb{E}X^* = \mathbb{E}X(X-1)/\mathbb{E}X.$$

Set $\mu_j = \lambda^{-1}\lambda_j$. Let Y be a r.v. such that $\mathbb{P}(Y=j) = \mu_j$, and set

$$Y' = Y^* + 1.$$

Then $\mathbb{P}(Y'=m) = m\mathbb{P}(Y=m)/\mathbb{E}Y$.

Denote by g a solution to the Stein equation (F.12):

$$\sum_{j\ge1} j\lambda_j g(j+k) - kg(k) = f(k) - \mathbb{E}f(Y) \qquad (k \in \mathbb{Z}_+).$$

Note that $g(0) = 0,\ g(j) = \Delta g(0) + \cdots + \Delta g(j-1)$. Then

$$\begin{aligned}
&\lambda \mathbb{E} Y g(Y+S_n) - \mathbb{E} S_n g(S_n) \\
&\quad= \lambda \mathbb{E} Y \mathbb{E} g(S_n+Y') - \mathbb{E} S_n \mathbb{E} g(S_n') = \mathbb{E} S_n \mathbb{E}(g(S_n+Y') - g(S_n')) \\
&\quad= \mathbb{E} S_n \sum_{j=1}^{\infty} g(j)(\mathbb{P}(S_n + Y' = j) - \mathbb{P}(S_n' = j)) \\
&\quad= \mathbb{E} S_n \sum_{j=1}^{\infty} \sum_{k=0}^{j-1} \Delta g(k)(\mathbb{P}(S_n + Y' = j) - \mathbb{P}(S_n' = j)) \\
&\quad= \mathbb{E} S_n \sum_{k=0}^{\infty} \Delta g(k)(\mathbb{P}(S_n + Y' \geq k+1) - \mathbb{P}(S_n' \geq k+1)).
\end{aligned}$$

Hence

$$\begin{aligned}
d_{TV}(S_n; \tilde{S}_n) &\leq \|\Delta g\|(\mathbb{E} S_n) \sum_{k=0}^{\infty} |\mathbb{P}(S_n + Y' > k) - \mathbb{P}(S_n' > k)| \\
&\leq \mathcal{M}(\lambda)(\mathbb{E} S_n) \sum_{k=0}^{\infty} |\mathbb{P}(S_n + Y' > k) - \mathbb{P}(S_n' > k)|. \qquad (6.37)
\end{aligned}$$

To prove the first statement of Theorem 6.9 we need the following lemma.

Lemma 6.13 [99] *Suppose that $X_1, \ldots, X_n$ are negatively associated r.v.s, and let Y' be as above. Then for any non-decreasing function f*

$$\mathbb{E} f(S_n') \leq E f(Y' + S_n).$$

Since $\mathbb{I}\{j > k\}$ is non-decreasing in j,

$$\begin{aligned}
&|\mathbb{P}(S_n + Y' > k) - \mathbb{P}(S_n' > k)| = |\mathbb{E}\mathbb{I}\{S_n + Y' > k\} - \mathbb{E}\mathbb{I}\{S_n' > k\}| \\
&= \mathbb{E}\mathbb{I}\{S_n + Y' > k\} - \mathbb{E}\mathbb{I}\{S_n' > k\} = \mathbb{P}(S_n + Y' > k) - \mathbb{P}(S_n' > k)
\end{aligned}$$

by Lemma 6.13. Changing the order of summation, we get

$$\begin{aligned}
&\sum_{k=0}^{\infty} |\mathbb{P}(S_n + Y' > k) - \mathbb{P}(S_n' > k)| = \sum_{k=0}^{\infty} (\mathbb{P}(S_n + Y' > k) - \mathbb{P}(S_n' > k)) \\
&\quad= \sum_{k=1}^{\infty} (\mathbb{P}(S_n + Y' \geq k) - \mathbb{P}(S_n' \geq k)) = \sum_{k=1}^{\infty} \sum_{j=k}^{\infty} (\mathbb{P}(S_n + Y' = j) - \mathbb{P}(S_n' = j)) \\
&\quad= \sum_{j=1}^{\infty} j(\mathbb{P}(S_n + Y' = j) - \mathbb{P}(S_n' = j)) = \mathbb{E}(S_n + Y') - \mathbb{E} S_n' = \mathbb{E}(Y' + S_n - S_n').
\end{aligned}$$

Therefore,

$$d_{TV}(S_n; \tilde{S}_n) \leq \mathcal{M}(\lambda) \mathbb{E}(Y' + S_n - S_n') \mathbb{E} S_n. \qquad (6.38)$$

Observe that

$$\mathbb{E} S_n = \sum_{j \geq 1} j \lambda_j = \lambda \mathbb{E} Y, \quad \mathbb{E} \sum_{j=1}^{n} X_j (X_j + \hat{X}_j) = \sum_{j \geq 1} j^2 \lambda_j = \lambda \mathbb{E} Y^2.$$

It follows from (6.36) that $\mathbb{E}S_n' = \mathbb{E}S_n^2/\mathbb{E}S_n$,

$$\mathbb{E}Y' = \frac{\mathbb{E}Y^2}{\mathbb{E}Y} = \frac{\sum_{i\ge1} i^2\lambda_i}{\sum_{i\ge1} i\lambda_i} = \frac{\mathbb{E}\sum_{j=1}^n X_j(X_j+\hat{X}_j)}{\mathbb{E}S_n},$$

where $\hat{X}$ is defined by (6.30). Substituting these estimates into (6.38), we get

$$d_{TV}(S_n,\tilde{S}_n) \le \mathcal{M}(\lambda)\Big(\mathbb{E}\sum_{j=1}^n X_j(X_j+\hat{X}_j) - \operatorname{var} S_n\Big),$$

which proves the first statement of Theorem 6.9.

For the proof of the second statement we use the following lemma.

Lemma 6.14 [99] *Suppose that $X_1,\dots,X_n$ are associated r.v.s. Let V be a random variable independent of all other r.v.s such that $\mathbb{P}(V=i) = \mathbb{E}X_i/\mathbb{E}S_n$ for all $i=1,\dots,n$. Then for all non-decreasing functions f*

$$\mathbb{E}f(S_n^*) \ge \mathbb{E}f(Y^* - S_n - X_V - \hat{X}_V).$$

Since $\mathbb{I}\{j>k\}$ is non-decreasing in j, arguing similarly to the proof of the first statement of the theorem and applying Lemma 6.14, we get

$$|\mathbb{P}(Y^*+S_n-X_V-\hat{X}_V>k)-\mathbb{P}(S_n^*>k)| = \mathbb{P}(S_n^*>k) - \mathbb{P}(Y^*+S_n-X_V-\hat{X}_V>k).$$

Taking into account that X_V and $\hat{X}_V$ are non-negative r.v.s, we obtain from (6.37) that

$$\begin{aligned}
d_{TV}(S_n;\tilde{S}_n) &\le \mathcal{M}(\lambda)(\mathbb{E}S_n)\Big\{\sum_{k=0}^\infty |\mathbb{P}(Y^*+S_n>k) - \mathbb{P}(Y^*+S_n-X_V-\hat{X}_V>k)| \\
&+ \sum_{k=0}^\infty |\mathbb{P}(Y^*+S_n-X_V-\hat{X}_V>k) - \mathbb{P}(S_n^*>k)|\Big\} \\
&= \mathcal{M}(\lambda)(\mathbb{E}S_n)\sum_{k=0}^\infty \big(\mathbb{P}(Y^*+S_n>k) - 2\mathbb{P}(Y^*+S_n-X_V-\hat{X}_V>k) \\
&+ \mathbb{P}(S_n^*>k)\big) \\
&= \mathcal{M}(\lambda)(\mathbb{E}S_n)(\mathbb{E}(Y^*+S_n) - 2\mathbb{E}(Y^*+S_n-X_V-\hat{X}_V) + \mathbb{E}S_n^*) \\
&= \mathcal{M}(\lambda)(\mathbb{E}S_n)(2\mathbb{E}(X_V+\mathbb{E}\hat{X}_V) + \mathbb{E}S_n^* - \mathbb{E}S_n - \mathbb{E}Y^*) \\
&= \mathcal{M}(\lambda)\Big\{2(\mathbb{E}S_n)\mathbb{E}(X_V+\hat{X}_V) + \operatorname{var} S_n - \mathbb{E}\sum_{i=1}^n X_i(X_i+\hat{X}_i)\Big\}. \qquad (6.39)
\end{aligned}$$

By the definition of V and $\hat{X}_j$ we have

$$(\mathbb{E}S_n)\mathbb{E}(X_V+\hat{X}_V) = \sum_{i=1}^n \mathbb{E}^2X_i + \sum_{i=1}^n\sum_{j\in\mathcal{I}(i)} \mathbb{E}X_j\mathbb{E}X_i = \sum_{i=1}^n\sum_{j\in\mathcal{I}(i)\cup\{i\}} \mathbb{E}X_j\mathbb{E}X_i$$

and

$$\mathbb{E}\sum_{i=1}^n X_i(X_i+\hat{X}_i) = \sum_{i=1}^n \mathbb{E}X_i^2 + \sum_{i=1}^n\sum_{j\in\mathcal{I}(i)} \mathbb{E}X_jX_i = \sum_{i=1}^n\sum_{j\in\mathcal{I}(i)\cup\{i\}} \mathbb{E}X_jX_i.$$

Substituting these estimates into (6.39) completes the proof of the second statement of Theorem 6.9. □

Exercises

6.1. Prove Lemma 6.1.

6.2. Prove (6.24).

6.3. Let $\mathcal{L}(X_i) = (1-p_i)I + p_i\mathcal{L}(X_i')$, where r.v.s $\{X_i'\}$ take values in $\mathbb{Z}_+$ and have finite means. Assume that $p^* \leq 1/4$. Prove that for any fixed natural number s there exists an absolute constant C such that

$$d_K(\mathcal{L}(S_n); H_{n,s}) \leq C(s+1)^{-1}Q_{\boldsymbol{\Pi}'}(1/\pi)\sum_{i=1}^{n}\mathbb{E}X_i(2p_i)^{s+1},$$

where $\boldsymbol{\Pi}'$ is a compound Poisson distribution with characteristic function

$$\widehat{\boldsymbol{\Pi}}'(t) = \exp\left(\frac{1}{2}\sum_{i=1}^{n}p_i(Ref_i(t)-1)\right),$$

$Ref_i(t)$ denotes the real part of the characteristic function of $\mathcal{L}(X_i')$ and

$$H_{n,s} = \exp\left(\sum_{i=1}^{n}\sum_{j=1}^{s}p_i^j(-1)^{j+1}(\mathcal{L}(X_i')-I)^{*j}/j\right).$$

Recall that $Q_{\boldsymbol{\Pi}'}$ denotes the concentration function.

6.4. Consider the sum $S_n = X_1 + \cdots + X_n$ of independent r.v.s, where distribution of each summand is a mixture of compound Poisson distributions:

$$\mathcal{L}(X_i) = \frac{1}{2}\exp(\alpha_i(F_{1i}-I)) + \frac{1}{2}\exp(\beta_i(F_{i2}-I)),$$

F_{1i}, F_{2i} $(i = 1, 2, \ldots, n)$ are arbitrary distributions and $\alpha_i, \beta_i > 0$.

One would like to approximate $\mathcal{L}(S_n)$ by a compound Poisson distribution $\mathcal{L}(Z_n^Y)$, where $Z_n^Y = Y_1 + \cdots + Y_n$, $\{Y_i\}$ are independent compound Poisson r.v.s,

$$\mathcal{L}(Y_i) = \exp\left(\frac{\alpha_i}{2}(F_{1i} - I) + \frac{\beta_i}{2}(F_{2i} - I)\right).$$

Prove that

$$d_{TV}(S_n; Z_n^Y) \leq \frac{3}{2}\sum_{i=1}^{n}(\alpha_i^2 + \beta_i^2).$$

Notes

Inequality (6.6) in Lemma 6.1 is formula (30) in Kolmogorov [179].
Lemma 6.2 is from [221], Lemma 5.4.
Theorem 6.5 is from [74], Theorem 3.1 and Corollary 3.1.
Theorem 6.6 has been proved by Hipp [152].
Theorem 6.8 is Theorem 5.2 in [221].
Theorem 6.9 is from Daly [99]. Theorem 6.4 is a partial case of Proposition 2.1 from [74].

7

Accuracy of CP approximation: discrete random variables

In this chapter, we deal with discrete random variables that are no longer assumed to take value 0 with "large" probability.

7.1 Independent Bernoulli random variables

Let $X, X_1, ..., X_n$ be independent Bernoulli $\mathbf{B}(p)$ r.v.s, $S_n = X_1 + ... + X_n$.

Though Poisson distribution is a natural proxy to $\mathcal{L}(S_n) = \mathbf{B}(n,p)$, compound Poisson approximation can be more accurate: Presman [237] has shown that

$$\sup_{n^{-2/3} \le p \le 1/2} d_{TV}(\mathbf{B}(n,p); P_{n,p}) = O(n^{-2/3}), \tag{7.1}$$

where $P_{n,p}$ is a shifted compound Poisson distribution (the approximating distribution $P_{n,p}$ is known explicitly; if $p \le n^{-2/3}$, the Binomial distribution $\mathbf{B}(n,p)$ is well approximated by the Poisson $\mathbf{\Pi}(np)$ law, see Chapter 3). A similar statement in terms of d_K is due to Meshalkin [206].

The Meshalkin–Presman result demonstrates what rate of the accuracy of approximation one can expect to achieve in the general case (cf. Kolmogorov's problem, see (10.3)).

We present Presman's result in Theorem 7.1.

Denote by $\lceil x \rceil$ the integer number that is the nearest to x from above, and let

$$\gamma = \lceil 3np^2 - 2np^3 \rceil, \quad \beta = \gamma - 3np^2 + 2np^3, \quad q = 1-p.$$

Note that $\beta \in [0;1)$.

Let η_1, η_2, η_3 be independent r.v.s with distributions

$$\mathcal{L}(\eta_1) = \mathbf{\Pi}(pq^2 - \beta/n), \;\; \mathcal{L}(\eta_2) = \mathbf{\Pi}(p^2q + \beta/3n), \;\; \mathcal{L}(\eta_3) = \mathbf{\Pi}(\beta/6n).$$

Set

$$Y = \gamma/n + \eta_1 - \eta_2 + 2\eta_3 .$$

Note that $Y - \gamma/n$ is a compound Poisson random variable. One can check that

$$\mathbb{E}Y = p, \;\; \mathbb{E}(Y-p)^2 = pq, \;\; \mathbb{E}(Y-p)^3 = pq(q-p),$$

matching the first three moments of $X - p$.

Let $P_{n,p} = \mathcal{L}(Y_1 + ... + Y_n)$, where $\{Y_i\}$ are independent copies of Y. Thus,

$$P_{n,p} = \mathcal{L}(\gamma + \pi_1 - \pi_2 + 2\pi_3),$$

DOI: 10.1201/9781003478164-7

where π_1, π_2, π_3 are independent r.v.s with distributions

$$\mathcal{L}(\pi_1) = \mathbf{\Pi}(npq^2 - \beta), \quad \mathcal{L}(\pi_2) = \mathbf{\Pi}(np^2q + \beta/3), \quad \mathcal{L}(\pi_3) = \mathbf{\Pi}(\beta/6).$$

Clearly, it suffices considering only $p \in [0; 1/2]$: if $\mathcal{L}(S_n) = \mathbf{B}(n,p)$, then $\mathcal{L}(n - S_n) = \mathbf{B}(n, 1-p)$.

Theorem 7.1 [237] *There exists an absolute constant* C *such that*

$$d_{TV}(\mathbf{B}(n,p); \mathbf{\Pi}(np)) \wedge d_{TV}(\mathbf{B}(n,p); P_{n,p}) \le C\varepsilon_{n,p} \qquad (0 \le p \le 1/2), \tag{7.2}$$

where $\varepsilon_{n,p} = \min\{np^2; p; \max\{1/(np)^2; 1/n\}\}$.

Bound (7.1) follows after noticing that

$$\sup_{0 \le p \le 1/2} \varepsilon_{n,p} = O(n^{-2/3}).$$

Note that $\varepsilon_{n,p} = O(n^{-1})$ if $p \ge 1/\sqrt{n}$.

Presman [237] has shown also that it is impossible to construct a compound Poisson or infinitely divisible distribution approximating $\mathbf{B}(n,p)$ with the accuracy better than $\varepsilon_{n,p}$: there exists an absolute constant $c > 0$ such that for any infinitely divisible distribution P

$$d_{TV}(\mathbf{B}(n,p); P) \ge d_K(\mathbf{B}(n,p); P) \ge c\varepsilon_{n,p} \qquad (1/n \le p \le 1/2). \tag{7.3}$$

Non-i.i.d. Bernoulli r.v.s.

Presman's result can be extended to the case of independent non-identically distributed Bernoulli r.v.s.

Assume that $\mathcal{L}(X_i) = \mathbf{B}(p_i)$, where $0 < p_i \le c_0 < 1$ $(\exists c_0 \in (0,1),\ i = 1, 2, \ldots, n)$. Let

$$\bar{\gamma} = \left\lceil (3p_1^2 - 2p_1^3) + \ldots + (3p_n^2 - 2p_n^3) \right\rceil, \quad \bar{\beta} = \bar{\gamma} - \sum_{i=1}^{n} (3p_i^2 - 2p_i^3).$$

Then $\bar{\beta} \in [0; 1)$. Denote $q_i = 1 - p_i$, and let $\bar{\eta}_{1i}, \bar{\eta}_{2i}, \bar{\eta}_{3i}$ be independent r.v.s with distributions

$$\mathcal{L}(\bar{\eta}_{1i}) = \mathbf{\Pi}(p_i q_i^2 - \bar{\beta}/n), \ \mathcal{L}(\bar{\eta}_{2i}) = \mathbf{\Pi}(p_i^2 q_i + \bar{\beta}/3n), \ \mathcal{L}(\bar{\eta}_{3i}) = \mathbf{\Pi}(\bar{\beta}/6n).$$

Set

$$\bar{Y}_i = \bar{\gamma}/n + \bar{\eta}_{1i} - \bar{\eta}_{2i} + 2\bar{\eta}_{3i} \quad (i = 1, 2, \ldots, n), \quad S_n^{\bar{Y}} = \bar{Y}_1 + \cdots + \bar{Y}_n.$$

Then $S_n^{\bar{Y}} - \bar{\gamma}$ is a compound Poisson r.v.,

$$\mathbb{E}\bar{Y}_i = p_i, \quad \mathbb{E}(\bar{Y}_i - p_i)^2 = p_i q_i, \quad \mathbb{E}(\bar{Y}_i - p_i)^3 = p_i q_i (q_i - p_i).$$

Denote

$$\sigma^2 = \sum_{i=1}^{n} p_i q_i, \quad \lambda = \sum_{i=1}^{n} p_i, \quad \lambda_k = \sum_{i=1}^{n} p_i^k \qquad (k \in \mathbb{N}). \tag{7.4}$$

As usual, C denotes (possibly different) absolute positive constants.

Theorem 7.2 *If* $\sum_{i=1}^n p_i q_i^2 \ge 1$, $0 < p_i \le \exists c_0 < 1$ $(i = 1, 2, \ldots, n)$, *then*

$$d_o(\mathcal{L}(S_n); \mathcal{L}(S_n^{\bar{Y}})) \le C(\lambda_2 + 1)/\lambda^2\sqrt{\lambda}, \tag{7.5}$$

$$d_{TV}(\mathcal{L}(S_n); \mathcal{L}(S_n^{\bar{Y}})) \le C(\lambda_2 + 1)/\lambda^2, \tag{7.6}$$

$$d_G(\mathcal{L}(S_n); \mathcal{L}(S_n^{\bar{Y}})) \le C(\lambda_2 + 1)/\lambda\sqrt{\lambda}. \tag{7.7}$$

Moreover, for any $k\in\mathbb{Z}$

$$(1+(k-\lambda)^2/\lambda)|\mathbb{P}(S_n=k)-\mathbb{P}(S_n^{\bar{Y}}=k)| \;\le\; C(\lambda_2+1)/\lambda^2\sqrt{\lambda}, \tag{7.8}$$
$$(1+(k-\lambda)^2/\lambda)|\mathbb{P}(S_n\le k)-\mathbb{P}(S_n^{\bar{Y}}\le k)| \;\le\; C(\lambda_2+1)/\lambda^2. \tag{7.9}$$

Recall that $\sum_{k\in\mathbb{Z}}\left(1+(k-a)^2/b^2\right)^{-1}\le 1+b\pi$ $(a\in\mathbb{R}, b>0)$ by (I.17). Set $b=\sqrt{\lambda}$. Estimates (7.6) and (7.7) follow from (7.8) and (7.9).

If $p_i=p$ $(\forall i)$ and $p\ge 1/n$, then (7.5)–(7.7) become

$$d_{\rm o}(\mathcal{L}(S_n);\mathcal{L}(S_n^{\bar{Y}})) \;\le\; C(p(np)^{-3/2}+(np)^{-5/2}) \tag{7.5*}$$
$$d_{TV}(\mathcal{L}(S_n);\mathcal{L}(S_n^{\bar{Y}})) \;\le\; C(1/n+1/(np)^2), \tag{7.6*}$$
$$d_G(\mathcal{L}(S_n);\mathcal{L}(S_n^{\bar{Y}})) \;\le\; C(\sqrt{p/n}+1/np\sqrt{np}). \tag{7.7*}$$

Recall that

$$d_{\rm o}(S_n;\pi_\lambda)\le np^2\min\{\sqrt{e/\pi}\,(npq)^{-3/2};1\}$$

by (3.3). It follows from (3.3) and (7.5*) that

$$\sup_{0\le p\le 1/2} d_{\rm o}(\mathbf{B}(n,p);\mathbb{CP})\le Cn^{-5/6}\,, \tag{7.10}$$

where $\mathbb{CP}$ is the class of (shifted) compound Poisson distributions.

Using (3.24) if $p\le n^{-2/3}$ and (7.7*) if $n^{-2/3}<p\le 1/2$, we get

$$\sup_{0\le p\le 1/2} d_G(\mathbf{B}(n,p);\mathbb{CP})\le Cn^{-1/2}\,. \tag{7.11}$$

Thus, the rate of the accuracy of uniform in p compound Poisson approximation in terms of $d_{\rm o}$, d_{TV}, d_G is n to the power of $-5/6$, $-4/6$, $-3/6$.

Symmetrised Bernoulli random variables.

Consider independent Bernoulli $\mathbf{B}(p_i)$ random variables $\{X_i\}$, where $p_i\in[0,1]$, $q_i=1-p_i$ $(i=1,\ldots,n)$. Let X_j' denote an independent copy of X_j, and set

$$S_n^\star=(X_1-X_1')+\ldots+(X_n-X_n').$$

The characteristic function of $\mathcal{L}(X_j-X_j')$ is $1-2p_jq_j\sin^2(t/2)\ge 0$. The characteristic function of $S_n^\star$ is

$$\mathbb{E}e^{itS_n^\star}=\prod_{j=1}^n\left(1-2p_jq_j\sin^2(t/2)\right)\ge 0.$$

It is natural to approximate $\mathcal{L}(S_n^\star)$ by a symmetrised Poisson distribution.

Let us consider the sum $\tilde{S}_n^\star$ of accompanying r.v.s, see (1.9):

$$\tilde{S}_n^\star\stackrel{d}{=}\pi_{\sigma^2}-\pi_{\sigma^2}'\,,$$

where π_{σ^2} and π_{σ^2}' are independent Poisson $\mathbf{\Pi}(\sigma^2)$ random variables, $\sigma^2=\sum_{i=1}^n p_iq_i$. Obviously, $\tilde{S}_n^\star$ is a Skellam compound Poisson r.v. with characteristic function

$$\mathbb{E}e^{it\tilde{S}_n^\star}=\exp(2\sigma^2(\cos t-1)).$$

Presman [238] has proved that

$$d_{TV}(S_n^\star;\tilde{S}_n^\star)\le\min\left\{0.7225\sum_{j=1}^n p_j^2q_j^2/(\sigma^2-p_jq_j)^2;\,4\sum_{j=1}^n(p_jq_j)^2\right\}. \tag{7.12}$$

If all $\{p_j\}$ are uniformly bounded away from 0 and 1, then the RHS of (7.12) is $O(1/n)$ as $n\to\infty$. The same is true if $p_j = p$ $(\forall j)$.

A multivariate version of (7.12) has been given by Kruopis & Čekanavičius [187].

If $p_i \equiv p \le 1/2$, then (Presman [237])

$$C_1\varepsilon^*_{n,p} \le d_{K}(S^\star_n, \tilde S^\star_n) \le d_{TV}(S^\star_n, \tilde S^\star_n) \le C_2\varepsilon^*_{n,p}\,, \tag{7.13}$$

where $\varepsilon^*_{n,p} = \min\{np^2; n^{-1}\}$. Thus, in the case of symmetrised r.v.s one can expect that the correct rate of the accuracy of compound Poisson approximation is n^{-1}.

The following gives an idea why the accuracy of approximation is much better in the case of symmetric distributions.

The behaviour of characteristic functions is usually a good indicator on the possible rate of the accuracy of approximation. For an arbitrary a we have

$$\begin{aligned}(1+a)^n - \mathrm{e}^{na} &= \mathrm{e}^{n\ln(1+a)} - \mathrm{e}^{na} = \mathrm{e}^{na-na^2/2+\dots} - \mathrm{e}^{na}\\ &= \mathrm{e}^{na}\left(1 - \frac{na^2}{2} + \dots\right) - \mathrm{e}^{na} = -\frac{na^2}{2}\,\mathrm{e}^{na} + \dots\end{aligned}$$

Thus, the accuracy of approximation is controlled by $na^2\mathrm{e}^{na}$.

Let us consider the approximation of Binomial distribution $\mathbf{B}(n;p)$ by Poisson distribution $\mathbf{\Pi}(np)$. The ch.f. of $\mathbf{B}(p)$ is $1+a$, where $a = p(\mathrm{e}^{it}-1)$, the ch.f. of $\mathbf{\Pi}(p)$ is e^a. Clearly, $|a| = |p(\mathrm{e}^{it}-1)| = 2p|\sin(t/2)|$.

Note that

$$\left|na^2\,\mathrm{e}^{na}\right| \le 4np^2\sin^2(t/2)\mathrm{e}^{-2np\sin^2(t/2)} = 2py\mathrm{e}^{-y}\,,$$

where $y = 2np\sin^2(t/2)$. Since $\sup_y y\mathrm{e}^{-y} = 1/\mathrm{e}$, the rate of the accuracy of pure Poisson approximation to the Binomial distribution in terms of d_K should be $O(p)$. This is true indeed, see Section 3.1.

Next, consider the Skellam approximation to a symmetrised Binomial distribution. The ch.f. of a symmetrised Bernoulli distribution is $1+a$, where $a = -4pq\sin^2(t/2)$. Clearly,

$$\left|na^2\,\mathrm{e}^{na}\right| = 16np^2q^2\sin^4(t/2)\mathrm{e}^{-4npq\sin^2(t/2)} = n^{-1}y^2\mathrm{e}^{-y}\,,$$

where $y = 4npq\sin^2(t/2)$. Note that $\sup_y y^2\mathrm{e}^{-y} = 4\mathrm{e}^{-2}$. Hence $|na^2\,\mathrm{e}^{na}| \le 4/n\mathrm{e}^2$, which is in line with Presman's estimate (7.12).

SCP approximation.

Let $\mathcal{L}(X_i) = \mathbf{B}(p_i)$ $(i=1,\dots,n)$, $q_j = 1-p_j$, $\theta = \lambda_2/\lambda$, cf. (7.4). Denote by

$$G_2 = \mathcal{L}(\pi_{\lambda+\lambda_2} + 2\pi_{-\lambda_2/2}) \tag{7.14}$$

the "distribution" of $\pi_{\lambda+\lambda_2} + 2\pi_{-\lambda_2/2}$, where $\pi_{\lambda+\lambda_2}$ and $\pi_{-\lambda_2/2}$ are independent "r.v.s", i.e., G_2 is a convolution of $\mathbf{\Pi}(\lambda+\lambda_2)$ and a compound Poisson unit measure $\mathbf{\Pi}(-\lambda_2/2, 2)$:

$$G_2 = \exp\bigl(\lambda(I_1 - I) - \lambda_2(I_1 - I)^{*2}/2\bigr)\,.$$

Presman [237] approximated the Binomial distribution by G_2 in terms of the total variation distance and has derived an estimate of order $\min\{np^3; p^{3/2}n^{-1/2}\}$.

Kruopis [185] has extended Presman's result to the case of non-identically distributed Bernoulli r.v.s. In the case of $\mathbf{B}(n,p)$ his result states that

$$d_{TV}(\mathbf{B}(n,p); G_2) \le 5\sqrt{\mathrm{e}}\,\lambda_3\min\{1.2\sigma^{-3} + 4.2\lambda_2\sigma^{-6}; 2 + \sigma^2 + 3.4\lambda_2\}. \tag{7.15}$$

Kruopis ([185], Theorem 7) has also derived a lower bound. Thus,

$$C_3 \min\left\{\frac{p\sqrt{p}}{\sqrt{n}}; np^3\right\} \le d_K(\mathbf{B}(n,p); G_2) \le d_{TV}(\mathbf{B}(n,p); G_2) \le C_4 \min\left\{\frac{p\sqrt{p}}{\sqrt{n}}; np^3\right\}. \quad (7.16)$$

Constants in (7.15) have been improved by Barbour & Xia [25] under the assumption that $\theta<1/2$. The following result is due to Zacharovas & Hwang [321].

Theorem 7.3 [321] If $\theta<1$, then for any $m\in\mathbb{N}$

$$d_{TV}(\mathcal{L}(S_n); G_2) \le \frac{\lambda_3}{2\lambda^{3/2}}\left(\frac{\sqrt{6}C_+}{(1-\theta)^2} + \frac{\sqrt{3\theta}}{2\sqrt{2}(1-\theta)^{5/2}}\right),$$

$$d_G(\mathcal{L}(S_n); G_2) \le \frac{\lambda_3}{\lambda}\left(\frac{\sqrt{2}C_+}{(1-\theta)^{3/2}} + \frac{\sqrt{3\theta}}{4\sqrt{2}(1-\theta)^2}\right),$$

$$|\mathbb{P}(S_n\le m) - G_2((-\infty,m])| \le \frac{\lambda_3}{\lambda^{3/2}}\left(\frac{\sqrt{6}C_+}{(1-\theta)^2} + \frac{\sqrt{3\theta}}{2\sqrt{2}(1-\theta)^{5/2}}\right) g_\lambda(m),$$

$$|\mathbb{P}(S_n=m) - G_2(\{m\})| \le \frac{\lambda_3}{\lambda^2}\left(\frac{2\sqrt{6}C_+}{(1-\theta)^{5/2}} + \frac{\sqrt{15\theta}}{2\sqrt{2}(1-\theta)^6}\right) g_\lambda(m),$$

where $C_+ = 0.3706,\ g_\lambda(m) = \mathrm{e}^{-(m-\lambda)^2/4(m+\lambda)}$.

The rate of the accuracy of approximation by G_2 is sharper than that of pure Poisson approximation. In some cases the rate of the RHS of (7.16) is sharper than that of (7.2) (consider, for example, $p = n^{-2/3}$). On the other hand, in some cases (e.g., $p = 1/2$), Presman's approximation (7.2) is more accurate.

Does there exist a SCP approximation that always provides a sharper rate of the accuracy of approximation than any compound Poisson approximation? The answer is affirmative for "moderately small" $\{p_i\}$.

Consider the Bernoulli distributions $\mathbf{B}(p_j)$ $(1\le j\le 1)$ and let

$$H^*_{n,k} = \exp\left(\sum_{i=1}^{n}\sum_{j=1}^{k}(-1)^{j+1}p_i^j(I_1-I)^{*j}/j\right) = \exp\left(\sum_{j=1}^{k}(-1)^{j+1}(I_1-I)^{*j}\lambda_j/j\right),$$

where $k\in\mathbb{N}$, $\lambda_j = \sum_{i=1}^n p_i^j$.

$H^*_{n,k}$ can be viewed as a "distribution" of the sum $\sum_{m=1}^k m\pi_{\lambda_{m,k}}$ of k independent Poisson "random variables", where

$$\lambda_{m,k} = \sum_{i=1}^{n}\sum_{j=m}^{k}\frac{(-1)^{2j-m+1}p_i^j}{j}\binom{j}{m} = (-1)^{m+1}\sum_{j=m}^{k}\frac{\lambda_j}{j}\binom{j}{m}. \quad (7.17)$$

Note that

$$H^*_{n,1} = \mathbf{\Pi}(\lambda), \quad H^*_{n,2} = \exp\left(\lambda(I_1-I) - \lambda_2(I_1-I)^{*2}/2\right).$$

Assume that $\max_{i\le n} p_i \le 1/4$. It follows from bound (12.13) by Roos that

$$d_{TV}(\mathcal{L}(S_n); H^*_{n,k}) \le C_k\lambda_{k+1}\min\{1; \lambda^{-(k+1)/2}\}. \quad (7.18)$$

If $p_i \equiv p$, then

$$d_{TV}(\mathbf{B}(n,p), H^*_{n,k}) \le C_k \min\left\{np^{k+1}; p^{(k+1)/2}n^{-(k-1)/2}\right\}. \quad (7.19)$$

Thus, if $p \le 1/4$, then there exists a SCP measure, which can approximate the Binomial distribution with the accuracy $O(n^{-(k-1)/2})$. In particular, for any $p \le 1/4$ the accuracy of approximation by $H^*_{n,3}$ is $O(n^{-1})$. According to (7.17), $H^*_{n,3}$ can be viewed as "distribution" of independent "random variables" $\pi_{\lambda_1+\lambda_2+\lambda_3} + 2\pi_{-\lambda_2/2-\lambda_3} + 3\pi_{\lambda_3/3}$.

Let

$$H_m(x) = (-1)^m \mathrm{e}^{x^2/2} \frac{\mathrm{d}^m}{\mathrm{d}x^m} \mathrm{e}^{-x^2/2} \qquad (m \in \mathbb{Z}_+,\ x \in \mathbb{R})$$

denote the Hermite polynomial. In particular, $H_0(x) = 1$, $H_1(x) = x$.

It is easy to check that $H^*_{n,2}$ can be expressed through Hermite polynomials:

$$H^*_{n,2}(m) = \mathrm{e}^{-\lambda-\lambda_2/2} \frac{\lambda_2^{m/2}}{m!} H_m\left(\frac{\lambda+\lambda_2}{\sqrt{\lambda_2}}\right).$$

$H^*_{n,2}$ satisfies the recurrent formula

$$(\lambda_1+\lambda_2)H^*_{n,2}(m-1) - \lambda_2 H^*_{n,2}(m-2) = mH^*_{n,2}(m) \qquad (m \ge 2) \qquad (m \in \mathbb{Z}_+),$$

cf. [71]. No such simple formulas exist for $H^*_{n,k}$ if $k \ge 3$.

Let

$$G^*_3 = H^*_{n,2} * (I+\lambda_3(I_1-I)^{*3}/3).$$

Then

$$G^*_3(k) = H^*_{n,2}(k) - \frac{\lambda_3}{3}\Delta^3 H^*_{n,2}(k-3) \qquad (k \in \mathbb{Z}_+)$$

(the difference operator Δ was defined in Chapter 1). Therefore, measure G^*_3 can also be expressed through Hermite polynomials.

According to Kruopis [185],

$$d_K(\mathcal{L}(S_n); G^*_3) \le \min\{2.3\sqrt{\mathrm{e}}\lambda_4\sigma^{-4} + 7.1\lambda_3^2\sigma^{-6}; 3\sqrt{\mathrm{e}}\lambda_4 + 3.3\mathrm{e}\lambda_3^2\}. \tag{7.20}$$

If all $p_i \le C < 1$, then

$$d_K(\mathcal{L}(S_n); G^*_3) \le C\lambda_4 \min\{1; \lambda^{-2}\}, \tag{7.21}$$

that is, the rate of the accuracy of approximation is the same as the one ensured by $H^*_{n,3}$, see (7.18).

An analogue of estimate (7.21) in terms of d_{TV} follows from [88], Theorem 3.2, if $\max_i p_i \le 1/100$. Nonuniform estimates for SCP to integer-valued r.v.s. were considered in [64].

SCP measure $H^*_{n,k}$ is not the only possible SCP approximation. For instance, SCP

$$\mathcal{L}(\pi_{\lambda-\lambda_2/2} - \pi_{-\lambda_2/2})$$

was used in [78], though the rate of the accuracy of approximation was worse than that provided by $H^*_{n,2}$.

Large deviations.

A large deviations result involving SCP approximation to the Binomial distribution $\mathbf{B}(n,p)$ has been suggested in [71].

Assume that $n^{-1} \le p \le 1/3$, $n \ge 4$, $x \in [np; n(1+p)^2/5] \cup \mathbb{N}$. Then

$$\frac{\mathbb{P}(S_n = k)}{H^*_{n,2}(\{k\})} = \mathrm{e}^{\Lambda(k)}\left\{1 + \frac{\theta_k A(k)}{1-\theta A(k)}\right\} \qquad (k \in \mathbb{N}), \tag{7.22}$$

where $|\theta_k|\le 1$,

$$A(y) = 14.4e^2 y(y-p+\sqrt{y/n}), \quad y=k/n,$$

$$\Lambda(y) = -n(1-p)\sum_{k=3}^{\infty}\left(\frac{y-p}{1-p}\right)^k \frac{1}{k(k-1)}\left\{1-\sum_{j=0}^{k-2}\binom{j+k-2}{j}(1-p)^{-j}\right\}.$$

A similar result holds also for distribution functions [71].

Open problem

7.1. Is the order of approximation in (7.10) and (7.11) optimal? Establish lower bounds.

7.2 Independent discrete random variables

A lattice random variable is a linear transform of an integer-valued random variable. In this section, we deal with integer-valued r.v.s.

Integer-valued random variables with finite 3rd moments.

Let non-degenerate r.v.s $X, X_1, X_2, ...$ take values in the set $\mathbb{Z}$ of integer numbers. Assume that $\mathbb{E}|X|^3<\infty$. Consider a SCP measure $\Psi_{n,k}$ with Fourier transform

$$\widehat{\Psi}_{n,k}(t) = \exp\left(\varkappa_1(e^{it}-1)+\varkappa_2(e^{it}-1)^2/2+\cdots+\varkappa_k(e^{it}-1)^k/k!\right),$$

where $\{\varkappa_j\}$ denote factorial cumulants (for example, $\varkappa_1=\mathbb{E}S_n$, $\varkappa_2=\text{var}\,S_n-\mathbb{E}S_n$). The case $k=1$ corresponds to Poisson approximation.

In the case of Bernoulli distribution $\mathbf{B}(p)$ we have $\mathbb{E}e^{itX}=1+p(e^{it}-1)$, hence

$$\varkappa_j=(-1)^{j+1}p^j(j-1)!$$

so that $\Psi_{n,k}$ coincides with Hipp's approximation $H^*_{n,k}$, see Section 7.1.

Note that $\Psi_{n,2}=\mathcal{L}(\pi_a+2\pi_b)$, where

$$a=2\mathbb{E}S_n-\text{var}\,S_n, \quad b=(\text{var}\,S_n-\mathbb{E}S_n)/2.$$

If $b\ge 0$, then $\mathcal{L}(\pi_a+2\pi_b)$ is a compound Poisson distribution, otherwise we have a SCP approximation. In the case of Bernoulli distribution $\mathbf{B}(p)$ one has $a=p+p^2$, $b=-p^2<0$.

Below we present a compound Poisson approximation involving first three moments of $\mathcal{L}(X_i)$.

Assume that factorial cumulants $\varkappa_j>0$ $(1\le j\le k)$, $\mathbb{E}|X_i|^3<\infty$ $(\forall i)$. Denote $S_{n,i}=S_n-X_i$,

$$d_+^{(i)}=d_{TV}(S_{n,i};S_{n,i}+1), \quad \tilde\theta=|\mathbb{E}S_n-\text{var}\,S_n|/\mathbb{E}S_n,$$

$$\hat\psi_i=\mathbb{E}|X_i(X_i-1)(X_i-2)|+|\mathbb{E}X_i|\mathbb{E}|X_i(X_i-1)|+2\mathbb{E}|X_i||\text{var}\,X_i-\mathbb{E}X_i|.$$

Theorem 7.4 *If* $\frac{2}{3}\text{var}\,S_n<\mathbb{E}S_n<2\text{var}\,S_n$, *then*

$$d_{TV}(S_n;\pi_a+2\pi_b)\le\frac{1}{(1-2\tilde\theta)}\left\{\frac{1}{\mathbb{E}S_n}\sum_{i=1}^n\hat\psi_i d_+^{(i)}+\mathbb{P}(S_n<0)\right\}. \quad (7.23)$$

According to (C.14), $d_+^{(i)} \leq \sqrt{2/\pi}\Big(1/4+\sum_{j=1,j\neq i}^n(1-d_{TV}(X_j;X_j+1))\Big)^{-1/2}$. The term $\mathbb{P}(S_n<0)$ can be evaluated using Chebyshev's inequality $\mathbb{P}(S_n<0)\leq 3/2\mathbb{E}S_n$, cf. (4.15).

If $\{X_i\}$ are Bernoulli $\mathbf{B}(p)$ r.v.s and $2/n<p<1/3$, then

$$a = np+np^2\,, \quad b = -np^2/2, \quad \hat\psi_i = 2p^3\,,$$

SCP measure $\mathcal{L}(\pi_a+2\pi_b)$ is $G_2 = \mathcal{L}(\pi_{\lambda+\lambda_2}+2\pi_{-\lambda_2/2})$, cf. (7.14), and (7.23) becomes

$$d_{TV}(S_n;\pi_a+2\pi_b) \leq 12\sqrt{2}p\sqrt{p/n}\,, \tag{7.23*}$$

cf. (7.16).

If $X_i \stackrel{d}{=} \pi_1$, then $a=n$, $b=0$, $\hat\psi_i=2$, $d_+^{(i)} \leq 1/\sqrt{2\mathrm{e}(n-1)}$, the LHS of (7.23) equals zero, the RHS of (7.23) is bounded by $\sqrt{2/\mathrm{e}(n-1)}$.

Similar approximations involving $\pi_a+2\pi_b$ or $\pi_a-\pi_b$ (with possibly negative b) have been suggested in [29]. Other examples showing that the rate of the accuracy of approximation in (7.23) can be $O(n^{-1/2})$ are given by Barbour & Xia [25].

Integer-valued random variables with finite 4th moments.

Let $\{X_i\}$ be independent integer-valued r.v.s. A natural question is if Presman's estimate (7.2) can be generalised to the case of non-identically distributed integer-valued r.v.s with finite 4th moments.

Denote $(i\geq 1)$

$$\mu_i = \mathbb{E}X_i,\ \mu = \mathbb{E}S_n,\ \sigma_i^2 = \operatorname{var} X_i,\ \sigma^2 = \operatorname{var} S_n\,,\ \beta_{3i} = \mathbb{E}(X_i-\mu_i)^3,\ \beta_3 = \mathbb{E}(S_n-\mu)^3\,.$$

Given $m\in\mathbb{N}$, let

$$\begin{aligned}\widehat\lambda_1 &= \sigma^2 - (\beta_3-\sigma^2+2m\delta)/(m-1),\ \ \delta = \widehat\gamma-\mu+\sigma^2-(\beta_3-\sigma^2)/m,\\ \widehat\lambda_2 &= m\delta/2(m-2),\ \ \widehat\lambda_m = (\beta_3-\sigma^2-2\delta m/(m-2))/m^2(m-1).\end{aligned}$$

Here m and $\widehat\gamma$ are chosen according to the following rules:

(a) If $\beta_3<\sigma^2$, then $\widehat\gamma = \lceil\mu-\sigma^2+m^{-1}(\beta_3-\sigma^2)\rceil$, $\quad m=-\max\{1;\lceil 8(1-\beta_3/\sigma^2)\rceil\}$;

(b) If $\sigma^2 \leq \beta_3 < \sigma^2+3$, then $\widehat\gamma = \lceil\mu-\sigma^2+m^{-1}(\beta_3-\sigma^2)\rceil+3$, $\quad m=-2$;

(c) If $\beta_3 \geq \sigma^2+3$, then $\widehat\gamma = \lceil\mu-\sigma^2+m^{-1}(\beta_3-\sigma^2)\rceil$, $\quad m=\max\{6;\lceil 8(\beta_3/\sigma^2-1)\rceil\}$.

Set

$$Y_m = \widehat\gamma + \pi_{\widehat\lambda_1} + 2\pi_{\widehat\lambda_2} + m\pi_{\widehat\lambda_m}\,.$$

The characteristic function of Y_m is

$$\mathbb{E}\exp(itY_m) = \exp\Big(\mathrm{i}t\widehat\gamma + \widehat\lambda_1(\mathrm{e}^{\mathrm{i}t}-1) + \widehat\lambda_2(\mathrm{e}^{2\mathrm{i}t}-1) + \widehat\lambda_m(\mathrm{e}^{\mathrm{i}tm}-1)\Big)\,.$$

The choice of parameters $\widehat\lambda_1\,,\widehat\lambda_2\,,\widehat\lambda_m$ ensures matching the first three moments of $\mathcal{L}(S_n)$:

$$\mathbb{E}Y_m = \mu,\ \operatorname{var} Y_m = \sigma^2,\ \mathbb{E}(Y_m-\mathbb{E}Y_m)^3 = \beta_3.$$

Denote $S_{n,i} = S_n - X_i\,,$

$$\begin{aligned}\kappa &= \max\{8;\lceil 8|1-\beta_3/\sigma^2|\rceil\},\ \ U = \sum_{i=1}^n(1-d_{TV}(X_i;X_i+1)),\\ \psi_i &= \mathbb{E}|(X_i-1)^{(3)}||\mu_i-\sigma_i^2+(\beta_{3i}-\sigma_i^2)/m|\\ &+ \mathbb{E}|X_i^{(3)}||\sigma_i^2-(\beta_{3i}-\sigma_i^2)/(m-1)|\\ &+ |\mathbb{E}(X_i+m-1)^{(3)}||(\beta_{3i}-\sigma_i^2)/(m(m-1))| + \mathbb{E}|X_i^{(4)}|.\end{aligned}$$

Theorem 7.5 *If* $\sigma^2 \geq 24$, *then*

$$d_{TV}(S_n; Y_m) \leq \frac{17.1}{3U\sigma^2}\Big\{\sum_{i=1}^{n}\psi_i + 2\kappa\Big\} + \frac{5}{3}\exp\left(-\frac{5\sigma^2}{48\kappa}\right) + 21\sigma^{-4} + 7\sigma^{-8}\sum_{i=1}^{n}\mathbb{E}(X_i - \mathbb{E}X_i)^4. \quad (7.24)$$

If $X_i \stackrel{d}{=} X$ $(\forall i)$ and $\mathcal{L}(X)$ does not depend on n, then the RHS of (7.24) $\leq C_X/n$.

Example 7.1 If $\mathcal{L}(S_n)$ is Binomial $\mathbf{B}(n,p)$, then $\mu_i = p$, $\sigma_i^2 = pq$, $\beta_{3i} = pq(1-2p)$, $\sigma^2 = npq$, $\beta_3 = npq(1-2p)$, $\mathbb{E}|X_i^{(3)}| = \mathbb{E}|X_i^{(4)}| = 0$, $|\mathbb{E}(X_i - 1)^{(3)}| = 6q$, $|\mathbb{E}(X_i - 2)^{(3)}| \leq 24$, $U = np$, $\kappa = 8$. If $p \leq 1/16$, then

$$m = -1, \ \widehat{\lambda}_1 = npq^2 - \delta, \ \widehat{\lambda}_2 = \delta/6, \ \widehat{\lambda}_{-1} = np^2q + \delta/3,$$

and $\mathcal{L}(Y_m)$ coincides with $P_{n,p}$ from Presman's Theorem 7.1 since $\delta = \beta$ and $\widehat{\gamma} = \gamma$.

If $npq \geq 24$ and $p \leq 1/16$, then $\psi_i \leq 42p^2q^2$, hence

$$d_{TV}(S_n; Y_m) \leq \frac{240}{n} + \frac{113}{(npq)^2} + \frac{5}{3}e^{-npq/80} = O(n^{-1} + (np)^{-2})),$$

cf. (7.2). □

Example 7.2 Since Theorem 7.5 aims at approximating r.v.s which can take on negative values, a natural question is if estimate (7.24) is sharp when $\mathcal{L}(-S_n) = \mathbf{B}(n,p)$ or $\mathbf{\Pi}(np)$?

Let $S_n \stackrel{d}{=} -\pi_\lambda$, where π_λ is a Poisson $\mathbf{\Pi}(\lambda)$ r.v. Then $n = 1$ (obviously, one can present $-\pi_\lambda$ as a sum of n i.i.d.r.v.s), $\mu_1 = \beta_{31} = -\lambda$, $\sigma_1^2 = \lambda$,

$$m = -16, \ \mu = -\lambda, \ \hat{\gamma} = \lceil -1.75\lambda \rceil, \ \delta = \gamma + 1.75\lambda, \ \beta_3 = -\lambda < \sigma^2 = \lambda, \ \kappa = 16,$$
$$\widehat{\lambda}_1 = 15\lambda/17 - 32\delta/17, \ \widehat{\lambda}_2 = 4\delta/9, \ \widehat{\lambda}_{-16} = \lambda/2176 + \delta/2448,$$

$U = 1 - d_{TV}(\pi_\lambda; \pi_\lambda - 1) \sim 1$ as $\lambda \to \infty$. Hence $Y_m = \lceil -1.75\lambda \rceil + \pi_{\widehat{\lambda}_1} + 2\pi_{\widehat{\lambda}_2} - 16\pi_{\widehat{\lambda}_m}$,

$$\psi_1 = \mathbb{E}|(-\pi_\lambda - 1)^{(3)}||\mu_1 - \sigma_1^2 - (\beta_{31} - \sigma_1^2)/16| + \mathbb{E}|(-\pi_\lambda)^{(3)}||\sigma_1^2 - (\beta_{31} - \lambda)/(m-1)| + |\mathbb{E}(-\pi_\lambda - 17)^{(3)}||(\beta_{31} - \lambda)/(m(m-1))| \leq C\lambda^4.$$

Thus, $\psi_1/U\sigma^2$ does not tend to 0 as $\lambda \to \infty$. Similarly, if Theorem 7.5 is applied to approximate $\mathcal{L}(S_n) = -\mathbf{B}(n,p)$, then the estimate of the accuracy of CP approximation is $O(1/np)$. □

Čekanavičius [68, 73] has suggested a SCP measure D_k that matches the first k moments of $\mathcal{L}(S_n)$, $k > 2$. In the case of i.i.d. lattice r.v.s that approximation ensures $d_{TV}(\mathcal{L}(S_n); D_k) \leq C_X n^{-(k-2)/2}$ assuming that $\mathcal{L}(X)$ is fixed.

Negative Binomial approximation.

Recall that Negative Binomial distribution is a particular compound Poisson distribution. Negative Binomial distribution is a natural choice when $\mathcal{L}(S_n)$ has a positive second factorial cumulant.

Let $S_n = X_1 + ... + X_n$, where $\{X_i\}$ are independent non-negative integer-valued random variables with finite third moments. Let

$$U = \sum_{i=1}^{n} (1 - d_{TV}(X_i; X_i + 1)).$$

Assume that

$$\mathbb{E}S_n > \mathrm{var}\, S_n .$$

Let Y denote a Negative Binomial $\mathbf{NB}(r, q)$ r.v., where

$$r = \frac{(\mathbb{E}S_n)^2}{\mathrm{var}\, S_n - \mathbb{E}S_n}, \quad q = \frac{\mathbb{E}S_n}{\mathrm{var}\, S_n}.$$

Then $\mathbb{E}Y = \mathbb{E}S_n = rp/q$, $\mathrm{var}\, Y = \mathrm{var}\, S_n = rp/q^2$, where $p = 1 - q$.

Theorem 7.6 *If* $\mathrm{var}\, S_n > \mathbb{E}S_n$, *then*

$$d_{TV}(S_n; Y) \leq \tag{7.25}$$
$$\frac{17.1}{Urp} \sum_{i=1}^{n} \sum_{k=1}^{\infty} k\left(\frac{k-1}{2} + \mathbb{E}X_i\right) |(q\mathbb{E}X_i + pk)\mathbb{P}(X_i = k) - (k+1)\mathbb{P}(X_i = k+1)|$$

Note that (7.25) is not applicable if $\{X_i\}$ are Poisson r.v.s, nor if $\{X_i\}$ are Bernoulli r.v.s.

Example 7.3 Let independent r.v.s $\{X_i\}$ have geometric distributions

$$\mathbb{P}(X_i = k) = q_i p_i^k \quad (k \geq 0),$$

where $q_i = 1 - p_i > 1/2$ $(i \geq 1)$. Then

$$\mathbb{E}S_n = \sum_{k=1}^{n} p_k/q_k, \quad \mathrm{var}\, S_n = \sum_{k=1}^{n} p_k/q_k^2,$$

and (7.25) yields

$$d_{TV}(S_n; Y) \leq 48 \left(\sum_{j=1}^{n} p_j\right)^{-1/2} \left(\mathbb{E}S_n\right)^{-1} \sum_{i=1}^{n} \left|\frac{1}{q_i} - \frac{1}{q}\right| \left(\frac{p_i}{q_i}\right)^2. \tag{7.26}$$

The approximation is exact if $p_i = p$ $(\forall i)$.

For instance, let $p_i = p/3$ if i is odd, $p_i = 2p/3$ if i is even, then the RHS of (7.26) is $O(\sqrt{p/n})$. □

Negative binomial approximation to conditional distributions was considered in [123].

7.3 Symmetric integer-valued random variables

In the case of symmetric (around 0) r.v.s the accuracy of compound Poisson approximation to $\mathcal{L}(S_n)$ is expected to be significantly better than $O(n^{-2/3})$.

Recall that any distribution satisfies (1.12*):

$$\mathcal{L}(X) = (1-p)I + p\mathcal{L}(X'),$$

where $\mathcal{L}(X') = \mathcal{L}(X|X\neq 0)$, $p = \mathbb{P}(X\neq 0)$, $\mathbb{P}(X'=k) = \mathbb{P}(X=k|X\neq 0)$ $(k\neq 0)$. In other words,

$$X \stackrel{d}{=} \tau X',$$

where τ is independent of X', $\mathcal{L}(\tau) = \mathbf{B}(p)$.

Below we extend Presman's estimate (7.12) to the case of symmetric integer-valued r.v.s. Denote

$$\lambda = \sum_{i=1}^{n} p_i\,, \quad \sigma^2 = \sum_{i=1}^{n} p_i q_i\,, \quad \lambda_k = \sum_{i=1}^{n} p_i^k \qquad (k\in\mathbb{N}).$$

Recall that $S_n = \sum_{j=1}^{n} \tau_i X_i'$, where $\mathcal{L}(\tau_i) = \mathbf{B}(p_i)$, $p_i = \mathbb{P}(X_i'\neq 0)$.

We assume that $X', X_1', \ldots, X_n'$ are i.i.d. *symmetric* r.v.s taking values in $\mathbb{Z}\setminus\{0\}$,

$$\operatorname{var} X' = \sigma'^2 < \infty\,, \qquad \theta := \lambda_2/\lambda < 1.$$

Theorem 2.7 in [83] states that

$$d_{TV}(\mathcal{L}(S_n); \mathbf{\Pi}(\lambda, X')) \;\le\; \frac{0.58\lambda_2\sqrt{\sigma'+1}}{\lambda^2(1-\sqrt{\theta})^{5/2}}\,, \tag{7.27}$$

$$d_{\mathrm{o}}(\mathcal{L}(S_n); \mathbf{\Pi}(\lambda, X')) \;\le\; \frac{0.82\lambda_2}{\lambda^2\sqrt{\lambda}(1-\sqrt{\theta})^{5/2}}\,. \tag{7.28}$$

It is easy to check that (7.27) is a generalisation of (7.12).

In the case of symmetrised Bernoulli $\mathbf{B}(p_i)$ r.v.s we have $S_n = \sum_{i=1}^{n}(X_{1i}-X_{2i})$, where

$$\mathcal{L}(X_{1i}-X_{2i}) = (q_i I + p_i I_1) * (q_i + p_i I_{-1}) = (1-2p_i q_i)I + p_i q_i(I_{-1} + I_1),$$

$$\lambda = 2\sum_{i=1}^{n} p_i q_i\,, \quad \lambda_2 = 4\sum_{i=1}^{n} p_i^2 q_i^2\,, \quad \sigma' = 1, \quad \theta \le \lambda_2/\lambda_1 \le 2\max_i p_i q_i \le 1/2.$$

If $p_i = p$ $(\forall i)$, then $\theta = 2pq$, and (7.27)–(7.28) become

$$d_{TV}(S_n; \pi_{npq} - \pi'_{npq}) \;\le\; \frac{0.58\sqrt{2}}{n(1-\sqrt{2pq}\,)^{5/2}}\,,$$

$$d_{\mathrm{o}}(S_n; \pi_{npq} - \pi'_{npq}) \;\le\; \frac{0.82}{n\sqrt{np}(1-\sqrt{2pq}\,)^{5/2}}\,.$$

where π_{npq}, π'_{npq} are independent Poisson $\mathbf{\Pi}(npq)$ r.v.s.

We now consider approximations by SCP measures defined by (6.22). We assume that $F_i \equiv F$ $(\forall i)$, where $F = \mathcal{L}(X')$.

Denote

$$H_{n,k} = \exp\left(\sum_{i=1}^{n}\sum_{j=1}^{k}(-1)^{j+1}p_i^j\,(F-I)^{*j}/j\right) = \exp\left(\sum_{j=1}^{k}(-1)^{j+1}\lambda_j(F-I)^{*j}/j\right),$$

where $k\in\mathbb{N}$. Observe that

$$H_{n,k} = \mathbf{\Pi}(\lambda_{1,k}; F) * \mathbf{\Pi}(\lambda_{2,k}, F^{*2}) * \cdots * \mathbf{\Pi}(\lambda_{m,k}, F^{*m}) * \cdots * \mathbf{\Pi}(\lambda_{k,k}, F^{*k}),$$

where $\lambda_{m,k} = (-1)^{m+1}\sum_{j=m}^{k} \frac{\lambda_j}{j}\binom{j}{m}$ are given by (7.17). In particular,

$$H_{n,1} = \mathcal{L}(\tilde{S}_n) = \mathbf{\Pi}(\lambda, F), \quad H_{n,2} = \exp(\lambda(F-I) - \lambda_2(F-I)^{*2}/2).$$

In the following theorem assumption $\theta < 1$ is dropped.

Theorem 7.7 *Suppose that* $\max_i p_i \le c_0 < 1$, $\lambda \ge 1$. *Then there exists an absolute constant* C *such that*

$$d_o(\mathcal{L}(S_n); \mathbf{\Pi}(\lambda, F)) \le \frac{C}{(1-c_0)^{5/2}} \frac{\lambda_2}{\lambda^2\sqrt{\lambda}}, \tag{7.29}$$

$$d_{TV}(\mathcal{L}(S_n); \mathbf{\Pi}(\lambda, F)) \le \frac{C\sqrt{\sigma'+1}}{(1-c_0)^{5/2}} \frac{\lambda_2}{\lambda^2}, \tag{7.30}$$

$$d_G(\mathcal{L}(S_n); \mathbf{\Pi}(\lambda, F)) \le \frac{C(\sigma'+1)^{3/2}}{(1-c_0)^2} \frac{\lambda_2}{\lambda^{3/2}}. \tag{7.31}$$

Moreover, if $c_0 \le 1/2$, $k \ge 2$, *then*

$$d_o(\mathcal{L}(S_n); H_{n,k}) \le \frac{C(k)\lambda_{k+1}}{(1-2c_0)\lambda^{k+3/2}}, \tag{7.32}$$

$$d_{TV}(\mathcal{L}(S_n); H_{n,k}) \le \frac{C(k)\lambda_{k+1}\sqrt{\sigma'+1}}{(1-2c_0)\lambda^{k+1}}. \tag{7.33}$$

In particular, bound (7.31) gives an estimate in terms of the Gini-Kantorovich distance of order $O\left(\left(\sum_{j=1}^{n} p_j q_j\right)^{-3/2} \sum_{i=1}^{n} p_i^2 q_i^2\right)$, which is $O(n^{-1/2})$ if all $\{p_i\}$ are uniformly bounded away from 1.

If $\lambda < 1$, then one can use estimates (6.6), (6.7), (6.10).

In the case of symmetrised Bernoulli r.v.s $\mathbb{P}(X'=1) = \mathbb{P}(X'=-1) = 1/2$, $c_0 = 1/2$. Theorem 7.7 can be applied with p_i replaced by $2p_i q_i$. If $p_i \ge n^{-1}$ for all r.v.s, then (7.30) has the same order as (7.12).

Open problem

7.2. Estimate constants in Theorem 7.7.

7.4 Discrete non-lattice distributions

Any discrete distribution can include zero in its support after a proper shift. W.l.o.g., we will assume that in this section.

The situation for discrete non-lattice distributions is very different from the situation for discrete lattice distributions. For example, if $X_1, \dots, X_n$ are independent r.v.s taking values

$$x_0 = 0, x_1, x_2, \dots, x_N.$$

Then $S_n = X_1 + \ldots + X_n$ takes values in the set

$$\{j_1 x_1 + j_2 x_2 + \cdots + j_N x_N : j_1 + j_2 + \cdots + j_N \le n, j_i \in \{0, 1, \dots, n\}\}.$$

Of course, if probabilities $\mathbb{P}(X_i=0)$ are close to one, then Khintchine's decomposition (1.12*) and results of Chapter 6 can be applied. Moreover, due to (6.6) (see also (A.3)) estimates such as (7.16) can be applied in the case of identically distributed r.v.s. However, in the general case results are scarce. We present one estimate for the case of a sum of non-identically distributed r.v.s.

Assume that $X_1,\dots,X_n$ are independent r.v.s taking on values $x_0=0, x_1, x_2,\dots,x_N$, where $\{x_j\}$ and N are fixed numbers. Denote

$$p_{ki} = \mathbb{P}(X_k=x_i) \quad (k=1,\dots,n,\ i=0,1,\dots N),$$

$$G_4 = \exp\left(\sum_{k=1}^{n}(\mathcal{L}(X_k)-I) - \frac{1}{2}\sum_{k=1}^{n}(\mathcal{L}(X_k)-I)^{*2}\right).$$

Theorem 7.8 [82] *Suppose that there exists an absolute constant* $\tilde{C}\in(0;1)$ *such that* $\mathbb{P}(X_k=0)\geq\tilde{C}>0$ $(k=1,\dots,n)$. *Then there exists constant* C_N *depending only on* N *such that*

$$d_K(\mathcal{L}(S_n);G_4) \leq C_N\frac{\max_{1\leq j\leq n}|x_j|}{\min_{1\leq j\leq n}|x_j|}\left(\left(\sum_{k=1}^{n}(1-p_{k0})\right)^{-1/2}\ln n+1\right)$$
$$\times\sum_{j=1}^{N}\left(\sum_{k=1}^{n}p_{kj}^3\right)\left(\sum_{m=1}^{n}p_{mj}\right)^{-3/2}+\mathrm{e}^{-n}.$$

If N is fixed, $\max_j|x_j|/\min_j|x_j|<C$, $C_3<p_{kj}<C_4$, $d_K(\mathcal{L}(S_n);G_4)=O(n^{-1/2})$. Similar results have been given in [81, 87].

Example 7.4 Let $X,X_1,\dots,X_n$ be i.i.d. *symmetric* r.v.s with the distribution

$$\mathcal{L}(X)=p_0I+p_1\mathcal{L}(X')+p_2\mathcal{L}(X'') \qquad (0<p_0,p_1,p_2<1),$$

where X' is a symmetric integer-valued r.v. with finite variance σ_1^2, X'' is a symmetric r.v. taking values in the set $\{0,\pm a,\pm 2a,\dots\}$ with finite variance σ_2^2. Note that

$$X\stackrel{d}{=}\tau(\tau'X'+(1-\tau')X''),$$

where $\mathcal{L}(\tau)=\mathbf{B}(p_0)$, $\mathcal{L}(\tau')=\mathbf{B}(p_1/(1-p_0))$, random variables τ,X',X'' are independent.

Proposition 7.9 *If* $p_0\geq 1/2$, *then there exists an absolute constant* $C>0$ *such that*

$$d_{TV}(\mathcal{L}(S_n);\mathbf{\Pi}(n;X))\leq C\sqrt{\sigma_1+\sigma_2+1}\left(\sqrt{\min\{p_1;p_2\}/n}+1/n\right). \tag{7.34}$$

For instance, if $\mathbb{P}(X=0)=\mathbb{P}(X=-\sqrt{2})=\mathbb{P}(X=\sqrt{2})=\mathbb{P}(X=-1)=\mathbb{P}(X=1)=1/5$, then $\mathbb{P}(X'=-\sqrt{2})=\mathbb{P}(X'=\sqrt{2})=1/2$, $\mathbb{P}(X''=-1)=\mathbb{P}(X'=1)=1/2$, $p_1=p_2=2/5$, hence the accuracy of approximation in (7.34) is $O(n^{-1/2})$.

We cannot apply (7.30) directly since we assume in (7.30) that $\mathcal{L}(X)$ is lattice.

It follows from Theorem 1 in [111] that $d_K(\mathcal{L}(S_n);\mathbf{\Pi}(n;X))=O(1/n)$ if p_1,p_2 are constants indicating that the accuracy in (7.34) is not optimal.

Open problems

7.3. Does an analogue of the estimate of Theorem 7.8 hold in terms of d_{TV}?

7.4. Can the order of approximation in (7.34) be improved?

7.5 Applications

An urn model with overflow

Suppose that n balls are distributed into m urns, and each ball is equally likely to be assigned to any urn. Each urn can hold at most k–1 balls, where $k \geq 2$ is a fixed number. If a ball is assigned to an urn that is already full, that ball is placed in an additional "overflow urn" of unlimited capacity.

Let W be the number of balls allocated to the overflow urn. More precisely, let

$$W = X_1 + \cdots + X_m\,,$$

where $X_j = (S_j - k + 1)\mathbb{I}\{S_j \geq k\}$ and S_j is the number of balls allocated to urn j. Set

$$\begin{aligned}\lambda_i &= \binom{n}{i+k-1}\left(\frac{1}{m}\right)^{i+k-2}\left(1-\frac{1}{m}\right)^{n-i-k+1} \qquad (i=1,\dots,n-k+1),\\ \lambda &= \lambda_1 + \cdots + \lambda_{n-k+1}\,, \quad Y_m = \sum_{i=1}^{n-k+1} i\pi_{\lambda_i}\,.\end{aligned}$$

Daly [99] has shown that

$$\begin{aligned}d_{TV}(W;Y_n) \;\leq\; & \mathcal{M}(\lambda)\Bigg\{m^2\Bigg(\sum_{i=k}^{n}(i-k+1)\binom{n}{i}\left(\frac{1}{m}\right)^i\left(1-\frac{1}{m}\right)^{n-i}\Bigg)^2\\ & - m(m-1)\sum_{i=k}^{n-k}\sum_{j=k}^{n-i}(i-k+1)(j-k+1)\frac{n!}{i!j!(n-i-j)!}\left(\frac{1}{m}\right)^{i+j}\left(1-\frac{2}{m}\right)^{n-i-j}\Bigg\}.\end{aligned}$$

Factor $\mathcal{M}(\lambda)$ obeys (6.16) with

$$\lambda_{n,1} - 2\lambda_{n,2} = \binom{n}{k}p^{k-1}(1-p)^{n-k-1}(1-p-2(n-k)p/(k+1))$$

(Daly [99]). The RHS of this estimate converges to zero as $n\to\infty$ if m is fixed (Boutsikas & Koutras [52]).

An overview of compound Poisson approximation results obtained via Stein's method can be found in [27, 28]. One must be aware that Stein's method may not always lead to a correct rate of compound Poisson approximation (consider, for instance, the problem of the length of the longest head run [112, 221]).

Compound Poisson approximation to the distribution of the number of k-out-of-n isolated vertices of a rectangular lattice on a torus has been presented in [255].

Compound Poisson approximation to the number of overlapping and non-overlapping occurrences of word patterns has been suggested in [129]. Compound Poisson approximation to the number of overlapping sequences has been studied in [59].

A Negative Binomial approximation to the number of parasites has been suggested in [36].

A review on the topic of compound Poisson approximation to the number of dependent claims has been given by Genest et al. [126].

7.6 Proofs

Proof of Theorem 7.1 (sketch). In view of (3.9) we only need to show that

$$d_{TV}(\mathbf{B}(n,p); P_{n,p}) \le C\max\{1/(np)^2; 1/n\} \qquad (n^{-2/3}<p\le 1/2).$$

The proof uses the method of characteristic functions. The key tool is Presman's Lemma D.7 or D.8.

W.l.o.g. we can assume that $npq^2 \ge 1$. Denote

$$M = \mathcal{L}(S_n - np) - P_{n,p} * I_{-np}\,.$$

Then $\widehat{M}(t) = f^n(t) - g^n(t)$, where $f(t) = \mathbb{E}e^{it(X-p)}$, $g(t) = \mathbb{E}e^{it(Y-p)}$.

Recall that $Y = \gamma/n + \eta_1 - \eta_2 + 2\eta_3$, where

$$\mathcal{L}(\eta_1) = \mathbf{\Pi}(pq^2 - \beta/n),\ \mathcal{L}(\eta_2) = \mathbf{\Pi}(p^2q + \beta/3n),\ \mathcal{L}(\eta_3) = \mathbf{\Pi}(\beta/6n),$$

$\gamma = \lceil 3np^2 - 2np^3 \rceil$, $\beta/n = \gamma/n - 3p^2 + 2p^3 \in [0; 1/n)$.

One can check that $f(t) = qe^{-itp} + pe^{iqt}$,

$$\begin{aligned} g(t) = \mathbb{E}e^{it(Y-p)} &= \exp\Big((3p^2 - 2p^3 - p)it + pq^2(e^{it}-1) + p^2q(e^{-it}-1) \\ &+ \frac{\beta}{n}\Big(it - \frac{2}{3}(e^{it}-1) + \frac{1}{3}(e^{-it}-1) + \frac{1}{6}(e^{it}-1)^2\Big)\Big). \end{aligned}$$

It follows from (D.6) that

$$|f(t)| \le \exp(-2pq\sin^2(t/2)) \le \exp(-2pqt^2/\pi^2) \qquad (|t|\le\pi).$$

Taking into account that $|e^{it}-1|^2 = 4\sin^2(t/2)$, we get

$$|g(t)| \le \exp(-2pq\sin^2(t/2))\exp(4\beta\sin^2(t/2)/3n) \le e^{4/3n}\exp(-2pqt^2/\pi^2)$$

as $|t|\le\pi$. Therefore,

$$|\widehat{M}(t)| \le n\max\{|f(t)|; |g(t)|\}^{n-1}|f(t)-g(t)| \le ne^{4/3}e^{-\kappa t^2}|f(t)-g(t)|$$

as $|t|\le\pi$, where $\kappa = 2(n-1)pq/\pi^2$.

It follows from Taylor's formula (D.3) that

$$|f(t)-g(t)| \le \sum_{j=1}^{5}|t|^j\,|\mathbb{E}(X-p)^j - \mathbb{E}(Y-p)^j|/j! + t^6(\mathbb{E}(X-p)^6 + \mathbb{E}(Y-p)^6)/6!$$

Note that $\mathbb{E}(X-p)^j = q(-p)^j + pq^j$. Expanding $g(t)$ in powers of (it), one obtain expressions for $\mathbb{E}(Y-p)^j$. We get $\mathbb{E}(Y-p)^k = \mathbb{E}(X-p)^k$ if $k=1,2,3$,

$$|\mathbb{E}(X-p)^4 - \mathbb{E}(Y-p)^4| \le 7p^2q^2 + 2\beta/n,\quad |\mathbb{E}(X-p)^5 - \mathbb{E}(Y-p)^5| \le 12p^2q^2 + 4\beta/n,$$
$$\mathbb{E}(Y-p)^6 \le \beta(10+30pq)/n + 10p^2q^2(q-p)^2 + 15p^2q^2(1+pq) + pq(p^5+q^5),$$

$\mathbb{E}(X-p)^6 \le pq$. Therefore,

$$\begin{aligned} |f(t)-g(t)| &\le \frac{(7p^2q^2+2\beta/n)t^4}{4!} + \frac{(12p^2q^2+4\beta/n)|t|^5}{5!} + \frac{\beta(10+30pq)t^6}{6!n} \\ &+ t^6(pq + 10p^2q^2(q-p)^2 + 15p^2q^2 + 15p^3q^3 + pq(p^5+q^5))/6! \\ &\le \frac{(7p^2q^2+2/n)t^4}{4!} + \frac{(12p^2q^2+4/n)|t|^5}{5!} + \frac{7t^6}{288n} + \frac{3pqt^6}{16}. \end{aligned}$$

In order to derive an estimate in terms of d_K, one can apply Tsaregradskii's inequality (D.14):

$$d_K(\mathbf{B}(n,p);P_{n,p}) = \|M\|_K \le \frac{1}{4}\int_{-\pi}^{\pi} |\widehat{M}(t)||t|^{-1}\mathrm{d}t.$$

Since $1/(n-1)\le 7/6n$ as $n\ge 7$ and

$$\int_0^{\pi} \mathrm{e}^{-ct^2}t^3\mathrm{d}t \le 1/2c^2\,, \quad \int_0^{\pi} \mathrm{e}^{-ct^2}t^4\mathrm{d}t \le 3\sqrt{\pi}/8c^{2.5}\,, \quad \int_0^{\pi} \mathrm{e}^{-ct^2}t^5\mathrm{d}t \le 1/c^3$$

for any $c>0$, we get

$$\begin{aligned} d_K(\mathbf{B}(n,p);P_{n,p}) &\le \frac{49\mathrm{e}^{4/3}\pi^2}{2304(n-1)} + \frac{7\mathrm{e}^{4/3}\pi^{5.5}}{1280\sqrt{2}(n-1)\sqrt{(n-1)pq}} + \frac{\mathrm{e}^{4/3}\pi^2}{192((n-1)pq)^2} \\ &+ \frac{21\mathrm{e}^{4/3}\pi^6}{1536((n-1)pq)^2} + \frac{\mathrm{e}^{4/3}\pi^{5.5}}{640\sqrt{2}((n-1)pq)^{5/2}} + \frac{7\mathrm{e}^{4/3}\pi^6}{2304((n-1)pq)^3} \\ &\le 11.01n^{-1} + 87.45(npq)^{-2}\,. \end{aligned}$$

Similarly one evaluates $\int_{-\pi}^{\pi} |\widehat{M}'(t)|^2 dt$ or $\int_{-\pi}^{\pi} |\widehat{M}''(t)|dt$. We then apply Lemma D.7 or D.8 in order to evaluate $d_{TV}(\mathbf{B}(n,p);P_{n,p})$. □

Proof of Theorem 7.2. The proof goes along the lines of the proof of Theorem 7.1. Recall that

$$\bar{Y}_i = \bar{\gamma}/n + \bar{\eta}_{1i} - \bar{\eta}_{2i} + 2\bar{\eta}_{3i} \quad (i=1,2,\ldots,n), \quad S_n^{\bar{Y}} = \bar{Y}_1 + \cdots + \bar{Y}_n\,,$$

where

$$\mathcal{L}(\bar{\eta}_{1i}) = \mathbf{\Pi}(p_i q_i^2 - \bar{\beta}/n),\ \mathcal{L}(\bar{\eta}_{2i}) = \mathbf{\Pi}(p_i^2 q_i + \bar{\beta}/3n),\ \mathcal{L}(\bar{\eta}_{3i}) = \mathbf{\Pi}(\bar{\beta}/6n),$$

$$\bar{\gamma} = \lceil (3p_1^2 - 2p_1^3) + \ldots + (3p_n^2 - 2p_n^3) \rceil\,, \quad \bar{\beta} = \bar{\gamma} - \sum_{i=1}^{n}(3p_i^2 - 2p_i^3).$$

Let φ_n and φ denote the characteristic functions of $\mathcal{L}(S_n - \lambda)$ and $\mathcal{L}(S_{\bar{Y}} - \lambda)$, respectively. Set

$$\begin{aligned} f_k(t) &= (1 - p_k + p_k\mathrm{e}^{\mathrm{i}t})\mathrm{e}^{-\mathrm{i}tp_k} = \mathbb{E}e^{\mathrm{i}t(X_k - \mathbb{E}X_k)}, \quad \varphi_\lambda(t) = \exp\big(-(1-c_0)\lambda t^2/\pi^2\big)\,, \\ b(t) &= \big(\mathrm{i}t - (\mathrm{e}^{\mathrm{i}t} - 1) + (\mathrm{e}^{-\mathrm{i}t} - 1)/3 + (\mathrm{e}^{2\mathrm{i}t} - 1)/6\big)\,\bar{\beta}, \\ g_k(t) &= \exp\big((3p_k^2 - 2p_k^3 - p_k)\mathrm{i}t + p_k q_k^2(\mathrm{e}^{\mathrm{i}t} - 1) + p^2 q_k(\mathrm{e}^{-\mathrm{i}t} - 1)\big)\,. \end{aligned}$$

Then

$$\varphi(t) = \mathrm{e}^{b(t)}\prod_{k=1}^{n} g_k(t), \quad \varphi_n(t) = \prod_{k=1}^{n} f_k(t).$$

Clearly, $\mathbb{E}e^{\mathrm{i}t(\bar{Y}_k - \mathbb{E}\bar{Y}_k)}$ equals

$$\exp\big((\bar{\gamma}/n - p_k)\mathrm{i}t + (p_k q_k^2 - \bar{\beta}/n)(\mathrm{e}^{\mathrm{i}t} - 1) + (p^2 q_k + \bar{\beta}/3n)(\mathrm{e}^{-\mathrm{i}t} - 1) + (\mathrm{e}^{2\mathrm{i}t} - 1)\bar{\beta}/6n\big)\,.$$

Note that

$$\varphi_n(t) - \varphi(t) = \prod_{k=1}^{n} f_k(t) - \mathrm{e}^{b(t)}\prod_{k=1}^{n} g_k(t) = \prod_{k=1}^{n} f_k(t) - \prod_{k=1}^{n} g_k(t) - (\mathrm{e}^{b(t)} - 1)\prod_{k=1}^{n} g_k(t).$$

By Taylor's formula (I.8),

$$|b(t)| \le Ct^4, \quad |b'(t)| \le C|t|^3, \quad |b''(t)| \le Ct^2. \tag{7.35}$$

Note that

$$|f_k(t)| \vee |g_k(t)| \le \exp\big(-2p_kq_k\sin^2(t/2)\big), \tag{7.36}$$
$$|f_k'(t)| \vee |g_k'(t)| \le Cp_k|t|, \quad |f_k''(t)| \vee |g_k''(t)| \le Cp_k \tag{7.37}$$

as $k=1,2,\dots,n$. Estimate (7.36) follows from (D.6). Indeed,

$$|f_k'(t) = |p_kq_k(\mathrm{e}^{\mathrm{i}tq_k} - \mathrm{e}^{-\mathrm{i}tp_k})| \le p_kq_k|t|$$

by Taylor's formula (I.8). Estimates (7.37) can be verified similarly.

Bounds (7.36)–(7.37) allow us to prove that

$$\left|\prod_{k=1}^n g_k(t)\right| \le \varphi_\lambda^2(t), \quad \left|\Big(\prod_{k=1}^n g_k(t)\Big)'\right| \le C\lambda|t|\varphi_\lambda(t), \quad \left|\Big(\prod_{k=1}^n g_k(t)\Big)''\right| \le C\lambda\varphi_\lambda(t). \tag{7.38}$$

Indeed, note that $\sigma^2 \ge (1-c_0)\lambda$. Since $|\sin(t/2)| \ge |t|/\pi$ as $|t| \le \pi$, we have

$$\prod_{k=1}^n |g_k(t)| = \exp\big(-2\sigma^2\sin^2(t/2)\big) \le \exp\big(-2(1-c_0)\lambda t^2/\pi^2\big) = \varphi_\lambda^2(t).$$

Obviously, $\lambda t^2\varphi_\lambda(t) = \lambda t^2\exp\big(-(1-c_0)\lambda t^2/\pi^2\big) \le \pi^2/\mathrm{e}(1-c_0)$. Moreover,

$$\prod_{m\ne k,j} |g_m(t)| \le \varphi_\lambda^2(t)\exp\big(2(p_kq_k+p_jq_j)\sin^2(t/2)\big) \le \mathrm{e}^4\varphi_\lambda^2(t). \tag{7.39}$$

Therefore,

$$\begin{aligned}\left|\Big(\prod_{k=1}^n g_k(t)\Big)''\right| &\le \sum_{k=1}^n |g_k''(t)|\prod_{j\ne k}^n |g_j(t)| + \sum_{k=1}^n |g_k'(t)|\sum_{j\ne k}^n |g_j'(t)|\prod_{m\ne j,k}^n |g_m(t)| \\ &\le C\lambda\varphi_\lambda^2(t) + C(\lambda t^2\varphi_\lambda(t))\lambda\varphi_\lambda(t) \le C\lambda\varphi_\lambda(t).\end{aligned}$$

The second estimate in (7.38) can be proved similarly.

Note that $|\mathrm{e}^{b(t)}| \le \mathrm{e}^2$ and $|\mathrm{e}^{b(t)} - 1| \le C|b(t)|$. Combining these estimates with (7.38) and (7.35), it is not difficult to prove that

$$\left|(\mathrm{e}^{b(t)}-1)\prod_{k=1}^n g_k(t)\right| \le Ct^4\varphi_\lambda(t); \quad \left|\Big((\mathrm{e}^{b(t)}-1)\prod_{k=1}^n g_k(t)\Big)'\right| \le C|t|^3\varphi_\lambda(t),$$
$$\left|\Big((\mathrm{e}^{b(t)}-1)\prod_{k=1}^n g_k(t)\Big)''\right| \le Ct^2\varphi_\lambda(t). \tag{7.40}$$

Expanding $f_k(t)$ and $g_k(t)$ in powers of $(\mathrm{i}t)$, we get

$$|f_k(t)-g_k(t)| \le Cp_k^2(t^4+|t|^5)+Cp_kt^6 \le Cp_k^2t^4(1+\pi)+Cp_kt^6,$$
$$|f_k'(t)-g_k'(t)| \le Cp_k^2|t|^3 + Cp_k|t|^5, \quad |f_k''(t)-g_k''(t)| \le Cp_k^2t^2 + Cp_kt^4 \tag{7.41}$$

as $|t|\le\pi$. Estimates (7.36)–(7.38), (7.41) combined with (I.1)–(I.2), (7.39) and the estimate $\lambda t^2\varphi_\lambda(t)\le C$ lead to

$$\left|\prod_{k=1}^{b} f_k(t)-\prod_{k=1}^{n} g_k(t)\right| \le C(\lambda_2 t^4+\lambda t^6)\varphi_\lambda(t),$$

$$\left|\left(\prod_{k=1}^{b} f_k(t)-\prod_{k=1}^{n} g_k(t)\right)'\right| \le C(\lambda_2|t|^3+\lambda|t|^5)\varphi_\lambda(t),$$

$$\left|\left(\prod_{k=1}^{b} f_k(t)-\prod_{k=1}^{n} g_k(t)\right)''\right| \le C(\lambda_2 t^2+\lambda t^4)\varphi_\lambda(t).$$

Combining the last estimates with (7.40), we obtain

$$|\varphi_n(t)-\varphi(t)| \le C((\lambda_2+1)t^4+\lambda t^6)\varphi_\lambda(t),$$
$$|((\varphi_n(t)-\varphi(t))\mathrm{e}^{-\mathrm{i}\lambda t})'| \le C((\lambda_2+1)|t|^3+\lambda|t|^5)\varphi_\lambda(t),$$
$$|((\varphi_n(t)-\varphi(t))\mathrm{e}^{-\mathrm{i}\lambda t})''| \le C((\lambda_2+1)t^2+\lambda t^4)\varphi_\lambda(t).$$

The proof of (7.5), (7.8), (7.9) now follows from Lemmas D.8 and D.9.

For example, applying (7.42) and (D.12), we get

$$\begin{aligned} d_{\mathrm{o}}(S_n;S_n^{\bar{Y}}) &\le C\int_{-\pi}^{\pi}((\lambda_2+1)t^4+\lambda t^6)\varphi_\lambda(t)\mathrm{d}t \\ &\le \frac{C(\lambda_2+1)}{\lambda^2\sqrt{\lambda}}+\frac{C}{\lambda^2\sqrt{\lambda}}\le\frac{C(\lambda_2+1)}{\lambda^2\sqrt{\lambda}}. \end{aligned}$$

Similarly one can derive remaining estimates. □

Proof of Theorem 7.4. According to Stein's method (cf. (F.12)), we need to evaluate

$$\mathbb{E}\left(ag(S_n+1)+2bg(S_n+2)-S_ng(S_n)\right),$$

where $a=2\mathbb{E}S_n-\operatorname{var}S_n$, $2b=\operatorname{var}S_n-\mathbb{E}S_n$, function g is a solution of Stein's equation

$$\sum_{i\ge1} i\lambda_i g(i+k)-kg(k)=f(k)-\mathbb{E}f(Y).$$

We plan to express $g(S_n+l)$, where $l=1,2$, via $g(S_{n,i}+1)$.

Clearly,

$$\mathbb{E}(ag(S_n+1)+2bg(S_n+2)-S_ng(S_n))=\sum_{i=1}^{n}\mathbb{E}\left(2b_ig(S_n+2)+a_ig(S_n+1)-X_ig(S_n)\right),$$

where $a_i=2\mathbb{E}X_i-\operatorname{var}X_i$, $2b_i=\operatorname{var}X_i-\mathbb{E}X_i$ $(i\in\{1,\dots,n\})$. We apply (I.5):

$$g(\omega+l)=g(\omega+1)+(l-1)\Delta g(\omega+1)+R(g,\omega,l), \tag{I.5*}$$

where

$$R(g,\omega,l)=\begin{cases}\sum_{i=1}^{l-2}(l-i-1)\Delta^2 g(\omega+i) & (l\ge3),\\ 0 & (l=1,2),\\ \sum_{j=0}^{-l}(-l-j+1)\Delta^2 g(\omega-j) & (l\le0).\end{cases}$$

It is easy to check that

$$|R(g,\omega,l)| \le \tfrac{1}{2}(l-1)(l-2)\|\Delta^2 g\|.$$

Recall that

$$|\mathbb{E}\Delta^k g(X)| \le 2\|\Delta^{k-1}g\| d_{TV}(X;X+1) \qquad (k\ge 1)$$

according to (F.5), where

$$\|\Delta g\| = \sup_k |\Delta g|.$$

In view of (I.5*) and (F.5), for any bounded function g

$$\begin{aligned} & |\mathbb{E}g(S_{n,i}+l) - \mathbb{E}g(S_{n,i}+1) - (l-1)\mathbb{E}\Delta g(S_{n,i}+1)| \\ \le\ & (l-1)(l-2)\sup_{m\in\mathbb{Z}} \mathbb{E}|\Delta^2 g(S_{n,i}+m)| \le (l-1)(l-2)\|\Delta g\| d_+^{(i)}\,. \end{aligned}$$

Note that $S_n = S_{n,i}+X_i$. Therefore,

$$g(S_n) = g(S_{n,i}+X_i) = g(S_{n,i}+1) + (X_i-1)\Delta g(S_{n,i}+1) + R(g,S_{n,i},X_i).$$

We conclude that

$$|\mathbb{E}X_i g(S_n) - \mathbb{E}X_i\mathbb{E}g(S_{n,i}+1) - \mathbb{E}X_i(X_i-1)\mathbb{E}\Delta g(S_{n,i}+1)| \le \|\Delta g\| d_+^{(i)}\mathbb{E}|X_i^{(3)}|.$$

Similarly, we express $g(S_n+1)$ and $g(S_n+2)$ via $g(S_{n,i}+1)$:

$$\begin{aligned} \mathbb{E}g(S_n+1) &= \mathbb{E}X_i\mathbb{E}g(S_{n,i}+2) - \mathbb{E}(X_i-1)\mathbb{E}g(S_{n,i}+1) + \mathbb{E}R(g,S_{n,i},X_i+1), \\ \mathbb{E}g(S_n+2) &= \mathbb{E}(X_i+1)\mathbb{E}g(S_{n,i}+2) - \mathbb{E}X_i\mathbb{E}g(S_{n,i}+1) + \mathbb{E}R(g,S_{n,i},X_i+2). \end{aligned}$$

Therefore,

$$\begin{aligned} & |2b_i\mathbb{E}g(S_n+2) + a_i\mathbb{E}g(S_n+1) - \mathbb{E}X_i g(S_n)| \\ \le\ & \Big|2b_i\mathbb{E}R(g,S_{n,i},X_i+2) + a_i\mathbb{E}R(g,S_{n,i},X_i+1)\Big| + \|\Delta g\| d_+^{(i)}\mathbb{E}|X_i^{(3)}|. \end{aligned}$$

Applying (F.5), we get

$$\begin{aligned} & \Big|2b_i\mathbb{E}R(g,S_{n,i},X_i+2) + a_i\mathbb{E}R(g,S_{n,i},X_i+1)\Big| \\ \le\ & \sum_{j=2}^{\infty} \mathbb{P}(X_i=j)\Big|2b_i\sum_{s=1}^{j}\mathbb{E}\Delta^2 g(S_{n,i}+s) + (2b_i+a_i)\sum_{s=1}^{j-1}\mathbb{E}\Delta^2 g(S_{n,i}+s)(j-s)\Big| \\ +\ & \sum_{j=-\infty}^{-2} \mathbb{P}(X_i=j)\Big|2b_i\Big(-\mathbb{E}\Delta^2 g(S_{n,i}+j+1) + \sum_{s=0}^{-j-2}\mathbb{E}\Delta^2 g(S_{n,i}-s)\Big) \\ +\ & (2b_i+a_i)\sum_{s=0}^{-j-1}(-j-s)\mathbb{E}\Delta^2 g(S_{n,i}-s)\Big| + \mathbb{P}(X_i=1)|2b_i\mathbb{E}\Delta^2 g(S_{n,i}+3)| \\ +\ & \mathbb{P}(X_i=-1)|(a_i+2b_i)\mathbb{E}\Delta^2 g(S_{n,i}) - 2b_i| \\ \le\ & \|\Delta g\| d_+^{(i)}\Big(4|b_i|\mathbb{E}|X_i| + |a_i+2b_i|\,\mathbb{E}|X_i(X_i-1)|\Big). \end{aligned}$$

We apply (F.13) with $\lambda_1 = 2\mathbb{E}S_n - \operatorname{var} S_n$, $2\lambda_2 = \operatorname{var} S_n - \mathbb{E}S_n$, $\lambda = \mathbb{E}S_n$, $\delta_1 = 0$, $\delta_2 = \sum_{i=1}^n \hat{\psi}_i d_+^{(i)}$ and $\varepsilon_\lambda = 0$ to complete the proof of Theorem 7.4. □

Proof of Theorem 7.5. The proof is similar to the proof of Theorem 7.4. According to (F.13) in example F.5, we need to estimate $\mathbb{E}(\mathcal{A}_{\lambda,\widehat{\gamma}}g)(S)$, where

$$(\mathcal{A}_{\lambda,\widehat{\gamma}}g)(j) = \widehat{\lambda}_1 g(j+1) + 2\widehat{\lambda}_2 g(j+2) + m\widehat{\lambda}_m g(j+m) - (j-\widehat{\gamma})g(j) \qquad (j\in\mathbb{Z}).$$

Recall that $\widehat{\gamma} = \lceil \mu-\sigma^2+m^{-1}(\beta_3-\sigma^2)\rceil$ or $\lceil \mu-\sigma^2+m^{-1}(\beta_3-\sigma^2)\rceil+3$,

$$\begin{aligned}\widehat{\lambda}_1 &= \sigma^2 - (\beta_3-\sigma^2+2m\delta)/(m-1), \quad \delta = \widehat{\gamma}-\mu+\sigma^2-(\beta_3-\sigma^2)/m,\\ \widehat{\lambda}_2 &= m\delta/2(m-2), \quad \widehat{\lambda}_m = (\beta_3-\sigma^2-2\delta m/(m-2))/m^2(m-1).\end{aligned}$$

Note that

$$\begin{aligned}(\mathcal{A}_{\lambda,\widehat{\gamma}}g)(S_n) &= \sum_{i=1}^{n}\Bigg\{\left(\sigma_i^2 - \frac{\beta_{3i}-\sigma_i^2}{m-1}\right)g(S_n+1) + \frac{\beta_{3i}-\sigma_i^2}{m(m-1)}g(S_n+m) - X_i g(S_n)\\ &+ \left(\mu_i-\sigma_i^2+\frac{\beta_{3i}-\sigma_i^2}{m}\right)g(S_n)\Bigg\}\\ &+ \delta\left(\frac{-2m}{m-1}g(S_n+1) + \frac{m}{m-2}g(S_n+2) - \frac{2g(S_n+m)}{(m-1)(m-2)} + g(S_n)\right).\end{aligned}$$

According to (I.6*),

$$g(\omega+l) = g(\omega+1) + (l-1)\Delta g(\omega+1) + \binom{l-1}{2}\Delta^2 g(\omega+1) + R_2(g,\omega,l),$$

where

$$R_2(g,\omega,l) = \begin{cases} \sum_{i=1}^{l-3}\binom{l-i-1}{2}\Delta^3 g(\omega+i) & (l\geq 4),\\ 0 & (1\leq l\leq 3),\\ -\sum_{j=0}^{-l}\binom{-l-j+2}{2}\Delta^3 g(\omega-j) & (l\leq 0).\end{cases}$$

Note that

$$\begin{aligned}\sum_{i=2}^{l-2}\binom{l-i}{2} &= (l-1)(l-2)(l-3)/6 \qquad (l\geq 4),\\ \sum_{i=0}^{-l}\binom{-l-i+2}{2} &= |(l-1)(l-2)(l-3)|/6 \qquad (l\leq 0),\end{aligned}$$

hence

$$|R_2(g,\omega,l)| \leq \tfrac{1}{6}|(l-1)(l-2)(l-3)|\,\|\Delta^3 g\|.$$

Therefore,

$$\begin{aligned}&\left|\mathbb{E}g(S_{n,i}+l+1) - \mathbb{E}g(S_{n,i}+1) - l\mathbb{E}\Delta g(S_{n,i}+1) - \binom{l}{2}\mathbb{E}\Delta^2 g(S_{n,i}+1)\right|\\ &\leq |l^{(3)}|\,\|\Delta g\| d'/6\end{aligned}$$

by (F.6), where $d' = \max_{1\leq i\leq n}\|\mathcal{L}(S_{n,i}) * (I_1-I)^{*2}\|$. Thus,

$$\begin{aligned}\mathbb{E}g(S_n+1) &= \mathbb{E}g(S_{n,i}+1) - \mu_i\mathbb{E}\Delta g(S_{n,i}+1) - \tfrac{1}{2}(\sigma_i^2-\mu_i+\mu_i^2)\mathbb{E}\Delta^2 g(S_{n,i}+1)|\\ &= \vartheta d'\|\Delta g\|\mathbb{E}|X_i^{(3)}|/6,\end{aligned}$$

where $\vartheta \in [0;1]$. Similarly,

$$\begin{aligned}\mathbb{E}g(S_n+k) &= \mathbb{E}g(S_{n,i}+1) + (\mu_i+m-1)\mathbb{E}\Delta g(S_{n,i}+1)\\ &+ \tfrac{1}{2}(\sigma_i^2+\mu_i^2+\mu_i(2m-3)+(k-1)(k-2))\mathbb{E}\Delta^2 g(S_{n,i}+1)\\ &+ \vartheta d'\|\Delta g\|\mathbb{E}|(X_i+k-1)^{(3)}|/6 \qquad (k\in\mathbb{Z}) \end{aligned} \tag{7.42}$$

and

$$\begin{aligned}\mathbb{E}X_i g(S_n) &= \mu_i\mathbb{E}g(S_{n,i}+1) + (\sigma_i^2+\mu_i^2-\mu_i)\mathbb{E}\Delta g(S_{n,i}+1)\\ &+ (\beta_{3i}+3\mu_i\sigma_i^2+\mu_i^3-3\sigma_i^2-3\mu_i^2+2\mu_i)\mathbb{E}\Delta^2 g(S_{n,i}+1)\\ &+ \vartheta d'\|\Delta g\|\mathbb{E}|X_i^{(4)}|/6.\end{aligned}$$

In particular, (7.42) yields expressions for $\mathbb{E}g(S_n)$ and $\mathbb{E}g(S_n+2)$.

Observe that $g(S_n+m)-g(S_n+2)$ equals

$$\begin{aligned}&= \sum_{j=2}^{m-1}\Delta g(S_n+j) = \sum_{j=2}^{m-1}\Big(\Delta g(S_n+1) + \sum_{l=1}^{j-1}\Delta^2 g(S_n+l)\Big)\\ &= (m-2)\Delta g(S_n+1) + \sum_{j=2}^{m-1}\sum_{l=1}^{j-1}\Big(\Delta^2 g(S_n) + \sum_{r=0}^{l-1}\Delta^3 g(S_n+r)\Big)\\ &= (m-2)\Delta g(S_n+1) + (m-1)(m-2)\Delta^2 g(S_n)/2 + \sum_{j=2}^{m-1}\sum_{l=1}^{j-1}\sum_{r=0}^{l-1}\Delta^3 g(S_n+r)\end{aligned}$$

$(m\geq 2)$. Substituting the above expressions into $\mathbb{E}(\mathcal{A}_{\lambda,\gamma}g)(S_n)$, we arrive at

$$\Big|\mathbb{E}(\mathcal{A}_{\lambda,\gamma}g)(S_n)\Big| \leq \frac{d'}{6}\sum_{i=1}^{n}\hat{\psi}_i\|\Delta g\| + \frac{2\delta}{(m-1)(m-2)}\left|\sum_{j=2}^{m-1}\sum_{l=1}^{j-1}\sum_{r=0}^{l-1}\mathbb{E}\Delta^3 g(S_n+r)\right|.$$

By (F.6),

$$\sup_i |\mathbb{E}\Delta^3 g(S_n+i)| \leq d'\|\Delta g\|.$$

Applying estimate (F.13), we get

$$d_{TV}(S_n;Y_m) \leq \frac{1}{1-2\theta}\left\{\frac{d'}{6\widehat{\lambda}}\left(\sum_{i=1}^{n}\hat{\psi}_i + 2|m|\delta\right) + \widehat{\eta} + (1+\widehat{\eta})\mathbb{P}(S_n<\widehat{\gamma})\right\},$$

where $Y_m = \widehat{\gamma} + \pi_{\widehat{\lambda}_1} + 2\pi_{\widehat{\lambda}_2} + m\pi_{\widehat{\lambda}_m}$,

$$\widehat{\lambda} = \widehat{\lambda}_1 + 2\widehat{\lambda}_2 + m\widehat{\lambda}_m,\ \ \widehat{\eta} = \mathbb{P}(Y_m<\widehat{\gamma}),\ \ \theta = \widehat{\lambda}^{-1}(2|\widehat{\lambda}_2| + |m(m-1)||\widehat{\lambda}_m|) < 1/5.$$

Concerning $(1+\widehat{\eta})(1-2\theta)^{-1}\mathbb{P}(S_n<\widehat{\gamma})$, observe that $\widehat{\eta}\leq 1$, $\mu-\widehat{\gamma}=\widehat{\lambda}\geq 5\sigma^2/6$. Indeed,

$$\widehat{\lambda} = \sigma^2 + (\sigma^2-\beta_3)/m - \delta. \tag{7.43}$$

Let us consider all three cases of choosing $\widehat{\gamma}$ and m.

(a) If $\beta_3<\sigma^2$, then $m= -\max\{1;\lceil 8(1-\beta_3/\sigma^2)\rceil\}$, and $\delta\leq 1\leq \sigma^2/24$, since $\sigma^2\geq 24$ by the assumption. If

$$1\geq \lceil 8(1-\beta_3/\sigma^2)\rceil \geq 8(1-\beta_3/\sigma^2),$$

i.e., $\beta_3 \geq 7\sigma^2/8$, then $m=-1$. It follows from (7.43) that

$$\widehat{\lambda} = \beta_3 - \delta \geq 7\sigma^2/8 - \sigma^2/24 = 5\sigma^2/6.$$

If $\lceil 8(1-\beta_3/\sigma^2)\rceil > 1$, then

$$\widehat{\lambda} = \sigma^2 + \frac{\sigma^2 - \beta_3}{\lceil 8(1-\beta_3/\sigma^2)\rceil} - \delta \geq 7\sigma^2/8 - \sigma^2/24 = 5\sigma^2/6.$$

(b) If $\sigma^2 \leq \beta_3 < \sigma^2+3$, then $m=-2$, $\delta \leq 4 \leq \sigma^2/6$ and

$$\widehat{\lambda} = 2\beta_3 - \sigma^2 - \delta \geq \sigma^2 - \sigma^2/6 = 5\sigma^2/6.$$

(c) Consider now the case where $\beta_3 \geq \sigma^2+3$. Then $m=\max\{6; \lceil 8\sigma^{-2}(\beta_3-\sigma^2)\rceil\}$ and $\delta \leq 1 \leq \sigma^2/24$. If $6 \geq \lceil 8\sigma^{-2}(\beta_3-\sigma^2)\rceil$, i.e., $\beta_3 \leq 7\sigma^2/4$, then $m=6$ and

$$\widehat{\lambda} = 7\sigma^2/6 - \beta_3/6 - \delta \geq 7\sigma^2/6 - 7\sigma^2/24 - \sigma^2/24 = 5\sigma^2/6.$$

If $6 < \lceil 8\sigma^{-2}(\beta_3-\sigma^2)\rceil$, then

$$\widehat{\lambda} \geq \sigma^2 - (\beta_3-\sigma^2)/8(\beta_3/\sigma^2-1) - \sigma^2/24 = 5\sigma^2/6.$$

Observe that

$$\begin{aligned} \frac{1+\widehat{\eta}}{1-2\theta}\mathbb{P}(S_n<\widehat{\gamma}) &\leq \frac{10\mathbb{E}(S_n-\mu)^4}{3\widehat{\lambda}^4} \leq \frac{10}{3(5\sigma^2/6)^2}\left(3\sigma^4 + \sum_{i=1}^n \mathbb{E}(X_i-\mu_i)^4\right) \\ &\leq 15 + 5\sum_{i=1}^n \mathbb{E}(X_i-\mu_i)^4/\sigma^4 . \end{aligned}$$

It follows from (F.14) that $\widehat{\eta} \leq \exp\Big(-\widehat{\lambda}/(8(|m|\vee 2)\Big) \leq \exp(-5\sigma^2/48\kappa)$.
According to (C.16),

$$d' = \max_{1\leq i\leq n} \|\mathcal{L}(S_{n,i}) * (I_1-I)^{*2}\|_{TV} \leq 17.1/U.$$

The result follows. □

Proof of Theorem 7.6. Let $S_{n,i} = S_n - X_i$, $\sigma_i^2 = \operatorname{var} X_i$, $\mu_i = \mathbb{E}X_i$, $p_{ij} = \mathbb{P}(X_i=j)$. Then

$$\begin{aligned} \mathbb{E}\Delta g(S_n+1) &= \sum_{j=0}^{\infty} p_{ij}\mathbb{E}\Delta g(S_{i,n}+j+1) \\ &= \mathbb{E}\Delta g(S_{i,n}+1) + \sum_{j=0}^{\infty} p_{ij} \sum_{l=1}^{j} \mathbb{E}\Delta^2 g(S_{i,n}+l). \end{aligned}$$

Denote $T_{ik} = q\mu_i p_{ik} + pkp_{ik} - (k+1)p_{i,k+1}$. Recall that

$$r = \frac{(\mathbb{E}S_n)^2}{\operatorname{var} S_n - \mathbb{E}S_n}, \quad q = \frac{\mathbb{E}S_n}{\operatorname{var} S_n}, \quad p=1-q. \tag{7.44}$$

Therefore,

$$\sum_{i=1}^n \sum_{k=1}^{\infty} kT_{ik} = \sum_{i=1}^n \left(q\mu_i^2 + p(\sigma_i^2+\mu_i^2) - \mathbb{E}X_i(X_i-1)\right) = \sum_{i=1}^n (\mu_i - q\sigma_i^2) = 0.$$

We want to evaluate $\mathbb{P}(S_n \in A) - \mathbb{P}(Y \in A)$ for any $A \subset \mathbb{Z}_+$. We apply Stein's method. According to (F.10), there exists function $g(\cdot) \equiv g_A(\cdot)$ such that

$$(r+j)pg(j+1) - jg(j) = \mathbb{I}\{j \in A\} - \mathbf{NB}(r,p)(A) \qquad (j \geq 1, A \subset \mathbb{Z}_+). \tag{F.10*}$$

Taking into account (F.10*), we derive that $\mathbb{P}(S_n \in A) - \mathbb{P}(Y \in A)$ equals

$$\sum_{i=1}^{n}\sum_{k=1}^{\infty} T_{ik} \sum_{s=1}^{k} \Big[-\sum_{j=0}^{\infty} p_{ij} \sum_{l=1}^{j} \mathbb{E}\Delta^2 g(S_{n,i}+l) + \sum_{m=1}^{s-1} \mathbb{E}\Delta^2 g(S_{n,i}+m) \Big]. \tag{7.45}$$

Recall that

$$\|g\| \leq 1.75/\sqrt{rpq}\,, \ \|\Delta g\| \leq 1/rp, \quad |\mathbb{E}\Delta^k g(X)| \leq 2\|\Delta^{k-1} g\| d_{TV}(X; X+1) \quad (k \geq 1)$$

by (F.5) and (F.11). The assertion of the theorem follows from (7.45) and (C.16). □

Proof of (7.26). Recall that independent r.v.s $\{X_i\}$ have geometric distributions

$$\mathbb{P}(X_i = k) = q_i p_i^k \quad (k \geq 0),$$

where $q_i = 1 - p_i > 1/2 \ \ (i \geq 1)$. Geometric distribution is unimodal, hence

$$d_{TV}(\mathcal{L}(X_i); \mathcal{L}(X_i+1)) = q_i\,.$$

In view of (7.44),

$$rp/q = \mathbb{E}S_n\,.$$

Observe that

$$\begin{aligned} &|q\mu_i p_{ik} + pkp_{ik} - (k+1)p_{i,k+1}| = p_i^k |qp_i + pkq_i - (k+1)q_i p_i| \\ =\ & p_i^k |k - (k+1)p_i|\,|q_i - q| = q_i p_i^k |k - p_i/q_i| |q_i - q|. \end{aligned}$$

Since $p_i < 1/2$, we have $p_i/q_i < 1$, hence

$$\begin{aligned} &\frac{1}{rq}\sum_{i=1}^{n} k\Big(\frac{k-1}{2} + \mu_i\Big)|q_i - q| q_i p_i^k \Big(k - p_i/q_i\Big) \\ =\ & \frac{1}{rq}\sum_{i=1}^{n} |q_i - q| \Big(\frac{1}{2}\mathbb{E}X_i(X_i-1)(X_i-\mu_i) + \mu_i \mathbb{E}X_i(X_i-\mu_i)\Big) \\ =\ & \frac{1}{rp}\sum_{i=1}^{n} |q_i - q| \Big(\frac{1}{2}\mathbb{E}X_i(X_i-1)(X_i-2) + \frac{1}{2}(2-\mu_i)\mathbb{E}X_i(X_i-1) + \mu_i\sigma_i^2\Big) \\ =\ & \frac{1}{rq}\sum_{i=1}^{n} |q_i - q| \Big(3\frac{p_i^3}{q_i^3} + \Big(2 - \frac{p_i}{q_i}\Big)\frac{p_i^2}{q_i^2} + \frac{p_i^2}{q_i^3}\Big) = \frac{3}{rp}\sum_{i=1}^{n} |q_i - q| \frac{p_i^2}{q_i^3} \\ =\ & \frac{3}{q\,\mathbb{E}S_n}\sum_{i=1}^{n} |q_i - q| \frac{p_i^2}{q_i^3} = \frac{3}{\mathbb{E}S_n}\sum_{i=1}^{n} \Big|\frac{1}{q_i} - \frac{1}{q}\Big| \frac{p_i^2}{q_i^2}\,. \end{aligned}$$

Combining these estimates and using (C.16), we derive (7.26). □

Proof of Theorem 7.7. Recall that $X', X_1', \ldots, X_n'$ are i.i.d. symmetric r.v.s taking values in $\mathbb{Z}\backslash\{0\}$, $\operatorname{var} X' = \sigma'^2 < \infty$, $\theta := \lambda_2/\lambda < 1/2$.

We aim at evaluating $d_o(\mathcal{L}(S_n); \mathbf{\Pi}(\lambda, F))$, $d_{TV}(\mathcal{L}(S_n); \mathbf{\Pi}(\lambda, F))$, $d_G(\mathcal{L}(S_n); \mathbf{\Pi}(\lambda, F))$.

Due to symmetry, $\widehat{F}(t)-1 = 2\sum_{j=1}^{\infty}(\cos(tj)-1)\mathbb{P}(X'=j)$, where $\widehat{F}$ denotes the characteristic function of $\mathcal{L}(X')$. Therefore,

$$\widehat{F}(t)-1 = -4\sum_{j=1}^{\infty}\sin^2(tj/2)\mathbb{P}(X'=j).$$

Since $\sin(tj) = 2\sin(tj/2)\cos(tj/2)$,

$$\begin{aligned}|\widehat{F}'(t)| &= \left|2\sum_{j=1}^{\infty} j\sin(tj)\mathbb{P}(X'=j)\right| \le 2\sigma'\Big(\sum_{j=1}^{\infty}\sin^2(tj)\mathbb{P}(X'=j)\Big)^{1/2} \\ &\le 2\sigma'\Big(4\sum_{j=1}^{\infty}\sin^2(tj/2)\mathbb{P}(X'=j)\Big)^{1/2} = 2\sigma'|\widehat{F}(t)-1|^{1/2}, \qquad (7.46)\end{aligned}$$

where we have applied the Bunyakovskiy-Cauchy-Schwartz inequality.

Let Y_F denote a r.v. with distribution $\mathbf{\Pi}(\lambda(1-c_0)/2, F))$, and let

$$\psi_F = \exp\Big(\lambda(1-c_0)(\widehat{F}-1)/4\Big)$$

denote the ch.f. of $\mathbf{\Pi}(\lambda(1-c_0)/4, F))$. Then $\psi_F^2(t)$ is the ch.f. of $\mathcal{L}(Y_F)$.

Recall that $\mathbb{P}(X'\in\mathbb{Z}\setminus\{0\}) = 1$. Applying (E.8), we get

$$\int_{-\pi}^{\pi}\psi_F^2(t)\,\mathrm{d}t \le 13Q_{Y_F}(1/\pi) \le C(\lambda(1-c_0))^{-1/2}. \qquad (7.47)$$

Since $\widehat{F}$ is real, $|\exp(\widehat{F}-1)| = \exp(\widehat{F}-1)$. Due to (D.6),

$$\big|(1-p_i) + p_i\widehat{F}\big| \le \exp\Big(p_i(1-p_i)(\widehat{F}-1)\Big) \le \exp\Big(p_i(1-c_0)(\widehat{F}-1)\Big).$$

Therefore,

$$|\mathbb{E}\exp(\mathrm{i}tS_n)| \le \psi_F^4(t) \qquad (t\in\mathbb{R}).$$

By (I.8),

$$\big|1+p_i(\widehat{F}-1) - \exp(p_i(\widehat{F}-1))\big| \le p_i^2|\widehat{F}-1|^2/2.$$

We repeatedly apply the trivial estimate

$$\prod_{j\neq i}^{n}\exp\Big(p_i(1-c_0)(\widehat{F}-1)\Big) \le \mathrm{e}^{1-c_0}\exp\Big(\lambda(1-c_0)(\widehat{F}-1)\Big) \le 3\psi_F^4 \qquad (7.48)$$

since $\exp\big(p_i(1-c_0)(1-\widehat{F})\big) \le \mathrm{e}^{1-c_0}$.

Let $M_F = \mathcal{L}(S_n) - \mathbf{\Pi}(\lambda; F)$. Then

$$\widehat{M_F}(t) = \mathbb{E}\mathrm{e}^{\mathrm{i}tS_n} - \widehat{\mathbf{\Pi}}(\lambda; F)(t) = \prod_{i=1}^{n}(1+p_i(\widehat{F}-1)) - \prod_{i=1}^{n}\mathrm{e}^{p_i(\widehat{F}-1)}.$$

By (I.1), (7.48), and (I.16),

$$\begin{aligned}|\widehat{M_F}(t)| &\le \psi_F(t)\sum_{i=1}^{n}\big|1+p_i(\widehat{F}-1) - \exp(p_i(\widehat{F}-1))\big| \\ &\le \lambda_2\psi_F(t)|\widehat{F}(t)-1|^2.\end{aligned}$$

Note that

$$|\widehat{F}(t)-1|^2\psi_F(t) \le C\lambda^{-2}.$$

Thus,

$$|\widehat{M}_F(t)| \le C(1-c_0)^{-2}\lambda_2\lambda^{-2}\psi_F(t).$$

Substituting this estimate into (D.12) and applying (7.47), we derive (7.29).

By (7.46),

$$\Big|\Big(1+p_i(\widehat{F}-1) - e^{p_i(\widehat{F}-1)}\Big)'\Big| = p_i|\widehat{F}'|\Big|1-e^{p_i(\widehat{F}-1)}\Big| \le Cp_i^2\sigma'|\widehat{F}-1|^{3/2}.$$

Denote

$$\widehat{V}_i = \prod_{j=1}^{i-1}(1-p_j+p_j\widehat{F}) \prod_{j=i+1}^{n} e^{p_j(\widehat{F}-1)}. \tag{7.49}$$

Then

$$|\widehat{V}_i'| \le C\lambda|\widehat{F}'|\prod_{j\ne i}\exp\Big(p_j(1-c_0)(\widehat{F}-1)\Big) \le C\lambda\sigma'\sqrt{|\widehat{F}-1|}\,\psi_F^4(t)$$

by (7.46). Applying (I.16), we get

$$\begin{aligned}|M_F'(t)| &\le \sum_{i=1}^n\Big|\Big(1+p_i(\widehat{F}-1) - e^{p_i(\widehat{F}_-1)}\Big)'\Big|\psi_F^4(t) + \frac{1}{2}\sum_{i=1}^n p_i^2|\widehat{F}-1|^2|\widehat{V}_i'|\\ &\le C\lambda_2\lambda^{-3/2}(1-c_0)^{-5/2}\psi_F^2(t).\end{aligned}$$

Taking into account (7.47), we prove that

$$\int_{-\pi}^{\pi}|\widehat{M}_F(t)|^2\,dt \le C\Big(\frac{\lambda_2}{\lambda^2(1-c_0)^2}\Big)^2\frac{1}{\sqrt{\lambda(1-c_0)}},$$

$$\int_{-\pi}^{\pi}|\widehat{M}_F'(t)|^2\,dt \le C\Big(\frac{\lambda_2\sigma'}{\lambda^{3/2}(1-c_0)^{5/2}}\Big)^2\frac{1}{\sqrt{\lambda(1-c_0)}}.$$

Applying (D.10) with $b=\sqrt{\lambda}\max\{\sigma';1\}(1-c_0)^{-1/2}$, we derive (7.30).
The proof of (7.31) is more involved. Observe that

$$\begin{aligned}F-I &= \sum_{k=1}^{\infty}\mathbb{P}(X'=k)(I_k+I_{-k}-2I) = \sum_{k=1}^{\infty}\mathbb{P}(X'=k)*(I_k-I)(I-I_{-k})\\ &= (I_1-I)*P',\end{aligned}$$

where $P'=\sum_{k=1}^{\infty}\mathbb{P}(X'=k)(I_{k-1}+\ldots+I)*(I-I_{-k})$. Clearly,

$$e^{p_i(F-I)} - I - p_i(F-I) = \sum_{k=2}^{\infty}\frac{p_i^k(F-I)^{*k}}{k!} = p_i^2(F-I)^{*2}*\sum_{k=0}^{\infty}\frac{p_i^k(F-I)^{*k}}{(k+2)!}.$$

Let V_i be a distribution with a characteristic function given by (7.49), i.e.,

$$V_i = \prod_{j=1}^{*(i-1)}((1-p_j)I+p_jF)*\prod_{j=i+1}^{*n}\exp(p_j(F-I)),$$

where $\prod_{j=s}^{*m} = I$ if $m<s$. By an analogue of (I.1) for convolutions of measures we get

$$\begin{aligned} M_F &= \mathcal{L}(S_n) - \mathbf{\Pi}(\lambda; F) = \sum_{i=1}^{n} \big((1-p_i)I+p_iF) - \exp(p_i(F-I))\big) * V_i \\ &= -(F-I)^{*2} * \sum_{i=1}^{n} p_i^2 V_i * \sum_{m=0}^{\infty} \frac{p_i^k (F-I)^{*m}}{(m+2)!} \\ &= (I_1 - I) * U_F, \end{aligned}$$

where

$$U_F = -(F-I) * P' * \sum_{i=1}^{n} p_i^2 V_i * \sum_{m=0}^{\infty} \frac{p_i^m (F-I)^{*m}}{(m+2)!}.$$

By (A.9), $\|M_F\|_{G} = \|U_F\|_{TV}$. We need to estimate $|\widehat{U}_F(t)|$ and $|\widehat{U}_F'(t)|$.
As in the proof of (7.46), we obtain

$$\begin{aligned} |\widehat{U}_F(t)| &\le C \sum_{k=1}^{\infty} \mathbb{P}(X'=k)|\sin(tk/2)||\widehat{F}-1|\lambda_2 \psi_F^4(t) \\ &\le C\sigma' |\widehat{F}-1|^{3/2} \psi_F^4 \le \frac{C\lambda_2 \sigma' \psi_F^2(t)}{\lambda^{3/2}(1-c_0)^{3/2}}. \end{aligned}$$

Employing (7.46) and (I.16), we get

$$\begin{aligned} |U_F'(t)| &\le C \sum_{k=1}^{\infty} \mathbb{P}(X'=k)|k-1+k-2+\cdots+1||\widehat{F}(t)-1|\lambda_2 \psi_F^4(t) \\ &+ C(\sigma')^2 |\widehat{F}(t)-1|\lambda_2 \psi_F^4(t) + \sum_{k=1}^{\infty} \mathbb{P}(X'=k)k||\sin(tk/2)|\sigma' \sqrt{|\widehat{F}(t)-1|}\,\lambda_2 \psi_F^4(t) \\ &+ \sum_{k=1}^{\infty} \mathbb{P}(X'=k)k|\sin(tk/2)||\widehat{F}(t)-1|\lambda_2 \sigma' \lambda \sqrt{\widehat{F}(t)-1}\,\psi_F^4(t) \\ &\le C(\sigma')^2 |\widehat{F}(t)-1|\lambda_2 \psi_F^4(t) + C(\sigma')^2 \lambda_2 |\widehat{F}(t)-1|\psi_F(t)(\lambda|\widehat{F}-|\psi_F^2(t))\psi_F(t) \\ &\le C(\sigma')^2 (1-c_0)^{-2} \lambda_2 \lambda^{-1} \psi_F(t). \end{aligned}$$

In order to complete the proof of (7.31), we apply (D.10) with $b = \sqrt{\lambda/(1-c_0)}\,\max\{1;\sigma'\}$.

The proof of (7.32) and (7.33) is simpler since, due to the assumptions, we can write $\mathcal{L}(S_n)$ as an exponent of a convergent series:

$$\mathcal{L}(S_n) = \prod_{i=1}^{n}((1-p_i)I+p_iF) = \prod_{i=1}^{n}(I+p_i(F-I)) = \exp(A(F)+B(F)),$$

where

$$A(F) = \sum_{i=1}^{n}\sum_{j=1}^{k} \frac{(-1)^{j+1}p_i^j}{j}(F-I)^{*j}, \quad B(F) = \sum_{i=1}^{n}\sum_{j=k+1}^{\infty} \frac{(-1)^{j+1}p_i^j}{j}(F-I)^{*j}.$$

Note that $H_{n,k} = \exp(A_k(F))$. Since $\widehat{F} \le 1$,

$$\widehat{A}(F)(t) = \lambda(\widehat{F}-1) - \sum_{i=1}^{n}\sum_{j=2}^{k} \frac{p_i^j}{j}(1-\widehat{F})^j \le \lambda(\widehat{F}-1), \quad \widehat{A}(F)(t)+\widehat{B}(F)(t) \le \lambda(\widehat{F}-1).$$

Besides,

$$|\widehat{B}(F)(t)| \leq \sum_{i=1}^{n} p_i^{k+1}|\widehat{F}-1|^{k+1} \sum_{j=k+1}^{\infty} (2p_i)^{j-k-1} \leq \frac{C(k)\lambda_{k+1}}{1-2c_0}|\widehat{F}-1|^{k+1}.$$

Employing (7.46), we prove that

$$|\widehat{B}(F)'(t)| \leq \lambda_{k+1}|\widehat{F}'||\widehat{F}-1|^{k}(1-2c_0)^{-1} \leq C(k)\lambda_{k+1}|\widehat{F}-1|^{k+1/2}\sigma'(1-2c_0)^{-1}$$

and $|A'| \leq C(k)\lambda|\widehat{F}-1|^{1/2}\sigma'$. Thus,

$$\begin{aligned} &\left|\mathrm{e}^{\widehat{A}(F)(t)}(\mathrm{e}^{\widehat{B}(F)(t)}-1)\right| \leq \mathrm{e}^{\lambda(\widehat{F}-1)}|\widehat{B}(F)(t)| \\ &\leq \frac{C(k)\lambda_{k+1}|\widehat{F}-1|^{k+1}}{1-2c_0}\mathrm{e}^{\lambda(\widehat{F}-1)} \leq \frac{C(s)\lambda_{k+1}}{(1-2c_0)\lambda^{k+1}}\mathrm{e}^{\lambda(\widehat{F}-1)/2}. \end{aligned} \tag{7.50}$$

An analogue of (7.47) is

$$\int_{-\pi}^{\pi} \mathrm{e}^{\lambda(\widehat{F}-1)/2}\,\mathrm{d}t \leq C\lambda^{-1/2}.$$

From (7.50) and (D.12) we derive (7.32). Similarly,

$$\begin{aligned} \left|\left(\mathrm{e}^{\widehat{A}(F)(t)}(\mathrm{e}^{\widehat{B}(F)(t)}-1)\right)'\right| &\leq |\widehat{B}(F)'(t)|\exp\Big(\widehat{A}(F)+\widehat{B}(t)\Big) \\ &+ |\widehat{A}'(t)||\mathrm{e}^{\widehat{A}(F)(t)}(\mathrm{e}^{\widehat{B}(F)(t)}-1)| \\ &\leq C(k)\lambda_{k+1}|\widehat{F}-1|^{k+1/2}\sigma'(1-2c_0)^{-1}\exp\Big(\lambda(\widehat{F}-1)/2\Big) \\ &+ C(k)|\widehat{F}-1|^{1/2}\sigma'\lambda_{k+1}(1-2c_0)^{-1}\lambda^{-k}\exp\Big(\lambda(\widehat{F}-1)/2\Big) \\ &\leq C(k)\sigma'\lambda_{k+1}(1-2c_0)^{-1}\lambda^{-k-1/2}\exp\Big(\lambda(\widehat{F}-1)/3\Big). \end{aligned}$$

Therefore,

$$\int_{-\pi}^{\pi}\left|\mathrm{e}^{\widehat{A}(F)(t)}(\mathrm{e}^{\widehat{B}(F)(t)}-1)\right|^2\mathrm{d}t \leq \frac{C(k)\lambda_{k+1}^2}{(1-2c_0)^2\lambda^{2k+5/2}},$$

$$\int_{-\pi}^{\pi}\left|\left(\mathrm{e}^{\widehat{A}(F)(t)}(\mathrm{e}^{\widehat{B}(F)(t)}-1)\right)'\right|^2\mathrm{d}t \leq \frac{C(k)\lambda_{k+1}^2}{(1-2c_0)^2\lambda^{2k+3/2}}.$$

To finish the proof of (7.33), it suffices to apply (D.10) with $b=\sqrt{\lambda}\max\{1;\sigma'\}$. □

Proof of Theorem 7.8. We use the method of characteristic functions and Lemma D.16. We need to evaluate the difference of characteristic functions twice: for all t and in a neighbourhood of zero. Moreover, the support of the distribution must be truncated to a set containing no more than Cn points, where C is an absolute constant. Lemma D.17 allows us to choose such a truncation.

Recall that $X_1,\dots,X_n$ are independent r.v.s taking on values $x_0=0,x_1,x_2,\dots,x_N$, where $\{x_j\}$ and N are fixed numbers,

$$G_4 = \exp\left(\sum_{k=1}^{n}(\mathcal{L}(X_k)-I) - \frac{1}{2}\sum_{k=1}^{n}(\mathcal{L}(X_k)-I)^{*2}\right).$$

Let $\widehat{F}_j$ denote the characteristic function of $\mathcal{L}(X_j)$, and let U_0 denote a distribution with non-negative characteristic function

$$\widehat{U}_0 = \exp\Big(\tilde{C}\sum_{j=1}^{n}(Re\widehat{F}_j-1)/10\Big)$$

We apply Lemma D.16 with $U=U_0$, $\mathbf{u}=(0,x_1,x_2,\dots,x_N)$, $m{=}90n$, $M=\mathcal{L}(S_n)-G_4$.

Observe that

$$\ln(m+1)\le C\ln n,\quad \|\ln G_4\|_{TV}\le 2n,\quad \operatorname{supp}\mathcal{L}(S_n)\subset K_a(\mathbf{u}),\quad \operatorname{supp}\mathcal{L}(H_{N,2})\subset K_{3a}(\mathbf{u}).$$

Applying Lemma D.17 with $b_1{=}8n$, $y{=}30$, we derive that $\delta(M,90n,\mathbf{u})\le \mathrm{e}^{-n}$.

It remains to estimate the difference of characteristic functions in a neighbourhood of zero. Note that

$$\widehat{F}_j = p_{j0}+(1-p_{j0})\widehat{V}_j,\quad \widehat{V}_j=\sum_{k=1}^{n}\mathrm{e}^{\mathrm{i}tx_k}p_{jk}\,.$$

Applying (D.6), we get

$$|\widehat{F}_j|\le \exp\Big(p_{j0}(1-p_{j0})(Re\,\widehat{V}_j-1)\Big)\le \exp\Bigg(-2\tilde{C}\sum_{k=1}^{N}p_{jk}\sin^2(tx_k/2)\Bigg)$$

and

$$\begin{aligned}
Re\,\widehat{F}_j-1+|\widehat{F}_j-1|^2/2 &\le (1-p_{j0})(Re\,\widehat{V}_j-1)+(1-p_{j0})^2|\widehat{V}_j-1|^2/2\\
&\le (1-p_{j0})(Re\,\widehat{V}_j-1)+(1-p_{j0})^2(1-Re\widehat{V}_j)\\
&\le \tilde{C}(1-p_{j0})(Re\,\widehat{V}_j-1)=-2\tilde{C}\sum_{k=1}^{N}p_{jk}\sin^2(tx_k/2).
\end{aligned}$$

Denote

$$\widehat{G}_{j4}=\exp\Big((\widehat{F}_j-1)-(\widehat{F}_j-1)^2/2\Big).$$

Then

$$|\widehat{F}_j-\widehat{G}_{j4}|\le C|\widehat{F}_j-1|^2\le C(N)\sum_{k=1}^{N}p_{jk}^3|\sin(tx_k/2)|^3\,.$$

Therefore,

$$\begin{aligned}
\Big|\prod_{j=1}^{n}\widehat{F}_j-\widehat{G}_4\Big| &\le C(N)\widehat{U}_0\exp\Bigg(-\tilde{C}\sum_{l=1}^{n}\sum_{k=1}^{N}p_{lk}\sin^2(tx_k/2)\Bigg)\sum_{j=1}^{n}|\widehat{F}_j-\widehat{G}_{j4}|\\
&\le C(N)\widehat{U}_0\exp\Bigg(-\tilde{C}\sum_{l=1}^{n}\sum_{k=1}^{N}p_{lk}\sin^2(tx_k/2)\Bigg)\sum_{j=1}^{n}\sum_{k=1}^{N}p_{jk}^3|\sin(tx_k/2)|^{3/2}\\
&\le C(N)\widehat{U}_0\sum_{j=1}^{n}\sum_{k=1}^{N}p_{jk}^3\Bigg(|\sin(tx_k/2)|^{3/2}\exp\Bigg(-\tilde{C}\sum_{l=1}^{n}p_{lk}\sin^2(tx_k/2)\Bigg)\Bigg)\\
&\le C(N)\widehat{U}_0\sum_{j=1}^{n}\sum_{k=1}^{N}p_{jk}^3\Bigg(\sum_{l=1}^{n}p_{lk}\Bigg)^{-3/2}.
\end{aligned}$$

Notice that $|\sin(tx_k/2)| \le |t| \max_j |x_j|/2$, so that

$$\Big|\prod_{j=1}^{n}\widehat{F}_j(t)-\widehat{G}_4(t)\Big| \le C(N)|t|\max_j|x_j|\widehat{U}_0(t)\sum_{j=1}^{n}\sum_{k=1}^{N}p_{jk}^3\Big(\sum_{l=1}^{n}p_{lk}\Big)^{-1}$$

if $|t| \le \pi/\max_j |x_j|$.

By (E.8),

$$\int\limits_{|t|<\pi/2\max|x_j|}\widehat{U}_0(t)\,\mathrm{d}t \le CQ_{U_0}(2\max_j|x_j|) \le \frac{C\max|x_j|}{\min|x_j|}Q_{U_0}(\min_j|x_j|/2).$$

Note that $Re\,\widehat{F}_j=(\widehat{F}_j+\widehat{F}_j^-)/2$, where $\widehat{F}_j^-(t)=\widehat{F}_j(-t)$. Moreover, $F_j^-(\{x\colon|x|>h\})=F_j(\{x\colon|x|>h\})$. By the property (E.7) of concentration functions,

$$Q_{U_0}(h) \le \Big(\frac{96}{95}\Big)^2 h\int\limits_{|t|<1/h}\widehat{U}_0(t)\,\mathrm{d}t \le \Big(\frac{96}{95}\Big)^2 h\int\limits_{|t|<1/h}\exp\Big(\tilde{C}\sum_{j=1}^{n}(Re\widehat{F}_j(t)-1)/10\Big)dt$$

for any $h>0$. By (E.6), $Q_{U_0}(h) \le C\big(\sum_{j=1}^{n}F_j(|x|\ge h)\big)^{-1/2}$.

We want h to be so small that $F_j(\{x\colon|x|\ge h\})=\mathbb{P}(|X_j|>0)$ for all j. Since $\{X_j\}$ are discrete and concentrated at $0, x_j, \dots$, the choice $h=\min|x_j|/2$ fits.

We apply (E.1) in order to switch from $Q_{U_0}(2\max|x|_j/\pi)$ to $Q_{U_0}(\min|x|_j/2)$:

$$\begin{aligned}Q_{U_0}(2\max|x|_j/\pi) &\le \frac{8\max|x|_j}{\pi\min|x|_j}Q_{U_0}(\min_j|x_j|/2)\\ &\le C\Big(\sum_{j=1}^{n}\mathbb{P}(|X_j|>0)\Big)^{-1/2} \le C\Big(\sum_{j=1}^{n}\sum_{k=1}^{N}p_{jk}\Big)^{-1/2}.\end{aligned}$$

An application of Lemma D.16 completes the proof. □

Proof of Proposition 7.9. Let $P_1=\mathcal{L}(X')$, $P_2=\mathcal{L}(X'')$,

$$\tilde{F}=\frac{p_0}{1-p_2}I+\frac{p_1}{1-p_2}P_1.$$

Then

$$\mathcal{L}(X)=(1-p_2)\tilde{F}+p_2P_2,\quad (1-p_2)(\tilde{F}-I)=p_1(P_1-I).$$

Observe that

$$d_{TV}(\mathcal{L}(S_n);\mathbf{\Pi}(n;X))=\frac{1}{2}\Big\|\Big((1-p_2)I+p_2\tilde{F}\Big)^{*n}-\exp\Big(n(1-p_2)(\tilde{F}-I)\Big)\Big\|_{TV} \le \mathcal{S}_1+\mathcal{S}_2+\mathcal{S}_3,$$

where

$$\begin{aligned}\mathcal{S}_1 &= \sum_{j=1}^{n}\binom{n}{j}(1-p_2)^jp_2^{n-j}\Big\|\tilde{F}^{*j}-\exp\Big(j(\tilde{F}-I)\Big)\Big\|_{TV}\|P_2^{*n-j}\|_{TV},\\ \mathcal{S}_2 &= \sum_{j=1}^{n}\binom{n}{j}(1-p_2)^jp_2^{n-j}\Big\|\exp\Big(j(\tilde{F}-I)\Big)-\exp\Big(n(1-p_2)(\tilde{F}-I)\Big)\Big\|_{TV}\|P_2^{*n-j}\|_{TV},\\ \mathcal{S}_3 &= \Big\|\exp\Big(n(1-\bar{p}_2)(\tilde{F}-I)\Big)\Big\|_{TV}\Big\|((1-p_2)I+p_2P_2)^{*n}-\exp(np_2(P_2-I))\Big\|_{TV}.\end{aligned}$$

The total variation of any distribution equals 1. Therefore,

$$\left\|P_2^{*(n-j)}\right\|_{TV} = \left\|\exp\Big(n(1-p_2)(\tilde{F}-I)\Big)\right\|_{TV} = 1.$$

We apply (7.30) and (I.3) to the remaining norm in $\mathcal{S}_1$ in order to prove that $\mathcal{S}_1 \le C\sqrt{\sigma_1+1}/n$.

Similarly, (C.30) and (I.4) allow us to derive $\mathcal{S}_2 \le C\sqrt{\sigma_2+1}\sqrt{p_2}/\sqrt{n}$.

Since the total variation norm is invariant to scale transform, we can apply (7.30) to prove that $\mathcal{S}_3 \le C\sqrt{\sigma_2+1}/n$. Collecting these estimates, we get (7.34) if $p_2 \le p_1$.

Observe that

$$\mathcal{L}(X) = (1-p_1)\hat{F} + p_1 P_1\,, \quad \hat{F} = \frac{p_0}{1-p_1} I + \frac{p_2}{1-p_1} P_2\,.$$

Repeating the previous argument, we obtain (7.34) in the case $p_2 > p_1$. □

Exercises

7.1. Let S_{2n} be a Binomial $\mathbf{B}(2n, 1/2)$ r.v., and let π_n, π'_n be independent Poisson $\mathbf{\Pi}(n/4)$ r.v.s. Show that there exists an absolute constant C such that

$$d_{TV}(S_{2n}; \pi_n - \pi'_n + n) \le C/n \qquad (n \in \mathbb{N}).$$

7.2. Let $X, X_1, X_2, \ldots$ be i.i.d.r.v.s,

$$\mathbb{P}(X=1)=p_1\,, \quad \mathbb{P}(X=2)=p_2\,, \quad \mathbb{P}(X=0)=1-p_1-p_2\,,$$

where p_1, $p_2 \in [0;1]$. Prove that there exists an absolute constant C and a compound Poisson distribution $P_n(p_1, p_2)$ such that

$$\sup_{0\le p_1, p_2 \le 1} d_{TV}(\mathcal{L}(S_n); P_n(p_1, p_2)) \le C n^{-2/3}.$$

7.3. Let $X, X_1, \ldots, X_n$ be i.i.d.r.v.s taking values in $\mathbb{Z}_+$. Assume that $\mathbb{E}X^3 < \infty$ and Franken's condition

$$a = \mathbb{E}X - \mathbb{E}^2X - \mathbb{E}X^{(2)} > 0$$

holds. Let $Z_n = \pi_{nu} + 2\pi_{nv}$, where π_{nu}, π_{nv} are independent Poisson r.v.s, $u = \mathbb{E}X - \mathbb{E}X^{(2)} + \mathbb{E}^2X$, $v = (\mathbb{E}X^{(2)} - \mathbb{E}^2X)/2$ (note that $v<0$ is allowed). Prove that

$$d_K(S_n; \pi_{nu} + 2\pi_{nv}) \le \frac{C(\mathbb{E}^3X + \mathbb{E}X^{(3)})}{a\sqrt{na}}.$$

7.4. Let S_n be the sum of r.v.s from Theorem 7.7. Suppose that $p := \max_i p_i \le 1/3$. Let

$$G_{n,p} = \mathbf{\Pi}(X', np) * (I - np^2(F-I)^{*2}/2), \quad \sigma'^2 = \operatorname{var} X'.$$

Prove that

$$d_{TV}(\mathcal{L}(S_n); G_{n,p}) \le C\sqrt{\sigma'+1}\,/n^2.$$

7.5. Let $X, X_1, \dots, X_n$ be i.i.d.r.v.s with

$$\mathbb{P}(X=-\sqrt{2}) = \mathbb{P}(X=-1) = \mathbb{P}(X=0) = \mathbb{P}(X=1) = \mathbb{P}(X=\sqrt{2}) = 1/5.$$

Let $S_n = X_1+\dots+X_n$, $\mathcal{L}(X) = F$,

$$D_n = D^{*n} = \exp\big(n(F-I) - n(F-I)^{*2}/2\big) = \mathbf{\Pi}(2n; F) * \mathbf{\Pi}(-n/2; F^{*2}).$$

Prove that

$$d_K(\mathcal{L}(S_n); D_n) \le C/n^2 .$$

Notes

Theorem 7.2 is a generalisation of Theorem 2.1 from [75] to the case of independent non-identically distributed Bernoulli r.v.s with an addition of non-uniform estimates and an estimate in terms of the Gini-Kantorovich distance.

Theorem 7.4 is Corollary 4.4 from [25].

Theorem 7.5 is Theorem 4.3 from [29].

Theorem 7.7 generalises Corollary 1 from [62].

Theorems 7.6 and 7.8 were proved in [307] and [82] respectively.

8

Sums of m-dependent integer-valued random variables

This chapter presents results on the accuracy of compound Poisson approximation to the distribution of a sum $S_n = X_1 + \cdots + X_n$ of m-dependent integer-valued random variables.

8.1 Compound Poisson approximation

Let $X_1, \ldots, X_n$ be m-dependent random variables, $S_n = X_1 + \cdots + X_n$. By grouping consecutive summands one can rewrite S_n as a sum of 1-dependent r.v.s. In this section, we deals with the sum of 1-dependent non-negative integer-valued random variables.

Given a sequence $\{Y_k\}_{k\geq 1}$ of real or complex-valued random variables, we define

$$\widehat{\mathbb{E}}(Y_1) = \mathbb{E}Y_1\,,\ \widehat{\mathbb{E}}(Y_1, Y_2) = \mathbb{E}Y_1Y_2 - \mathbb{E}Y_1\mathbb{E}Y_2\,,$$

$$\widehat{\mathbb{E}}(Y_1, Y_2, \cdots, Y_k) = \mathbb{E}Y_1Y_2\cdots Y_k - \sum_{j=1}^{k-1}\widehat{\mathbb{E}}(Y_1, \cdots, Y_j)\mathbb{E}Y_{j+1}\cdots Y_k. \qquad (k\geq 2).$$

If $\{Y_k\}_{k\geq 1}$ are independent, then $\widehat{\mathbb{E}}(Y_1, Y_2, \cdots, Y_k) = 0 \ (k>1)$.

Recall that the jth factorial moment $\mathbb{E}X(X-1)\cdots(X-j+1)$ of $\mathcal{L}(X)$ is denoted by $\mathbb{E}X^{(j)}$ $(j\geq 1)$. Set

$$\Gamma_1 = \mathbb{E}S_n\,,\quad \Gamma_2 = \frac{1}{2}(\mathrm{var}\,S_n - \mathbb{E}S_n) = \frac{1}{2}\sum_{k=1}^{n}(\mathbb{E}X_k^{(2)} - \mathbb{E}^2X_k) + \sum_{k=2}^{n}\widehat{\mathbb{E}}(X_{k-1}, X_k),$$

$$\Gamma_3 = \frac{1}{6}\sum_{k=1}^{n}\left(\mathbb{E}X_k^{(3)} - 3\mathbb{E}X_k\mathbb{E}X_k^{(2)} + 2\mathbb{E}^3X_k\right) - \sum_{k=2}^{n}\left(\mathbb{E}X_{k-1} + \mathbb{E}X_k\right)\widehat{\mathbb{E}}(X_{k-1}, X_k)$$
$$+\frac{1}{2}\sum_{k=2}^{n}\left(\widehat{\mathbb{E}}(X_{k-1}^{(2)}, X_k) + \widehat{\mathbb{E}}(X_{k-1}, X_k^{(2)})\right) + \sum_{k=3}^{n}\widehat{\mathbb{E}}(X_{k-2}, X_{k-1}, X_k).$$

For the sake of convenience we assume that $\mathbb{E}_k^{(j)} = 0$ if $k\leq 0$; $\sum_k^n = 0$ if $k>n$.
Denote

$$\mathbf{F_n} = \mathcal{L}(S_n),\quad \widehat{\mathbf{F}}_\mathbf{n}(t) = \mathbb{E}e^{itS_n}\,.$$

Note that Γ_1, $2\Gamma_2$, and $6\Gamma_3$ are factorial cumulants of $\mathcal{L}(S_n)$:

$$\widehat{\mathbf{F}}_\mathbf{n}(t) = \exp\left(\Gamma_1(e^{it}-1) + \Gamma_2(e^{it}-1)^2 + \Gamma_3(e^{it}-1)^3 + \ldots\right).$$

DOI: 10.1201/9781003478164-8

Set $\mathbf{D}=\exp\big(\Gamma_1(I_1-I)+\Gamma_2(I_1-I)^{*2}\big)$. Then

$$\mathbf{D}=\exp\big((I_1-I)\mathbb{E}S_n+(I_1-I)^{*2}(\operatorname{var}S_n-\mathbb{E}S_n)/2\big)=\exp((I_1-I)\mu+(I_2-I)\lambda)$$

is a SCP measure corresponding to

$$\pi_\mu+2\pi_\lambda\,,$$

where Poisson random variables π_μ, π_λ are independent,

$$\mu=\Gamma_1-2\Gamma_2=2\mathbb{E}S_n-\operatorname{var}S_n\,,\quad \lambda=\Gamma_2=(\operatorname{var}S_n-\mathbb{E}S_n)/2.$$

Clearly, $\mathbb{E}(\pi_\mu+2\pi_\lambda)=\mathbb{E}S_n$, $\operatorname{var}(\pi_\mu+2\pi_\lambda)=\operatorname{var}S_n$.

Note that $\Gamma_2<0$ in the case of the Binomial distribution.

Set

$$\tilde{\mathbf{D}}=(I_1-I)^{*3}*\Gamma_3*\mathbf{D}.$$

The results of this section are derived assuming the following assumptions $(k=1,\dots,n)$:

$$\mathbb{E}X_k\le 1/100,\quad \mathbb{E}X_k^{(2)}\le\mathbb{E}X_k\,,\quad \mathbb{E}X_k^{(4)}<\infty,\tag{8.1}$$

$$\sum_{k=1}^n\mathbb{E}X_k^{(2)}\le\mathbb{E}S_n/20,\quad \sum_{k=2}^n|\operatorname{cov}(X_{k-1},X_k)|\le\mathbb{E}S_n/20.\tag{8.2}$$

These assumptions hold, e.g., in the case of independent Bernoulli r.v.s taking zero with large probabilities.

Assuming that $X_k\equiv 0$, $k\le 0$, we define

$$\begin{aligned}R_1 &= \sum_{k=1}^n\Big\{\mathbb{E}^3X_k+\mathbb{E}X_k\mathbb{E}X_k^{(2)}+\mathbb{E}X_k^{(3)}+\widehat{\mathbb{E}}_2^+(X_{k-1},X_k)\\ &\quad +(\mathbb{E}X_{k-2}+\mathbb{E}X_{k-1}+\mathbb{E}X_k)\mathbb{E}X_{k-1}X_k+\widehat{\mathbb{E}}^+(X_{k-2},X_{k-1},X_k)\Big\},\\ R_2 &= \sum_{k=1}^n\Big\{\mathbb{E}^4X_k+\mathbb{E}^2X_k^{(2)}+\mathbb{E}X_k^{(4)}+\mathbb{E}^2X_{k-1}X_k\\ &\quad +(\mathbb{E}X_{k-2}+\mathbb{E}X_{k-1}+\mathbb{E}X_k)(\mathbb{E}X_k^{(3)}+\widehat{\mathbb{E}}_2^+(X_{k-1},X_k))\\ &\quad +\sum_{l=0}^3\mathbb{E}X_{k-l}\widehat{\mathbb{E}}^+(X_{k-2},X_{k-1},X_k)+\widehat{\mathbb{E}}_2^+(X_{k-2},X_{k-1},X_k)\\ &\quad +\widehat{\mathbb{E}}_3^+(X_{k-1},X_k)+\widehat{\mathbb{E}}^+(X_{k-3},X_{k-2},X_{k-1},X_k)\Big\},\end{aligned}$$

where

$$\begin{aligned}\widehat{\mathbb{E}}^+(X_1) &= \mathbb{E}X_1,\qquad \widehat{\mathbb{E}}^+(X_1,X_2)=\mathbb{E}X_1X_2+\mathbb{E}X_1\mathbb{E}X_2,\\ \widehat{\mathbb{E}}^+(X_1,\dots,X_k) &= \mathbb{E}X_1\dots X_k+\sum_{j=1}^{k-1}\widehat{\mathbb{E}}^+(X_1,X_2,\dots,X_j)\mathbb{E}X_{j+1}X_{j+2}\cdots X_k,\\ \widehat{\mathbb{E}}_2^+(X_{k-1},X_k) &= \widehat{\mathbb{E}}(X_{k-1}^{(2)},X_k)+\widehat{\mathbb{E}}(X_{k-1},X_k^{(2)}),\\ \widehat{\mathbb{E}}_2(X_{k-2},X_{k-1},X_k) &= \widehat{\mathbb{E}}(X_{k-2}^{(2)},X_{k-1},X_k)+\widehat{\mathbb{E}}(X_{k-2},X_{k-1}^{(2)},X_k)+\widehat{\mathbb{E}}(X_{k-2},X_{k-1},X_k^{(2)}),\\ \widehat{\mathbb{E}}_3(X_{k-1},X_k) &= \widehat{\mathbb{E}}(X_{k-1}^{(3)},X_k)+\widehat{\mathbb{E}}(X_{k-1}^{(2)},X_k^{(2)})+\widehat{\mathbb{E}}(X_{k-1},X_k^{(3)}).\end{aligned}$$

For instance, let $\{X_i\}$ be Bernoulli $\mathbf{B}(p_i)$ random variables. If $\{X_i\}$ are 1-dependent, then in general we can expect $R_1\le C\sum_1^n p_k^{3/2}$. However, if $\{X_i\}$ are independent, then $R_1\sim\sum_1^n p_i^3$, $R_2\sim\sum_{i=1}^n p_i^4$, and the results of this section are comparable to the results from Section 7.1.

Theorem 8.1 *Let conditions (8.1) and (8.2) be satisfied. There exists an absolute constant C such that for all $n \in \mathbb{N}$*

$$d_{TV}(\mathbf{F_n};\mathbf{D}) \le CR_1 \min\{1;(\mathbb{E}S_n)^{-3/2}\}, \tag{8.3}$$
$$d_{TV}(\mathbf{F_n};\mathbf{D}+\tilde{\mathbf{D}}) \le CR_1^2 \min\{1;(\mathbb{E}S_n)^{-3}\} + CR_2 \min\{1;(\mathbb{E}S_n)^{-2}\}. \tag{8.4}$$

If, in addition, $\{X_k\}$ are bounded:

$$|X_k| \le \tilde{C} \qquad (k=1,2,\ldots,n), \tag{8.5}$$

then

$$d_G(\mathbf{F_n};\mathbf{D}) \le CR_1/\mathbb{E}S_n\,. \tag{8.6}$$

If $\{\mathcal{L}(X_i)\}$ do not depend on n and $\{X_i\}$ are bounded i.i.d.r.v.s, then estimates (8.3) and (8.4) are of order $O(n^{-1/2})$ and $O(n^{-1})$, respectively. Thus, the rate of the accuracy of approximation is comparable to that in the central limit theorem (CLT).

If r.v.s $\mathcal{L}(X_i) = \mathbf{B}(p_i)$ and $\{X_i\}$ are independent, then the rate of the accuracy of approximation in (8.3) is $C\sum_1^n p_i^3(1 \vee \sum_1^n p_i)^{-3/2}$, cf. (7.15).

Estimates in terms of the point metric typically provide better rate of the accuracy of approximation by factor $(\mathbb{E}S_n)^{-1/2}$.

Theorem 8.2 *Let conditions (8.1) and (8.2) be satisfied. Then for arbitrary $n \in \mathbb{N}$*

$$d_o(\mathbf{F_n};\mathbf{D}) \le CR_1 \min\{1;(\mathbb{E}S_n)^{-2}\}, \tag{8.7}$$
$$d_o(\mathbf{F_n};\mathbf{D}+\tilde{\mathbf{D}}) \le CR_1^2 \min\{1;(\mathbb{E}S_n)^{-7/2}\} + CR_2 \min\{1;(\mathbb{E}S_n)^{-5/2}\}. \tag{8.8}$$

If $\operatorname{var} S_n > \mathbb{E}S_n$, then the Negative Binomial approximation can be applied.

Let $\mathbf{NB}(r,\bar{q})$ be defined by (1.4), where r and $\bar{q}$ obey

$$r\bar{q}/(1-\bar{q}) = \Gamma_1\,, \quad r\big(\bar{q}/(1-\bar{q})\big)^2 = 2\Gamma_2\,. \tag{8.9}$$

Let

$$\widetilde{\mathbf{NB}}(r,\bar{q}) = \big(\Gamma_3 - 4\Gamma_2^2(3\Gamma_1)^{-1}\big)\,\mathbf{NB}(r,\bar{q}) * (I_1 - I)^{*3}\,,$$

so that $\mathbf{NB}(r,\bar{q}) + \widetilde{\mathbf{NB}}(r,\bar{q})$ is the first-order asymptotic expansion of $\mathbf{NB}(r,\bar{q})$.

Theorem 8.3 *Assume that conditions (8.1) and (8.2) are satisfied and $\Gamma_2 > 0$. Then ($n \in \mathbb{N}$)*

$$d_{TV}(\mathbf{F_n};\mathbf{NB}(r,\bar{q})) \le C \min\big\{1;\Gamma_1^{-3/2}\big)\Big(R_1 + \Gamma_2^2\Gamma_1^{-1}\Big), \tag{8.10}$$
$$d_{TV}(\mathbf{F_n};\mathbf{NB}(r,\bar{q})+\widetilde{\mathbf{NB}}(r,\bar{q})) \le C\Big(R_1^2 \min\big\{1;\Gamma_1^{-3}\big\} + R_2 \min\big\{1;\Gamma_1^{-2}\big\}$$
$$+\Gamma_2^3\Gamma_1^{-2}\min\big\{1;\Gamma_1^{-2}\big\} + \Gamma_2^2\Gamma_1^{-1}|\Gamma_3 - 4\Gamma_2^2(3\Gamma_1)^{-1}|\min\big\{1;\Gamma_1^{-3}\big\}\Big). \tag{8.11}$$

If, in addition, (8.5) is satisfied, then

$$d_G(\mathbf{F_n};\mathbf{NB}(r,\bar{q})) \le C\Big(R_1\Gamma_1^{-1} + \Gamma_2^2\Gamma_1^{-2}\Big). \tag{8.12}$$

If $\mathcal{L}(X_i)$ does not depend on n and is of bounded support, then the accuracy of approximation in (8.10) and (8.11) is $O(n^{-1/2})$ and $O(n^{-1})$, respectively.

Estimate (8.12) is non-trivial if $R_1 = o(\Gamma_1)$.

The next theorem presents estimates in terms of the point metric.

Theorem 8.4 *Assume that conditions (8.1) and (8.2) are satisfied and* $\Gamma_2>0$. *Then for all* $n\in\mathbb{N}$

$$d_{\rm o}(\mathbf{F_n};\mathbf{NB}(r,\bar q)) \le C\min\{1;\Gamma_1^{-2}\}\Big(R_1+\Gamma_2^2\Gamma_1^{-1}\Big), \tag{8.13}$$

$$d_{\rm o}(\mathbf{F_n};\mathbf{NB}(r,\bar q)+\widetilde{\mathbf{NB}}(r,\bar q)) \le C\Big(R_1^2\min\{1;\Gamma_1^{-7/2}\}+R_2\min\{1;\Gamma_1^{-5/2}\}$$
$$+\Gamma_2^3\Gamma_1^{-2}\min\{1;\Gamma_1^{-5/2}\}+\Gamma_2^2\Gamma_1^{-1}|\Gamma_3-4\Gamma_2^2(3\Gamma_1)^{-1}|\min\{1;\Gamma_1^{-7/2}\}\Big). \tag{8.14}$$

If $\{X_i\}$ are independent geometric $\Gamma_0(q_i)$ r.v.s and $\min_{i\le n} q_i$ is bounded away from 0, then the accuracy of approximation in (8.13) and (8.14) is $O(n^{-1})$ and $O(n^{-3/2})$, respectively.

Similar non-uniform estimates can be obtained for probabilities and distribution functions. Denote $\mathbf{F_n}\{x\}=\mathbb{P}(S_n=x)$, $\mathbf{D}\{x\}:=\mathbf{D}(\{x\})$, $\mathbf{NB}(r,\bar q)\{x\}:=\mathbf{NB}(r,\bar q)(\{x\})$. Let $\mathbf{F_n}[x]:=\mathbb{P}(S_n\le x)$, $\mathbf{D}[x]=\mathbf{D}((-\infty,x])$, $\mathbf{NB}(r,\bar q)[x]:=\mathbf{NB}(r,\bar q)((-\infty,x])$.

Theorem 8.5 *Let conditions (8.1), (8.2), and (8.5) be satisfied. Then for any* $x\in\mathbb{N}$

$$\big(1+(x-\Gamma_1)^2/\Gamma_1\big)\,|\mathbf{F_n}[x]-\mathbf{D}[x]| \le CR_1\Gamma_1^{-3/2}, \tag{8.15}$$

$$\big(1+(x-\Gamma_1)^2/\Gamma_1\big)\,|\mathbf{F_n}\{x\}-\mathbf{D}\{x\}| \le CR_1\Gamma_1^{-2}. \tag{8.16}$$

If, in addition, $\Gamma_2>0$, *then*

$$\big(1+(x-\Gamma_1)^2/\Gamma_1\big)\,|\mathbf{F_n}[x]-\mathbf{NB}(r,\bar q)[x]| \le C\big(R_1\Gamma_1^{-3/2}+\Gamma_2^2\Gamma_1^{-5/2}\big), \tag{8.17}$$

$$\big(1+(x-\Gamma_1)^2/\Gamma_1\big)\,|\mathbf{F_n}\{x\}-\mathbf{NB}(r,\bar q)\{x\}| \le C\big(R_1\Gamma_1^{-2}+\Gamma_2^2\Gamma_1^{-3}\big). \tag{8.18}$$

Estimates in terms of the point metric are better by factor $\Gamma_1^{-1/2}$.

The next theorem presents lower bounds.

Theorem 8.6 *Let conditions (8.1), (8.2), and (8.5) be satisfied. Then for any* $n\ge1$ *and any* $\alpha\ge1$

$$d_K(\mathbf{F_n};\mathbf{D})\ge C\,\Gamma_1^{-3/2}\alpha^{-3}\big(|\Gamma_3|-R_1/\alpha\big). \tag{8.19}$$

If, in addition, $\Gamma_2>0$, *then*

$$d_K(\mathbf{F_n};\mathbf{NB}(r,\bar q))\ge\frac{C}{\alpha^3}\bigg(|\Gamma_3-4\Gamma_2^2(3\Gamma_1)^{-1}|\Gamma_1^{-3/2}-\frac{R_1\Gamma_1^{-2}}{\alpha}-\frac{\Gamma_2^3\Gamma_1^{-4}}{\alpha}\bigg). \tag{8.20}$$

Since $d_K(\mathbf{F_n};\mathbf{D})\le d_{TV}(\mathbf{F_n};\mathbf{D})$, Theorem 8.6 provides lower bounds in terms of the total variation distance as well. Note also that in some cases the right-hand sides of (8.19)–(8.20) may be negative.

8.2 Applications

2-runs statistic.

Let $\xi_1,\xi_2,\dots$ be independent Bernoulli $\mathbf{B}(p)$ random variables, where $0<p<1$. Set

$$S_{n,2}=\xi_0\xi_1+\xi_1\xi_2+\cdots+\xi_{n-1}\xi_n\,.$$

Then $\mathbb{E}S_n=np^2$, $\operatorname{var}S_n=np^2+2np^3-3np^4-2p^3(1-p)$.

Denote

$$\mathbf{D}_2 = \exp\bigl(np^2(I_1-I) + \Gamma_2^*(I_1-I)^{*2}\bigr)\,, \quad \mathbf{D}_3 = \exp\bigl(np^2(I_1-I)+\Gamma_2^*(I_1-I)^{*2}+\Gamma_3^*(I_1-I)^{*3}\bigr)\,,$$

where

$$\Gamma_2^* = np^3\Bigl(1-\tfrac{3}{2}p\Bigr) - p^3(1-p)\,, \quad \Gamma_3^* = np^4\Bigl(1-4p+\tfrac{10}{3}p^2\Bigr) - 2p^4(1-p)(1-2p)\,.$$

Note that

$$\mathbf{D}_2 = \mathcal{L}(\pi_{np^2-2\Gamma_2^*} + 2\pi_{\Gamma_2^*}), \quad \mathbf{D}_3 = \mathcal{L}(\pi_{np^2-2\Gamma_2^*+3\Gamma_3^*} + 2\pi_{\Gamma_2^*-3\Gamma_3^*} + 3\pi_{\Gamma_3^*}),$$

where all Poisson variables are independent. The choice of parameters guarantee that $\mathbf{D}_2$ (resp. $\mathbf{D}_3$) has the same two (resp. three) moments as S_n.

Theorem 8.7 *There exists an absolute constant C such that*

$$\begin{aligned} d_{TV}(\mathcal{L}(S_{n,2});\mathbf{D}_2) &\le C\min\{np^4; p/\sqrt{n}\}, \\ d_{TV}(\mathcal{L}(S_{n,2});\mathbf{D}_3) &\le C\min\{np^5; p/n\} \end{aligned}$$

if $p\le 1/5$ and $n\ge 3$. If, in addition, $np^2\ge 1$, then

$$|d_{TV}(\mathcal{L}(S_{n,2});\mathbf{D}_2) - \tilde{C}p/\sqrt{n}\,| \le C\bigl(p^2/\sqrt{n} + 1/n\bigr), \tag{8.21}$$

where $\tilde{C} = (1+4e^{-3/2})\sqrt{2/\pi}$.

Assumptions in Theorem 8.7 are weaker than those in Theorem 8.1.

If p is a constant, then the accuracy of approximation by $\mathbf{D}_3$ is of order $O(n^{-1})$. As follows from (8.21), the accuracy of approximation by $\mathbf{D}_2$ is of correct order. If $p \to 0$, $np^2 \to \infty$, then

$$d_{TV}(\mathcal{L}(S_{n,2});\mathbf{D}_2) \sim \tilde{C}/\sqrt{n}\,.$$

The 2-runs statistic can be extended in natural way to m-runs. when $X_j = \xi_1\xi_2\cdots\xi_m$. Daly [99] has proved that the accuracy of compound Poisson approximation to m-runs $S_{n,m}$ in terms of the total variation distance is of the order $mp^m/(1-5p)$ if $p_i \equiv p < 1/5$ $(\forall i)$. Further reading on the topic includes [59, 60, 88, 308, 90, 91].

Patterns: (k_1, k_2)-events.

Let $\{\xi_i\}$ be independent Bernoulli $\mathbf{B}(p_i)$ random variables $(0<p_i<1)$.

Given two natural numbers k_1, k_2, a (k_1, k_2)-event is a pattern of at least k_1 consecutive failures followed by at least k_2 consecutive successes.

Set

$$Y_j = (1-\xi_{j-m+1})\cdots(1-\xi_{j-k_2})\xi_{j-k_2+1}\cdots\xi_j\,,$$

where $j\in\{m, m+1, \ldots, n\}$, $m=k_1+k_2$, and denote

$$N_n(k_1,k_2) = Y_m + \cdots + Y_n\,, \quad a(p) = (1-p)^{k_1}p^{k_2}\,.$$

In the particular case where $k_1=1$, $k_2=k$ $N_n(1,k)$ is the number of success runs of length $\ge k$.

It is well known that the limiting distribution of $N_n(k_1,k_2)$ is Poisson, the accuracy of Poisson approximation is $O(a(p))$, see [161, 305]. It is proved in [295] that

$$\begin{aligned} \mathbb{E}N_n(k_1,k_2) &= (n-m+1)a(p), \\ \operatorname{var} N_n(k_1,k_2) &= (n-m+1)a(p) + (1-4m+3m^2-n(2m-1))a^2(p). \end{aligned}$$

Under mild assumptions $\operatorname{var} N(n;k_1,k_2) < \mathbb{E}N(n;k_1,k_2)$. Therefore, the Negative Binomial approximation is not applicable.

Assume that

$$k_1{>}0,\quad k_2{>}0,\quad m{\le}n/2,\quad ma(p){\le}0.01,\quad \Gamma_1=(n{-}m{+}1)a(p){\ge}1.$$

The direct consequence of these assumptions is that $n{-}m{+}1{\ge}100m$.

Clearly, $\{Y_1,Y_2,\dots\}$ are m-dependent r.v.s. One can group them in the following way:

$$(Y_m+Y_{m+1}+\dots+Y_{2m-1})+(Y_{2m}+Y_{2m+1}+\dots+Y_{3m-1})+\dots=X_1+X_2+\dots,$$

i.e., $X_1=Y_m+Y_{m+1}+\dots+Y_{2m-1}$, $X_2=Y_{2m}+Y_{2m+1}+\dots+Y_{3m-1}$, etc.

Thus, $N_n(k_1,k_2)$ is a sum of 1-dependent random variables. Each X_j (with exception of the last one) is a sum of m r.v.s.

Let $K=\lfloor\frac{n-m+1}{m}\rfloor$, so that $\frac{n-m+1}{m}=K+\delta$, where $0\le\delta<1$. Only one r.v. among $Y_1,\dots,Y_{2m-1}$ can be equal to 1, all others are equal to 0. Therefore,

$$\mathbb{E}X_j=ma(p),\quad \mathbb{E}X_{K+1}=\delta ma(p)\qquad (j=1,\dots,K).$$

Consequently,

$$\mathbb{E}X_j^{(2)}=\mathbb{E}X_j^{(3)}=\mathbb{E}X_j^{(4)}=\widehat{\mathbb{E}}_2^+(X_1,X_2)=\widehat{\mathbb{E}}_2^+(X_1,X_2,X_3)=\widehat{\mathbb{E}}_3(X_1,X_2)=0.$$

For calculation of $\mathbb{E}X_1X_2$ note that there are the following possibilities: the first summand of X_1 equals 1 ($Y_m=1$) and any of the summands of X_2 equals 1 (m variants); the second summand of X_1 equals 1 and any of the summands of X_2, beginning from the second one, equals 1 ($m{-}1$ cases); etc. Each event has probability $a^2(p)$. Therefore,

$$\mathbb{E}X_1X_2=a^2(p)(m+(m{-}1)+(m{-}2)+\dots+1)=a(p)^2m(m{+}1)/2.$$

Similarly, as $j=1,\dots,K$, $\delta{>}0$,

$$\begin{aligned}
\mathbb{E}X_j=ma(p),\quad \mathbb{E}X_{j-1}X_j&=m(m{+}1)a^2(p)/2,\quad \mathbb{E}X_{K+1}=\delta ma(p),\\
\widehat{\mathbb{E}}(X_{j-1},X_j)&=-m(m{-}1)a^2(p)/2,\\
\mathbb{E}X_{j-2}X_{j-1}X_j&=m(m{+}1)(m{+}2)a^3(p)/6,\\
\widehat{\mathbb{E}}(X_{j-2},X_{j-1},X_j)&=a^3(p)m(m{-}1)(m{-}2)/6,\\
\mathbb{E}X_KX_{K+1}&=\delta m(\delta m{+}1)a^2(p)/2,\\
\widehat{\mathbb{E}}(X_K,X_{K+1})&=a^2(p)\delta m(\delta m{+}1{-}2m)/2,\\
\mathbb{E}X_{K-1}X_KX_{K+1}&=\delta m(\delta m{+}1)(\delta m{+}2)a^3(p)/6,\\
\widehat{\mathbb{E}}X_{K-1}X_KX_{K+1}&=a^3(p)\delta m(9m^2{-}9m{+}2)/6.
\end{aligned}\tag{8.22}$$

Taking into account (8.22), we can calculate Γ_2:

$$\begin{aligned}
2\Gamma_2&=-\left(Km^2a^2(p)+\delta^2m^2a^2(p)\right)-(K{-}1)m(m{-}1)a^2(p)\\
&\quad+\delta ma^2(p)(\delta m{+}1{-}2m)=-a^2(p)m\left(2m(K{+}\delta)-(K{+}\delta)-(m{-}1)\right)\\
&=-a^2(p)\left((n{-}m{+}1)(2m{-}1)-m(m{-}1)\right).
\end{aligned}$$

Similarly,

$$\Gamma_3=a^3(p)\left((n{-}m{+}1)(3m{-}1)(3m{-}2)-4m(2m{-}1)(m{-}1)\right)/6.$$

Noting that $m \geq 2$, we derive

$$\begin{aligned} R_1 &\leq K(ma(p))^3 + 3ma(p)\left((K-2)m(m+1)a^2(p)/2 + \delta m(\delta m+1)a^2(p)/2\right) \\ &\quad +(\delta ma(p))^3 + C(K+\delta)m^3a^3(p) \leq Cm^3a^3(p)(K+\delta) \\ &\leq C(n-m+1)m^2a^3(p). \end{aligned}$$

Similarly,

$$\widehat{\mathbb{E}}^+(X_1,X_2,X_3,X_4) \leq Cm^4a^4(p), \quad R_2 \leq C(n-m+1)m^3a^4(p),$$

and $\Gamma_3 \geq C(n-m+1)m^2a^3(p)$.

Next, we check the conditions (8.2). We have already noted that $\kappa_2(j) = 0$. Note that

$$\begin{aligned} (K-1)|\widehat{\mathbb{E}}(X_1,X_2)| + |\widehat{\mathbb{E}}(X_{K-1},X_K)| &\leq Km(m-1)a^2(p)/2 + \delta m 2ma^2/2 \\ &\leq (K+\delta)2m^2a^2(p) \leq \frac{2ma^2}{n-m+1} = 2ma\Gamma_1 \leq 0.02\Gamma_1 . \end{aligned}$$

It follows from (8.3) and (8.19) that there exist absolute constants C and c such that

$$\frac{ca^{3/2}(p)m^2}{\sqrt{n-m+1}} \leq d_K(\mathcal{L}(N(n;k_1,k_2));\mathbf{D}) \leq d_{TV}(\mathcal{L}(N(n;k_1,k_2));\mathbf{D}) \leq \frac{Ca^{3/2}(p)m^2}{\sqrt{n-m+1}}. \quad (8.23)$$

Thus, upper and lower estimates are of the same order.

Various other approximations to $\mathcal{L}(N_n(k_1,k_2))$ have been suggested in [88, 296, 305, 308]. We present an analogue of (7.2) established in [308]. Let $\mathcal{S} = \{p\colon m(1-p)^{k_1}p^{k_2} \leq 0.01\}$. Recall that $\mathcal{D}$ denotes the set of infinitely divisible distributions. If $p \in \mathcal{S}$, then there exist constants $c(m) < C(m)$ such that

$$c(m)n^{-2/3} \leq \sup_{p\in\mathcal{M}} \inf_{D\in\mathcal{D}} d_{TV}(\mathcal{L}(N_n(k_1,k_2));D) \leq C(m)n^{-2/3} \quad (8.24)$$

as $n \geq C(m)$. It is shown in [308] that $\mathcal{D}$ in (8.24) can be replaced by the class of compound Poisson distributions.

Symmetric three-point distributions.

It is well-known that symmetry radically improves compound Poisson approximation – cf. (7.12), (7.30). A natural question is if symmetry has a similar effect on m-dependent variables? By considering a 1-dependent analogue of (7.12) with explicit dependency structure we show that the answer is affirmative.

We consider sums of 1-dependent symmetric three-point random variables.

Let $\{\xi_j\}_{j\geq 1}$ be independent (not necessarily identically distributed) symmetric r.v.s,

$$\mathbb{P}(\xi_j = -1) = p_j\,, \quad \mathbb{P}(\xi_j = 0) = 1-2p_j\,, \quad \mathbb{P}(\xi_j = 1) = p_j\,.$$

Set $S_n = X_1 + \cdots + X_n$, where $X_j = \xi_j\xi_{j+1}$. Then

$$\mathbb{P}(X_j = 0) = 1-4p_jp_{j+1}\,, \quad \mathbb{P}(X_j = -1) = \mathbb{P}(X_j = 1) = 2p_jp_{j+1}\,.$$

Obviously,

$$\mathbb{E}X_j = 0, \quad \operatorname{var} X_j = 4p_jp_{j+1}\,, \quad \mathbb{E}X_1X_2 = \mathbb{E}\xi_1\xi_2^2\xi_3 = \mathbb{E}\xi_1\mathbb{E}\xi_2^2\mathbb{E}\xi_3 = 0\,.$$

Hence $\operatorname{var}(X_1+X_2) = 4p_1p_2 + 4p_2p_3$.

Let $\tilde{S}_n$ denote the sum of accompanying $\{X_j\}_{j\geq 1}$ independent r.v.s defined by (1.9). Then

$$\mathcal{L}(\tilde{S}_n) = \exp\left(2\sum_{j=1}^{n} p_j p_{j+1}(I_{-1}+I_1-2I)\right).$$

Denote $\sigma^2 = \operatorname{var} S_n = 4\sum_{j=1}^{n} p_j p_{j+1}$.

Theorem 8.8 *Assume that* $\max_{j\leq n} p_j \leq 1/24$. *Then for all* $n\geq 2$

$$d_{TV}(S_n; \tilde{S}_n) \leq \sum_{k=2}^{n-1}(p_{k-1}p_k p_{k+1} + (p_k p_{k+1})^2)\min\{\sigma^{-4}; 1\}.$$

In the case of identically distributed r.v.s with $p_k \equiv p > 0$ the accuracy of approximation in Theorem 8.8 is of order $np^3\min\{1; (np)^{-2}\} = \min\{np^3; p/n\}$.

If p is bounded away from 0, then the accuracy of approximation is $O(n^{-1})$. Moreover, the accuracy of approximation is always at least $O(n^{-1/2})$.

8.3 Proofs

8.3.1 Auxiliary results

Let $X_1,\dots,X_n$ be 1-dependent non-negative integer-valued random variables. . Set

$$z = \mathrm{e}^{\mathrm{i}t}-1, \quad Z_j = \exp(\mathrm{i}tX_j) - 1.$$

We shall omit argument t below (for example, we write φ_k instead of $\varphi_k(t)$). We also assume that $X_m \equiv 0$ if $m \leq 0$. In this section, symbol ϑ denotes real or complex number such that $|\vartheta|\leq 1$.

The following Lemma was proved by Heinrich [149].

Lemma 8.9 *Let* $Z_1, Z_2, \dots, Z_k$ *be 1-dependent complex-valued random variables. If* $\mathbb{E}|Z_j|^k < \infty$ $(j = 1, 2, \dots, k)$, *then for any complex numbers* $a_1, a_2, \dots, a_k$

$$\widehat{\mathbb{E}}(Z_1 + a_1, Z_2 + a_2, \dots, Z_k + a_k) = \widehat{\mathbb{E}}(Z_1, Z_2, \dots, Z_k). \tag{8.25}$$

If $\mathbb{E}|Z_m|^2 < \infty$ $(1 \leq m \leq k)$, *then*

$$|\widehat{\mathbb{E}}(Z_1, \cdots, Z_k)| \leq 2^{k-1}\prod_{m=1}^{k}(\mathbb{E}|Z_m|^2)^{1/2}.$$

The next lemma can be easily proved by induction.

Lemma 8.10 *For all* $t\in\mathbb{R}$ *and* $k\geq 2$

$$\widehat{\mathbb{E}}^+(|Z_1|, \dots, |Z_k|) \leq 4\widehat{\mathbb{E}}^+(|Z_1|, \dots, |Z_{k-1}|). \tag{8.26}$$

The essence of Heinrich's method used below is the possibility to write down the characteristic function of S_n as a product of functions.

Lemma 8.11 *Let* $\max_{1\le k\le n}\mathbb{E}X_k \le 0.01$. *Then*

$$\widehat{\mathbf{F}}_{\mathbf{n}} = \varphi_1\varphi_2\cdots\varphi_n\,,$$

$$\varphi_k = 1 + \mathbb{E}Z_k + \sum_{j=1}^{k-1}\frac{\widehat{\mathbb{E}}(Z_j, Z_{j+1},\dots,Z_k)}{\varphi_j\varphi_{j+1}\cdots\varphi_{k-1}} \qquad (k=2,\dots,n). \tag{8.27}$$

where $\varphi_1 = \mathbb{E}\mathrm{e}^{\mathrm{i}tX_1}$. *Moreover, for* $t\in\mathbb{R}$,

$$|\varphi_k - 1| \le 1/10, \quad 1/|\varphi_k| \le 10/9 \qquad (k=1,\dots,n), \tag{8.28}$$

$$|\varphi_k - 1| \le |z|(0.66\mathbb{E}X_{k-1} + 4.13\mathbb{E}X_k) \quad (k=2,\dots,n), \tag{8.29}$$

$$\begin{aligned}|\varphi_k - 1 - \mathbb{E}Z_k| &\le \sin^2(t/2)(0.38\mathbb{E}X_k + 0.29\mathbb{E}X_{k-1}\\ &+ 15.6\mathbb{E}X_{k-1}X_k + 0.1\mathbb{E}X_{k-2}X_{k-1}) \quad (k=3,\dots,n).\end{aligned} \tag{8.30}$$

Proof of Lemma 8.11. By the symmetry property

$$\widehat{\mathbb{E}}(\mathrm{e}^{\mathrm{i}tX_1},\mathrm{e}^{\mathrm{i}tX_2},\dots,\mathrm{e}^{\mathrm{i}tX_k}) = \widehat{\mathbb{E}}(\mathrm{e}^{\mathrm{i}tX_k},\dots,\mathrm{e}^{\mathrm{i}tX_1}).$$

Therefore, setting $\mathbb{E}\mathrm{e}^{\mathrm{i}tS_0} := 1$, we obtain

$$\begin{aligned}\mathbb{E}\mathrm{e}^{\mathrm{i}tS_k} &= \widehat{\mathbb{E}}(\mathrm{e}^{\mathrm{i}tX_1},\dots,\mathrm{e}^{\mathrm{i}tX_k}) + \sum_{j=1}^{k-1}\mathbb{E}\mathrm{e}^{\mathrm{i}tS_j}\widehat{\mathbb{E}}(\mathrm{e}^{\mathrm{i}tX_{j+1}},\dots,\mathrm{e}^{\mathrm{i}tX_k})\\ &= \mathbb{E}\mathrm{e}^{\mathrm{i}tX_k}\mathbb{E}\mathrm{e}^{\mathrm{i}tS_{k-1}} + \sum_{j=1}^{k-1}\mathbb{E}\mathrm{e}^{\mathrm{i}tS_{j-1}}\widehat{\mathbb{E}}(\mathrm{e}^{\mathrm{i}tX_j},\dots,\mathrm{e}^{\mathrm{i}tX_k})\\ &= \mathbb{E}\mathrm{e}^{\mathrm{i}tX_k}\mathbb{E}\mathrm{e}^{\mathrm{i}tS_{k-1}} + \sum_{j=1}^{k-1}\mathbb{E}\mathrm{e}^{\mathrm{i}tS_{j-1}}\widehat{\mathbb{E}}(Z_j,\dots,Z_k).\end{aligned} \tag{8.31}$$

For the last equality we applied (8.25). Let $\varphi_j(t) = \mathbb{E}\mathrm{e}^{\mathrm{i}tS_j}/\mathbb{E}\mathrm{e}^{\mathrm{i}tS_{j-1}}$. Then (8.27) follows from (8.31) if we divide it by $\mathbb{E}\mathrm{e}^{\mathrm{i}tS_{k-1}}$, provided $|\varphi_1|,\dots|\varphi_{k-2}| > 0$, which will follow from (8.28). For the proof of (8.28) and (8.30) we use mathematical induction and the following trivial inequalities:

$$|z| \le 2, \quad |Z_k| \le 2, \quad |Z_k| \le X_k|z|. \tag{8.32}$$

The first estimate in (8.28) follows from (8.29) and (8.32) and by the assumption of the lemma. The second estimate in (8.28) follows from the first estimate:

$$|\varphi_k| \ge |1 - |\varphi_k - 1|| \ge 9/10.$$

Let us assume that (8.28) holds for $h_1, \dots, h_{k-1}$. From (8.27) and equation (8.28),

$$\begin{aligned}
|\varphi_k - 1 - \mathbb{E}Z_k| &\le \frac{|\widehat{\mathbb{E}}(Z_{k-1}, Z_k)|}{|\varphi_{k-1}|} + \frac{|\widehat{\mathbb{E}}(Z_{k-2}, Z_{k-1}, Z_k)|}{|\varphi_{k-3}\varphi_{k-1}|} \\
&\quad + \frac{|\widehat{\mathbb{E}}(Z_{k-3}, \dots, Z_k)|}{|\varphi_{k-3}\cdots\varphi_{k-1}|} + \frac{|\widehat{\mathbb{E}}(Z_{k-4}, \dots, Z_k)|}{|\varphi_{k-4}\cdots\varphi_{k-1}|} \\
&\quad + \frac{|\widehat{\mathbb{E}}(Z_{k-5}, \dots, Z_k)|}{|\varphi_{k-5}\cdots\varphi_{k-1}|} + \sum_{j=1}^{k-6} \frac{|\widehat{\mathbb{E}}(Z_j, \dots, Z_k)|}{|\varphi_j\varphi_{j+1}\cdots\varphi_{k-1}|} \\
&\le \left(\frac{10}{9}\right)|\widehat{\mathbb{E}}(Z_{k-1}, Z_k)| + \left(\frac{10}{9}\right)^2 |\widehat{\mathbb{E}}(Z_{k-2}, Z_{k-1}, Z_k)| \\
&\quad + \left(\frac{10}{9}\right)^3 |\widehat{\mathbb{E}}(Z_{k-3}, \dots, Z_k)| + \left(\frac{10}{9}\right)^4 |\widehat{\mathbb{E}}(Z_{k-4}, \dots, Z_k)| \\
&\quad + \left(\frac{10}{9}\right)^5 |\widehat{\mathbb{E}}(Z_{k-5}, \dots, Z_k)| + \sum_{j=1}^{k-6} \left(\frac{10}{9}\right)^{k-j} |\widehat{\mathbb{E}}(Z_j, \dots, Z_k)|. \qquad (8.33)
\end{aligned}$$

By (8.32) and Lemma 8.9,

$$\begin{aligned}
\mathbb{E}|Z_j| &\le \mathbb{E}X_j|z| \le 0.02|\sin(t/2)|, \\
\mathbb{E}|Z_j|^2 &\le 2\mathbb{E}|Z_j| \le 2\mathbb{E}X_j|z| = 4\mathbb{E}X_j|\sin(t/2)|, \qquad (8.34) \\
|\widehat{\mathbb{E}}(Z_j, \dots Z_k)| &\le 2^{k-j}2^{(k-j+1)/2}|z|^{(k-j+1)/2}\prod_{l=j}^{k}\sqrt{\mathbb{E}X_l} \\
&\le 2^{2(k-j)-1}|z|^2\sqrt{\mathbb{E}X_k\mathbb{E}X_{k-1})}/10^{k-j-1} \\
&\le 4^{k-j}\,(\mathbb{E}X_k + \mathbb{E}X_{k-1})\sin^2(t/2)/10^{k-j-1} \\
&= 10\sin^2(t/2)(\mathbb{E}X_k + \mathbb{E}X_{k-1})0.4^{k-j}. \qquad (8.35)
\end{aligned}$$

Consequently,

$$\begin{aligned}
\sum_{j=1}^{k-6}\left(\frac{10}{9}\right)^{k-j}|\widehat{\mathbb{E}}(Z_j, \dots, Z_k)| &\le 10\sin^2(t/2)(\mathbb{E}X_k + \mathbb{E}X_{k-1})\sum_{j=1}^{k-6}\left(\frac{4}{9}\right)^{k-j} \\
&\le 2 \times 0.07\sin^2(t/2)(\mathbb{E}X_k + \mathbb{E}X_{k-1}). \qquad (8.36)
\end{aligned}$$

By 1-dependence, (8.32) and Hölder's inequality,

$$\begin{aligned}
|\mathbb{E}Z_{k-j}\cdots Z_k| &\le \prod_{i=k-j}^{k}\sqrt{\mathbb{E}|Z_i|^2} \le \prod_{i=k-j}^{k}\sqrt{2\mathbb{E}X_i|z|} \\
&\le 2^{(j+1)/2}|z|^{(j+1)/2}\sqrt{\mathbb{E}X_{k-1}\mathbb{E}X_k}/10^{j-1} \\
&\le 2^{j-1}|z|^2\frac{\mathbb{E}X_{k-1} + \mathbb{E}X_k}{2}/10^{j-1} \\
&= 2^j\sin^2(t/2)(\mathbb{E}X_{k-1} + \mathbb{E}X_k)/10^{j-1} \qquad (8.37)
\end{aligned}$$

as $j \ge 3$. Moreover, for any j

$$\begin{aligned}
|\mathbb{E}Z_{j-1}Z_j| &\le 2\mathbb{E}|Z_j| \le 4|\sin(t/2)|\mathbb{E}X_j, \\
|\mathbb{E}Z_{j-1}Z_j| &\le |z|^2\mathbb{E}X_{j-1}X_j = 4\sin^2(t/2)\mathbb{E}X_{j-1}X_j \qquad (8.38)
\end{aligned}$$

and

$$|\mathbb{E}Z_{j-2}Z_{j-1}Z_j| \le 2\mathbb{E}|Z_{j-1}Z_j| \le 8\sin^2(t/2)\mathbb{E}X_{j-1}X_j\,. \tag{8.39}$$

Therefore, we derive from (8.34) that

$$\begin{aligned} |\widehat{\mathbb{E}}(Z_{j-1},Z_j)| &\le \mathbb{E}|Z_{j-1}Z_j| + \mathbb{E}X_{j-1}\mathbb{E}X_j|z|^2 \\ &\le 2.02|z|\mathbb{E}X_j \le 0.0404|\sin(t/2)|. \end{aligned} \tag{8.40}$$

Similarly, applying (8.37), (8.38), (8.39), and (8.42), we obtain the following rough estimates:

$$\begin{aligned} |\widehat{\mathbb{E}}(Z_{j-2},Z_{j-1},Z_j)| &\le |z||\sin(t/2)|(0.2\mathbb{E}X_{j-1} + 0.2804\mathbb{E}X_j) \\ &\le 0.01\sin^2(t/2), \\ |\widehat{\mathbb{E}}(Z_{j-3},\dots,Z_j)| &\le |z||\sin(t/2)|(0.044\mathbb{E}X_{j-1} + 0.1348\mathbb{E}X_j) \\ &\le 0.0036\sin^2(t/2), \\ |\widehat{\mathbb{E}}(Z_{j-4},\dots,Z_j)| &\le |z||\sin(t/2)|(0.0169\mathbb{E}X_{j-1} + 0.0405\mathbb{E}X_j) \\ &\le 0.00115\sin^2(t/2). \end{aligned} \tag{8.41}$$

Taking into account that $\mathbb{E}X_{k-1} \le 0.01$, we get

$$\begin{aligned} |\widehat{\mathbb{E}}(Z_{k-1},Z_k)| &\le \mathbb{E}|Z_{k-1}Z_k| + \mathbb{E}X_{k-1}\mathbb{E}X_k|z|^2 \\ &\le \sin^2(t/2)(4\mathbb{E}X_{k-1}X_k + 0.04\mathbb{E}X_k). \end{aligned} \tag{8.42}$$

Similarly, taking into account (8.37)–(8.41), we get

$$\begin{aligned} |\widehat{\mathbb{E}}(Z_{k-2},Z_{k-1},Z_k)| &\le \sin^2(t/2)(8.08\mathbb{E}X_{k-1}X_k + 0.08\mathbb{E}X_{k-2}X_{k-1} + 0.0008\mathbb{E}X_{k-1}), \\ |\widehat{\mathbb{E}}(Z_{k-3},\dots,Z_k)| &\le \sin^2(t/2)(0.3216\mathbb{E}X_{k-1}X_k + 0.08\mathbb{E}X_{k-1} + 0.1\mathbb{E}X_k), \\ |\widehat{\mathbb{E}}(Z_{k-4},\dots,Z_k)| &\le \sin^2(t/2)(0.3632\mathbb{E}X_{k-1}X_k + 0.0176\mathbb{E}X_{k-1} + 0.0248\mathbb{E}X_k\,, \\ |\widehat{\mathbb{E}}(Z_{k-5},\dots,Z_k)| &\le \sin^2(t/2)(0.0944\mathbb{E}X_{k-1}X_k + 0.0068\mathbb{E}X_{k-1} + 0.0091\mathbb{E}X_k)). \end{aligned} \tag{8.43}$$

Combining (8.36), (8.42)–(8.43) with (8.33), we derive (8.30).

For the proof of (8.29) we apply the mathematical induction.

First, we assume that $k \ge 6$. If (8.28) holds for the first $k-1$ functions, then we expand φ_k as in (8.33):

$$|\varphi_k - 1| \le \mathbb{E}|Z_k| + \left(\frac{10}{9}\right)|\widehat{\mathbb{E}}(Z_{k-1},Z_k)| + \cdots + \sum_{j=1}^{k-4}\left(\frac{10}{9}\right)^{k-j}|\widehat{\mathbb{E}}(Z_j,\dots,Z_k)|.$$

Applying (8.37), (8.34) and (8.40)–(8.41), we complete the proof of (8.29) in the case $k \ge 6$. If $k = 1,\dots,6$, then expansion of φ_k in powers of z is shorter and estimates are smaller. □

Lemma 8.12 *Assume that (8.1) – (8.2) hold. Then for all* $t \in \mathbb{R}$

$$|\varphi_k| \le 1 - \lambda_k \sin^2(t/2) \le \exp\bigl(-\lambda_k\sin^2(t/2)\bigr)\,,$$

$$|\widehat{\mathbf{F}}_{\mathbf{n}}| \le \prod_{k=1}^{n}|\varphi_k| \le \exp\bigl(-0.26\Gamma_1\sin^2(t/2)\bigr)\,,$$

where $\lambda_k = 1.606\mathbb{E}X_k - 0.288\mathbb{E}X_{k-1} - 2\mathbb{E}X_k^{(2)} - 0.1\mathbb{E}X_{k-2}X_{k-1} - 15.58\mathbb{E}X_{k-1}X_k\,.$

Proof of Lemma 8.12. It follows from Lemma D.1 that

$$1+\mathbb{E}Z_k = \mathbb{E}\exp(\mathrm{i}tX_k) = 1+\mathbb{E}X_k z + C_k\vartheta\mathbb{E}(X_k-1)|z|^2/2,$$

where $z=\mathrm{e}^{\mathrm{i}t}-1$, C_k is an absolute constant. Therefore,

$$|\varphi_k| \le |1+\mathbb{E}Z_k| + |\varphi_k - 1 - \mathbb{E}Z_k| \le |1+\mathbb{E}X_k z| + \mathbb{E}X_k(X_k-1)/2|z|^2 + |\varphi_k - 1 - \mathbb{E}Z_k|.$$

Note that

$$|1+c(\mathrm{e}^{\mathrm{i}t}-1)|^2 = (1-c+c\cos t)^2 + (c\sin t)^2 = 1-4c(1-c)\sin^2(t/2)$$

if $0<c<1$. Consequently,

$$|1+\mathbb{E}X_k z| \le \sqrt{1-4\mathbb{E}X_k(1-\mathbb{E}X_k)\sin^2(t/2)} \le 1-2\mathbb{E}X_k(1-\mathbb{E}X_k)\sin^2(t/2).$$

Combining the last estimate with (8.30), we get the first estimate of the lemma. The second estimate follows from the fact that

$$\mathbb{E}X_{k-1}X_k \le |\mathrm{cov}(X_{k-1},X_k) + \mathbb{E}X_{k-1}\mathbb{E}X_k| \le |\mathrm{cov}(X_{k-1},X_k)| + 0.01\mathbb{E}X_k\,.$$

Therefore,

$$\sum_{k=1}^{n}\mathbb{E}X_{k-1} - 1.52\sum_{k=1}^{n}\mathbb{E}X_k(X_k-1) - 12\sum_{k=2}^{n}\mathbb{E}X_{k-1}X_k > 0.2\Gamma_1\,.$$

□

In order to expand φ_k in powers of z and to estimate the remainder terms we use the following notation:

$$\begin{aligned}
\gamma_2^*(k) &= \frac{\mathbb{E}X_k^{(2)}}{2} + \widehat{\mathbb{E}}(X_{k-1},X_k),\\
\gamma_3^*(k) &= \frac{\mathbb{E}X_k^{(3)}}{6} + \frac{\widehat{\mathbb{E}}_2(X_{k-1},X_k)}{2} + \widehat{\mathbb{E}}(X_{k-2},X_{k-1},X_k) - \mathbb{E}X_{k-1}\widehat{\mathbb{E}}(X_{k-1},X_k),\\
r_0(k) &= \mathbb{E}X_k^{(2)} + \sum_{l=0}^{3}\mathbb{E}^2X_{k-l} + \mathbb{E}X_{k-1}X_k,\\
r_1(k) &= \mathbb{E}X_k^{(3)} + \sum_{l=0}^{5}\mathbb{E}^3X_{k-l} + \mathbb{E}X_{k-1}\mathbb{E}X_{k-1}X_k + \widehat{\mathbb{E}}_2^+(X_{k-1},X_k)\\
&\quad +\widehat{\mathbb{E}}^+(X_{k-2},X_{k-1},X_k),\\
r_2(k) &= \mathbb{E}X_k^{(4)} + \sum_{l=0}^{7}\mathbb{E}^4X_{k-l} + \mathbb{E}^2X_k^{(2)} + \mathbb{E}^2X_{k-1}^{(2)} + \mathbb{E}^2X_{k-1}X_k + \mathbb{E}^2X_{k-2}X_{k-1}\\
&+ (\mathbb{E}X_{k-2}+\mathbb{E}X_{k-1})\widehat{\mathbb{E}}_2^+(X_{k-1},X_k) + \sum_{l=0}^{3}\mathbb{E}X_{k-l}\widehat{\mathbb{E}}^+(X_{k-2},X_{k-1},X_k)\\
&+ \widehat{\mathbb{E}}_3^+(X_{k-1},X_k) + \widehat{\mathbb{E}}_2^+(X_{k-2},X_{k-1},X_k) + \widehat{\mathbb{E}}^+(X_{k-3},X_{k-2},X_{k-1},X_k).
\end{aligned}$$

Lemma 8.13 *Let condition (8.1) be satisfied* $(k=1,\dots,n)$. *Then there exists an absolute constant* C *such that for all* $t\in\mathbb{R}$ *and* $k=1,\dots,n$

$$\varphi_k = 1+\mathbb{E}X_k z+\vartheta C|z|^2 r_0(k), \tag{8.44}$$

$$\varphi_k = 1+\mathbb{E}X_k z+\gamma_2^*(k)z^2+\vartheta C|z|^3 r_1(k), \tag{8.45}$$

$$\varphi_k = 1+\mathbb{E}X_k z+\gamma_2^*(k)z^2+\gamma_3^*(k)z^3+\vartheta C|z|^4 r_2(k), \tag{8.46}$$

$$(\varphi_k-1)^2 = \mathbb{E}^2X_k z^2+C\vartheta|z|^3(\mathbb{E}X_{k-1}+\mathbb{E}X_k)\Big\{\mathbb{E}X_k^{(2)}+\mathbb{E}X_{k-1}X_k+\sum_{l=0}^{3}\mathbb{E}^2X_{k-l}\Big\}, \tag{8.47}$$

$$(\varphi_k-1)^3 = \mathbb{E}^3X_k z^3+C\vartheta|z|^4\Big\{\mathbb{E}^2X_k^{(2)}+\sum_{l=0}^{3}\mathbb{E}^4X_{k-l}+\mathbb{E}^2X_{k-1}X_k\Big\}. \tag{8.48}$$

For $k=2,\dots,n$

$$\frac{1}{\varphi_{k-1}} = 1+C\vartheta|z|\{\mathbb{E}X_{k-2}+\mathbb{E}X_{k-1}\}, \tag{8.49}$$

$$\frac{1}{\varphi_{k-1}} = 1-\mathbb{E}X_{k-1}z+C\vartheta|z|^2\{\mathbb{E}X_{k-1}^{(2)}+\sum_{l=1}^{4}\mathbb{E}^2X_{k-1}+\mathbb{E}X_{k-2}X_{k-1}\}, \tag{8.50}$$

$$\begin{aligned}\frac{1}{\varphi_{k-1}} &= 1-\mathbb{E}X_{k-1}z-\Big(\mathbb{E}X_{k-1}^{(2)}/2-\mathbb{E}^2X_{k-1}+\widehat{\mathbb{E}}(X_{k-2},X_{k-1})\Big)z^2\\ &+ C\vartheta|z|^3\Big\{\mathbb{E}X_{k-1}^{(3)}+(\mathbb{E}X_{k-2}+\mathbb{E}X_{k-1})(\mathbb{E}X_{k-1}^{(2)}+\mathbb{E}X_{k-2}X_{k-1})\\ &+ \sum_{l=1}^{6}\mathbb{E}^3X_{k-l}+\widehat{\mathbb{E}}_2^+(X_{k-2},X_{k-1})+\widehat{\mathbb{E}}^+(X_{k-3},X_{k-2},X_{k-1})\Big\}.\end{aligned} \tag{8.51}$$

For $k=3,\dots,n$

$$\frac{1}{\varphi_{k-1}\varphi_{k-2}} = 1+C\vartheta|z|\{\mathbb{E}X_{k-3}+\mathbb{E}X_{k-2}+\mathbb{E}X_{k-1}\}. \tag{8.52}$$

Proof of Lemma 8.13. We present the proof in the case $k\geq 7$ (the same argument applies if $k<7$).

The lemma is proved in four steps. First, we prove (8.44), (8.45), (8.49), and (8.52). Second, we derive (8.50) and (8.47). Then we prove (8.51) and (8.48). The final step is to prove (8.46). At each step, we employ results from the previous step. All proofs are similar.

Due to (8.28), we have

$$1/\varphi_{k-1} = \frac{1}{1-(1-\varphi_{k-1})} = 1+\sum_{j=1}^{\infty}(1-\varphi_{k-1})^j = 1+C\vartheta|1-\varphi_{k-1}|.$$

Therefore, (8.49) and (8.52) will follow if we prove (8.29) and (8.28).

From Lemmas 8.11, D.1, 8.10, equation (8.28) and second estimate in (8.35),

$$
\begin{aligned}
|\varphi_k| &= 1+\mathbb{E}Z_k+\frac{|\widehat{\mathbb{E}}(Z_{k-1},Z_k)|}{|\varphi_{k-1}|}+\frac{|\widehat{\mathbb{E}}(Z_{k-2},Z_{k-1},Z_k)|}{|\varphi_{k-2}\varphi_{k-1}|}+\sum_{j=1}^{k-3}\frac{|\widehat{\mathbb{E}}(Z_j,\dots,Z_k)|}{|\varphi_j\varphi_{j+1}\cdots\varphi_{k-1}|}\\
&\leq 1+\mathbb{E}X_kz+C\vartheta|z|^2\mathbb{E}X_k^{(2)}+C\vartheta\widehat{\mathbb{E}}^+(|Z_{k-1}|,|Z_k|)\\
&\quad +C\vartheta|z|^2\sqrt{\mathbb{E}X_{k-3}\mathbb{E}X_{k-2}\mathbb{E}X_{k-1}\mathbb{E}X_k}\\
&= 1+\mathbb{E}X_kz+C\vartheta|z|^2\Big\{\mathbb{E}X_k^{(2)}+\mathbb{E}X_{k-1}X_k+\mathbb{E}X_{k-1}\mathbb{E}X_k+\sum_{l=0}^{3}\mathbb{E}^2X_{k-l}\Big\}\\
&= 1+\mathbb{E}X_kz+C\vartheta|z|^2r_0(k),
\end{aligned}
$$

which proves (8.44).

The proof of (8.45) is almost identical. We take a longer expansion in (8.27) and note that $Z_k=X_kz+\vartheta X_k^{(2)}|z|^2/2$ due to (B.9). Therefore,

$$
\begin{aligned}
\widehat{\mathbb{E}}(Z_{k-1},Z_k) &= \widehat{\mathbb{E}}(X_{k-1}z+\vartheta|z|^2X_{k-1}^{(2)},Z_k)=z\widehat{\mathbb{E}}X_{k-1}Z_k+C\vartheta|z|^3\widehat{\mathbb{E}}^+(X_{k-1}^{(2)},X_k)\\
&= z^2\widehat{\mathbb{E}}(X_{k-1},X_k)+C\vartheta|z|^3\widehat{\mathbb{E}}_2^+(X_{k-1},X_k).
\end{aligned}
$$

The other proofs follow the same lines with the only exception that the results obtained at the previous steps are used. For example, for the proof of (8.46) we apply Lemma 8.11:

$$
|\varphi_k|=1+\mathbb{E}Z_k+\sum_{j=1}^{k-1}\frac{|\widehat{\mathbb{E}}(Z_j,\dots,Z_k)|}{|\varphi_j\varphi_{j+1}\cdots\varphi_{k-1}|}=1+\mathbb{E}Z_k+\left(\sum_{j=k-2}^{k-1}+\sum_{j=k-6}^{k-3}+\sum_{j=1}^{k-7}\right)\frac{|\widehat{\mathbb{E}}(Z_j,\dots,Z_k)|}{|\varphi_j\varphi_{j+1}\cdots\varphi_{k-1}|}.
$$

By (8.35),

$$
\sum_{j=1}^{k-7}\frac{|\widehat{\mathbb{E}}(Z_j,\dots,Z_k)|}{|\varphi_j\varphi_{j+1}\cdots\varphi_{k-1}|}\leq\sum_{j=1}^{k-7}C|z|^4\sqrt{\mathbb{E}X_{k-7}\mathbb{E}X_k}\leq C|z|^4\sum_{l=0}^{7}\mathbb{E}^4X_{k-l}.
$$

By (8.26),

$$
\begin{aligned}
\left|\sum_{j=k-6}^{k-3}\frac{|\widehat{\mathbb{E}}(Z_j,\dots,Z_k)|}{|\varphi_j\varphi_{j+1}\cdots\varphi_{k-1}|}\right| &\leq C\sum_{j=k-6}^{k-3}\widehat{\mathbb{E}}^+(|Z_j|,\dots,|Z_k|)\leq C\widehat{\mathbb{E}}^+(|Z_{k-3}|,\dots,|Z_k|)\\
&\leq C|z|^4\widehat{\mathbb{E}}^+(X_{k-3},\dots,X_k).
\end{aligned}
$$

For other summands, we apply Lemma D.1 and use the previous estimates. □

In order to derive estimates of the accuracy of compound Poisson approximation in terms of the total variation distance by the method of characteristic functions we will need bounds involving derivatives of characteristic functions. Below the prime denotes a derivative with respect to t.

Lemma 8.14 *Assume condition (8.1). Then for all* $t\in\mathbb{R}$

$$
\begin{aligned}
(\widehat{\mathbb{E}}(Z_j,\dots,Z_k))' &= \sum_{i=j}^{k}\widehat{\mathbb{E}}(Z_j,\dots,Z_i',\dots,Z_k),\\
|\widehat{\mathbb{E}}(Z_j,\dots,Z_i',\dots,Z_k)| &\le 2^{[3(k-j)+1]/2}|z|^{(k-j)/2}\prod_{l=j}^{k}\sqrt{\mathbb{E}X_l}\,,\\
|\widehat{\mathbb{E}}(Z_j,\dots,Z_m',\dots,Z_i',\dots Z_k)| &\le 4^{k-j}2^{-1}|z|\prod_{l=j}^{k}\left(\sqrt{\mathbb{E}X_l}\right).
\end{aligned}
$$

If, in addition, (8.5) is satisfied, then there exists an absolute constant C *such that*

$$
|\widehat{\mathbb{E}}(Z_j,\dots,Z_i'',\dots,Z_k)| \le C4^{k-j}\prod_{l=j}^{k}\sqrt{\mathbb{E}X_l}\,.
$$

Proof of Lemma 8.14. The first identity of the lemma is due to Heinrich [149]); it is obvious. Applying (8.37), we obtain

$$
|\widehat{\mathbb{E}}(Z_j,\dots,Z_i',\dots,Z_k)| \le 2^{k-j}\sqrt{\mathbb{E}|Z_i'|^2}\prod_{l\ne i}^{k}\sqrt{\mathbb{E}|Z_l|^2}\,.
$$

Due to assumption (8.1), $\mathbb{E}X_1(X_1-1)\le\mathbb{E}X_1$. Therefore,

$$
\mathbb{E}|Z_l'|^2=\mathbb{E}|\mathrm{i}\mathrm{e}^{\mathrm{i}tX_l}X_l|^2=\mathbb{E}X_l^2=\mathbb{E}X_l(X_l-1+1)=\mathbb{E}X_l(X_l-1)+\mathbb{E}X_l\le 2\mathbb{E}X_l\,.
$$

Combining the last estimate with the bound $\mathbb{E}|Z_l|^2\le 2\mathbb{E}|Z_l|\le 2|z|\mathbb{E}X_l$, the second statement follows. Similarly,

$$
\begin{aligned}
&|\widehat{\mathbb{E}}(Z_j,\dots,Z_m',\dots,Z_i',\dots Z_k)| \le 2^{k-j}\sqrt{\mathbb{E}|Z_m'|^2\mathbb{E}|Z_i'|^2}\prod_{l=j,l\ne m,i}(2|z|\mathbb{E}X_1)^{1/2}\\
&\le 2^{k-j+1}\sqrt{\mathbb{E}X_m\mathbb{E}X_i}2^{(k-j-1)/2}|z|^{(k-j-1)/2}\prod_{l=j,l\ne m,i}^{k}\sqrt{\mathbb{E}X_l}\le 4^{k-j}2^{-1}|z|\prod_{l=j}^{k}\sqrt{\mathbb{E}X_l}\,,
\end{aligned}
$$

and

$$
\begin{aligned}
|\widehat{\mathbb{E}}(Z_j,\dots,Z_i'',\dots,Z_k)| &\le 2^{k-j}\sqrt{\mathbb{E}|Z_i''|^2}\prod_{l=j,l\ne i}^{k}\sqrt{2|z|\mathbb{E}X_l}\\
&\le 2^{k-j}\sqrt{\mathbb{E}X_i^4}2^{(k-j)/2}|z|^{(k-j)/2}\prod_{l=j,l\ne i}^{k}\sqrt{\mathbb{E}X_l}\\
&\le 4^{k-j}2^{-1}|z|\tilde{C}\sqrt{\mathbb{E}X_i^2}\prod_{l=j,l\ne i}^{k}\sqrt{\mathbb{E}X_l}\le 4^{k-j}2^{-1/2}\tilde{C}\prod_{l=j}^{k}\sqrt{\mathbb{E}X_l}\,.
\end{aligned}
$$

The proof is complete. □

Lemma 8.15 *Let condition (8.1) be satisfied* ($k=1,\dots,n$), *and let* φ_k *be defined as in Lemma 8.11. Then for all* $t\in\mathbb{R}$

$$\varphi_k' = 33\vartheta[\mathbb{E}X_k+\mathbb{E}X_{k-1}], \tag{8.53}$$

$$\varphi_k' = \mathbb{E}X_k z' + \vartheta C|z|(r_0(k)+\widehat{\mathbb{E}}^+(X_{k-2},X_{k-1})), \tag{8.54}$$

$$\varphi_k' = \mathbb{E}X_k z' + \gamma_2^*(k)(z^2)' + \vartheta C|z|^2\Big([\mathbb{E}X_{k-2}+\mathbb{E}X_k]\mathbb{E}X_{k-1}X_k + \widehat{\mathbb{E}}^+(X_{k-4},X_{k-3},X_{k-2}) + \widehat{\mathbb{E}}^+(X_{k-3},X_{k-2},X_{k-1}) + r_1(k)\Big), \tag{8.55}$$

$$|\varphi_k' - \mathbb{E}X_k z' - \gamma_2^*(k)(z^2)' - \gamma_3^*(k)(z^3)'| \leq C|z|^3\Big(r_2(k)+\widehat{\mathbb{E}}^+(X_{k-4},\dots,X_{k-1})+\widehat{\mathbb{E}}^+(X_{k-5},\dots,X_{k-2})\Big). \tag{8.56}$$

If, in addition, (8.5) is satisfied, then there exists an absolute constant C *such that*

$$\begin{aligned}\varphi_k'' &= C(\mathbb{E}X_k+\mathbb{E}X_{k-1}),\\ \varphi_k'' &= \mathbb{E}X_k z'' + \gamma_2^*(k)(z^2)'' + C|z|r_1(k),\end{aligned}$$

where $r_1(k), r_2(k)$ *are defined before Lemma 8.13.*

Proof of Lemma 8.15. Note that

$$\left(\frac{\widehat{\mathbb{E}}(Z_j,\dots,Z_k)}{\varphi_j\cdots\varphi_{k-1}}\right)' = \frac{(\widehat{\mathbb{E}}(Z_j,\dots,Z_k))'}{\varphi_j\cdots\varphi_k} - \frac{\widehat{\mathbb{E}}(Z_j,\dots,Z_k)}{\varphi_j\cdots\varphi_k}\sum_{m=j}^{k-1}\frac{\varphi_m'}{\varphi_m}.$$

The proof by mathematical induction is just a repetition of the proof of Lemma 8.13. For example, (8.53) is easily verifiable for $k=0,1$. Assume that (8.53) holds for $\varphi_1,\varphi_2,\dots\varphi_{k-1}$. From Lemmas 8.11, (D.1) and (8.28) we derive

$$\begin{aligned}|\varphi_k'| &\leq \mathbb{E}X_k + \sum_{j=1}^{k-1}\frac{|(\widehat{\mathbb{E}}(Z_j,\dots,Z_k))'|}{|\varphi_j\cdots\varphi_{k-1}|} + \sum_{j=1}^{k-1}\frac{|\widehat{\mathbb{E}}(Z_j,\dots,Z_k)|}{|\varphi_j\cdots\varphi_{k-1}|}\sum_{m=j}^{k-1}\frac{|\varphi_m'|}{|\varphi_m|}\\ &\leq \mathbb{E}X_k + \sum_{j=1}^{k-1}\left(\frac{10}{9}\right)^{k-j}\sum_{i=j}^{k}|\widehat{\mathbb{E}}(Z_j,\dots,Z_i',\dots,Z_k)|\\ &+ \sum_{j=1}^{k-1}\left(\frac{10}{9}\right)^{k-j+1}|\widehat{\mathbb{E}}(Z_j,\dots,Z_k)|(k-j)33\cdot 0.02.\end{aligned}$$

By Lemma 8.14,

$$|\widehat{\mathbb{E}}(Z_j,\dots,Z_i',\dots,Z_k)| \leq [\mathbb{E}X_{k-1}+\mathbb{E}X_k](0.04)^{k-j}10/\sqrt{2}.$$

Combining the last two estimates and (8.35), we complete the proof of (8.53) for φ_k.

We omit the proofs of remaining expressions. Note only that due to (I.8)

$$\begin{aligned}(\mathrm{e}^{\mathrm{i}tX}-1)' &= \mathrm{i}X\mathrm{e}^{\mathrm{i}tX} = \mathrm{i}\mathrm{e}^{\mathrm{i}t}X\mathrm{e}^{\mathrm{i}t(X-1)}\\ &= z'X\Big(1+(X-1)z+\vartheta(X-1)(X-2)|z|^2/2\Big),\\ (\mathrm{e}^{\mathrm{i}tX})'' &= \mathrm{i}^2X^2\mathrm{e}^{\mathrm{i}tX} = \mathrm{i}^2X(X-1)(\mathrm{e}^{\mathrm{i}t})^2\mathrm{e}^{\mathrm{i}t(X-2)} + \mathrm{i}^2\mathrm{e}^{\mathrm{i}t}X\mathrm{e}^{\mathrm{i}t(X-1)}\\ &= \mathrm{i}^2(\mathrm{e}^{\mathrm{i}t})^2X(X-1)[1+\vartheta(X-2)|z|]\\ &\quad+\mathrm{i}^2\mathrm{e}^{\mathrm{i}t}X[1+(X-1)z+\vartheta(X-1)(X-2)|z|^2/2]\\ &= Xz''+X(X-1)(z^2)''/2+\vartheta 2X(X-1)(X-2)|z|\end{aligned}$$

and

$$\varphi_k'' = (\mathbb{E}Z_k)'' + \sum_{j=1}^{k-1} \frac{\Psi_{j,k}''}{\varphi_j\cdots\varphi_{k-1}} - 2\sum_{j=1}^{k-1} \frac{\Psi_{j,k}'}{\varphi_j\cdots\varphi_{k-1}} \sum_{i=j}^{k-1} \frac{\varphi_i'}{\varphi_i}$$
$$+\sum_{j=1}^{k-1} \frac{\Psi_{j,k}}{\varphi_j\cdots\varphi_{k-1}} \left(\sum_{i=j}^{k-1} \frac{\varphi_i'}{\varphi_i}\right)^2 + \sum_{j=1}^{k-1} \frac{\Psi_{j,k}}{\varphi_j\cdots\varphi_{k-1}} \sum_{i=j}^{k-1} \left(\frac{\varphi_i'}{\varphi_i}\right)^2$$
$$-\sum_{j=1}^{k-1} \frac{\Psi_{j,k}}{\varphi_j\cdots\varphi_{k-1}} \sum_{i=j}^{k-1} \frac{\varphi_i''}{\varphi_i},$$

where $\Psi_{j,k} = \widehat{\mathbb{E}}(Z_j, Z_{j+1}, \dots, Z_k)$. □

Let $(j=1,\dots,n,\ l=2,\dots,n)$

$$g_j(t) = \exp\Big(\mathbb{E}X_j z + \big((\mathbb{E}X_j^{(2)} - \mathbb{E}^2 X_j)/2 + \widehat{\mathbb{E}}(X_{j-1}, X_j)\big) z^2\Big).$$

Lemma 8.16 *Let conditions in (8.1) be satisfied, $k=1,2,\dots,n$. Then for all $t\in\mathbb{R}$*

$$g_k = 1 + C\vartheta|z|(\mathbb{E}X_{k-1} + \mathbb{E}X_k), \tag{8.57}$$
$$g_k' = C\vartheta([\mathbb{E}X_{k-1} + \mathbb{E}X_k), \tag{8.58}$$
$$g_k = 1 + \mathbb{E}X_k z + \gamma_2^*(k) z^2 + C\vartheta|z|^3\big\{\mathbb{E}^3 X_{k-1} + \mathbb{E}^3 X_k + \mathbb{E}X_k \mathbb{E}X_k^{(2)} + (\mathbb{E}X_{k-1} + \mathbb{E}X_k)\mathbb{E}X_{k-1}X_k\big\}, \tag{8.59}$$
$$g_k' = \mathbb{E}X_k z' + \gamma_2^*(k)(z^2)' + C\vartheta|z|^2\big\{\mathbb{E}^3 X_{k-1} + \mathbb{E}^3 X_k + \mathbb{E}X_k \mathbb{E}X_k^{(2)} + (\mathbb{E}X_{k-1} + \mathbb{E}X_k)\mathbb{E}X_{k-1}X_k\big\}, \tag{8.60}$$
$$g_k = 1 + \mathbb{E}X_k z + \gamma_2^*(k) z^2 + \tilde{\gamma}_3^*(k) z^3 + C\vartheta|z|^4\big\{\mathbb{E}^4 X_{k-1} + \mathbb{E}^4 X_k + \mathbb{E}^2 X_k^{(2)} + \mathbb{E}^2 X_{k-1}X_k\big\}, \tag{8.61}$$
$$g_k' = \mathbb{E}X_k z' + \gamma_2^*(k)(z^2)' + \tilde{\gamma}_3^*(k)(z^3)' + C\vartheta|z|^3\big\{\mathbb{E}^4 X_{k-1} + \mathbb{E}^4 X_k + \mathbb{E}^2 X_k^{(2)} + \mathbb{E}^2 X_{k-1}X_k\big\}, \tag{8.62}$$
$$|g_k| \le \exp\big(-\lambda_k \sin^2(t/2)\big), \tag{8.63}$$

where λ_k is as in Lemma 8.12 and

$$\tilde{\gamma}_3^*(k) = \frac{\mathbb{E}X_k \mathbb{E}X_k^{(2)} - \mathbb{E}^3 X_k}{2} + \mathbb{E}X_k \widehat{\mathbb{E}}(X_{k-1}, X_k) + \frac{\mathbb{E}^3 X_k}{6}.$$

Proof of Lemma 8.16. For any complex number z we have

$$\mathrm{e}^z = 1 + z + \frac{z^2}{2} + \dots + \frac{z^m}{m!} + \vartheta \frac{|z|^{m+1}}{(m+1)!}\mathrm{e}^{|z|} \qquad (m\in\mathbb{N}).$$

Due to (8.1), $\mathbb{E}X_j^{(2)} \le \mathbb{E}X_j$. Therefore,

$$\mathbb{E}X_{j-1}X_j \le \sqrt{\mathbb{E}X_{j-1}^2 \mathbb{E}X_j^2} \le \sqrt{(\mathbb{E}X_{j-1}(X_{j-1}-1) + \mathbb{E}X_{j-1}(\mathbb{E}X_j(X_j-1) + \mathbb{E}X_j)}$$
$$\le 2\sqrt{\mathbb{E}X_{j-1}\mathbb{E}X_j} \le 2(\mathbb{E}X_{j-1} + \mathbb{E}X_j). \tag{8.64}$$

Hence $\exp(g_k)$ is bounded by an absolute constant C, and (8.57) and (8.58) easily follow.

We have

$$\begin{aligned} g_k &= 1+\mathbb{E}X_kz+\gamma_2^*(k)z^2+C\vartheta|z|^3\big(\mathbb{E}^3X_k+\mathbb{E}^2X_k^{(2)}+\mathbb{E}X_k\mathbb{E}X_k^{(2)} \\ &\quad +\mathbb{E}X_k\widehat{\mathbb{E}}^+(X_{k-1},X_k)+\mathbb{E}X_{k-1}\mathbb{E}^2X_k+(\widehat{\mathbb{E}}^+(X_{k-1},X_k))^2\big). \end{aligned}$$

Moreover,

$$\mathbb{E}^2X_k^{(2)}\le\mathbb{E}X_k\mathbb{E}X_k^{(2)},\qquad \mathbb{E}X_{k-1}\mathbb{E}^2X_k\le\mathbb{E}^3X_{k-1}+\mathbb{E}^3X_k$$

and

$$\begin{aligned} \widehat{\mathbb{E}}^+(X_{k-1},X_k) &\le 2(\mathbb{E}X_{k-1}X_k)^2+2\mathbb{E}^2X_{k-1}\mathbb{E}^2X_k \\ &\le 2(\mathbb{E}X_{k-1}+\mathbb{E}X_k)\mathbb{E}X_{k-1}X_k+2\mathbb{E}^3X_{k-1}+2\mathbb{E}^3X_k. \end{aligned}$$

Thus, (8.59) follows. Estimates (8.60)–(8.62) can be proved similarly.

For the proof of (8.63) note that

$$\widehat{\mathbb{E}}^+(X_{k-1},X_k)\le\mathbb{E}X_{k-1}X_k+0.01\mathbb{E}X_k,\quad \mathbb{E}^2X_k\le 0.01\mathbb{E}X_k$$

and

$$\begin{aligned} |g_k| &\le \exp\Big(-2\mathbb{E}X_k\sin^2(t/2)+2(\mathbb{E}X_k^{(2)}+\mathbb{E}^2X_k+2\widehat{\mathbb{E}}^+(X_{k-1},X_k))\sin^2(t/2)\Big) \\ &\le \exp\Big(-1.98\mathbb{E}X_k\sin^2(t/2)+2\mathbb{E}X_k^{(2)}\sin^2(t/2)+4\mathbb{E}X_{k-1}X_k\sin^2(t/2)\Big), \end{aligned}$$

which completes the proof of (8.63). □

Set $E=I_1-I$.

Lemma 8.17 *Let conditions (8.1) and (8.2) be satisfied, and let M be a finite signed measure. Then for any $0\le\alpha\le 1$ there exists an absolute constant C_α such that*

$$\|M*\exp\big(\Gamma_1E+\alpha\Gamma_2E^{*2}\big)\|_{TV}\le C_\alpha\|M*\exp(0.18\Gamma_1E)\|_{TV}.$$

Note that $\exp(0.18\Gamma_1E)$ is a Poisson distribution. The term $\exp(\alpha\Gamma_2E^{*2})$ is responsible for the first-order asymptotic expansion.

Proof of Lemma 8.17. Due to (8.1) and (8.2), $0.82\Gamma_1-3.1|\Gamma_2|\ge 0$. Thus,

$$\begin{aligned} &\|M*\exp\big(\Gamma_1E+\alpha\Gamma_2E^{*2}\big)\|_{TV} \\ &\qquad\le \|M*\exp((\Gamma_1-3.1|\Gamma_2|)E)\|_{TV}\,\|\exp\big(3.1|\Gamma_2|E+\alpha\Gamma_2E^{*2}\big)\|_{TV} \\ &\qquad\le \|M*\exp(0.18\Gamma_1E)\|_{TV}\,\|\exp\big(3.1|\Gamma_2|E+\alpha\Gamma_2E^{*2}\big)\|_{TV}. \end{aligned}$$

It remains to check that $\|\exp\big(3.1|\Gamma_2|E+\alpha\Gamma_2E^{*2}\big)\|\le C$ by Lemma B.1. □

Recall that

$$\mathbf{NB}(r,\bar q)=\exp\left(r\sum_{j=1}^{\infty}\frac{\bar q^j}{j}(I_j-I)\right),$$

i.e., $\mathcal{L}(\sum_{j=1}^{\infty}j\pi_{r\bar q^j/j})=\mathbf{NB}(r,\bar q)$.

Lemma 8.18 *Let conditions (8.1) and (8.2) be satisfied, and let $r, \bar{q}$ be defined by (8.9). Then there exists an absolute constant C such that*

$$
\begin{aligned}
\mathbf{NB}(r,\bar{q}) &= \exp\left(\sum_{j=1}^{\infty}\frac{r}{j}\left(\frac{1-\bar{q}}{\bar{q}}\right)^{j}E^{*j}\right)\\
&= \exp\left(\Gamma_1 E+\Gamma_2 E^{*2}+\frac{4\Gamma_2^2}{3\Gamma_1}E^{*3}+\frac{2\Gamma_2^3}{\Gamma_1^2}E^{*4}*\frac{\Theta}{0.7}\right)\\
&= \exp\left(\Gamma_1 E+\Gamma_2 E^{*2}+\frac{4\Gamma_2^2}{3\Gamma_1}E^{*3}*\frac{\Theta}{0.7}\right)=\exp\left(\Gamma_1 E+\Gamma_2 E^{*2}*\frac{\Theta}{0.7}\right)\\
&= \exp\left(\frac{3}{28}\Gamma_1 E+\Gamma_1 E^{*2}*\Theta\right)=C\exp(\Gamma_1 E/2)*\Theta. \qquad (8.65)
\end{aligned}
$$

Relation (8.65) means that Negative Binomial distribution can be represented as a convolution of a Poisson distribution and measure Θ such that $\|\Theta\|_{TV}\leq 1$.
Proof of Lemma 8.18. Due to (8.2),

$$
\begin{aligned}
\Gamma_2 &= \frac{1}{2}\sum_{k=1}^{n}(\mathbb{E}^2X_k^{(2)}-\mathbb{E}^2X_k)+\sum_{k=1}^{n}\operatorname{cov}(X_{k-1},X_k)\\
&\leq \frac{1}{2}\sum_{k=1}^{n}\mathbb{E}X_k^{(2)}+\sum_{k=1}^{n}|\operatorname{cov}(X_{k-1},X_k)|\leq\frac{3}{40}\Gamma_1.
\end{aligned}
$$

Therefore,

$$
\frac{\bar{q}}{1-\bar{q}}=\frac{2\Gamma_2}{\Gamma_1}\leq 0.15,\qquad \frac{\bar{q}}{1-\bar{q}}\|E\|_{TV}\leq 0.15(\|I_1\|_{TV}+\|I\|_{TV})=0.3.
$$

Due to assumptions,

$$
\widehat{\mathbf{NB}}(r,\bar{q})(t)=\left(\frac{1-\bar{q}}{1-\bar{q}\mathrm{e}^{\mathrm{i}t}}\right)^{r}=\left(1-\frac{\bar{q}(\mathrm{e}^{\mathrm{i}t}-1)}{1-\bar{q}}\right)^{-r}=\exp\left(\sum_{j=1}^{\infty}\frac{r}{j}\left(\frac{\bar{q}}{1-\bar{q}}\right)^{j}(\mathrm{e}^{\mathrm{i}t}-1)^{j}\right),
$$

the series in the exponent converge absolutely. Due to the correspondence between characteristic functions and measures we can write

$$
\mathbf{NB}(r,\bar{q})=\exp\left(\sum_{j=1}^{\infty}\frac{r}{j}\left(\frac{\bar{q}}{1-\bar{q}}\right)^{j}E^{*j}\right).
$$

Observe that

$$
\sum_{j=4}^{\infty}\frac{r}{j}\left(\frac{\bar{q}}{1-\bar{q}}\right)^{j}E^{*j}=\frac{r}{4}\left(\frac{\bar{q}}{1-\bar{q}}\right)^{4}E^{*4}*\sum_{j=4}^{\infty}\frac{4}{j}\left(\frac{\bar{q}}{1-\bar{q}}\right)^{j-4}E^{*(j-4)}
$$

and

$$
\sum_{j=4}^{\infty}\frac{4}{j}\left(\frac{\bar{q}}{1-\bar{q}}\right)^{j-4}\|E\|_{TV}^{j-4}\leq\sum_{j=4}^{\infty}(0.3)^{j-4}=\frac{1}{0.7}.
$$

Therefore,

$$\begin{aligned}\mathbf{NB}(r,\bar q) &= \exp\left(\Gamma_1 E + \Gamma_2 E^{*2} + r\left(\frac{\bar q}{1-\bar q}\right)^3\frac{E^{*3}}{3} + r\left(\frac{\bar q}{1-\bar q}\right)^4\frac{E^{*4}}{4} * \frac{\Theta}{0.7}\right)\\ &= \exp\left(\Gamma_1 E + \Gamma_2 E^{*2} + \frac{4\Gamma_2^2}{3\Gamma_1}E^{*3} + \frac{2\Gamma_2^3}{\Gamma_1^2}E^{*4} * \frac{\Theta}{0.7}\right),\end{aligned}$$

which proves the second equation in (8.65).

Since

$$\left\| I + \frac{3\Gamma_2 E}{2\Gamma_1} * \frac{\Theta}{0.7}\right\|_{TV} \leq 1 + \frac{3\cdot 2\cdot 1}{2\cdot 40\cdot 0.7} \leq \frac{1}{0.7},$$

we can write

$$\frac{4\Gamma_2^2}{3\Gamma_1}E^{*3} + \frac{2\Gamma_2^3}{\Gamma_1^2}E^{*4} * \frac{\Theta}{0.7} = \frac{4\Gamma_2^2}{3\Gamma_1}E^{*3} * \left(I + \frac{3\Gamma_2 E}{2\Gamma_1} * \frac{\Theta}{0.7}\right) = \frac{4\Gamma_2^2}{3\Gamma_1}E^{*3} * \frac{\Theta}{0.7}.$$

Thus, the third expression follows. Similarly we prove the forth expression. Finally, observe that $\Gamma_2 = \Gamma_1(\Gamma_2/\Gamma_1) \leq 3\Gamma_1/40$, which allows us to prove the last expression in (8.65). □

8.3.2 Proofs of Theorems 8.1–8.8

In the proofs of Theorems 8.1–8.6 and Theorem 8.8 we omit dependence of Fourier transforms on t, and write $\widehat{D}$ instead of $\widehat{D}(t)$, etc. We retain notation from the previous subsection.

Proof of Theorem 8.1. Recall that $\widehat{\mathbf{F}}_{\mathbf{n}} = \varphi_1\varphi_2\cdots\varphi_n$, $\widehat{\mathbf{D}} = g_1 g_2 \dots g_n$, where $\varphi_j = \mathbb{E}e^{itS_j}/\mathbb{E}e^{itS_{j-1}}$,

$$g_j(t) = \exp\Big(\mathbb{E}X_j z + \Big((\mathbb{E}X_j^{(2)} - \mathbb{E}^2 X_j)/2 + \widehat{\mathbb{E}}(X_{j-1}, X_j)\Big) z^2\Big).$$

Denote $D = \mathbf{F}_{\mathbf{n}} - \mathbf{D} - \tilde{\mathbf{D}}$. Then $D = J_1 + J_2$, where

$$\begin{aligned}\widehat{J}_1 &= \prod_{j=1}^n \varphi_j - \prod_{j=1}^n g_j - \sum_{m=1}^n (\varphi_m - g_m)\prod_{j\neq m}^n g_j,\\ \widehat{J}_2 &= \prod_{j=1}^n g_j + \sum_{m=1}^n(\varphi_m - g_m)\prod_{j\neq m}^n g_j - \prod_{j=1}^n g_j(1+\Gamma_3 z^3).\end{aligned}$$

We estimate $\widehat{J}_1$ and $\widehat{J}_2$ separately. Note that

$$\begin{aligned}\prod_{j=1, j\neq m,l}^n \exp(-\lambda_j\sin^2(t/2)) &\leq \exp(-0.26\Gamma_1\sin^2(t/2))\exp((\lambda_m+\lambda_l)\sin^2(t/2))\\ &\leq \exp(\lambda_m+\lambda_l)\exp(-0.26\Gamma_1\sin^2(t/2)). \qquad (8.66)\end{aligned}$$

This bound is valid for any $t\in\mathbb{R}$, $m,l\in\{1,2,\dots,n\}$ since all $\{\lambda_j\}\le C$ (λ_j are defined in Lemma 8.12).

Applying Lemmas B.3, (8.66), 8.12, 8.13, 8.16, and I.1, we obtain

$$\begin{aligned}
|\widehat{J}_1| &= \Big|\sum_{l=2}^{n}(\varphi_l-g_l)\prod_{j=l+1}^{n}\varphi_j\sum_{m=1}^{l-1}(\varphi_m-g_m)\prod_{j=1,j\neq m}^{l-1}g_j\Big| \\
&\le C\sum_{l=2}^{n}|\varphi_l-g_l|\sum_{m=1}^{l-1}|\varphi_m-g_m|\prod_{j=1,j\neq m,l}^{n}\exp(-1.3\lambda_j\sin^2(t/2)) \\
&\le C\exp(-0.26\Gamma_1\sin^2(t/2))\Big(\sum_{k=1}^{n}|\varphi_k-g_k|\Big)^2 \le C\exp(-0.2\Gamma_1\sin^2(t/2))\,R_1^2|z|^6 \\
&\le C\exp(-0.1\Gamma_1\sin^2(t/2))\,R_1^2\min\{1;\Gamma_1^{-3}\}.
\end{aligned}$$

Similarly, taking into account (8.46), (8.57), (8.61), and (8.63), we get

$$\begin{aligned}
|\widehat{J}_2| &\le \Big|\prod_{j=1}^{n}g_j(1+\Gamma_3z^3)-\prod_{j=1}^{n}g_j-\prod_{j=1}^{n}g_j\sum_{m=1}^{n}(\varphi_m-g_m)\Big| \\
&\quad+\Big|\sum_{m=1}^{n}(\varphi_m-g_m)\Big(\prod_{j=1}^{n}g_j-\prod_{j\neq m}^{n}g_j\Big)\Big| \\
&= \Big|\prod_{j=1}^{n}g_j\Big(\sum_{m=1}^{n}(\varphi_m-g_m)-\Gamma_3z^3\Big)\Big|+\Big|\sum_{m=1}^{n}(\varphi_m-g_m)\prod_{j\neq m}g_j(g_m-1)\Big| \\
&\le CR_2|z|^4\exp(-0.26\Gamma_1\sin^2(t/2)) \le C\exp(-0.2\Gamma_1\sin^2(t/2))\,R_2\min\{\{1;\Gamma_1^{-2}\}.
\end{aligned}$$

Therefore,

$$|\widehat{D}|\le C\exp(-0.2\Gamma_1\sin^2(t/2))\,(R_1^2\min\{1;\Gamma_1^{-3}\}+R_2\min\{1;\Gamma_1^{-2}\}). \tag{8.67}$$

Let

$$\tilde{\varphi}_k=\varphi_k\exp(-\mathrm{i}t\mathbb{E}X_k)\,,\quad \tilde{g}_k=g_k\exp(-\mathrm{i}t\mathbb{E}X_k)\,.$$

Observe that $|\tilde{\varphi}_l'-\tilde{g}_l'|\le C(|\varphi_l'-g_l'|+\mathbb{E}X_k|\varphi_l-g_l|)$. Moreover, taking into account (8.54), (8.55), and (8.64), we get

$$\begin{aligned}
|\tilde{\varphi}_l'| &\le |\varphi_l'-\mathbb{E}X_lz'|+\mathbb{E}X_l|\mathrm{e}^{\mathrm{i}t}-\varphi_l| \\
&\le |\varphi_l'-\mathbb{E}X_lz'|+\mathbb{E}X_l|z|+\mathbb{E}X_l|1-\varphi_l|\le C|z|\sum_{j=0}^{3}\mathbb{E}X_{l-j}\,.
\end{aligned}$$

A similar estimate holds for $|\tilde{g}_l'|$.

Taking into account (8.66), Lemmas 8.12, 8.13, 8.15, and 8.16, we prove that

$$\begin{aligned}
\Big|(\mathrm{e}^{-\mathrm{i}\Gamma_1 t}J_1)'\Big| &\le \sum_{l=2}^{n}|\tilde\varphi_l'-\tilde g_l'|\prod_{j=l+1}^{n}|\tilde\varphi_l|\sum_{m=1}^{l-1}|\tilde\varphi_m-\tilde g_m|\prod_{j=1,j\neq m}^{l-1}|\tilde g_j|\\
&\quad+\sum_{l=2}^{n}|\tilde\varphi_l-\tilde g_l|\sum_{j=l+1}^{n}|\tilde\varphi_j'|\prod_{i=l+1,i\neq j}^{n}|\tilde\varphi_j|\sum_{m=1}^{l-1}|\tilde\varphi_m-\tilde g_m|\prod_{j=1,j\neq m}^{l-1}|\tilde g_j|\\
&\quad+\sum_{l=2}^{n}|\tilde\varphi_l-\tilde g_l|\prod_{j=l+1}^{n}|\tilde\varphi_j|\sum_{m=1}^{l-1}|\tilde\varphi_m'-\tilde g_m'|\prod_{j=1,j\neq m}^{l-1}|\tilde g_j|\\
&\quad+\sum_{l=2}^{n}|\tilde\varphi_l-\tilde g_l|\prod_{j=l+1}^{n}|\tilde\varphi_j|\sum_{m=1}^{l-1}|\tilde\varphi_m-\tilde g_m|\sum_{j=1,j\neq m}^{l-1}|\tilde g_j'|\prod_{k=1,k\neq m,j}^{l-1}|\tilde g_k|\\
&\le C\exp\big(-0.26\Gamma_1\sin^2(t/2)\big)\Big(\sum_{l=2}^{n}|\tilde\varphi_l'-\tilde g_l'|\sum_{m=1}^{l-1}|\tilde\varphi_m-\tilde g_m|\\
&\quad+\Big(\sum_{l=1}^{n}|\tilde\varphi_l-\tilde g_l|\Big)^2\sum_{j-1}^{n}(|\tilde\varphi_j'|+|\tilde g_j'|)\Big)\\
&\le C\exp\big(-0.26\Gamma_1\sin^2(t/2)\big)\,(R_1^2|z|^5(1+\Gamma_1|z|^2)+R_1^2|z|^7\Gamma_1)\\
&\le C\exp\big(-0.2\Gamma_1\sin^2(t/2)\big)\,R_1^2\min\{1;\Gamma_1^{-5/2}\}. \qquad (8.68)
\end{aligned}$$

Similarly, we prove that

$$|(\mathrm{e}^{-\mathrm{i}\Gamma_1 t}J_2)'|\le C|z|^5\Gamma_1R_2\exp\big(-0.26\Gamma_1\sin^2(t/2)\big)\,.$$

Combining this estimate with (8.68), we obtain

$$|(\mathrm{e}^{-\mathrm{i}\Gamma_1 t}\widehat{D})'| \;\le\; C\exp\big(-0.1\Gamma_1\sin^2(t/2)\big)\,(R_1^2\min\{1;\Gamma_1^{-5/2}\}+R_2\min\{1;\Gamma_1^{-3/2})\}.$$

For the proof of (8.4) we use (8.67), (D.10) with $v=\Gamma_1$ and $u=\max\{1;\Gamma_1\}$. For the proof of (8.3) we use identity

$$\prod_{j=1}^{n}\varphi_j-\prod_{j=1}^{n}g_j=\sum_{j=1}^{n}(\varphi_j-g_j)\prod_{l=j+1}^{n}\varphi_l\prod_{l=1}^{j-1}g_l\,.$$

The rest of the proof is very similar to the proof of (8.4): we obtain estimates

$$|(\widehat{\mathbf{F}}_{\mathbf{n}}\exp(-\mathrm{i}t\Gamma_1)-\widehat{\mathbf{D}}\exp(-\mathrm{i}t\Gamma_1))| \;\le\; CR_1|z|^3\exp\big(-0.1\Gamma_1\sin^2(t/2)\big)\,, \qquad (8.69)$$

$$|(\widehat{\mathbf{F}}_{\mathbf{n}}\exp(-\mathrm{i}t\Gamma_1)-\widehat{\mathbf{D}}\exp(-\mathrm{i}t\Gamma_1))'| \;\le\; CR_1|z|^2\exp\big(-0.1\Gamma_1\sin^2(t/2)\big)\,. \qquad (8.70)$$

It remains to apply Lemma D.8. To get estimate (8.6) it suffices to sum estimates (8.15) and apply (I.17). □

Proof of Theorem 8.2. From (D.12) with (8.69) and Lemma I.1 we get

$$d_{\mathrm{o}}(\mathbf{F}_{\mathbf{n}};\mathbf{D})\le\frac{1}{2\pi}\int_{-\pi}^{\pi}|\widehat{\mathbf{F}}_{\mathbf{n}}-\widehat{\mathbf{D}}|\,\mathrm{d}t\le CR_1\int_{-\pi}^{\pi}|z|^3\exp\big(-0.1\Gamma_1\sin^2(t/2)\big)\,\mathrm{d}t\le CR_1\min\{1;\Gamma_1^{-2}\}.$$

Similarly, applying (D.12) and Lemma I.1 to (8.67) we prove (8.8). □

Proof of Theorem 8.3. Recall that $E = I_1 - I$. Applying (8.65) and (C.10), we obtain

$$\begin{aligned}
\|\mathbf{D} - \mathbf{NB}(r,\bar q)\|_{TV} &= \Big\|\mathbf{D} - \mathbf{D} * \exp\Big(\frac{4\Gamma_2^2}{\Gamma_1}E^{*3} * \frac{\Theta}{0.7}\Big)\Big\|_{TV} \\
&= \Big\|\mathbf{D} * \int_0^1 \Big(\exp\Big(\tau\frac{4\Gamma_2^2}{\Gamma_1}E^{*3} * \frac{\Theta}{0.7}\Big)\Big)' \mathrm{d}\tau\Big\|_{TV} \\
&\leq C\int_0^1 \frac{\Gamma_2^2}{\Gamma_1}\Big\|E^{*3}\exp\Big(\Gamma_1 E + \Gamma_2 E^{*2} + \tau\frac{4\Gamma_2^2}{\Gamma_1}E^{*3} * \frac{\Theta}{0.7}\Big)\Big\|_{TV}\mathrm{d}\tau \\
&\leq C\frac{\Gamma_2^2}{\Gamma_1}\|E^{*3} * \exp(0.5\Gamma_1 E)\|_{TV} \leq C\frac{\Gamma_2^2}{\Gamma_1}\min\{1;\Gamma_1^{-3/2}\}.
\end{aligned}$$

Combining the last estimate with (8.3), we prove (8.10).

Let

$$M_1 = \frac{4\Gamma_2^2}{3\Gamma_1}E^{*3}, \quad M_2 = \frac{20\Gamma_2^3}{7\Gamma_1^2}E^{*4}, \quad M_3 = \Gamma_3 E^{*3} - M_1\,. \tag{8.71}$$

Using (I.9), Lemma 8.18 and (C.10), we obtain

$$\begin{aligned}
\mathbf{NB}(r,\bar q) &= \mathbf{D} * \exp(M_1 + \Theta * M_2) \\
&= \mathbf{D} * \Big(I + M_1 + M_1^{*2} * \int_0^1 (1-s)\mathrm{e}^{sM_1}\mathrm{d}s\Big) * \Big(I + M_2 * \Theta * \int_0^1 \mathrm{e}^{xM_2*\Theta}\mathrm{d}x\Big) \\
&= \mathbf{D} * (I + M_1) + M_1^{*2} * \int_0^1 (1-s)\mathbf{D} * \mathrm{e}^{sM_1}\mathrm{d}s \\
&+ \int_0^1\int_0^1 M_2 * \Theta * (I + M_1 + M_1^{*2}(1-s)) * \mathbf{D} * \mathrm{e}^{sM_1 + xM_2}\mathrm{d}s\mathrm{d}x \\
&= \mathbf{D} * (I + M_1) + C\exp(0.5\Gamma_1 E) * \big(M_1^{*2} * \Theta + C[M_2 + M_1 * M_2 * \Theta] * \Theta \\
&+ CM_1^{*2} * M_2 * \Theta\big) = \mathbf{D} * (I + M_1) + C\Gamma_2^3\Gamma_1^{-2}\exp(0.25\Gamma_1 E) * E^{*4} * \Theta,
\end{aligned}$$

where C is an absolute constant.

By the triangle inequality,

$$\begin{aligned}
&\|\mathbf{F_n} - \mathbf{NB}(r,\bar q) * (I + M_3)\|_{TV} \\
&\leq \|\mathbf{F_n} - \mathbf{D} * (I + \Gamma_3 E^{*3})\|_{TV} + \|\mathbf{D} * (I + \Gamma_3 E^{*3}) - \mathbf{D} * (I + M_1) * (I + M_3)\|_{TV} \\
&+ C\|\Gamma_2^3\Gamma_1^{-1}\exp(0.25\Gamma_1 E) * E^{*4} * (I + M_3)\|_{TV} := J_{31} + J_{32} + J_{33}\,.
\end{aligned}$$

By Lemma 8.17 and (C.10)

$$\begin{aligned}
J_{32} &\leq C\|\exp(0.9\lambda E) * \Gamma_2^2\Gamma_1^{-1}(\Gamma_3 - 4\Gamma_2^2(3\Gamma_1)^{-1})E^{*6}\|_{TV} \\
&\leq \Gamma_2^2\Gamma_1^{-1}|\Gamma_3 - 4\Gamma_2^2(3\Gamma_1)^{-1}|\min\{1;\Gamma_1^{-3}\}.
\end{aligned}$$

Similarly

$$\begin{aligned}
J_{33} &\leq C\|\exp(0.25\Gamma_1 E) * \Gamma_2^3\Gamma_1^{-2}E^{*4}\|_{TV} \\
&\quad + C\|\exp(0.25\Gamma_1 E) * \Gamma_2^3\Gamma_1^{-2}(\Gamma_3 - 4\Gamma_2^2(3\Gamma_1)^{-1})E^{*7}\|_{TV} \\
&\leq C\Gamma_2^3\Gamma_1^{-2}\min(1,\Gamma_1^{-2}) + C\Gamma_2^3\Gamma_1^{-2}|\Gamma_3 - 4\Gamma_2^2(3\Gamma_1)^{-1}|\min\{1;\Gamma_1^{-7/2}\}.
\end{aligned}$$

Combining the last two estimates and applying (8.4) in order to evaluate J_{31}, we derive (8.11).

To prove estimate (8.12) it suffices to sum estimates (8.17) and apply (I.17). □

Proof of Theorem 8.4 is similar to the proof of Theorem 8.3. It suffices to apply (A.7). For example, arguing as in the proof of Theorem 8.3, we get

$$\begin{aligned}\|\mathbf{D}-\mathbf{NB}(r,\bar{q})\|_\circ &\leq \|E*(\mathbf{D}-\mathbf{NB}(r,\bar{q}))\|_{TV}\\ &\leq C\frac{\Gamma_2^2}{\Gamma_1}\Big\|E^{*4}*\exp(0.5\Gamma_1 E)\Big\|_{TV} \leq C\frac{\Gamma_2^2}{\Gamma_1}\min\{1;\Gamma_1^{-2}\}.\end{aligned}$$

All other proofs are very similar and, therefore, omitted. □

Proof of Theorem 8.5. Recall that $z=\mathrm{e}^{it}-1$. We apply Lemma D.9. Similarly to the proof of (8.69) and (8.70) we can check that

$$|(\widehat{\mathbf{F}}_{\mathbf{n}}\exp(-it\Gamma_1)-\widehat{\mathbf{D}}\exp(-it\Gamma_1))''|\leq CR_1|z|\exp\big(-0.1\Gamma_1\sin^2(t/2)\big)\,.$$

Combining this estimate with (8.69) and applying Lemma D.9, we derive (8.15) and (8.16).

For the proof of (8.17) and (8.18) observe that

$$\Gamma_2\leq\frac{3}{40}\Gamma_1,\quad \frac{\bar{q}}{1-\bar{q}}=\frac{2\Gamma_2}{\Gamma_1}\leq 0.15$$

due to the assumptions, see also [88]. Therefore, $\widehat{\mathbf{NB}}(t)\exp(-\Gamma_1 it)=\exp(A)$, where

$$A=\Gamma_1(z-\mathrm{i}t)+\Gamma_2 z^2+\sum_{j=3}^{\infty}r\bar{q}^j z^j/j(1-\bar{q})^j=\Gamma_1(z-\mathrm{i}t)+\Gamma_2 z^2+\vartheta C\Gamma_2^2\Gamma_1^{-1}|z|^3\,.$$

Denote

$$\psi=\exp\big(-0.1\Gamma_1\sin^2(t/2)\big)\,.$$

To use (D.16) we need estimates for derivatives. We have

$$|A'|\leq C\Gamma_1|z|,\quad |A''|\leq C\Gamma_1,\quad |\mathrm{e}^A|\leq\psi^2\,.$$

Let

$$B=\Gamma_1(z-\mathrm{i}t)+\Gamma_2 z^2,\quad u_1(t)=(\mathrm{e}^A-\mathrm{e}^B)/z,$$

so that $\widehat{\mathbf{D}}(t)\exp(-\Gamma_1 it)=\exp(B)$. Note that $|\mathrm{e}^B|\leq\psi^2$ and

$$|\mathrm{e}^A-\mathrm{e}^B|=\left|(A-B)\int_0^1\mathrm{e}^{\tau A+(1-\tau)B}\mathrm{d}\tau\right|\leq\psi^2|A-B|\leq C\psi^2\Gamma_2^2|z|^3/\Gamma_1\,.$$

Then

$$|u_1(t)|\leq\frac{|\mathrm{e}^A-\mathrm{e}^B|}{|z|}\leq\psi^2\frac{|A-B|}{|z|}\leq C\psi^2\frac{\Gamma_2^2|z|^2}{\Gamma_1},\quad \int_{-\pi}^{\pi}|u_1(t)|\mathrm{d}t\leq C\frac{\Gamma_2^2}{\Gamma_1^2\sqrt{\Gamma_1}}\,. \tag{8.72}$$

Also,

$$\begin{aligned}|(\mathrm{e}^A-\mathrm{e}^B)''| &\le |A''||\mathrm{e}^A-\mathrm{e}^B|+|(A')^2||\mathrm{e}^A-\mathrm{e}^B|+|A''-B''||\mathrm{e}^B|+|(A')^2-(B')^2||\mathrm{e}^B|\\ &\le C\psi^2\Big\{\Gamma_1\frac{\Gamma_2^2}{\Gamma_1}|z|^3+\Gamma_1^2|z|^2\frac{\Gamma_2^2}{\Gamma_1}|z|^3+\frac{\Gamma_2^2}{\Gamma_1}|z|+\Gamma_1|z|\frac{\Gamma_2^2}{\Gamma_1}|z|^2\Big\}\le C\psi|z|\frac{\Gamma_2^2}{\Gamma_1}.\end{aligned}$$

Similarly,

$$|(\mathrm{e}^A-\mathrm{e}^B)'|\le|A'||\mathrm{e}^A-\mathrm{e}^B|+|\mathrm{e}^B||A'-B'|\le C\psi|z|^2\Gamma_2^2/\Gamma_1,$$

and we obtain

$$|u_1(t)''|\le C\psi\Gamma_2^2/\Gamma_1,\quad \int_{-\pi}^{\pi}|u_1(t)''|\mathrm{d}t\le C\Gamma_2^2/\Gamma_1\sqrt{\Gamma_1}. \tag{8.73}$$

Estimates (8.72) and (8.73) and Lemma D.9 allow us to write

$$\sup_x\big(1+(x-\Gamma_1)^2/\Gamma_1\big)\,|\mathbf{D}[x]-\mathbf{NB}(r,\bar q)[x]|\le C\Gamma_2^2/\Gamma_1^2\sqrt{\Gamma_1}.$$

This together with (8.15) proves (8.18). The proof of estimate (8.17) is similar and is therefore omitted. □

Proof Theorem 8.6. Note that

$$\begin{aligned}\mathrm{e}^{-\mathrm{i}t\Gamma_1}\Gamma_3\widehat{\mathbf{D}}z^3 &= \Gamma_3(\mathrm{i}t)^3+\Gamma_3(z^3-(\mathrm{i}t)^3)+\Gamma_3z^3(\mathrm{e}^{-\mathrm{i}t\Gamma_1}\widehat{\mathbf{D}}-1)\\ &= \Gamma_3(\mathrm{i}t)^3+\vartheta C|\Gamma_3||t|^3(|t|+\Gamma_{1n}t^2),\\ \mathrm{e}^{-\mathrm{i}t\Gamma_1}(\widehat{\mathbf{F}}_{\mathbf{n}}-\widehat{\mathbf{D}}) &= \mathrm{e}^{-\mathrm{i}t\Gamma_1}(\Gamma_3\widehat{\mathbf{D}}z^3+\widehat{\mathbf{F}}_{\mathbf{n}}-\widehat{\mathbf{D}}(1+\Gamma_3z^3))\\ &= \mathrm{e}^{-\mathrm{i}t\Gamma_1}\Gamma_3\widehat{\mathbf{D}}z^3+\vartheta C(R_1^2|z|^6+R_2|z|^4).\end{aligned}$$

We apply Lemma D.15 with $a=\Gamma_1$, $b=\alpha\sqrt{\Gamma_1}$, $t\exp(-t^2/2)$ when α is sufficiently large:

$$\begin{aligned}d_K(\mathbf{F_n};\mathbf{D}) &\ge \frac{C|\Gamma_3|}{\alpha^3\Gamma_1^{3/2}}-\frac{C|\Gamma_3|}{\alpha^4\Gamma_1^{3/2}}-\frac{C|\Gamma_3|}{\alpha^5\Gamma_1^{3/2}}-\frac{CR_1^2}{\alpha^6\Gamma_1^3}-\frac{CR_2}{\alpha^4\Gamma_1^2}\\ &\ge \frac{C}{\alpha^3}\Big(\frac{|\Gamma_3|}{\Gamma_1^{3/2}}\Big(1-\frac{C}{\alpha}-\frac{C}{\alpha^2}\Big)-\frac{C}{\alpha}\Big(\frac{R_1^2}{\Gamma_{1n}^3}+\frac{R_2}{\Gamma_1^2}\Big)\Big)\\ &\ge \frac{C}{\alpha^3}\Big(\frac{|\Gamma_3|}{\Gamma_1^{3/2}}-\frac{C}{\alpha}\Big(\frac{R_1^2}{\Gamma_1^3}+\frac{R_2}{\Gamma_1^2}\Big)\Big),\end{aligned}$$

leading to (8.7).

Let M_1,M_2,M_3 be defined by (8.71). Note that

$$|1+\Gamma_3z^3-(1+\widehat{M}_1)(1+\widehat{M}_3)|\le\frac{C\Gamma_2^2}{\Gamma_1}\Big|\Gamma_3-\frac{4\Gamma_2^2}{3\Gamma_1}\Big||z|^6,$$

$$\begin{aligned}\mathrm{e}^{-\Gamma_1\mathrm{i}t}\widehat{\mathbf{NB}}(r,\bar q)z^3 &= (\mathrm{i}t)^3+(z^3-(\mathrm{i}t)^3)+(\mathrm{e}^{-\Gamma_1\mathrm{i}t}\widehat{\mathbf{NB}}(r,\bar q)-1)z^3\\ &= (\mathrm{i}t)^3+\vartheta C|t|^3(|t|+\Gamma_1t^2).\end{aligned}$$

Rewriting estimates of the proof of Theorem 8.3 for characteristic functions, we get

$$\begin{aligned}
\widehat{\mathbf{F}}_{\mathbf{n}} - \widehat{\mathbf{NB}}(r,\bar{q}) &= \widehat{\mathbf{NB}}(r,\bar{q})\widehat{M}_3 + [\widehat{\mathbf{F}}_{\mathbf{n}} - \widehat{\mathbf{D}}(1+\Gamma_3 z^3)] \\
&\quad +(\widehat{\mathbf{D}}(1+\Gamma_3 z^3) - \widehat{\mathbf{D}}(1+\widehat{M}_1)(1+\widehat{M}_3)) + (1+\widehat{M}_3)[\widehat{\mathbf{D}}(1+\widehat{M}_1) - \widehat{\mathbf{NB}}(r,\bar{q})] \\
&= \widehat{\mathbf{NB}}(r,\bar{q})\widehat{M}_3 + \vartheta C\Big\{ R_1^2|z|^6 + R_2|z|^4 + \frac{\Gamma_2^2}{\Gamma_1}\Big|\Gamma_3 - \frac{4\Gamma_2^2}{3\Gamma_1}\Big||z|^6 \\
&\quad + \Big(1 + \Big|\Gamma_3 - \frac{4\Gamma_2^2}{3\Gamma_1}\Big||z|^3\Big)\frac{\Gamma_2^3}{\Gamma_1^2}|z|^4(1+\Gamma_2|z|^2)\Big\} \\
&= \widehat{\mathbf{NB}}(r,\bar{q})\widehat{M}_3 + \vartheta C\Big\{ R_1^2|z|^6 + R_2|z|^4 + \frac{\Gamma_2^3|z|^4}{\Gamma_1^2}(1+\Gamma_2|z|^2) \\
&\quad + \Big|\Gamma_3 - \frac{4\Gamma_2^2}{3\Gamma_1}\Big|\frac{\Gamma_2^2}{\Gamma_1}|z|^6\Big(\frac{|z|(1+\Gamma_2|z|^2)}{\Gamma_1} + 1\Big)\Big\} \\
&= \widehat{\mathbf{NB}}(r,\bar{q})z^3\Big(\Gamma_3 - \frac{4\Gamma_2^2}{3\Gamma_1}\Big) + \vartheta C\Big\{ R_1^2|z|^6 + R_2|z|^4 \\
&\quad + \frac{\Gamma_2^3|z|^4}{\Gamma_1^2}(1+\Gamma_2|z|^2) + \Big|\Gamma_3 - \frac{4\Gamma_2^2}{3\Gamma_1}\Big|\frac{\Gamma_2^2}{\Gamma_1}|z|^6\Big\}.
\end{aligned}$$

The proof of (8.20) is completed by an application of Lemma D.15 with $a = \Gamma_1$, $b = \alpha\sqrt{\Gamma_1}$, and sufficiently large constant α. □

Proof of Theorem 8.7 is similar to the proof of Theorem 8.1, therefore, we only give a sketch. The main difference in the proof is that the explicit nature of dependence allows us to prove by induction that

$$\widehat{\mathbb{E}}(Z_1, Z_2, \dots, Z_k) = z^k p^{k+1}(1-p)^{k-1} \qquad (0<p<1, k\ge 1). \tag{8.74}$$

Since $p\le 1/5$, one can check that

$$|\varphi_k| \le 1 - \frac{16}{25}p^2\sin^2\frac{t}{2} \le \exp\Big(-\frac{16}{25}p^2\sin^2\frac{t}{2}\Big).$$

Repeating the argument of the proofs of Lemma 8.13 and Lemma 8.15, we obtain

$$\begin{aligned}
\varphi_1 &= 1+p^2z, \quad \varphi_2 = 1+p^2z+p^3(1-p)z^2 - p^5(1-p)z^3 + C\vartheta p^5|z|^4, \\
\varphi_k &= 1+p^2z+p^3(1-p)z^2+p^4(1-p)(1-2p)z^3 + C\vartheta p^5|z|^4 \qquad (k=3,4,\dots), \\
\varphi_1' &= p^2z', \quad \varphi_2' = p^2z'+p^3(1-p)(z^2)' - p^5(1-p)(z^3)' + C\vartheta p^5|z|^3, \\
\varphi_k' &= p^2z'+p^3(1-p)(z^2)'+p^4(1-p)(1-2p)(z^3)' + C\vartheta p^5|z|^3 \qquad (k=3,4,\dots).
\end{aligned}$$

Proceeding as in the proof of Theorem 8.1 and applying Lemma D.8 with $v = np^2$ and $u = \max\{1; np^2\}$, we obtain the third estimate in Theorem 8.7.

By Lemma B.1, for any (signed) finite measure M concentrated on integers we have

$$\begin{aligned}
&\|M * \exp\big(np^2E + \alpha\Gamma_2^* E^{*2} + \tau\Gamma_3^* E^{*3}\big)\|_{TV} \\
&\qquad \le C\|M*\exp\big(np^2E + np^2E^{*2} * (p(1-3p/2) + p^2(1-4p+10p^2/3)E)\big)\|_{TV} \\
&\qquad \le C\|M*\exp\big(0.4np^2E\big)\|_{TV},
\end{aligned} \tag{8.75}$$

where C is an absolute constant, $E=I_1-I$, $\alpha,\tau\in[0,1]$. Recall that

$$\mathbf{D}_2 = \exp\big(np^2E + \Gamma_2^*E^{*2}\big), \quad \mathbf{D}_3 = \exp\big(np^2E+\Gamma_2^*E^{*2}+\Gamma_3^*E^{*3}\big).$$

For the proof of the first estimate in Theorem 8.7 we apply (C.10) and the following bound:

$$\begin{aligned}
\|\mathbf{D}_3-\mathbf{D}_2\|_{TV} &= \|\mathbf{D}_2*(\exp(\ln\mathbf{D}_3-\ln\mathbf{D}_2)-I)\| = \Big\|\mathbf{D}_2*\int_0^1(\exp(\tau(\ln\mathbf{D}_3-\ln\mathbf{D}_2)))'\mathrm{d}\tau\Big\|_{TV}\\
&= \Big\|\mathbf{D}_2*(\ln\mathbf{D}_2-\ln\mathbf{D}_3)*\int_0^1\exp(\tau(\ln\mathbf{D}_3-\ln\mathbf{D}_2))\,\mathrm{d}\tau\Big\|_{TV}\\
&= \Big\|(\ln\mathbf{D}_2-\ln\mathbf{D}_3)*\int_0^1\exp(\tau\ln\mathbf{D}_3+(1-\tau)\ln\mathbf{D}_2)\,\mathrm{d}\tau\Big\|_{TV}\\
&\le C\|(\ln\mathbf{D}_3-\ln\mathbf{D}_2)*\exp\big(0.4np^2E\big)\|_{TV}\le C\|np^4E^{*3}*\exp\big(0.4np^2E\big)\|_{TV}\\
&\le Cnp^4\min\{1;(np^2)^{-3/2}\}.
\end{aligned}$$

Similarly,

$$\begin{aligned}
&\mathbf{D}_3-\mathbf{D}_2*(I+\gamma_3^*E^3)=\mathbf{D}_2*\big(\exp\big(\gamma_3^*E^{*3}\big)-I-\gamma_3^*E^{*3}\big)\\
&=\mathbf{D}_2*(\gamma_3^*E^{*3})^{*2}*\int_0^1(1-s)\exp\big(s\gamma_3^*E^{*3}\big)\,\mathrm{d}s.
\end{aligned}$$

Applying (8.75) and (C.10), we get

$$\|\mathbf{D}_3-\mathbf{D}_2*(I+\gamma_3^*E^{*3})\|_{TV}\le C\|(\gamma_3^*)^2E^{*6}*\exp\big(0.4np^2E\big)\|_{TV}\le C(np^4)^2\min\{1;(np^2)^{-3}\}.$$

This bound and the triangle inequality yield

$$d_{TV}(\mathcal{L}(S_{n,2});\mathbf{D}_3)\le C\min\{np^5;p/n\}.$$

Set $\mathbf{D}_1=\exp\big(np^2E\big)$. Next we prove that

$$\begin{aligned}
\|(\mathbf{D}_2-\mathbf{D}_1)*E^{*3}\|_{TV} &= \Big\|E^{*3}*\mathbf{D}_1\int_0^1\gamma_2^*E^{*2}*\exp\big(\tau\gamma_2^*E^{*2}\big)\,\mathrm{d}\tau\Big\|_{TV}\\
&\le Cnp^3\int_0^1\Big\|\mathbf{D}_1*\mathrm{e}^{\tau\gamma_2^*E}*E^{*5}\mathrm{d}\tau\Big\|_{TV}\le Cnp^3\Big\|E^{*5}*\mathrm{e}^{0.4np^2E}\Big\|_{TV}\le\frac{Cp}{\sqrt{n}}. \qquad (8.76)
\end{aligned}$$

By the triangle inequality, for any finite measures M, V and any constant $\tilde{C}$

$$\Big|\|M\|_{TV}-\tilde{C}\Big|\le\Big|\|M\|_{TV}-\|V\|_{TV}\Big|+\Big|\|V\|_{TV}-\tilde{C}\Big|\le\|M-V\|_{TV}+\Big|\|V\|_{TV}-\tilde{C}\Big|.$$

Applying the second estimate of Theorem 8.7, (C.8), (8.76) and Lemma C.3, we obtain

$$\begin{aligned}
\Big|\|\mathcal{L}(S_{n+1,2})-\mathbf{D}_2\|_{TV}-\frac{\tilde{C}p}{\sqrt{n}}\Big| &\le \|\mathcal{L}(S_{n+1,2})-\mathbf{D}_2*(I+\gamma_3^*E^{*3})\|_{TV}+\Big|\|\gamma_3^*\mathbf{D}_2*E^{*3}\|_{TV}-\frac{\tilde{C}p}{\sqrt{n}}\Big|\\
&\le \frac{Cp}{n}+\|\mathbf{D}_2*E^{*3}(\gamma_3^*-np^4)\|_{TV}+\Big|np^4\|\mathbf{D}_2*E^{*3}\|_{TV}-\frac{\tilde{C}p}{\sqrt{n}}\Big|\\
&\le \frac{Cp^2}{\sqrt{n}}+np^4\|(\mathbf{D}_2-\mathbf{D}_1)*E^{*3}\|_{TV}+\Big|np^4\|\mathbf{D}_1*E^{*3}\|_{TV}-\frac{\tilde{C}p}{\sqrt{n}}\Big|\\
&\le Cp^2/\sqrt{n}+np^4\Big|\|\mathbf{D}_1*E^{*3}\|_{TV}-\tilde{C}/(np^2)^{3/2}\Big|\\
&\le Cp^2/\sqrt{n}+C/n.
\end{aligned}$$

The proof is complete. □

Proof of Theorem 8.8 (sketch). The proof is similar to the proof of Theorem 8.1, therefore we only outline the main differences. First, note that

$$\mathbb{E}Y_1Y_2\dots Y_k = 2p_1p_2\dots p_{k+1}(z+\bar{z})^k = 2p_1p_2\dots p_{k+1}(-1)^k(z\bar{z})^k \tag{8.77}$$

for each $k\in\mathbb{N}$, where $z=\mathrm{e}^{\mathrm{i}t}-1$, $\bar{z}=\mathrm{e}^{-\mathrm{i}t}-1$, $Y_j=\mathrm{e}^{\mathrm{i}tX_j}-1$.

Indeed, if $k=1$, then

$$\begin{aligned}\mathbb{E}Y_1 &= \mathbb{E}(\mathrm{e}^{\mathrm{i}t\xi_1\xi_2}-1) = p_1\mathbb{E}(\mathrm{e}^{\mathrm{i}t\xi_2}-1)+p_1\mathbb{E}(\mathrm{e}^{-\mathrm{i}t\xi_2}-1)\\ &= p_1(p_2z+p_2\bar{z})+p_1(p_2\bar{z}+p_2z) = 2p_1p_2(z+\bar{z}).\end{aligned}$$

We use mathematical induction. Assume that (8.77) is true for a particular $k\in\mathbb{N}$. We need to prove that the formula is true for $k+1$. Indeed,

$$\begin{aligned}\mathbb{E}Y_1Y_2\dots Y_{k+1} &= \mathbb{E}Y_1\dots Y_k(\mathrm{e}^{\mathrm{i}t\xi_{k+1}}-1)p_{k+2}+\mathbb{E}Y_1\dots Y_k(\mathrm{e}^{-\mathrm{i}t\xi_{k+1}}-1)p_{k+2}\\ &= \mathbb{E}Y_1\dots Y_{k-1}(\mathrm{e}^{\mathrm{i}t\xi_k\xi_{k+1}}-1)(\mathrm{e}^{\mathrm{i}t\xi_{k+1}}-1)p_{k+2}\\ &\quad+\mathbb{E}Y_1\dots Y_{k-1}(\mathrm{e}^{\mathrm{i}t\xi_k\xi_{k+1}}-1)(\mathrm{e}^{-\mathrm{i}t\xi_{k+1}}-1)p_{k+2}\\ &= \mathbb{E}Y_1\dots Y_{k-1}(\mathrm{e}^{\mathrm{i}t\xi_k}-1)zp_{k+1}p_{k+2}\\ &\quad+\mathbb{E}Y_1\dots Y_{k-1}(\mathrm{e}^{-\mathrm{i}t\xi_k}-1)\bar{z}p_{k+1}p_{k+2}\\ &\quad+\mathbb{E}Y_1\dots Y_{k-1}(\mathrm{e}^{\mathrm{i}t\xi_k}-1)\bar{z}p_{k+1}p_{k+2}\\ &\quad+\mathbb{E}Y_1\dots Y_{k-1}(\mathrm{e}^{-\mathrm{i}t\xi_k}-1)zp_{k+1}p_{k+2}\\ &= p_{k+1}p_{k+2}(z+\bar{z})\Big(\mathbb{E}Y_1\dots Y_{k-1}(\mathrm{e}^{\mathrm{i}t\xi_k}-1)+\mathbb{E}Y_1\dots Y_{k-1}(\mathrm{e}^{-\mathrm{i}t\xi_k}-1)\Big)\\ &= p_{k+2}(z+\bar{z})\mathbb{E}Y_1\dots Y_k = 2p_1\dots p_{k+1}p_{k+2}(z+\bar{z})^{k+1}.\end{aligned}$$

It is easy to check that

$$-z\bar{z} = -(\mathrm{e}^{\mathrm{i}t}-1)(\mathrm{e}^{-\mathrm{i}t}-1) = -2+\mathrm{e}^{\mathrm{i}t}+\mathrm{e}^{-\mathrm{i}t} = z+\bar{z}.$$

This completes the proof of (8.77).

Similarly, we prove by mathematical induction that

$$\begin{aligned}\mathbb{E}(Y_1,Y_2\dots Y_k) &= 2p_1\dots p_{k+1}(1{-}!2p_2)(1-2p_3)\dots(1-2p_k)(z+\bar{z})^k\\ &= 2p_1\dots p_{k+1}(1-2p_2)(1-2p_3)\dots(1-2p_k)(-1)^k(z\bar{z})^k\end{aligned}$$

as $k\geq 2$.

The following steps are similar to the proof of Theorem 8.1. Observe that

$$\mathbb{E}\mathrm{e}^{\mathrm{i}tS_n} = \varphi_1(t)\varphi_2(t)\dots\varphi_n(t).$$

Here $\varphi_1(t)=1+2p_1p_2(z+\bar{z})$,

$$\begin{aligned}\varphi_k(t) &= 1+2p_kp_{k+1}(z+\bar{z})\\ &\quad+\sum_{j=1}^{k-1}\frac{2p_jp_{j+1}\dots p_{k+1}(1-2p_{j+1})\dots(1-2p_k)(z+\bar{z})^{k-j+1}}{\varphi_j(t)\varphi_{j+1}(t)\dots\varphi_{k-1}(t)}\qquad (k=2,\dots,n).\end{aligned}$$

The following estimates can be proved by induction for all t, $k=2,\dots,n$, $n\geq 3$:

$$\begin{gathered}|\varphi_k(t)-1|\leq 3p^2|z|^2,\quad |\varphi_k(t)|\leq 1-6p_kp_{k+1}\sin^2(t/2),\\ |\varphi_k(t)-1-2p_kp_{k+1}(z+\bar{z})|\leq 3|z|^4p_{k-1}p_kp_{k+1}.\end{gathered}$$

Note that from assumption $p=\max p_i\leq 1/24$ it follows that $|\varphi_k(t)-1|\leq 3p^2|z|^2\leq 1/48$ and $1/|\varphi_k(t)|\leq 1/(1-|1-\varphi_k(t)|)\leq 48/47$.

All the other steps of the proof are very similar to the proof of Theorem 8.1 and, therefore, are omitted. □

Exercises

Let $\xi, \xi_0, \xi_1, \dots, \xi_n$ be i.i.d.r.v.s, $\mathcal{L}(\xi) = \mathbf{B}(p)$. Let $S_n = X_1 + \cdots + X_n$, $X_i = \xi_i(1-\xi_{i-1})$ $(i=1,\dots,n)$, $a=p(1-p)$, $z=\mathrm{e}^{\mathrm{i}t}-1$, $Z_j=\exp(\mathrm{i}tX_j) - 1$.

Assume that $p \le 1/10$ (the latter assumption is much weaker than (8.1)).

8.1. Prove that $\widehat{\mathbb{E}}(Z_1, Z_2, \dots, Z_k) = (-1)^{k+1} a^k z^k$ $(k = 2, \dots, n)$.

8.2. Prove that $\mathbb{E}\exp(\mathrm{i}tS_n) = \varphi_1\varphi_2 \dots \varphi_n$, where

$$\varphi_k = 1 + az + \sum_{j=1}^{k} \frac{(-1)^{k-j} a^{k-j+1} z^{k-j+1}}{\varphi_j \cdots \varphi_{k-1}} \qquad (k=2,\dots,n),$$

$$|\varphi_k - 1| \le 1/4, \quad 1/|\varphi_k| \le 4/3.$$

8.3. Prove that $|\varphi_k| \le \exp\bigl(-a\sin^2(t/2)\bigr)$ $(k = 1, 2, \dots, n)$.

8.4. Prove that

$$\varphi_k = 1 + az - az^2 + \vartheta C|a|^3|z|^3 \qquad (k>1),$$

where C is an absolute constant, $|\vartheta| \le 1$.

8.5. Let $Y_n = \pi_b + 2\pi_u$, where $b = na - (3n-2)a^2$, $u = -(3n-2)a^2/2 < 0$. Prove that

$$d_{\mathrm{o}}(S_n; Y_n) \le Cp/n, \quad d_{K}(S_n; Y_n) \le Cp^{3/2} n^{-1/2}.$$

Note that this is a SCP approximation.

Notes

Estimates (8.3), (8.4), (8.10), (8.11) have been proved in [88].
Estimates (8.6), (8.12), (8.15), (8.16), (8.17), (8.18) have been obtained in [306].
Theorem 8.6 is from [91]. Theorem 8.7 is from [229].
Estimate (8.21) has been proved in [229].
Symmetric three-point runs have been investigated in [192].

9

Markov Binomial distribution

The sum $S_n = \xi_1 + ... + \xi_n$ of 0-1 random variables $\xi_1, ..., \xi_n$ plays important role in extreme value theory, mathematical biology, etc. This chapter is devoted to the case where r.v.s $\{\xi_k\}$ form a Markov chain with the initial distribution

$$\mathbb{P}(\xi_0 = 1) = p_0, \quad \mathbb{P}(\xi_0 = 0) = 1 - p_0 \quad (p_0 \in [0,1])$$

and transition probabilities

$$\begin{aligned} \mathbb{P}(\xi_i = 1 \mid \xi_{i-1} = 1) &= \beta, \qquad & \mathbb{P}(\xi_i = 0 \mid \xi_{i-1} = 1) &= 1 - \beta, \\ \mathbb{P}(\xi_i = 1 \mid \xi_{i-1} = 0) &= \alpha, \qquad & \mathbb{P}(\xi_i = 0 \mid \xi_{i-1} = 0) &= 1 - \alpha, \end{aligned}$$

where $\alpha, \beta \in (0,1)$, $i \in \mathbb{N}$. If $p_0 = \alpha/(1 - \beta + \alpha)$, then the chain is stationary.

The distribution of the sum

$$S_n = \xi_1 + \cdots + \xi_n$$

is sometimes called the Markov Binomial (MB) distribution.

The Markov Binomial distribution is a generalisation of the Binomial distribution to the case of 0-1 r.v.s forming a Markov chain. If $p_0 = 0$ and $\alpha = \beta = p$, then $\mathcal{L}(S_n)$ has the Binomial $\mathbf{B}(n,p)$ distribution.

It is known that a centreed normalised Binomial distribution can have either Normal, Poisson or degenerate weak limit (Kozulyaev [183]). In the case of Markov Binomial distribution the class of limit laws for a centreed normalised sum has seven different elements (see Dobrushin [108]). One of these limit distributions is a compound Poisson law with a geometric compounding distribution. A compound Poisson limit theorem for the Markov Binomial distribution can be found in [125, 182, 311].

9.1 Accuracy of approximation

Note that S_n depends on ξ_0. Below, we employ the following approach that takes into account the initial distribution $\mathcal{L}(\xi_0)$:

$$\mathcal{L}(S_n) \approx \{\text{approximation of } \mathcal{L}(\xi_0)\} * \{\text{CP or SCP approximation}\}.$$

Below p_0, α, β may depend on n. We usually assume that

$$\beta \le 1/2, \quad \alpha \le 1/30. \tag{9.1}$$

Markov chain $\{\xi_k\}_{k \ge 0}$ may be non-stationary.

DOI: 10.1201/9781003478164-9

Denote

$$\begin{aligned}
H &= I+\varkappa_2(K_\beta-I), \quad K_\beta = (1-\beta)\sum_{j=0}^{\infty}\beta^j I_{j+1},\\
\gamma &= \lfloor b\rfloor, \quad \{b\} = b-\lfloor b\rfloor, \qquad \gamma_2 = -\frac{(1-\beta)\alpha^2}{(1-\beta+\alpha)^2}\Big(\beta+\frac{1-\beta}{1-\beta+\alpha}\Big)-\frac{\gamma_1^2}{2},\\
\gamma_3 &= \frac{\gamma_1^3}{3}+\frac{\gamma_1^2}{(1-\beta)(1-\beta+\alpha)}\Big\{\beta^2\alpha+\frac{\beta(1-\beta)(2\alpha-1+\beta)}{1-\beta+\alpha}+\frac{2\alpha(1-\beta)^2}{(1-\beta+\alpha)^2}\Big\}\\
&\quad+\frac{\gamma_1^2\alpha}{1-\beta+\alpha}\Big(\beta+\frac{1-\beta}{1-\beta+\alpha}\Big),\\
\varkappa_1 &= \gamma_1\Big(\frac{\alpha-\beta}{1-\beta+\alpha}-p_0\Big), \quad \varkappa_2 = p_0\frac{\beta(1-\beta)}{1-\beta+\alpha},\\
D_{0n} &= \exp((n-p_0)\gamma_1(K_\beta-I)), \quad D_j = \exp\Big(\sum_{i=1}^{j}\gamma_1(K_\beta-I)^{*i}\Big) \quad (1\le j\le 3).
\end{aligned}$$

Note that K_β is a geometric distribution: $K_\beta(k)=(1-\beta)\beta^{k-1}$ $(k=1,2,\dots)$, hence H is a CP distribution. Set

$$b=n(\gamma_1+4\gamma_2+3\gamma_3), \quad \gamma=\lfloor b\rfloor,$$

$$\lambda_1 = n(\gamma_1+4\gamma_2+3\gamma_3)-\{b\}, \quad \lambda_2=\{b\}/6, \quad \lambda_{-1}=n(2|\gamma_2|-3\gamma_3)+\{b\},$$

and let

$$\begin{aligned}
D_{4n} &= K_\beta^{*\gamma}*\exp\big(\lambda_1(K_\beta-I)+\lambda_2(K_\beta^{*2}-I)+\lambda_{-1}(I_{-1}-I)/(1-\beta)\big),\\
A_0 &= H*D_{0n}, \qquad A_{01} = H*D_{0n}*(I+n\gamma_2(K_\beta-I)^{*2}),\\
A_1 &= H*\exp(\varkappa_1(K_\beta-I))*D_{4n}, \quad A_2 = H*\exp(\varkappa_1(K_\beta-I))*D_2^{*n},\\
A_3 &= H*\exp(\varkappa_1(K_\beta-I))*D_3^{*n}.
\end{aligned}$$

Observe that $H\equiv I$ if $p_0=0$.

Approximations A_1–A_3 are constructed similarly to those in Chapters 7 and 9. A_0 is an analogue of the accompanying distribution $\mathcal{L}(\tilde S_n)$, A_{01} is a first-order asymptotic expansion (compare with $\mathbf{D}$ and $\mathbf{D}+\tilde{\mathbf{D}}$ from (8.3) and (8.4)). Approximation A_1 is an analogue of Presman's $P_{n,p}$ from (7.1). A_2 and A_3 are the analogues of $\hat H_{n,2}$ and $\hat H_{n,3}$ from (7.32). In order to ensure sufficient accuracy, the initial distribution $\mathcal{L}(\xi_0)$ is approximated by $H*\exp(\varkappa_1(K_\beta-I))$.

Theorem 9.1 *If $\beta\le 1/2$, $\alpha\le 1/30$, then there exist absolute constants $c,C\in(0;1)$ such that*

$$\begin{aligned}
d_{TV}(\mathcal{L}(S_n);A_0) &\le C\left(\alpha(\alpha+\beta)\min\{1;1/\sqrt{n\alpha}\}+\min\{\alpha;n\alpha^2\}+(\alpha+\beta)\mathrm{e}^{-cn}\right), &(9.2)\\
d_{TV}(\mathcal{L}(S_n);A_{01}) &\le C\left(\alpha^2+\alpha\beta\min\{1;1/\sqrt{n\alpha}\}+(\alpha+\beta)\mathrm{e}^{-cn}\right), &(9.3)\\
d_{\mathrm{o}}(\mathcal{L}(S_n);A_0) &\le C\alpha(\alpha+\beta)+C\min\left\{\sqrt{\alpha/n};n\alpha^2\right\}+C(\alpha+\beta)\mathrm{e}^{-cn}, &(9.4)\\
d_G(\mathcal{L}(S_n);A_0) &\le C\alpha(\alpha+\beta)+C\min\{\alpha\sqrt{n\alpha};n\alpha^2\}+C(\alpha+\beta)\mathrm{e}^{-cn}. &(9.5)
\end{aligned}$$

The remainder term e^{-cn} is due to the method of the proof. It can be dropped if $\alpha\ge 1/n$.

If $\alpha \geq 1/n$, then the RHS of (9.2), (9.3), and (9.5) are respectively $O(\alpha)$, $O(\alpha^2)$ and $O(\alpha\sqrt{n\alpha})$. If $p_0 = 0$, $\alpha = \beta$, then $\mathcal{L}(S_n)$ becomes the Binomial distribution $\mathbf{B}(n,\alpha)$, and the rate of the accuracy in Theorem 9.1 is the same as resp. in (3.7), (3.31), and (3.24). However, A_0 remains a compound Poisson law with a geometric compounding distribution, it does not reduce to Poisson approximation.

According to Theorem 4.2 in [76],

$$d_{TV}(\mathcal{L}(S_n); \mathbf{\Pi}(n\gamma_1, K_\beta)) \leq C\max\{\alpha; p_0\}\min\{1; 1/\sqrt{n\alpha}\} + C\min\{\alpha; n\alpha^2\} + C\mathrm{e}^{-cn} \quad (9.6)$$

in the assumptions of Theorem 9.1. The structure of the approximating distribution in (9.6) is simpler than A_0 but the accuracy of approximation in (9.6) can be significantly worse than that in (9.2). For example, if $p_0 = 1/2$ and $\alpha = n^{-2/3}$, then the accuracy of approximation in (9.2) is $O(n^{-2/3})$, while the accuracy the accuracy of approximation in (9.6) is $O(n^{-1/6})$.

In the next theorem we apply a compound Poisson approximation similar to $P_{n,p}$ in (7.2).

Theorem 9.2 *If $\beta \leq 1/4$, $\alpha \leq 1/30$ and $n\alpha \geq 3$, then*

$$d_{TV}(\mathcal{L}(S_n); A_1) \leq C\max\{n^{-1}; (n\alpha)^{-2}\}.$$

If $p_0 = 0$ and the assumptions of Theorem 9.2 are fulfilled, then A_1 is a compound Poisson distribution. If $\alpha > n^{-2/3}$, the rate of the accuracy of approximation in Theorem 9.2 is $n^{-2/3}$. If $\alpha \leq n^{-2/3}$, then approximation A_0 can be applied yielding the estimate $O(n^{-2/3})$.

The accuracy of approximation is sharper if one uses a SCP approximation.

Theorem 9.3 *Assume that $\beta \leq 1/2$, $\alpha \leq 1/30$, $n\alpha \geq 1$. There exist absolute constants c, C such that*

$$\begin{aligned}
d_{TV}(\mathcal{L}(S_n); A_2) &\leq C(\alpha+\beta)\left(\min\{\alpha; \sqrt{\alpha/n}\} + \mathrm{e}^{-cn}\right), \\
d_o(\mathcal{L}(S_n); A_2) &\leq C(\alpha+\beta)\left(\min\{\alpha; n^{-1}\} + \mathrm{e}^{-cn}\right), \\
d_G(\mathcal{L}(S_n); A_2) &\leq C(\alpha+\beta)\left(\alpha + \mathrm{e}^{-cn}\right).
\end{aligned}$$

Theorem 9.4 *If $\beta \leq 1/2$, $\alpha \leq 1/30$, $n\alpha \geq 1$, then there exist absolute constants c, C such that*

$$\begin{aligned}
d_{TV}(\mathcal{L}(S_n); A_3) &\leq C(\alpha+\beta)\left(\min\{\alpha; n^{-1}\} + \mathrm{e}^{-cn}\right), \\
d_o(\mathcal{L}(S_n); A_3) &\leq C(\alpha+\beta)\left(\min\{\alpha; 1/n\sqrt{n\alpha}\} + \mathrm{e}^{-cn}\right), \\
d_G(\mathcal{L}(S_n); A_3) &\leq C(\alpha+\beta)\left(\min\{\alpha; \sqrt{\alpha/n}\} + \mathrm{e}^{-cn}\right).
\end{aligned}$$

As expected, the rate of the accuracy of SCP approximations is much better than the rate in Theorem 9.1. For example, if $\alpha = const.$, then the rate of the accuracy of approximation in terms of the total variation distance in Theorem 9.4 is $O(n^{-1})$ vs $O(1)$ in Theorem 9.1.

The following theorem presents lower bounds. Recall that

$$A_0 = (I + \varkappa_2(K_\beta - I)) * \exp((n - p_0)\gamma_1(K_\beta - I)).$$

Theorem 9.5 *If $\beta\le 1/2$, $\alpha\le 1/30$, $n\alpha\ge 1$, then there exist absolute constants c, C such that*

$$d_{TV}(\mathcal{L}(S_n);A_0) \ \ge\ C\alpha\Big(1-c\big(\alpha+\beta/\sqrt{n\alpha}\big)\Big), \tag{9.7}$$

$$d_{\mathrm{o}}(\mathcal{L}(S_n);A_0) \ \ge\ C\Big(1-c\big(\alpha+\beta/\sqrt{n\alpha}\big)\Big)\sqrt{\alpha/n}, \tag{9.8}$$

$$d_{G}(\mathcal{L}(S_n);A_0) \ \ge\ C\alpha\sqrt{n\alpha}\Big(1-c\big(\alpha+\beta/\sqrt{n\alpha}\big)\Big), \tag{9.9}$$

$$d_{TV}(\mathcal{L}(S_n);A_2) \ \ge\ C\Big(|\gamma_3|-c(\alpha+\beta)/\sqrt{n\alpha}\Big)\sqrt{\alpha/n}, \tag{9.10}$$

$$d_{\mathrm{o}}(\mathcal{L}(S_n);A_2) \ \ge\ C\Big(|\gamma_3|-c(\alpha+\beta)/\sqrt{n\alpha}\Big)/n, \tag{9.11}$$

$$d_{G}(\mathcal{L}(S_n);A_2) \ \ge\ C\alpha\Big(|\gamma_3|-c(\alpha+\beta)/\sqrt{n\alpha}\Big). \tag{9.12}$$

It is obvious that estimates in Theorem 9.5 are non-trivial only when the expressions in the brackets are positive.

To ensure that $C\alpha(1-c\alpha-c\beta/\sqrt{n\alpha})>0$, usually one needs the assumption that $\alpha\to 0$, $n\alpha\to\infty$. If $\alpha\equiv\alpha(n)\to 0$, $\beta\le 1/2$, $n\alpha\to\infty$, then there exist absolute constants c, C such that

$$c\alpha\le d_{TV}(\mathcal{L}(S_n);A_0)\le C\alpha,\quad c\alpha\sqrt{n\alpha}\le d_G(\mathcal{L}(S_n);A_0)\le C\alpha\sqrt{n\alpha}$$

for all sufficiently large n. If, in addition, $\beta\equiv\beta(n)=o(1)$ as $n\to\infty$, then

$$d_{TV}(\mathcal{L}(S_n);A_0)\sim 6\alpha/\sqrt{2\pi\mathrm{e}}. \tag{9.13}$$

Assume that $\alpha\equiv\alpha(n)\to 0$, $n\alpha\to\infty$, $\beta\equiv\beta(n)\to\tilde\beta$ as $n\to\infty$, the right-hand sides of estimates of (9.10)–(9.12) are positive. Then there exist absolute constants c, C such that

$$c\sqrt{\alpha/n}\le d_{TV}(\mathcal{L}(S_n);A_2)\le C\sqrt{\alpha/n}\,,\quad c\alpha\le d_G(\mathcal{L}(S_n);A_2)\le C\alpha\,.$$

If β is "small", then the approximating measure can be simplified. Let

$$w=\frac{\alpha}{1-\beta+\alpha},\quad u=\frac{\alpha(1-\beta)(\beta-\alpha)}{(1-\beta+\alpha)^3}-\frac{w^2}{2}\,.$$

Denote $A_{w,u}=\mathcal{L}(\pi_{w-2u}+2\pi_u)$. In the case of the Binomial $\mathbf{B}(\alpha)$ distribution $w=\alpha$, $u=-\alpha^2/2$.

Theorem 9.6 *If $\beta\le 1/20$ and $\alpha\le 1/30$, then*

$$d_{TV}(\mathcal{L}(S_n);A^{*n}_{w,u})\le C(\alpha+\beta)^2\min\{n\alpha;(n\alpha)^{-1/2}\}+C|\alpha-\beta|\min\{1;(n\alpha)^{-1/2}\}. \tag{9.14}$$

If $\beta=O(\alpha)$, then the RHS of (9.14) is $O(\alpha n^{-1/2})$. Moreover, if $\alpha=\beta$, then the Markov Binomial distribution becomes the Binomial distribution, $A^{*n}_{w,u}$ coincides with G_2 from Section 7.1, the accuracy of approximation in Theorem 9.6 is the same as that of the upper bound in (7.16). Further results of this kind can be found in [79, 85].

9.2 Some other results

Negative Binomial approximation. Let $\xi_0, \xi_1, \ldots, \xi_n$ be a Markov chain with the initial distribution

$$\mathbb{P}(\xi_0=1) = 1 - \mathbb{P}(\xi_0=0) = p_0 \in [0,1]$$

and transition probabilities

$$\mathbb{P}(\xi_i=1 \,|\, \xi_{i-1}=0) = 1 - \mathbb{P}(\xi_i=0 \,|\, \xi_{i-1}=0) = \alpha,$$
$$\mathbb{P}(\xi_i=1 \,|\, \xi_{i-1}=1) = 1 - \mathbb{P}(\xi_i=0 \,|\, \xi_{i-1}=1) = \beta,$$

where $\alpha \in (0;1), \beta \in (0;1)$ $(i \in \mathbb{N})$.

Assume that $\mathbb{P}(\xi_0 = 1) = \alpha/(1-\beta+\alpha)$ (stationarity). Then $\mathbb{E}S_n = n\alpha/(1-\beta+\alpha)$,

$$\begin{aligned} \operatorname{var} S_n &= \frac{n\alpha(1-\beta)}{(1-\beta+\alpha)^2} + \frac{2n\alpha(1-\beta)(\beta-\alpha)}{(1-\beta+\alpha)^3} + \frac{2\alpha(1-\beta)(\beta-\alpha)((\beta-\alpha)^n-1)}{(1-\beta+\alpha)^4} \\ &= \frac{n\alpha(1-\beta)(1+\beta-\alpha)}{(1-\beta+\alpha)^3} + \frac{2\alpha(1-\beta)(\beta-\alpha)((\beta-\alpha)^n-1)}{(1-\beta+\alpha)^4}. \end{aligned}$$

If $\operatorname{var} S_n > \mathbb{E}S_n$, let Y denote a Negative Binomial $\mathbf{NB}(r,p)$ r.v. defined by (1.4), where

$$r = \frac{\mathbb{E}^2 S_n}{\operatorname{var} S_n - \mathbb{E}S_n}, \quad p = 1 - \frac{\mathbb{E}S_n}{\operatorname{var} S_n}.$$

Theorem 9.7 [318] If $\mathbb{P}(\xi_0 = 1) = \alpha/(1-\beta+\alpha)$ and $\operatorname{var} S_n > \mathbb{E}S_n$, then

$$\begin{aligned} d_{TV}(S_n;Y) &\le \frac{|\alpha-\beta|(5+43\max\{\alpha;\beta\})}{(1-\max\{\alpha;\beta\})^2}\left(\frac{2\sqrt{5}}{\sqrt{n}} \cdot \frac{\sqrt{1-\beta+\alpha}}{\sqrt{\alpha(1-\beta)\min\{1-\alpha;\beta;1/2\}}}\right. \\ &\quad \left. + \frac{360}{n} \cdot \frac{(1-\alpha)(1-\beta)^2+\alpha^2\beta}{\alpha(1-\beta)(1-\beta+\alpha)} + \beta^{\lfloor n/4 \rfloor}\right). \end{aligned} \tag{9.15}$$

If $\alpha \equiv \alpha(n)$ and $\beta \equiv \beta(n)$ are bounded away from 0 and 1, then the RHS of (9.15) is $O(n^{-1/2})$. The same rate of shifted Poisson approximation has been achieved by Barbour & Lindvall [32].

Symmetric three-state Markov chain.

The case of a symmetric three-state Markov chain has been investigated by Šliogerė & Čekanavičius [285]. Consider r.v.s $\xi_0, \xi_1, \ldots$ taking values in the set $\{a_1, a_2, a_3\}$. In this section, we assume that $\{\xi_i\}$ is a (possibly non-stationary) Markov chain.

Denote $S_n = \xi_1 + \cdots + \xi_n$, $\mathbb{P}(\xi_0=a_1) = p_{01}$, $\mathbb{P}(\xi_0=a_2) = p_{02}$, $\mathbb{P}(\xi_0=a_3) = p_{03}$. Let the transition probabilities be equal to

$$\begin{aligned} &\mathbb{P}(\xi_i=a_1 \,|\, \xi_{i-1}=a_1) = \mathbb{P}(\xi_i=a_3 \,|\, \xi_{i-1}=a_1) = a, \quad \mathbb{P}(\xi_i=a_2 \,|\, \xi_{i-1}=a_1) = 1-2a, \\ &\mathbb{P}(\xi_i=a_1 \,|\, \xi_{i-1}=a_2) = \mathbb{P}(\xi_i=a_3 \,|\, \xi_{i-1}=a_2) = b, \quad \mathbb{P}(\xi_i=a_2 \,|\, \xi_{i-1}=a_2) = 1-2b, \\ &\mathbb{P}(\xi_i=a_1 \,|\, \xi_{i-1}=a_3) = \mathbb{P}(\xi_i=a_3 \,|\, \xi_{i-1}=a_3) = a, \quad \mathbb{P}(\xi_i=a_2 \,|\, \xi_{i-1}=a_3) = 1-2a, \end{aligned}$$

where $a, b \in [0;1/2]$. Denote

$$\begin{aligned} \tilde{D} &= \exp\left(\frac{2b(1-2a)}{1-2a+2b}(\tilde{B}-I)\right), \qquad \tilde{B} = \frac{(1-2a)}{2}\sum_{j=0}^{\infty} a^i (I_{-1}+I_1)^{*(j+1)}, \\ \tilde{M} &= \frac{1-2(a-b)}{1+2b}\sum_{j=0}^{\infty}\left(\frac{a}{1+2b}\right)^j (I_{-1}+I_1)^{*j}, \qquad \tilde{E} = \left(1-\frac{2ap_{02}}{1-2a}\right)I + \frac{ap_{02}}{1-2a}(I_{-1}+I). \end{aligned}$$

Theorem 9.8 [285] *Assume that* $a_1=-1$, $a_2=0$, $a_3=1$. *If* $0<a,b\leq 1/30$, *then*

$$\begin{aligned} d_{TV}(\mathcal{L}(S_n); \tilde{E}*\tilde{M}*\tilde{D}^{*n}) &\leq C\left(\min\{1/n; b\} + 0.2^n|a-b|\right), \\ d_o(\mathcal{L}(S_n); \tilde{E}*\tilde{M}*\tilde{D}^{*n}) &\leq C\left(\min\{1/n\sqrt{nb}; b\} + 0.2^n|a-b|\right), \\ d_G(\mathcal{L}(S_n); \tilde{E}*\tilde{M}*\tilde{D}^{*n}) &\leq C\left(\min\{\sqrt{b/n}; b\} + 0.2^n|a-b|\right). \end{aligned}$$

Approximation $\tilde{E}*\tilde{M}*\tilde{D}^{*n}$ has three distinctive components:

(i) $\tilde{E}$ is a 3-points distribution, it is not infinitely divisible, it approximates $\mathcal{L}(\xi_0)$.

(ii) $\tilde{M}$ is a compound geometric distribution with symmetric compounding distribution $(I_{-1}+I_1)/2$.

(iii) $\tilde{D}$ is a compound Poisson distribution with a geometric compounding distribution.

The accuracy of approximation in Theorem 9.8 is $O(n^{-1})$.

Hsiau [158] has extended the compound Poisson limit theorem to the case of a stationary Markov chains with more than two states. A review of some results can be found in [228]. Further extensions of the Markov Binomial model were considered in [212, 313, 333]. Large deviations for Markov Binomial distribution have been studied by Jensen [166]. First-order asymptotic expansions can be found in [85, 86].

Open problems

9.1 Can the assumptions on α, β in Theorem 9.3 be weaken?

9.2 Generalise Theorems 9.3 and 9.7 to the case of a Markov chain with more than 3 states.

9.3 Proofs

9.3.1 Auxiliary results

In Lemmas 9.9–9.14 we set

$$U_\beta = K_\beta - I,$$

where K_β denotes geometric $\mathbf{\Gamma}(\beta)$ distribution:

$$K_\beta = (1-\beta)\sum_{j=0}^{\infty}\beta^j I_{j+1}, \quad \widehat{K}_\beta(t) = (1-\beta)e^{it}/(1-\beta e^{it}).$$

Lemma 9.9 *Let* $\alpha\leq 1/30$, $\beta\leq 1/4$, $|t|\leq\pi$. *Then*

$$Re\widehat{K}_\beta(t) \leq 1-\frac{2}{1+\beta}\sin^2(t/2) \leq -\frac{8}{9\pi^2}t^2, \quad |\widehat{K}_\beta(t)-1| \leq \frac{4}{3}|t|,$$

where Re *denotes the real part of the complex number.*

The statement of Lemma 9.9 follows from the fact that

$$|\sin t|/|t| \geq 2/\pi \qquad (|t|\leq\pi/2).$$

Lemma 9.10 *If $\beta\le 1/2$, then for any unit measure M*

$$\|U_\beta * M\|_G \ge \frac{2}{3}\|M\|_{TV}, \quad \|U_\beta * M\|_{TV} \ge \frac{2}{3}\|(I_1-I)*M\|_{TV}. \tag{9.16}$$

Proof of Lemma 9.10. It is not difficult to check that

$$\widehat{U}_\beta(t) = (\mathrm{e}^{\mathrm{i}t}-1)/(1-\beta\mathrm{e}^{\mathrm{i}t}),$$

that is,

$$U_\beta = (I_1-I) * \sum_{j=0}^{\infty}\beta^j I_j, \quad U_\beta * (I-\beta I_1) = I_1 - I, \quad (I-\beta I_1) * \sum_{j=0}^{\infty}\beta^j I_j = I.$$

By (A.6),

$$\|M * (I_1-I)\|_{TV} = \|M * U_\beta * (I-\beta I)\|_{TV} \le \|M * U_\beta\|_{TV}\|I-\beta I\|_{TV} \le \|M * U_\beta\|_{TV}(1+\beta).$$

Similarly,

$$\|U_\beta * M\|_G = \left\| M * (I_1-I) * \sum_{j=0}^{\infty}\beta^j I_j \right\|_G = \left\| M * \sum_{j=0}^{\infty}\beta^j I_j \right\|_{TV}$$

by (A.9). On the other hand,

$$\|M\|_{TV} = \left\| M * \sum_{j=0}^{\infty}\beta^j I_j * (I-\beta I_1) \right\|_{TV} \le (1+\beta)\left\| M * \sum_{j=0}^{\infty}\beta^j I_j \right\|_{TV}.$$

The statement of Lemma 9.10 now easily follows from the fact that $1+\beta\le 3/2$. □

Recall that

$$\varkappa_2 = p_0(\beta(1-\beta))/(1-\beta+\alpha), \; H = I + \varkappa_2(K_\beta - I).$$

Hence

$$\widehat{H}(t) = 1 + \frac{\varkappa_2(\mathrm{e}^{\mathrm{i}t}-1)}{1-\beta\mathrm{e}^{\mathrm{i}t}} = \frac{(1-\varkappa_2)(1-v\mathrm{e}^{\mathrm{i}t})}{1-\beta\mathrm{e}^{\mathrm{i}t}} = \exp\left(\sum_{j=1}^{\infty}\frac{\beta^j}{j}(1-v^j)(\mathrm{e}^{\mathrm{i}tj}-1)\right),$$

where $v=(1-\varkappa_2/\beta)/(1-\varkappa_2)$. Note that $1-v^j>0$, therefore H is CP distribution. Let

$$H^{(-1)} := \exp\left(-\sum_{j=1}^{\infty}\frac{\beta^j}{j}(1-v^j)(I_j-I)\right).$$

Then

$$\widehat{H}^{(-1)} = (1-\beta\mathrm{e}^{\mathrm{i}t})/(1-\varkappa_2)(1-v\mathrm{e}^{\mathrm{i}t}), \quad \widehat{H}(t)\widehat{H}^{(-1)}(t) = 1 \;\; (\forall t).$$

Since measures H, $H^{(-1)}$ are of bounded variation, this entails $H*H^{(-1)}=I$. □

Lemma 9.11 *Let (9.1) be satisfied. Then*

$$\|H^{(-1)}\|_{TV} \le \mathrm{e}^2. \tag{9.17}$$

For any (signed) finite measure M concentrated at integers

$$\|M * H\|_{TV} \ge \mathrm{e}^{-2}\|M\|_{TV}, \quad \|M * \exp(-p_0\gamma_1 U_\beta)\|_{TV} \ge \|M\|_{TV}. \tag{9.18}$$

Proof of Lemma 9.11. Since $\beta \le 1/2$, we have $v<1$. Estimate (9.17) easily follows from the property $\|e^M\|_{TV} \le e^{\|M\|_{TV}}$ and the trivial estimate $\|I_j - I\|_{TV} \le 2$.

Recall that $H * H^{(-1)} = I$. By (A.1),

$$\|M\|_{TV} = \|M * H * H^{(-1)}\|_{TV} \le \|M * H\|_{TV}\, \|H^{(-1)}\|_{TV} \le e^2 \|M * H\|_{TV}.$$

Since $\exp(p_0\gamma_1 U_\beta)$ is a distribution, its total variation is 1. Therefore,

$$\|M\|_{TV} = \|M * \exp(-p_0\gamma_1 U_\beta) * \exp(p_0\gamma_1 U_\beta)\,\|_{TV} \le \|M * \exp(-p_0\gamma_1 U_\beta)\,\|_{TV}.$$

The proof is complete. □

Set $F_n = \mathcal{L}(S_n)$. We need certain facts about $\widehat{F}_n$.

Lemma 9.12 *Let (9.1) be satisfied. Then*

$$\widehat{F}_n(t) = \widehat{\Lambda}_1^n(t)\, \widehat{W}_1(t) + \widehat{\Lambda}_2^n(t)\, \widehat{W}_2(t), \tag{9.19}$$

where

$$\begin{aligned}
\widehat{\Lambda}_1(t) &= (\beta e^{it} + 1-\alpha + \widehat{D}^{1/2}(t))/2, \qquad \widehat{\Lambda}_2(t) = (\beta e^{it} + 1-\alpha - \widehat{D}^{1/2}(t))/2,\\
\widehat{W}_1(t) &= \frac{p_0}{2}\left(1 + \frac{1-\beta+\alpha+\beta(e^{it}-1)}{\widehat{D}^{1/2}(t)}\right) + \frac{1-p_0}{2}\left(1 + \frac{1-\beta+\alpha+(2\alpha-\beta)(e^{it}-1)}{\widehat{D}^{1/2}(t)}\right),\\
\widehat{W}_2(t) &= \frac{p_0}{2}\left(1 - \frac{1-\beta+\alpha+\beta(e^{it}-1)}{\widehat{D}^{1/2}(t)}\right) + \frac{1-p_0}{2}\left(1 - \frac{1-\beta+\alpha+(2\alpha-\beta)(e^{it}-1)}{\widehat{D}^{1/2}(t)}\right),\\
\widehat{D}(t) &= (\beta e^{it}+1-\alpha)^2 + 4e^{it}(\alpha-\beta) = (1+\alpha-\beta e^{it})^2\left(1 + \frac{4\alpha(e^{it}-1)}{(1+\alpha-\beta e^{it})^2}\right)\\
&= \big(1-\beta+\alpha+\beta(e^{it}-1)\big)^2\left(1 + \frac{4(1-\beta)(\alpha-\beta)(e^{it}-1)}{(1-\beta+\alpha+\beta(e^{it}-1))^2}\right).
\end{aligned}$$

Obviously, $\widehat{D}(0) = (1-\beta+\alpha)^2 > 1$ if $\beta < \alpha$, thus, $\widehat{D}$ is not a characteristic function.

Proof of Lemma 9.12. Using the standard matrix notation, we have

$$\widehat{F}_n(t) = (p_0, 1-p_0)\, \widetilde{P}^n(t) \begin{pmatrix} 1 \\ 1 \end{pmatrix},$$

where

$$\widetilde{P}(t) = \begin{pmatrix} \beta e^{it} & 1-\beta \\ \alpha e^{it} & 1-\alpha \end{pmatrix},$$

cf. [314]. We apply the standard spectral decomposition of matrices:

$$\widetilde{P}^n(t) = \hat{\Lambda}_1^n(t)\, \vec{x}_1\, \vec{y}_1^{\,T} + \hat{\Lambda}_2^n(t)\, \vec{x}_2\, \vec{y}_2^{\,T},$$

where $\hat{\Lambda}_1(t)$, $\hat{\Lambda}_2(t)$ are eigenvalues of matrix $\widetilde{P}(t)$, $\vec{x}_1, \vec{x}_2, \vec{y}_1, \vec{y}_2$, are its left and right eigenvectors, T stands for transposition.

$\hat{\Lambda}_1(t)$, $\hat{\Lambda}_2(t)$ can be found by solving the equation

$$\widehat{\Lambda}^2(t) - \widehat{\Lambda}(t)(1-\alpha+\beta e^{it}) + e^{it}(\beta-\alpha) = 0.$$

To find $\widehat{W}_1(t)$ and $\widehat{W}_2(t)$, we use relations

$$\widetilde{P}(t)\,\vec{x}_j = \widehat{\Lambda}_j(t)\,\vec{x}_j, \qquad \vec{y}_j^{\,T}\,\widetilde{P}(t) = \widehat{\Lambda}_j(t)\,\vec{y}_j^{\,T}, \qquad \vec{y}_j^{\,T}\,\vec{x}_j = 1 \qquad (j=1,2),$$

where

$$\begin{aligned}\vec{x}_j^{\,T} &= ((1-\beta)(\widehat{\Lambda}_j(t)-1+\alpha),\,(1-\beta)\alpha e^{it}),\\ \vec{y}_j^{\,T} &= \left(\frac{(\widehat{\Lambda}_j(t)-1+\alpha)/(1-\beta)}{(1-\beta)\alpha e^{it}+(\widehat{\Lambda}_j(t)-1+\alpha)^2},\, \frac{1}{(1-\beta)\alpha e^{it}+(\widehat{\Lambda}_j(t)-1+\alpha)^2}\right).\end{aligned}$$

We have

$$\widehat{F}_n(t) = \Lambda_1^n(t)(p_0,1-p_0)\vec{x}_1\,\vec{y}_1^{\,T}\begin{pmatrix}1\\1\end{pmatrix} + \Lambda_2^n(t)(p_0,1-p_0)\vec{x}_2\,\vec{y}_2^{\,T}\begin{pmatrix}1\\1\end{pmatrix}.$$

It can be checked that

$$\begin{aligned}\widehat{W}_j(t) &= (p_0,1-p_0)\vec{x}_j\,\vec{y}_j^{\,T}\begin{pmatrix}1\\1\end{pmatrix} = p_0x_{j1}(y_{j1}+y_{j2}) + (1-p_0)x_{j2}(y_{j1}+y_{j2})\\ &= p_0\,\frac{(\widehat{\Lambda}_j(t)-1+\alpha)(\widehat{\Lambda}_j(t)+\alpha-\beta)}{(1-\beta)\alpha e^{it}+(\widehat{\Lambda}_j(t)-1+\alpha)^2} + (1-p_0)\frac{\alpha e^{it}(\widehat{\Lambda}_j(t)+\alpha-\beta)}{(1-\beta)\alpha e^{it}+(\widehat{\Lambda}_j(t)-1+\alpha)^2}.\end{aligned}$$

Substituting expressions for $\hat{\Lambda}_j(t)$ into the last equation, we complete the proof. □

For the proofs we will use smoothing properties of exponential measures, see Section C in Appendix. We express F_n in convolution powers of U_β. We will use Lemma 9.12 and the fact that there is a unique correspondence between a measure and its Fourier transform.

We denote by $\Lambda_1, \Lambda_2, W_1, W_2$ measures with Fourier transforms $\widehat{\Lambda}_1(t), \widehat{\Lambda}_2(t), \widehat{W}_1(t)$, $\widehat{W}_2(t)$, respectively. Set $\tilde{K} = \sum_{j=0}^{\infty}(\beta I_1 - \alpha I - 2\gamma_1 U_\beta)^{*j}$,

$$\tilde{L} = \frac{4\alpha^2}{(1-\beta+\alpha)^2}U_\beta^{*2} * \Big[(1-\beta)^2 I + \beta(1-\beta+\alpha) * (I-\beta I_1)\Big] * \tilde{K}^{*2}. \tag{9.20}$$

Recall that $U_\beta = K_\beta - I$, where $K_\beta = (1-\beta)\sum_{j=0}^{\infty}\beta^j I_{j+1}$.

Lemma 9.13 *Let assumption (9.1) hold. Then*

$$F_n = \Lambda_1^{*n} * W_1 + \Lambda_2^{*n} * W_2\,, \tag{9.19*}$$

where

$$\begin{aligned}\Lambda_1 &= I + \gamma_1 U_\beta + \tfrac{1}{2}\Big\{(1+\alpha)I - \beta I_1 + 2\gamma_1 U_\beta\Big\} * \sum_{j=1}^{\infty}\frac{(1/2)^{(j)}}{j!}(-1)^j\tilde{L}^{*j},\\ \Lambda_2 &= \beta I_1 + (1-\alpha)I - \Lambda_1, \quad \tilde{B} = \sum_{j=0}^{\infty}(\beta I_1 - \alpha I)^{*j},\\ W_1 &= \tfrac{1}{2}\Big\{I + ((1-\beta+\alpha)I + \beta(I_1-I)) * \tilde{K} * \sum_{j=0}^{\infty}\frac{(-1/2)^{(j)}}{j!}(-1)^j\tilde{L}^{*j}\Big\}\\ &\quad + (1-p_0)(\alpha-\beta)(I_1-I) * \tilde{K} * \sum_{j=0}^{\infty}\frac{(-1/2)^{(j)}}{j!}(-1)^j\tilde{L}^{*j},\\ W_2 &= \tfrac{1}{2}\Big\{I - ((1-\beta+\alpha)I + \beta(I_1-I)) * \tilde{K} * \sum_{j=0}^{\infty}\frac{(-1/2)^{(j)}}{j!}(-1)^j\tilde{L}^{*j}\Big\}\\ &\quad - (1-p_0)(\alpha-\beta)(I_1-I) * \tilde{K} * \sum_{j=0}^{\infty}\frac{(-1/2)^{(j)}}{j!}(-1)^j\tilde{L}^{*j}.\end{aligned}$$

We recall that C denotes (possibly different) absolute constants. Lemma 9.13 allows to use properties of convolutions instead of the characteristic functions method.

Denote $a_1 = \gamma_1$, $a_2 = \gamma_2 + a_1^2/2$, $a_3 = \gamma_3 + a_1 a_2 - a_1^3/3$. Given measure P, we use notation

$$\ln P = \sum_{j=1}^{\infty} (-1)^{j+1}(P-I)^{*j}/j,$$

for $\|P-I\|_{TV} \leq 1$.

Lemma 9.14 *Let condition (9.1) be satisfied. Then*

$$\begin{aligned}
\Lambda_1 &= I + \sum_{j=1}^{3} a_j U_\beta^{*j} + C\alpha^3(\beta+\alpha)U_\beta^{*4} * \Theta, && (9.21)\\
\ln \Lambda_1 &= \sum_{j=1}^{3} \gamma_j U_\beta^{*j} + C\alpha^3(\beta+\alpha)U_\beta^{*4} * \Theta, && (9.22)\\
\ln \Lambda_1 &= \gamma_1 U_\beta + \frac{19}{60}\gamma_1 U_\beta^{*2} * \Theta, && (9.23)\\
\|\Lambda_2\|_{TV} &\leq 19/30, \quad \|\Lambda_1 - I\|_{TV} \leq 0.1, && (9.24)\\
W_1 &= I + (\varkappa_1+\varkappa_2)U_\beta + C\alpha(\beta+\alpha)U_\beta^{*2} * \Theta, && (9.25)\\
W_2 &= C(\alpha+\beta)(I_1 - I) * \Theta, \qquad \|W_2\|_{TV} \leq 7. && (9.26)
\end{aligned}$$

Besides, for any finite signed measure M on $\mathbb{Z}$ and any $t>0$

$$\begin{aligned}
\|M * \exp(t\ln\Lambda_1)\|_{TV} &\leq C\|M * \exp((t\gamma_1/30)U_\beta)\|_{TV}, && (9.27)\\
\|M * D_{jt}\|_{TV} &\leq C\|M * \exp((t\gamma_1/30)U_\beta)\|_{TV} \quad (j=1,2,3), && (9.28)
\end{aligned}$$

where $D_{jt} = \exp(t\sum_{i=1}^{j}\gamma_1(K_\beta - I)^{*i})$.

Proof of Lemma 9.14. Recall that $U_\beta = K_\beta - I$, where

$$K_\beta = (1-\beta)\sum_{j=0}^{\infty} \beta^j I_{j+1},$$

$\tilde{L}$ is defined by (9.20), $\tilde{K} = \sum_{j=0}^{\infty}(\beta I_1 - \alpha I - 2\gamma_1 U_\beta)^{*j}$. Since $a_1 = \gamma_1 \leq 1/30$, $1/(1-\beta+\alpha) \leq 2$, we have

$$\|U_\beta\|_{TV} \leq \|K_\beta\|_{TV} + 1 = 2, \quad \|\tilde{K}\|_{TV} \leq \sum_{j=0}^{\infty}(\alpha+\beta+4a_1)^j \leq 3,$$

$$\|\tilde{L}\|_{TV} \leq 9 \cdot 4 \cdot \alpha^2 \cdot 4\big(1 + \beta(1+\beta)/(1-\beta+\alpha)\big) \leq 0.4.$$

We have $|(1/2)^{(2)}/2| = 1/8$, $|(1/2)^{(3)}/3!| = 1/16$, $|(1/2)^{(j)}/j!| \leq 5/128$, $(j \geq 4)$. Therefore,

$$\sum_{j=1}^{\infty} \frac{|(1/2)^{(j)}|}{j!}\|\tilde{L}\|_{TV}^{(j-1)} \leq \frac{1}{2} + \frac{0.4}{8} + \frac{(0.4)^2}{16} + \frac{5}{128}\frac{0.4^3}{0.6} \leq 0.5642$$

and

$$\begin{aligned}
(I(1+\alpha) - \beta I_1 + 2a_1 U_\beta) * \tilde{L} &= \frac{4\alpha^2}{(1-\beta+\alpha)^2}((1-\beta)^2 I + \beta(1-\beta+\alpha)(I - \beta I_1)) * U_\beta^{*2} * \tilde{K}\\
&= \frac{12\gamma_1\alpha}{(1-\beta)(1-\beta+\alpha)}((1-\beta)^2 + \beta(1-\beta+\alpha)(1+\beta))U_\beta^{*2} * \Theta\\
&= \gamma_1 U_\beta^{*2} * \Theta.
\end{aligned}$$

Consequently,

$$\Lambda_1 = I + \gamma_1 U_\beta + 0.2821\gamma_1 U_\beta^{*2} * \Theta = I + 1.2821\gamma_1 U_\beta * \Theta = I + 0.1\Theta,$$

$$\begin{aligned}\ln \Lambda_1 &= \Lambda_1 - I + \sum_{j=2}^{\infty} \frac{(-1)^{j+1}}{j}(\Lambda_1 - I)^{*j} = \gamma_1 U_\beta + 0.2821\gamma_1 U_\beta^{*2} * \Theta \\ &+ \frac{1}{2}1.2821\gamma_1^2 U_\beta^{*2} * \sum_{j=2}^{\infty}(0.1)^{j-2} * \Theta = \gamma_1 U_\beta + \frac{19}{60}\gamma_1 U_\beta^{*2} * \Theta.\end{aligned}$$

Moreover, $\|\Lambda_2\|_{TV} \leq \alpha + \beta + \|\Lambda_1 - I\|_{TV} \leq 19/30$. Thus, we have proved (9.23).

By the properties of the total variation norm and Lemma B.1,

$$\begin{aligned}\|M * e^{t\ln\Lambda_1}\|_{TV} &\leq \left\|M * \exp\left(\frac{t\gamma_1}{30}U_\beta\right)\right\|_{TV} \left\|\exp\left(\frac{29t\gamma_1}{30}U_\beta + \frac{19t\gamma_1}{60}U_\beta^{*2} * \Theta\right)\right\|_{TV} \\ &\leq \left\|M * \exp\left(\frac{t\gamma_1}{30}U_\beta\right)\right\|_{TV}.\end{aligned}$$

This leads to (9.27).

Estimate (9.28) can be proved similarly. For the proof of (9.21) we note that

$$\begin{aligned}\Lambda_1 &= I + \gamma_1 U_\beta - \frac{\alpha^2}{(1-\beta+\alpha)^2}U_\beta^{*2}((1-\beta)^2 I + \beta(1-\beta+\alpha)(I-\beta I_1)) * \tilde{K} \\ &\quad + C\alpha^4 U_\beta^{*4} * \Theta, \qquad (9.29)\\ I_1 - I &= U_\beta * (I - \beta I_1), \quad (1-\beta+\alpha)\tilde{B} = I + \beta(I_1 - I) * \tilde{B}, \\ I_1 - I &= 2U_\beta * \Theta, \quad (1-\beta+\alpha)(I-\beta I_1) * \tilde{B} = (1-\beta)I - \beta\alpha(I-\beta I_1) * \tilde{B} * U_\beta, \quad (9.30)\\ \tilde{B} &= \frac{1}{1-\beta+\alpha}I + \frac{\beta(1-\beta)}{(1-\beta+\alpha)^2}U_\beta - \frac{\beta^2\alpha}{(1-\beta+\alpha)^2}U_\beta * (I_1 - I) * \tilde{B}, \\ \tilde{K} &= \tilde{B} + 2\gamma_1 U_\beta * \tilde{K} * \tilde{B} = \tilde{B} - 2\gamma_1 \tilde{B}^{*2} * U_\beta + 4\gamma_1^2 \tilde{B}^{*2} * U_\beta^{*2} * \tilde{B}^{*2} * \tilde{K}. \quad (9.31)\end{aligned}$$

Substituting (9.30) and (9.31) into (9.29), we obtain (9.21). Taking into account (9.24), we prove that

$$\ln \Lambda_1 = \sum_{j=1}^{3} \frac{(-1)^{j+1}}{j}(\Lambda_1 - I)^{*j} + C(\Lambda_1 - I)^{*4} * \Theta.$$

For the proof of (9.22) it suffices to apply (9.21). From (9.31) and the first relation in (9.30) we get

$$\beta(I_1 - I) * \tilde{K} = \beta(I_1 - I) * \tilde{B} + C\beta\alpha U_\beta^{*2} * \Theta.$$

Moreover,

$$(I_1 - I) * \tilde{B} = \frac{(1-\beta)}{1-\beta+\alpha}U_\beta - \frac{\beta\alpha(I_1 - I) * \tilde{B} * U_\beta}{1-\beta+\alpha} = \frac{(1-\beta)U_\beta}{1-\beta+\alpha} + C\alpha U_\beta^{*2} * \Theta.$$

The last two equations allow us to prove (9.25).

Since $W_1 + W_2 = I$, we have the first relation in (9.26), while

$$\|W_2\|_{TV} \leq \frac{1}{2}\left(1 + (1-\beta+\alpha)3\sum_{j=0}^{\infty}\frac{|(-1/2)^{(j)}|\,0.4^j}{j!}\right) + |\alpha - p| \cdot 2 \cdot 3\sum_{j=0}^{\infty}\frac{|(-1/2)^{(j)}|\,0.4^j}{j!} < 7.$$

The proof of Lemma 9.14 is complete. □

9.3.2 Proofs of Theorems 9.1–9.6.

Proof of Theorem 9.1. We have

$$\begin{aligned} d_{TV}(\mathcal{L}(S_n);A_0) &\leq \|\Lambda_1^{*n}-D_1^{*n}\|_{TV}\|W_1\|_{TV}+\|D_1^{*n}*(W_1-H*\exp(-p_0\gamma_1U_\beta))\|_{TV}\\ &+ \|\Lambda_2\|_{TV}^n\|W_2\|_{TV}. \end{aligned}$$

Besides,

$$\begin{aligned} \|\Lambda_1^{*n}-D_1^{*n}\|_{TV} &\leq n\int_0^1\|(\ln\Lambda_1-\gamma_1U_\beta)*\exp(\tau n\ln\Lambda_1+(1-\tau)n\gamma_1U_\beta)\,\|_{TV}\,\mathrm{d}\tau\\ &\leq Cn\|(\ln\Lambda_1-\gamma_1U_\beta)*\exp(n\gamma_1U_\beta/30)\,\|_{TV}\\ &\leq Cn\alpha^2\|U_\beta^{*2}*\exp(n\gamma_1U_\beta/30)\,\|_{TV} \end{aligned}$$

by (B.1). According to Lemma 9.14, $W_1-H*\exp(-p_0\gamma_1U_\beta)$ equals

$$\begin{aligned} &(W_1-I-(\varkappa_1+\varkappa_2)*U_\beta)+(I+(\varkappa_1+\varkappa_2)U_\beta-(I-p_0\gamma_1U_\beta)*H)\\ &+(H*(I-p_0\gamma_1U_\beta-\exp(-p_0\gamma_1U_\beta)))=C\alpha(\alpha+\beta)U_\beta*\Theta. \end{aligned}$$

where C is an absolute constant. Applying (C.8), (9.26), (9.24), we derive (9.2).
Local estimates (9.4) and (9.5) can be proved similarly employing (A.7) and (A.9).

For the proof of (9.3) note that

$$\begin{aligned} d_{TV}(\mathcal{L}(S_n;A_{01}) &\leq \|\Lambda_2\|_{TV}^n\|W_2\|_{TV}+\|W_1\|_{TV}\,\|\Lambda_1^{*n}-D_2^{*n}\|_{TV}\\ &+\|W_1\|_{TV}\,\|D_1^{*n}*(\mathrm{e}^{n\gamma_2U_\beta^{*2}}-I-n\gamma_2U_\beta^{*2})\|_{TV}\\ &+\|D_1^{*n}*(I+n\gamma_2U_\beta^{*2})*(W_1-H*\mathrm{e}^{-p_0\gamma_1U_\beta})\|_{TV}. \end{aligned}$$

Using (9.28), we obtain

$$\begin{aligned} \|\Lambda_1^{*n}-D_2^{*n}\|_{TV} &\leq Cn\Big\|[\ln\Lambda_1-\gamma_1U_\beta-\gamma_2U_\beta^{*2}]\\ &* \int_0^1\exp\big(\tau n\ln\Lambda_1+(1-\tau)[n\gamma_1U_\beta+n\gamma_2U_\beta^{*2}]\big)\,\mathrm{d}\tau\Big\|_{TV}\\ &\leq Cn\|[\ln\Lambda_1-\gamma_1U_\beta-\gamma_2U_\beta^{*2}]*\exp((n\gamma_1/30)U_\beta)\,\|_{TV}\\ &\leq Cn\alpha^2(\alpha+\beta)\|U_\beta^{*3}*\exp((n\gamma_1/30)U_\beta)\,\|_{TV}. \end{aligned}$$

Similarly,

$$\begin{aligned} \Big\|D_1^{*n}*(\mathrm{e}^{n\gamma_2U_\beta^{*2}}-I-n\gamma_2U_\beta^{*2})\Big\|_{TV} &\leq \Big\|(n\gamma_2)^2U_\beta^{*4}\int_0^1D_1^{*n}*\mathrm{e}^{\tau n\gamma_2U_\beta^{*2}}(1-\tau)\mathrm{d}\tau\Big\|_{TV}\\ &\leq C(n\gamma_2)^2\|U_\beta^{*4}*\exp((n\gamma_1/30)U_\beta)\,\|_{TV}. \end{aligned}$$

By (C.8),

$$\|D_1^{*n}*(I+n\gamma_2U_\beta^{*2})*M\|_{TV}\leq\|D_1^{*n/2}*M\|_{TV}\,(1+n|\gamma_2|\|U_\beta^{*2}*D_1^{*n/2}\|_{TV})\leq C\|D_1^{*n/2}*M\|_{TV},$$

where $M=W_1-H*\exp(-p_0\gamma U_\beta)$. The rest of the proof is very similar to the proof of (9.2). □

Proof of Theorems 9.3–9.4 is similar to that of Theorem 9.1. Using Lemma 9.14, it is not difficult to show that

$$\begin{aligned} W_1 - \mathrm{e}^{\varkappa_1 U_\beta} * H &= (W_1 - I - (\varkappa_1 + \varkappa_2) * U_\beta) + (I + (\varkappa_1+\varkappa_2)U_\beta - (I+\varkappa_1 U_\beta) * H) \\ &+ \ H * (I + \varkappa_1 U_\beta - \mathrm{e}^{\varkappa_1 U_\beta}) = C\alpha(\beta+\alpha)U_\beta^{*2} * \Theta, \\ \|\Lambda_1^{*n} * W_1 - D_2^{*n} * H * \mathrm{e}^{\varkappa_1 U_\beta}\|_{TV} &\le \ \|\Lambda_1^{*n} - D_2^{*n}\|_{TV} \, \|W_1\|_{TV} \\ &+ \ \|D_2^{*n} * (W_1 - H * \mathrm{e}^{\varkappa_1 U_\beta})\|_{TV}, \end{aligned}$$

yielding the first estimate of Theorem 9.3. Other estimates can be proved similarly. □

Proof of Theorem 9.2. It is clear that SCP measure D_{4n} is infinitely divisible. We now check that it is a compound Poisson distribution.

Taking into account that $\beta \le 1/4$ and $\alpha \le 1/30$, we observe that

$$\frac{45}{47} \le \frac{1-\beta}{1-\beta+\alpha} \le 1, \quad \frac{30}{31} \le \frac{1}{1-\beta+\alpha} \le \frac{4}{3}, \quad \frac{\alpha}{1-\beta+\alpha} \le \frac{2}{47}.$$

Therefore,

$$\begin{aligned} \gamma_1 &\le \alpha, \quad \gamma_1 \ge \alpha 45/47 \ge 0.957\alpha, \\ |\gamma_2| &= \frac{(1-\beta)\alpha^2}{(1-\beta+\alpha)^2}\left(\frac{\beta}{2} + \frac{1-\beta}{1-\beta+\alpha} + \frac{1}{2}\right) \le \frac{(1-\beta)\alpha^2}{(1-\beta+\alpha)^2}\left(\frac{1}{8} + \frac{3}{2}\right) \le \frac{13}{6}\alpha^2 \le 0.073\alpha, \\ |\gamma_2| &\ge \frac{(1-\beta)\alpha^2}{(1-\beta+\alpha)^2}\left(\frac{45}{47} + \frac{1}{2}\right) \ge 1.457\frac{(1-\beta)\alpha^2}{(1-\beta+\alpha)^2} \ge 1.395\frac{\alpha^2}{1-\beta+\alpha} \ge 1.35\alpha^2, \\ |\gamma_3| &\le \gamma_1^2\left\{\frac{\gamma_1}{3} + \frac{\beta^2\alpha}{(1-\beta)(1-\beta+\alpha)} + \frac{2\beta\alpha}{(1-\beta+\alpha)^2}\right. \\ &\quad \left. + \frac{\beta(1-\beta)}{(1-\beta+\alpha)^2} + \frac{2(1-\beta)\alpha}{(1-\beta+\alpha)^3} + \frac{\beta\alpha}{(1-\beta+\alpha)} + \frac{(1-\beta)\alpha}{(1-\beta+\alpha)^2}\right\} \\ &\le \frac{\alpha^2}{1-\beta+\alpha}\left\{\frac{\alpha}{3} + \frac{\beta^2\alpha}{1-\beta+\alpha} + \frac{2\beta\alpha}{1-\beta+\alpha} + \beta + \frac{2\alpha}{1-\beta+\alpha} + \beta\alpha + \alpha\right\} \le 0.02\alpha, \\ |\varkappa_1| &\le \alpha \max\left\{\frac{\alpha}{1-\beta+\alpha}; \frac{\beta}{1-\beta+\alpha} + p_0\right\} \le \frac{4}{90}, \quad |\varkappa_2| \le \frac{1}{4}. \end{aligned}$$

Hence $\gamma = \lfloor n(\gamma_1 + 4\gamma_2 + 3\gamma_3 \rfloor \ge 6n|\gamma_2| - 6n\gamma_3 \ge 6n(|\gamma_2| - |\gamma_3|) \ge 4.8n\alpha^2 > 0$.

Recall that K_β is a geometric distribution: $K_\beta(\{k\}) = (1-\beta)\beta^{k-1}$ $(k=1,2,\dots)$. Hence

$$\widehat{K}_\beta(t) = (1-\beta)\mathrm{e}^{\mathrm{i}t}/(1-\beta\mathrm{e}^{\mathrm{i}t}).$$

For any $\gamma > 0$ we denote by $K_{\beta,\gamma}$ a compound Poisson distribution with the characteristic function

$$\widehat{K}_{\beta,\gamma}(t) = \exp\left(\mathrm{i}t\gamma + \gamma\sum_{j=1}^{\infty}\frac{\beta^j}{j}(\mathrm{e}^{\mathrm{i}tj} - 1)\right).$$

Since $n\alpha \ge 3$,

$$\lambda_1 + \varkappa_1 \ge n(\gamma_1 - 4|\gamma_2| - 3|\gamma_3|) - \beta - |\varkappa_1| \ge 0.605n\alpha - 1.045 > 0.2n\alpha > 0.$$

Therefore, $\exp((\lambda_1 + \varkappa_1)(K_\beta - I))$ is a compound Poisson distribution. Similarly,

$$\lambda_2 \ge 0, \quad \lambda_{-1} = n(2|\gamma_2| - 3\gamma_3) + \beta/3 \ge n(2|\gamma_2| - 3|\gamma_3|) \ge 1.05n\alpha^2 > 0.$$

Note that $\exp\Big(\lambda_2(K_\beta^{*2} - I) + \lambda_{-1}(I_{-1} - I)/(1-\beta)\Big)$ is a compound Poisson distribution.

Next, we apply the triangle inequality and Lemma D.6 in order to show that D_{4n} is close to D_3^{*n}. In general, D_3 is a signed compound Poisson measure, not a distribution.

First, we estimate $|\widehat{D}_3^n(t)|$ and $|\widehat{D}_{4n}(t)|$. By the property of characteristic functions (D.5),

$$|\widehat{K}_\beta(t)-1|^2 \le 2(1-Re\widehat{K}_\beta(t)).$$

Since $|\widehat{K}_\beta(t)-1|\le 2$, Lemma 9.9 yields

$$\begin{aligned} |\widehat{D}_3^n(t)| &\le \exp\Big(n\gamma_1(Re\widehat{K}_\beta(t)-1) + 2n|\gamma_2|(1-Re\,\widehat{K}_\beta(t)) + 4n|\gamma_3|(1-Re\,\widehat{K}_\beta(t))\Big) \\ &\le \exp\Big(-0.731\alpha(Re\widehat{K}_\beta(t)-1)\Big) \le \exp\big(-0.065\alpha t^2\big). \end{aligned} \tag{9.32}$$

Similarly,

$$\begin{aligned} |\widehat{D}_{4n}(t)| &\le \exp\Big(\lambda_1(Re\,\widehat{K}_\beta(t)-1)\Big) \le \exp\Big(0.605n\alpha(Re\,\widehat{K}_\beta(t)-1)\Big) \\ &\le \exp\big(-0.054n\alpha t^2\big). \end{aligned} \tag{9.33}$$

Next, we compare $\ln \widehat{D}_3^n(t)$ and $\ln \widehat{D}_{4n}(t)$. Let ϑ denote any number such that $|\vartheta|\le 1$. By (I.8),

$$\mathrm{e}^{\mathrm{i}x} = 1 + \mathrm{i}x + (\mathrm{i}x)^2/2 + (\mathrm{i}x)^3/6 + \vartheta|x|^4/24 \qquad (x\in\mathbb{R}).$$

One can prove that

$$\begin{aligned} \widehat{K}_\beta^\gamma(t) &= \exp\left(\gamma\frac{\mathrm{i}t}{1-\beta} + \frac{\gamma\beta(\mathrm{i}t)^2}{2(1-\beta)^2} + \gamma\frac{\beta(1+\beta)(\mathrm{i}t)^3}{6(1-\beta)^3} + \vartheta|\gamma|t^4/24\right), \\ \widehat{K}_\beta(t)-1 &= \frac{\mathrm{i}t}{1-\beta} + \frac{(1+\beta)(\mathrm{i}t)^2}{2(1-\beta)^2} + \frac{(1+4\beta+\beta^2)(\mathrm{i}t)^3}{6(1-\beta)^3} + \vartheta t^4/24, \\ (\widehat{K}_\beta(t)-1)^2 &= \frac{(\mathrm{i}t)^2}{(1-\beta)^2} + \frac{(1+\beta)(\mathrm{i}t)^3}{(1-\beta)^3} + \vartheta C t^4, \qquad (\widehat{K}_\beta(t)-1)^3 = \frac{(\mathrm{i}t)^3}{(1-\beta)^3} + \vartheta C t^4, \\ \widehat{K}_\beta^2(t)-1 &= \frac{2(\mathrm{i}t)}{1-\beta} + \frac{(2+\beta)(\mathrm{i}t)^2}{(1-\beta)^2} + \frac{(4+7\beta+\beta^2)(\mathrm{i}t)^3}{3(1-\beta)^2} + \vartheta C t^4, \\ \mathrm{e}^{-\mathrm{i}t}-1 &= -\mathrm{i}t - t^2/2 - (\mathrm{i}t)^3/6 + \vartheta C t^4 \end{aligned}$$

if $|t|\le\pi$. Consequently,

$$n\ln\widehat{D}_3(t) = n\gamma_1(\widehat{K}_\beta(t)-1) + n\gamma_2\frac{(\mathrm{i}t)^2}{(1-\beta)^2} + n(\gamma_2(1+\beta)+\gamma_3)\frac{(\mathrm{i}t)^3}{(1-\beta)^3} + \vartheta(n\alpha^2)t^4/24,$$

$$\ln\widehat{D}_{4n}(t) = n\gamma_1(\widehat{K}_\beta(t)-1) + n\gamma_2\frac{(\mathrm{i}t)^2}{(1-\beta)^2} + n(\gamma_2(1+\beta)+\gamma_3)\frac{(\mathrm{i}t)^3}{(1-\beta)^3} + \vartheta C(n\alpha^2+1)t^4,$$

Combining the last two expansions with (9.32), (9.33) and the trivial estimate $x\mathrm{e}^{-x}\le 1$ $(x>0)$, we prove that

$$\begin{aligned} |n\widehat{D}_3(t) - \widehat{D}_{4n}(t)| &\le \max\{|\widehat{D_3}(t)|; |\widehat{D}_{4n}(t)|\}|n\ln\widehat{D}_3(t) - \ln\widehat{D}_{4n}(t)| \\ &\le \exp\big(-0.054n\alpha t^2\big)\,(n\alpha^2+1)t^4 \\ &\le C\exp\big(-0.05n\alpha t^2\big)\,(n^{-1} + (n\alpha)^{-2}). \end{aligned} \tag{9.34}$$

Similarly,

$$
\begin{aligned}
(\gamma \ln \widehat{K}_\beta(t))' &= \widehat{K}'_\beta(t)\gamma\Big(1-\frac{\mathrm{i}t}{1-\beta}+\frac{(\mathrm{i}t)^2}{2(1-\beta)}\Big)+\vartheta C|\gamma||t|^3,\\
(\mathrm{e}^{-\mathrm{i}t}-1)' &= \widehat{K}'_\beta(t)\Big(-1+\frac{2(\mathrm{i}t)}{1-\beta}-\frac{(\mathrm{i}t)^2(1+1-\beta)}{(1-\beta)^2}\Big)+\vartheta C|t|^3,\\
(n\ln \widehat{D}_3(t))' &= \widehat{K}'_\beta(t)\Big(n\gamma_1+\frac{2n\gamma_2(\mathrm{i}t)}{1-\beta}+\frac{n(\gamma_2(1+\beta)+3\gamma_3)(\mathrm{i}t)^2}{(1-\beta)^2}\Big)+\vartheta Cn\alpha^2|t|^3,\\
(\ln \widehat{D}_{4n}(t))' &= \widehat{K}'_\beta(t)\Big(n\gamma_1+\frac{2n\gamma_2(\mathrm{i}t)}{1-\beta}+\frac{n(\gamma_2(1+\beta)+3\gamma_3)(\mathrm{i}t)^2}{(1-\beta)^2}\Big)+\vartheta C(n\alpha^2+1)|t|^3,
\end{aligned}
$$

$$
|(n\ln \widehat{D}_3(t)-\ln \widehat{D}_{4n}(t))'| \le C(n\alpha^2+1)|t|^3 \tag{9.35}
$$

if $|t|\le\pi$. Let $v=n\alpha/(1-\beta+\alpha)$. Then

$$
\begin{aligned}
(n\ln \widehat{D}_3(t)-\mathrm{i}tv)' &= n\gamma_1\widehat{K}'_\beta(t)-\mathrm{i}a+2n\gamma_2(\widehat{K}_\beta(t)-1)\widehat{K}'_\beta(t)+3n\gamma_3(\widehat{K}_\beta(t)-1)^3\widehat{K}'_\beta(t)\\
&= \frac{n\alpha\mathrm{i}((1-\beta)^2\mathrm{e}^{\mathrm{i}t}-(1-\beta\mathrm{e}^{\mathrm{i}t})^2)}{(1-\beta+\alpha)(1-\beta\mathrm{e}^{\mathrm{i}t})^2}+\vartheta Cn\alpha^2|\widehat{K}_\beta(t)-1| = \vartheta Cn\alpha|t|. \quad (9.36)
\end{aligned}
$$

Thus,

$$
\begin{aligned}
\Big(\widehat{D}_3^n(t)\mathrm{e}^{-\mathrm{i}tv}-\widehat{D}_{4n}(t)\mathrm{e}^{-\mathrm{i}tv}\Big)' &= \widehat{D}_3^n(t)\mathrm{e}^{-\mathrm{i}tv}(n\ln \widehat{D}_3(t)-\mathrm{i}tv)'-\widehat{D}_{4n}(t)\mathrm{e}^{-\mathrm{i}tv}(\ln \widehat{D}_{4n}(t)-\mathrm{i}t)'\\
&= (n\ln \widehat{D}_3(t)-\mathrm{i}tv)'\mathrm{e}^{-\mathrm{i}tv}(\widehat{D}_3^n(t)-\widehat{D}_{4n}(t))\\
&+ \widehat{D}_{4n}(t)\mathrm{e}^{-\mathrm{i}tv}(n\ln \widehat{D}_3(t)-\ln \widehat{D}_{4n}(t))'.
\end{aligned}
$$

Taking into account (9.35), (9.36), (9.34), we obtain

$$
\begin{aligned}
\Big|\Big(\widehat{D}_3^n(t)\mathrm{e}^{-\mathrm{i}tv}-\widehat{D}_{4n}(t)\mathrm{e}^{-\mathrm{i}tv}\Big)'\Big| &\le Cn\alpha|t|\exp\big(-0.05n\alpha t^2\big)\,(n^{-1}+(n\alpha)^{-2})\\
&+ C\exp\big(-0.054n\alpha t^2\big)\,(n\alpha^2+1)|t|^3\\
&\le C\sqrt{n\alpha}(n^{-1}+(n\alpha)^{-2})\exp\big(-0.04n\alpha t^2\big). \quad (9.37)
\end{aligned}
$$

Applying Lemma D.8 with $M=D_3^{*n}-D_{4n}$, $u=\sqrt{n\alpha}$, $v=n\alpha/(1-\beta+\alpha)$ and taking into account (9.34) and (9.37), we derive

$$
\|D_3^{*n}-D_{4n}\|_{TV} \le C(n^{-1}+(n\alpha)^{-2}). \tag{9.38}
$$

Recall that H is a distribution, hence $\|H\|_{TV}=1$. By the properties of the total variation norm and (9.38),

$$
\begin{aligned}
\|H*\exp(\varkappa_1(K_\beta-I))*(D_3^{*n}-D_{4n})\|_{TV} &\le \|H\|_{TV}\,\|\exp(\varkappa_1(K_\beta-I))\,\|_{TV}\|D_3^{*n}-D_{4n}\|_{TV},\\
&\le \exp(|\varkappa_1|(\|K_\beta\|_{TV}+\|I\|_{TV}))\,\|D_3^{*n}-D_{4n}\|_{TV}\\
&\le \exp(8/90)\,\|D_3^{*n}-D_{4n}\|_{TV}\\
&\le C(n^{-1}+(n\alpha)^{-2}).
\end{aligned}
$$

Applying the triangle inequality, (9.39) and Theorem 9.3, we complete the proof of Theorem 9.2. □

Proof of Theorem 9.5. Taking into account Theorem 3.6, (9.16) and (9.18), we get

$$
\begin{aligned}
d_{TV}(\mathcal{L}(S_n);A_0) &\ge n|\gamma_2|\|H*D_0*U_\beta^{*2}\|_{TV}-C\alpha\big(\alpha+p/\sqrt{n\alpha}\big)\\
&\ge Cn|\gamma_2|\|D_1^n(I_1-I)^{*2}\|_{TV}-C\alpha\big(\alpha+\beta/\sqrt{n\alpha}\big). \quad (9.39)
\end{aligned}
$$

Set $z=t/(h\sqrt{n\alpha})$, $\mu=n\gamma_1/(1-\beta)$; constant $h>1$ will be chosen later.

It is easy to check that for all $|t|\leq\pi$

$$\begin{aligned}|\exp\Big(n\varkappa_1(\widehat{U_\beta}(t)-\mathrm{i}t/(1-\beta))\Big)-1| &\leq Cn\alpha t^2,\\ |\widehat{D}_2^n(t)|\leq 1,\quad |\widehat{D}_2^n(t)\exp(-\mathrm{i}tn\gamma_1/(1-\beta))-1| &\leq Cn\alpha t^2.\end{aligned}$$

Hence

$$\tilde{J}=\Big|\int_{\mathbb{R}}\mathrm{e}^{-t^2/2}\widehat{D}_1^n(z)\mathrm{e}^{-\mathrm{i}z\mu}(\mathrm{e}^{\mathrm{i}z}-1)^2\mathrm{d}t\Big|\geq\Big|\int_{\mathbb{R}}\mathrm{e}^{-t^2/2}z^2\mathrm{d}t\Big|-\tilde{J}_1-\tilde{J}_2, \tag{9.40}$$

where

$$\begin{aligned}\tilde{J}_1&=\int_{\mathbb{R}}\mathrm{e}^{-t^2/2}z^2|\widehat{D}_1^n(z)\mathrm{e}^{-\mathrm{i}z\mu}-1|\mathrm{d}t\leq Cn\alpha\int_{\mathbb{R}}z^4\mathrm{e}^{-t^2/2}\mathrm{d}t=\frac{C}{h^4n\alpha},\\ \tilde{J}_2&=\int_{\mathbb{R}}\mathrm{e}^{-t^2/2}|\widehat{D}_1^n(z)\mathrm{e}^{-\mathrm{i}z\mu}||(\mathrm{e}^{\mathrm{i}z}-1)^2-(\mathrm{i}z)^2|\mathrm{d}t\leq\frac{C}{h^3n\alpha\sqrt{n\alpha}}\end{aligned}$$

for a particular absolute constant C. Combining the last two estimates with (9.40) and choosing h to be sufficiently large absolute constant, we obtain

$$\tilde{J}\geq\frac{C}{h^2n\alpha}\left(1-\frac{\tilde{C}}{h^2}-\frac{\hat{C}}{h\sqrt{n\alpha}}\right)\geq\frac{c}{n\alpha},$$

where $C,\tilde{C},\hat{C},c$ are absolute constants. Applying Lemma D.15 and substituting the result into (9.39), we get (9.7).

Estimate (9.9) can be proved similarly using (9.16). For the proof of remaining estimates one can apply Theorem 3.11 with $t\exp\big(-t^2/2\big)$ instead of $\exp\big(-t^2/2\big)$. The proof is almost identical to that of Theorem 9.5 and is therefore omitted. □

Proof of (9.13). Denote

$$A_{11}=\frac{4|\gamma_2|}{\gamma_1(1-\beta)\sqrt{2\pi\mathrm{e}}}.$$

Then

$$\begin{aligned}\Big|d_{TV}(\mathcal{L}(S_n);A_0)-A_{11}\Big| &\leq d_{TV}(\mathcal{L}(S_n);A_{01})+\Big\|(H*\mathrm{e}^{-p_0\gamma_1U_\beta}-I)*D_1^{*n}*U_\beta^{*2}n\gamma_2\Big\|_{TV}\\ &\quad+n|\gamma_2|\Big\|\Big(U_\beta^{*2}-\frac{1}{(1-\beta)^2}(I_1-I)^{*2}\Big)*D_1^{*n}\Big\|_{TV}\\ &\quad+\frac{n|\gamma_2|}{(1-\beta)^2}\Big\|(I_1-I)^{*2}*\Big(D_1^{*n}-\exp\Big(\frac{n\gamma_1}{1-\beta}(I_1-I)\Big)\Big)\Big\|_{TV}\\ &\quad+\left|\frac{n|\gamma_2|}{(1-\beta)^2}\Big\|(I_1-I)^{*2}*\exp\Big(\frac{n\gamma_1}{1-\beta}(I_1-I)\Big)\Big\|_{TV}-A_{11}\right|.\end{aligned}$$

We apply Theorem 3.6, (9.30), Lemma C.3, and the following easily verifiable relations:

$$U_\beta=\frac{(I_1-I)}{1-\beta}*\sum_{j=0}^{\infty}\Big(\frac{\beta}{1-\beta}\Big)^j(I_1-I)^{*j}=\frac{I_1-I}{1-\beta}+\frac{3\beta}{(1-\beta)^2}(I_1-I)^{*2}*\Theta$$

and

$$\begin{aligned}D_1-\exp\Big(\frac{\gamma_1}{1-\beta}(I_1-I)\Big) &= \exp\Big(\frac{\gamma_1}{1-\beta}(I_1-I)\Big)*\Big(\exp\Big(\frac{3\beta\gamma_1}{(1-\beta)^2}(I_1-I)^{*2}*\Theta\Big)-I\Big)\\ &= C\beta\alpha(I_1-I)^{*2}*\Theta.\end{aligned}$$

Note that

$$\Big\|(I_1-I)^{*2}*\Big(D_1^{*n}-\exp\Big(\frac{n\gamma_1}{1-\beta}(I_1-I)\Big)\Big)\Big\|_{TV}$$
$$=\Bigg\|(I_1-I)^{*2}*\Big(D_1-\exp\Big(\frac{\gamma_1}{1-\beta}(I_1-I)\Big)\Big)*\sum_{j=1}^{n}D_1^{*(n-j)}*\exp\Big((j-1)\frac{\gamma_1}{1-\beta}(I_1-I)\Big)\Bigg\|_{TV}$$
$$\le Cn\beta\alpha\Big(\Big\|(I_1-I)^{*4}*D_1^{*n/3}\Big\|_{TV}+\Big\|(I_1-I)^{*4}*\exp\Big(\frac{n\gamma_1}{3(1-\beta)}(I_1-I)\Big)\Big\|_{TV}\Big),$$

yielding (9.13). □

Proof of Theorem 9.6 (sketch). Expanding $\widehat{\Lambda}_1(t)$ from Lemma 9.12 in powers of $\mathrm{e}^{\mathrm{i}t}-1$, we obtain

$$\widehat{\Lambda}_1(t)=1+w(\mathrm{e}^{\mathrm{i}t}-1)+\frac{u}{2}(\mathrm{e}^{\mathrm{i}t}-1)^2+C\vartheta\alpha(\alpha+\beta)^2|\mathrm{e}^{\mathrm{i}t}-1|^3,$$
$$\widehat{\Lambda}_1'(t)=\mathrm{i}\mathrm{e}^{\mathrm{i}t}(w+u(\mathrm{e}^{\mathrm{i}t}-1))+C\vartheta\alpha(\alpha+\beta)^2|\mathrm{e}^{\mathrm{i}t}-1|^2.$$

After quite standard calculation we get

$$|\widehat{\Lambda}_1(t)-\widehat{A}_s(t)|\ \le\ C\alpha(\alpha+\beta)^2|t|^3,\quad |\widehat{\Lambda}_1'(t)-\widehat{A}_s'(t)|\le C\alpha(\alpha+\beta)^2t^2,$$
$$\big|\big(\mathrm{e}^{-\mathrm{i}tw}\widehat{\Lambda}_1(t)\big)'\big|\ \le\ C\alpha|t|,\quad \big|\big(\mathrm{e}^{-\mathrm{i}tw}\widehat{A}_s(t)\big)'\big|\le C\alpha|t|,$$

where C is an absolute constant. Moreover,

$$|\widehat{\Lambda}_1(t)|,|\widehat{A}_s(t)|\le\exp(-C\kappa_1\sin^2(t/2))\qquad(|t|\le\pi).$$

Applying Lemma D.8 with $u=\max\{1;\sqrt{n}\}$ and $v=nw$, we obtain

$$\|\Lambda_1^{*n}-A_s^{*n}\|_{TV}\le C(\alpha+\beta)^2\min\{(n\alpha)^{-1/2};n\alpha\}.$$

Combining the last estimate with (9.24) and (9.26), we complete the proof of the theorem. □

Exercises

9.1. Prove (9.6).

9.2. Let $\beta\le1/2$, $\alpha\le1/30$, $n\alpha\ge1$, $p_0=0$. Denote

$$A_{21}=A_2*(I+n\gamma_3(K_\beta-I)^{*3}).$$

Prove that

$$d_{TV}(F_n;A_{21})\le C(\alpha+\beta)/n, \tag{9.41}$$

where C is an absolute constant.

9.3. Let $\alpha,\beta\le1/3$, $p_0=1$, $m<n/2$. Let

$$\eta_{m,n}=\pi_{m\beta(1+\beta)}+2\pi_{-m\beta^2/2}+\pi_{(n-m-1)\alpha(1+\alpha)}+\pi_{-(n-m-1)\alpha^2/2}+\pi_{\alpha/(1-\beta+\alpha)}.$$

Prove that there exists an absolute constant C such that

$$\max_m |\mathbb{P}(S_n = m) - (1-\beta+\alpha)\mathbb{P}(\eta_{m,n} = m)| \leq \frac{C(\alpha^2 \vee \beta^2)}{n(\alpha \wedge \beta)}.$$

Hint. Use the following inversion formula (see [76]):

$$\mathbb{P}(S_n = m) = \frac{1}{2\pi} \int_{-\pi}^{\pi} (1-\beta+\alpha \mathrm{e}^{\mathrm{i}t})(1-\beta+\beta \mathrm{e}^{\mathrm{i}t})^m (1-\alpha+\alpha \mathrm{e}^{\mathrm{i}t})^{n-m-1} \mathrm{d}t.$$

Notes

Theorems 9.1 and Theorems 9.3–9.6 were proved in [86].
Theorem 9.2 is from [284].

10

Compound Poisson approximations related to Kolmogorov's problem

Let $X, X_1, \ldots, X_n$ be independent random variables, and let $S_n = X_1 + \ldots + X_n$.

Results of this chapter are closely related to the first uniform Kolmogorov's problem. For the sake of completeness we discuss the second Kolmogorov's problem as well.

10.1 Kolmogorov's first problem

Let $X_1, \ldots, X_n$ be independent random variables. Distributions of r.v.s $X_1, ..., X_n$ may depend on n, i.e., $\{X_1, \ldots, X_n\}_{n \geq 1}$ is a triangular array of r.v.s.

If $X_1, \ldots, X_n$ are infinitesimal and the limiting distribution of $S_n = X_1 + ... + X_n$ as $n \to \infty$ exists, then it is infinitely divisible (Khintchine [176]).

In early 1950s Kolmogorov has raised the problem of evaluating the accuracy of infinitely divisible approximation to $\mathcal{L}(S_n)$.

Kolmogorov's first problem is concerned with i.i.d.r.v.s, while Kolmogorov's second problem deals with independent but not necessarily identically distributed random variables.

The problem is call "uniform" since the estimate of the accuracy of approximation established by Kolmogorov [179] is *uniform* over the class $\mathcal{F}$ of all probability distributions.

Let $X, X_1, \ldots, X_n$ be i.i.d.r.v.s. Prokhorov [239, 241, 245] has proved that for any distribution $\mathcal{L}(X)$ there exists a sequence of infinitely divisible distributions that are "close" to $\mathcal{L}(S_n)$:

$$d_K(\mathcal{L}(S_n); \mathcal{D}) \equiv \inf_{P \in \mathcal{D}} d_K(\mathcal{L}(S_n); P) \to 0 \qquad (n \to \infty).$$

Kolmogorov [179] has derived an estimate of the accuracy of approximation that is uniform over $\mathcal{F}$ (the so-called first uniform Kolmogorov's theorem): there exists an absolute constant C such that

$$\sup_{\mathcal{L}(X) \in \mathcal{F}} d_K(\mathcal{L}(S_n); \mathcal{D}) \leq C n^{-1/5}. \tag{10.1}$$

Kolmogorov then improved (10.1) to

$$\sup_{\mathcal{L}(X) \in \mathcal{F}} d_K(\mathcal{L}(S_n); \mathcal{D}) \leq C n^{-1/3}. \tag{10.2}$$

Kolmogorov used the sum of normal and compound Poisson independent r.v.s in order to approximate $\mathcal{L}(S_n)$. Le Cam [191] proved that the rate $n^{-1/3}$ holds if one employs only compound Poisson r.v.s, see (10.5).

Observe the extreme generality of estimates (10.1), (10.2) – there are no moment or structural assumptions on $\mathcal{L}(X)$.

DOI: 10.1201/9781003478164-10

It took over 25 years of research before the correct rate of the accuracy of approximation in (10.1) has been established by Arak [6, 7]. A comprehensive history of the problem can be found in the monograph by Arak & Zaitsev [8].

Arak [6, 7] has shown that there exist absolute constants $0<c<C<\infty$ such that

$$cn^{-2/3} \leq \sup_{P\in\mathcal{F}} d_K(\mathcal{L}(S_n);\mathcal{D}) \leq Cn^{-2/3}, \tag{10.3}$$

where P stands for $\mathcal{L}(X)$. The lower bound in (10.3) sets a limit to the rate of the accuracy of compound Poisson approximation as well.

Denote by $\mathbb{CP}$ the class of (shifted) compound Poisson distributions. Arak [6] has proved that

$$\sup_{P\in\mathcal{F}} d_K(\mathcal{L}(S_n);\mathbb{CP}) \leq Cn^{-2/3}.$$

We now discuss the idea of the proof of (10.3) in more detail.

Arak employs the method of characteristic functions. The classical approach used to obtain (10.5) relies on relation (D.21) and is limited by the Kolmogorov-Rogozin inequality (E.2), which yields the accuracy $Cn^{-1/2}$ at best.

Arak has solved two problems: first, he has constructed an infinitely divisible characteristic function, which is "close" to the characteristic function of $\mathcal{L}(S_n)$ for all $t\in\mathbb{R}$; second, he has developed a method of evaluating $d_K(F;G)$ via characteristic functions of F,G.

Arak begins by proving that for any one-dimensional distribution F there exist vector $\mathbf{u}=(u_1,\dots,u_9)$, natural number $m\leq n^{5/3}$ and discrete distribution F_A such that

$$d_K(F^{*n};F_A^{*n})\leq Cn^{-2/3}, \quad \text{supp}F_A\subset K_m(\mathbf{u}),$$

where $K_m(\mathbf{u})$ is defined by (D.25). Therefore, it suffices constructing a compound Poisson approximation to F_A^{*n}.

F_A can be viewed as the distribution of a one-dimensional "projection" $u_1\xi_1+u_2\xi_2+\dots+u_9\xi_9$ of a 9-dimensional lattice random vector $\vec{\xi}=(\xi_1,\xi_2,\dots,\xi_9)$.

Let $\bar{t}=(t_1,t_2,\dots,t_9)$, and let $f(\bar{t})=\mathbb{E}\exp(\mathrm{i}(t_1\xi_1+t_2\xi_2+\dots+\xi_9t_9))$ denote the characteristic function of $\vec{\xi}$. Arak has constructed a characteristic function $g_n(\bar{t})$ of a 9-dimensional compound Poisson random vector, which is close to $f^n(\bar{t})$ for all $\bar{t}$.

Two facts are essential for the construction of $g_n(\bar{t})$: the periodicity of $f(\bar{t})=f(t_1+2\pi,t_2+2\pi,\dots,t_9+2\pi)$ and the fact that $\mathbf{\Pi}(1,2X-X'-X'')$ has the second and the third cumulants differ from those of those of $\mathcal{L}(X)$ by a factor of 6 if $\mathbb{E}X^4<\infty$:

$$\mathbb{E}\exp(it(2X-X'-X'')) = 1+6\sum_{k=2}^{3}\varkappa_k(it)^k/k! + 11\theta\mathbb{E}X^4t^4,$$

where X,X',X'' are i.i.d.r.v.s, $\varkappa_k$ denotes the kth cumulant of $\mathcal{L}(X)$, $|\theta|\leq1$, see Lemma D.2.

Indeed, recall that $\varkappa_1=\mathbb{E}X$, $\varkappa_2=\text{var}\,X$, $\varkappa_3=\mathbb{E}(X-\mathbb{E}X)^3$,

$$\ln\mathbb{E}e(itX) = \sum_{k=1}^{3}\varkappa_k(it)^k/k! + o(t^3) \qquad (t\to0).$$

Let Y denote a r.v. with the distribution $\mathcal{L}(Y)=\mathbf{\Pi}(1,2X-X'-X'')$. Then

$$\text{var}\,Y=\mathbb{E}(2X-X'-X'')^2=6\text{var}\,X, \quad \mathbb{E}(Y-\mathbb{E}Y)^3=6\mathbb{E}(X-\mathbb{E}X)^3.$$

A compound Poisson distribution is then used to approximate F_A^{*n}. Approximation is sharper when it involves more matching moments. Approximation of $f^n(\bar{t})$ by

$\exp(n\,(f(2\bar{t})\,f^2(-\bar{t})-1)/6)$ ensures closeness of the second and the third moments. Therefore, $\exp(n\,(f(2\bar{t})\,f^2(-\bar{t})-1)/6)$ is the main component of $g_n(\bar{t})$. The means of two distributions are equal thanks to suitable shifting. Convolution with specially constructed CP distributions ensures that lattice structure is retained (after shifting). The closeness of distributions is estimated with the help of Lemma D.16.

Arak's [6] proof is very complicated. To some extent its limitations were demonstrated in [81], where a first-order asymptotic expansion G_F was constructed with the accuracy

$$d_K(F^{*n}; G_F) \le C_\varepsilon n^{-1+\varepsilon},$$

where $0<\varepsilon\le 1/3$, C_ε is a constant depending only on ε. However, $C_\varepsilon\to\infty$ as $\varepsilon\to 0$. Therefore, a modification of Arak's original proof is needed if one seeks improving this asymptotic expansion.

So far there is no multidimensional analogue of Arak's inequality (10.3).

Zaitsev [330] has shown that d_K in (10.3) cannot, in general, be replaced by the total variation distance. Relations (7.2), (10.8) can be viewed as solutions of Kolmogorov's first problem for special classes of distributions.

Open problems

10.1 Derive a multidimensional analogue of Arak's inequality (10.3).

10.2. Is it true that

$$\sup_{\mathcal{L}(X)\in\mathcal{F}_s} d_K(\mathcal{L}(S_n); \mathcal{D}) \le C/n,$$

where C is an absolute constant?

10.2 Kolmogorov's second problem

Kolmogorov's second problem is concerned with independent but not identically distributed random variables $X_1, X_2, \ldots X_n$. The task is to evaluate the accuracy of infinitely divisible approximation to $\mathcal{L}(S_n)$.

The problem has no solution without additional assumptions on $\{X_i\}$. Indeed, let $\{X_1, X_2, \ldots\}$ be the sequence $\{X, 0, 0, \ldots\}$, where the first r.v. X is not infinitely divisible while all other r.v.s are zeros. Then $S_n \equiv X_1+\ldots+X_n = X$ cannot be approximated by any infinitely divisible r.v.

However, if random variables $\{X_i\}$ take with large probabilities values in a finite interval $[0;T]$, where $T>0$, then $\mathcal{L}(S_n)$ can be approximated by an infinitely divisible distribution in terms of the Lévy distance.

Let $X_1, \ldots, X_n$ be independent random variables ($\mathcal{L}(X_i)$ may depend on n). Given constants $a_1, \ldots, a_n$, we denote by $\tilde{X}_{1,a_1}, \ldots, \tilde{X}_{n,a_n}$ accompanying $X_1-a_1, \ldots, X_n-a_n$ independent random variables. Set

$$\tilde{S}_{n,a} = \tilde{X}_{1,a_1} + \ldots + \tilde{X}_{n,a_n}.$$

If $a_i=0$ $(\forall i)$, then $\tilde{S}_{n,0}=\tilde{S}_n$ is defined by (1.9).

Recall that $\mathcal{L}(X_i)$ can be presented as

$$\mathcal{L}(X_i) = (1-p_i)U_i + p_i V_i \qquad (0\le p_i\le 1),$$

cf. (6.1), where distributions $\{U_i\}$ may be chosen concentrated on finite intervals (say, $[b_i; c_i]$) of lengths $c_i - b_i \le T$ (one can choose $U_i = \mathcal{L}(X_i | X_i \in [b_i; c_i])$), $p_i = \mathbb{P}(X_i \notin [b_i; c_i]$.

Denote by d_L the Lévy distance, and let

$$a_i = \int_{\mathbb{R}} x U_i(\mathrm{d}x), \quad p_n^* = \max\{p_1, \dots, p_n\}.$$

According to Zaitsev & Arak [323] (see also [8], Chapter 6), there exists an absolute constant C such that

$$d_L(\mathcal{L}(S_n); \mathcal{D}) \le C(p_n^* + T(|\ln T| + 1)). \tag{10.4}$$

Letting $T \to 0$, we get

$$d_L(\mathcal{L}(S_n); \mathcal{D}) \le C p_n^*, \tag{10.4*}$$

cf. (6.13). Zaitsev & Arak [323] have proved that estimate (10.4) is of correct order.

The following result is cited from [8], Theorem 4. Let $\varepsilon \in (0; 1)$.

Theorem 10.1 *If there exist constants β_i $(i = 1, \dots, n)$ such that $d_L(\mathcal{L}(X_i); I_{\beta_i}) \le \varepsilon$, then there exist constants $a_1, a_2, \dots, a_n$ and an absolute constant C such that*

$$d_L(S_n; \tilde{S}_{n,a}) \le C\varepsilon(|\ln \varepsilon| + 1).$$

For any $\delta \in (0, 1]$ there exist i.i.d. random variables $X, X_1, ..., X_n$, absolute constant c and $n \in \mathbb{N}$ such that $d_L(\mathcal{L}(X); I) \le \delta$ and

$$d_L(S_n; \mathcal{D}) \ge c\delta(|\ln \delta| + 1).$$

The second inequality of Theorem 10.1 shows that the first one cannot be improved.

Zaitsev [327] dealt also with a multivariate version of the second uniform Kolmogorov's problem.

Let $F_1, F_2, \dots, F_n$ be k-dimensional distributions such that $d_L(F_i; I) \le \varepsilon$ $(i = 1, 2, \dots, n)$. Zaitsev [327] has shown that there is an accompanying compound Poisson distribution G such that

$$d_L(F_1 * \cdots * F_n; G) \le C_k \varepsilon(|\ln \varepsilon| + 1),$$

where positive constant C_k depends only on k.

A comprehensive overview of the results concerning the one-dimensional Kolmogorov's problem can be found in [8], including the results in terms of the Lévy-Prokhorov distance. The multivariate case is discussed in [140, 141, 142, 332].

10.3 Shifted compound Poisson approximation

In this section, we show that a suitably shifted sum of i.i.d.r.v.s can be well approximated by a sum of accompanying compound Poisson random variables.

Let $X, X_1, \dots, X_n$ be independent and identically distributed r.v.s. Given a constant a, we denote by $\tilde{X}_{1,a}, ..., \tilde{X}_{n,a}$ independent accompanying $X_1{+}a, \dots, X_n{+}a$ random variables. Let

$$\tilde{S}_{n,a} = \tilde{X}_{1,a} + ... + \tilde{X}_{n,a}\,.$$

Clearly, $\mathcal{L}(\tilde{S}_{n,a})$ is compound Poisson.

The accuracy of shifted compound Poisson approximation by accompanying laws is $n^{-1/3}$: there exist absolute constants $0<c\le C<\infty$ such that

$$cn^{-1/3} \le \sup_{\mathcal{L}(X)} \inf_{a\in\mathbb{R}} d_K(S_n; \tilde{S}_{n,a}-na) \le Cn^{-1/3}. \tag{10.5}$$

The upper bound in (10.5) has been given by Le Cam [191]; according to Ibragimov & Presman [163], $C \le 8$. The lower bound in (10.5) has been indicated by Ibragimov & Presman [163] and proved by Presman in [8], Section 8.4 (the lower bound is achieved when $\mathcal{L}(X)=\mathbf{B}(p)$).

Note that the rate of the accuracy of shifted Poisson approximation to the Binomial distribution is $n^{-1/2}$, see (3.41); the rate of the accuracy of shifted compound Poisson approximation to the Binomial distribution is $n^{-2/3}$, see (7.1).

Denote

$$F_a = \mathcal{L}(X+a), \quad \tilde{F}_a = \mathbf{\Pi}(1, F_a), \quad \Delta_{ja} = \binom{n}{j} \tilde{F}_a^{*(n-j)} * (F_a - \tilde{F}_a)^{*j}.$$

In particular, $\mathcal{L}(S_n+na) = F_a^{*n}$, $\Delta_{j0} = \mathbf{\Pi}(n, F_a)$.

Theorem 10.2 *If $k\ge 0$, $n\ge 1$, then for any $F\in\mathcal{F}$*

$$\inf_a d_K\Big(F_a^{*n}; \sum_{j=0}^{k} \Delta_{ja}\Big) \le C(k) n^{-(k+1)/(2k+3)}. \tag{10.6}$$

The upper bound in (10.5) follows from (10.6) if $k=0$.

Denote

$$A_{1a} = -n(F_a - I)^{*2}/2, \quad A_{2a} = n^2 (F_a - I)^{*4}/8.$$

Theorem 10.3 *If $n\ge 3$, then for any $F\in\mathcal{F}$*

$$\inf_a d_K(F_a^{*n}; \mathbf{\Pi}(n, F_a) * (I + A_{1a})) \le Cn^{-2/5},$$
$$\inf_a d_K(F_a^{*n}; \mathbf{\Pi}(n, F_a) * (I + A_{1a} + A_{2a})) \le Cn^{-3/7}.$$

The question about optimality of estimates of Theorem 10.3 remains open.

Theorem 10.4 *There exists an absolute constant C such that for any integer $n\ge 1$ and any distribution F*

$$d_K\Big(F_a^{*n}; \exp\big(n(F_a - I) - n(V - I)^{*2}/2\big)\Big) \le Cn^{-1/2} \tag{10.7}$$

for some $a\in\mathbb{R}$ and distribution $V=V(F,a)$.

Bound (10.7) states that for any X there exist constant a and distribution $V=V(F,a)$ such that $\mathcal{L}(S_n + na)$ can be approximated by $\mathcal{L}(\tilde{S}_n) * \mathbf{\Pi}(n, V) * \mathbf{\Pi}(-n/2, V^{*2})$ with the accuracy $Cn^{-1/2}$.

Open problems

10.3. Improve the accuracy of approximation in (10.6)–(10.7).

10.4. Evaluate constant C in (10.7).

10.4 Symmetric random variables

Recall that the distribution of r.v. X is symmetric if $\mathbb{P}(X \in B) = \mathbb{P}(X \in -B)$ for any Borel set B. The set of symmetric distributions is denoted by $\mathcal{F}_s$.

If $X, X_1, \ldots, X_n$ are i.i.d. *symmetric* random variables, then estimates of the accuracy of approximation typically can be improved.

Zaitsev [322] (upper bound) and Studnev [289] (lower bound) have shown that there exist absolute constants $0<c<C<\infty$ such that

$$cn^{-1/2} \le \sup_{\mathcal{L}(X)\in\mathcal{F}_s} d_{\scriptscriptstyle K}(S_n;\tilde{S}_n) \le Cn^{-1/2}, \tag{10.8}$$

where $\tilde{S}_n = \tilde{X}_1 + ... + \tilde{X}_n$ is the sum of independent accompanying random variables. Note that no moment assumption is imposed.

If $\mathbb{P}(X=-1) = \mathbb{P}(X=1)=1/2$, Studnev [289] has shown that

$$\mathbb{P}(S_n \le 0) - \mathbb{P}(\tilde{S}_n \le 0) = 1/2\sqrt{2\pi n}(1+o(1)) \qquad (n\to\infty).$$

The next theorem generalises (10.8) to the case of a distribution with a symmetric component.

Recall that any distribution $P = \mathcal{L}(X)$ admits representation

$$X \stackrel{d}{=} (1-\tau)\xi + \tau\eta, \tag{6.1*}$$

where $\mathcal{L}(\tau)=\mathbf{B}(p)$, $p\in[0;1]$, random variables τ, ξ, η are independent. In other words,

$$\mathcal{L}(X) = qU + pV, \tag{6.1*}$$

where $U=\mathcal{L}(\xi)$, $V=\mathcal{L}(\eta)$, $q=1-p$. Hence $\mathcal{L}(S_n) = (qU+pV)^{*n}$,

$$\mathcal{L}(\tilde{S}_n) = \exp(nq(U-I)+np(V-I)) = \mathbf{\Pi}(nq,U) * \mathbf{\Pi}(np,V).$$

Denote

$$L_1 = \exp\bigl(-p^2(V-I)^{*2}/2\bigr) = \mathbf{\Pi}(p^2,V) * \mathbf{\Pi}(-p^2/2, V^{*2}).$$

Theorem 10.5 *If $p \le \tilde{C} < 1$, then*

$$\sup_{U\in\mathcal{F}_s}\sup_{V\in\mathcal{F}} d_{\scriptscriptstyle K}(\mathcal{L}(S_n);\mathcal{L}(\tilde{S}_n)) \le C\bigl(n^{-1/2}+p\bigr), \tag{10.9}$$

$$\sup_{U\in\mathcal{F}_s}\sup_{V\in\mathcal{F}} d_{\scriptscriptstyle K}(\mathcal{L}(S_n);\mathcal{L}(\tilde{S}_n) * L_1^{*n}) \le Cn^{-1/2}. \tag{10.10}$$

Relation (10.9) with $p=0$ yields the upper bound in (10.8). Relation (10.10) is a first-order asymptotic expansion (SCP approximation).

Next we present an analogue of Theorem 7.7.

Recall that $X_i \stackrel{d}{=} \tau_i X_i'$, where r.v.s $X_1', X_2', ..., \tau_1, \tau_2, ...$ are independent, $\mathcal{L}(\tau_i)=\mathbf{B}(p_i)$, $\mathcal{L}(X_i')=\mathcal{L}(X_i|X_i\neq 0)$, $p_i = \mathbb{P}(X_i\neq 0)$. Then

$$S_n \stackrel{d}{=} \sum_{i=1}^n \tau_i X_i'.$$

Assume that $X', X_1', X_2', \ldots$ are i.i.d. *symmetric* r.v.s. Then $\mathcal{L}(\tau_i X_i')$ has a non-negative characteristic function if $\max_{i\le n} p_i \le 1/2$.

Denote $F = \mathcal{L}(X')$ $(i\ge 1)$,

$$\lambda = \sum_{i=1}^{n} p_i\,, \quad \lambda_k = \sum_{i=1}^{n} p_i^k \qquad (k\in\mathbb{N}),$$

and let $(k\in\mathbb{N})$

$$H_{n,k} = \exp\left(\sum_{i=1}^{n}\sum_{j=1}^{k}(-1)^{j+1}p_i^j\,(F-I)^{*j}/j\right) = \exp\left(\sum_{j=1}^{k}(-1)^{j+1}\lambda_j(F-I)^{*j}/j\right).$$

Theorem 10.6 *Suppose that* $\max_{i\le n} p_i \le 1/4$, $\lambda\ge 1$. *If* $\mathcal{L}(X')\in\mathcal{F}_s$, *then*

$$d_K(\mathcal{L}(S_n); H_{n,k}) \le C(k)\lambda_{k+1}\lambda^{-k-1} \qquad (k\in\mathbb{N}). \tag{10.11}$$

If $k=1$ and $p_1 = \ldots = p_n = p$, then (10.11) yields

$$d_K(\mathcal{L}(S_n); \mathbf{\Pi}(np, X')) < C/n. \tag{10.11*}$$

Estimate (7.33) holds under the assumption that $\{X_i'\}$ are integer-valued r.v.s with finite second moments. No such assumption is needed in Theorem 10.6. On the other hand, (10.11) holds in terms of a weaker metric.

10.5 Random variables with non-negative ch.f.s

Let $\mathcal{F}_+ \subset \mathcal{F}_s$ denote the set of distributions with non-negative characteristic functions. The requirement for characteristic function to be non-negative sounds very technical. However, note that $\mathcal{L}(X_1 - X_2)\in\mathcal{F}_+$ if X_1, X_2 are i.i.d.r.v.s. As another example, consider a symmetric distribution $\mathcal{L}(X)$ on $\mathbb{Z}$ with $\mathbb{P}(X=0)\ge 1/2$. Then $\mathcal{L}(X)\in\mathcal{F}_+$.

This section deals with the class $\mathcal{F}_+$ of distributions with non-negative ch.f.s.

Let $X, X_1, \ldots, X_n$ be i.i.d.r.v.s with a non-negative characteristic function. Recall that $\tilde{S}_n$ denotes the sum of accompanying r.v.s, see (1.9).

Arak [4, 5] has shown that there exist absolute constants $0<c<C<\infty$ such that

$$cn^{-1} \le \sup_{\mathcal{L}(X)\in\mathcal{F}_+} d_K(S_n; \tilde{S}_n) \le Cn^{-1}. \tag{10.12}$$

Some light on the difference between (10.8) and (10.12) can be shed by the following argument. Since $|\widehat{F}(t)| = \widehat{F}(t) \le \exp(\widehat{F}(t)-1)$, we have

$$\begin{aligned}\left|\widehat{F}^n(t) - \exp(n(\widehat{F}(t)-1))\right| &\le n\left|\widehat{F}(t) - \exp(\widehat{F}(t)-1)\right| \max\{|\widehat{F}(t)|^{n-1}; \exp((n-1)(\widehat{F}(t)-1))\} \\ &\le n(\widehat{F}(t)-1)^2 \exp\Big((n-1)(\widehat{F}(t)-1)\Big) \le 1/n \qquad (n\ge 4).\end{aligned}$$

No such estimate is possible for a general symmetric distribution, where $-1\le\widehat{F}(t)\le 1$.

On the other hand, if $F\in\mathcal{F}_s$, then $F^{*2}\in\mathcal{F}_+$. Therefore, it follows from (10.12) that

$$\sup_{F\in\mathcal{F}_s} d_{K}(F^{*2n};\exp(n(F^{*2}-I)))\le C/n,$$
$$\sup_{F\in\mathcal{F}_s} d_{K}(F^{*2n+1};F*\exp(n(F^{*2}-I)))\le C/n.$$

Zaitsev [327] (see also [8], Theorem 5.1) has shown that the upper bound in (10.12) holds for a more general class of distributions: if $\mathcal{L}(X)\in\mathcal{F}_\alpha$ $(\exists\alpha\in(0;2))$, then there exist an absolute constant $C<\infty$ such that

$$d_{K}(S_n;\tilde{S}_n)\le C\big(n^{-1}+\exp(-n\alpha+C\ln^3 n)\big). \tag{10.13}$$

A non-uniform analogue of (10.13) has been given by Zaitsev [329].

Example 10.1 Let $X,X_1,X_2,\ldots,X_n$ be independent and identically distributed symmetric r.v.s such that $q:=\mathbb{P}(X=0)>0$. Then there exist an absolute constant $C<\infty$ such that

$$d_{K}(S_n;\tilde{S}_n)\le Cn^{-1}+C\exp\big(-2nq+C\ln^3 n\big). \tag{10.13*}$$

Indeed, by Khintchine's formula (1.12) $\mathcal{L}(X)=qI+(1-q)F$, where $F\in\mathcal{F}_s$. Consequently,

$$\mathbb{E}e^{itX}=q+(1-q)\widehat{F}(t)\ge q-(1-q)=-1+2q>-1.$$

Therefore, (10.13) can be applied.

The accuracy of compound Poisson approximation can be $O(n^{-1})$ without the assumption that $\mathbb{P}(X=0)>0$. For instance, if $\mathbb{P}(X=-1)=\mathbb{P}(X=1)=1/2$, then $d_{K}(S_n;P_n)\le C/n$ for a particular compound Poisson distribution P_n, cf. (7.2). □

Assume that $X,X_1,\ldots,X_n$ are i.i.d.r.v.s,

$$\mathcal{L}(X)=(1-p)U+pV,$$

where $V\in\mathcal{F}$, $U\in\mathcal{F}_+$. Then

$$\mathcal{L}(S_n)=((1-p)U+pV)^{*n},\quad \mathcal{L}(\tilde{S}_n)=\exp(n(1-p)(U-I)+np(V-I)).$$

Set

$$\begin{aligned} L_0 &= I-n(\mathcal{L}(X)-I)^{*2}/2, \qquad L_2=-n(1-p)p(U-I)*(V-I)+np^3(V-I)^{*3}/3,\\ L_1 &= \exp\big(-p^2(V-I)^{*2}/2\big), \quad L_3=I-n(U-I)^{*2}/2+n(U-I)^{*3}/3+n^2(U-I)^{*4}/8. \end{aligned}$$

The next theorem is an analogue of Theorem 10.5 with $U\in\mathcal{F}_+$ while V is arbitrary.

Theorem 10.7 *Suppose that $p\le\tilde{C}<1$. Then there exist an absolute constant $C<\infty$ such that*

$$\sup_{U\in\mathcal{F}_+}\sup_{V\in\mathcal{F}} d_{K}(\mathcal{L}(S_n);\mathcal{L}(\tilde{S}_n))\le C\big(n^{-1}+p\big), \tag{10.14}$$

$$\sup_{U\in\mathcal{F}_+}\sup_{V\in\mathcal{F}} d_{K}(\mathcal{L}(S_n);\mathcal{L}(\tilde{S}_n)*L_0)\le C\big(n^{-1}+p\big)^2, \tag{10.15}$$

$$\sup_{U\in\mathcal{F}_+}\sup_{V\in\mathcal{F}} d_{K}(\mathcal{L}(S_n);\mathcal{L}(\tilde{S}_n)*L_1^{*n})\le C\big(n^{-1}+\sqrt{p/n}\,\big), \tag{10.16}$$

$$\sup_{U\in\mathcal{F}_+}\sup_{V\in\mathcal{F}} d_{K}(\mathcal{L}(S_n);\mathcal{L}(\tilde{S}_n)*L_1^{*n}*L_2)\le Cn^{-1}. \tag{10.17}$$

For any $U\in\mathcal{F}_+$

$$d_{K}(U^{*n};\exp(n(U-I))*L_0)\le Cn^{-2},\quad d_{K}(U^{*n};\exp(n(U-I))*L_3)\le Cn^{-3}. \tag{10.18}$$

Inequality (10.14) is a generalisation of the upper bound in (10.12). Estimates (10.16) and (10.17) demonstrate that approximations of order $O(n^{-1/2})$ or $O(n^{-1})$ can be achieved if $\mathcal{L}(X) = (1-p)U + pV$ for a particular $U \in \mathcal{F}_+$ and $p \le \tilde{C} < 1$.

Theorem 10.7 presents "short" asymptotic expansions. We now consider "long" asymptotic expansions.

Let $\mathcal{L}(X) = (1-p)U + pV$ (hence $\mathcal{L}(S_n) = ((1-p)U + pV)^{*n}$), and let $(j \in \mathbb{Z}_+)$

$$B_0 = \mathbf{\Pi}(1, X), \quad B_1 = B_0 * L_1,$$

$$\Delta_{0j} = \binom{n}{j} B_0^{*(n-j)} * (\mathcal{L}(X) - B_0)^{*j}, \quad \Delta_{1j} = \binom{n}{j} B_1^{*(n-j)} * (\mathcal{L}(X) - B_1)^{*j}.$$

In particular, $\Delta_{00} = \mathbf{\Pi}(n, X) = \mathcal{L}(\tilde{S}_n)$, $\Delta_{01} = n\mathbf{\Pi}(n-1, X) * (\mathcal{L}(X) - \mathbf{\Pi}(1, X))$.

Theorem 10.8 *If* $p \le c < 1$, $s \in \mathbb{N}$, $0 \le s \le n/2$, *then*

$$\sup_{U \in \mathcal{F}_+} \sup_{V \in \mathcal{F}} d_K\left(\mathcal{L}(S_n); \sum_{j=0}^{s} \Delta_{0j}\right) \le C(s)(n^{-1} + p)^{s+1},$$

where $\sup_{U \in \mathcal{F}_+} \sup_{V \in \mathcal{F}} \|\Delta_{0j}\|_K \le C(j)(n^{-1} + p)^j \quad (j = 1, 2, \ldots, s)$,

$$\sup_{U \in \mathcal{F}_+} \sup_{V \in \mathcal{F}} d_K\left(\mathcal{L}(S_n); \sum_{j=0}^{s} \Delta_{1j}\right) \le C(s)(n^{-1} + \sqrt{p/n})^{s+1},$$

where $\sup_{U \in \mathcal{F}_+} \sup_{V \in \mathcal{F}} \|\Delta_{1j}\|_K \le C(j)(n^{-1} + \sqrt{p/n})^j \quad (j = 1, 2, \ldots, s)$.

Open problems

10.5. Let $X, X_1, ..., X_n$ be i.i.d.r.v.s with a symmetric distribution. Will SCP measure

$$\exp(n(P_X - I) - n(P_X - I)^{*2}/2)$$

approximate P_X^n with the rate $O(n^{-1})$? with the rate $o(n^{-1/2})$?

10.6 Proofs

The general scheme of proving results like (10.5) is as follows.

Let $X, X_1, ..., X_n$ be i.i.d.r.v.s. A common step is to switch from the task of evaluating $d_K(S_n; Y)$, where Y is a CP or an infinitely divisible r.v., to the task of evaluating $d_K(S_n + \xi_h; Y + \xi_h)$, where r.v. ξ_h is independent of (S_n, Y) and has certain suitable properties. This is called "smoothing".

One can choose $\xi_h = h\xi_o$, where h is a constant, the ch.f. of $\mathcal{L}(\xi_o)$ vanishes outside of $[-1; 1]$, hence the ch.f. of $\mathcal{L}(\xi_h)$ vanishes outside of $[-1/h; 1/h]$.

Switching from $d_K(S_n; Y)$ to $d_K(S_n + \xi_h; Y + \xi_h)$ at a cost of adding an error term of order $Q_{S_n}(h) \asymp 1/\sqrt{np}$, where $p \in (0; 1)$ will be chosen later, is possible thanks to Lemma D.12.

The next step is to recall that distribution $F = \mathcal{L}(X)$ can be presented as

$$F = qA + pV, \tag{10.19}$$

where $A = \mathcal{L}(X|X\in B)$, $V = \mathcal{L}(X|X\notin B)$, B is a measurable set, $p = \mathbb{P}(X\notin B)$, $q = 1-p$, see (6.1). This can be rewritten as

$$X \stackrel{d}{=} (1-\tau)X' + \tau X'',$$

where $\mathcal{L}(X') = A$, $\mathcal{L}(X'') = V$, r.v.s τ, X', X'' are independent. Hence

$$S_n \stackrel{d}{=} \sum_{i=1}^{n-\nu_n} X_i' + \sum_{i=1}^{\nu_n} X_i'',$$

where $\nu_n, X', X_1', X_2', ..., X'', X_1'', X_2'', ...$ are independent r.v.s, $X_i' \stackrel{d}{=} X'$ $(\forall i)$, $X_i'' \stackrel{d}{=} X''$ $(\forall i)$, $\mathcal{L}(\nu_n) = \mathbf{B}(n,p)$. Obviously, $S_n + \xi_h \stackrel{d}{=} \sum_{i=1}^{n-\nu_n} X_i' + \sum_{i=1}^{\nu_n} X_i'' + \xi_h$.

Random variables $\sum_{i=1}^{n-\nu_n} X_i'$ and $\sum_{i=1}^{\nu_n} X_i''$ can be replaced by their independent copies at a cost of adding an error term of order p.

Thanks to (3.15) and (5.6), $d_{TV}(\sum_{i=1}^{\nu_n} X_i''; \sum_{i=1}^{\pi_{np}} X_i'') \leq d_{TV}(\nu_n; \pi_{np}) \leq p$.

Distribution $\mathcal{L}(\sum_{i=1}^{n-\nu_n} X_i' + \xi_h)$, where r.v. ξ_h is independent of $X_1, ..., X_n, \nu_n$ and has certain suitable properties, can be well approximated by a normal or an "accompanying" compound Poisson law thanks to the properties of the ch.f. of $\mathcal{L}(X')$, see Lemma 10.9.

With a proper choice of $h = h(p)$ the total bound is of order $p + 1/\sqrt{np}$. It is usually possible to find set B such that $p \equiv p(n) = n^{-1/3}$, leading to a bound of order $n^{-1/3}$.

If $\mathcal{L}(X)$ is absolutely continuous, then decomposition (10.19) is possible for any $p\in(0;1)$ with $B = [h_-; h_+]$. Indeed, it is known that $X \stackrel{d}{=} F^{-1}(\eta)$, where r.v. η is uniformly distributed on $[0;1]$. One can define

$$A(\cdot) = \mathbb{P}(F^{-1}(\eta)\in\cdot|\eta\in[p/2; 1-p/2]), \quad V(\cdot) = \mathbb{P}(F^{-1}(\eta)\in\cdot|\eta\notin[p/2; 1-p/2]).$$

Then

$$\text{supp}(A) = [F^{-1}(p/2); F^{-1}(1-p/2)], \quad \text{supp}(V) = \mathbb{R}\setminus(F^{-1}(p/2); F^{-1}(1-p/2)).$$

Denote

$$h_- = F^{-1}(p/2), \quad h_+ = F^{-1}(1-p/2).$$

Then $\text{supp}(A) = [h_-; h_+]$, $\text{supp}(V) = \mathbb{R}\setminus(h_-; h_+)$. One can apply (E.2).

If $\mathcal{L}(X)$ is not continuous, then one may have a situation where $A = \mathcal{L}(X)$ – consider, for example, $\mathcal{L}(X) = \mathbf{B}(p)$, then $F^{-1}(p/2) = 0$, $F^{-1}(1-p/2) = 1$.

Example 10.2 If $F = \mathbf{B}(p)$, then

$$h_- = 0, \quad h_+ = 1.$$

Hence $F((-\infty; h_-]) = q$, $F([h_+;\infty)) = p$, the length of interval $[h_-; h_+]$ is 1, $Q_F(1) = 1$, and we cannot immediately use (E.2), which is needed when applying Lemma D.12.

This example highlights the difficulty that arises when the d.f. is not continuous.

In order to overcome this difficulty, one can deal with $X - \varepsilon\xi$ instead of X, where r.v. ξ is independent of X and has a continuous d.f., $\varepsilon > 0$.

If ξ is a normal $\mathcal{N}(0;1)$ r.v., then $\mathbb{P}(X - \varepsilon\xi \leq 0) = q\mathbb{P}(\xi \geq 0) + p\mathbb{P}(\xi \geq 1/\varepsilon) \to q/2$ as $\varepsilon \to 0$.

Let ξ be an exponential $\mathbf{E}(1)$ r.v. Then $\mathbb{P}(X - \varepsilon\xi \leq 0) = q + p\mathbb{P}(\xi \geq 1/\varepsilon) \to q$ as $\varepsilon \to 0$.

Instead of an exponential r.v. one can use a uniform $\mathbf{U}[0;1]$ r.v., etc. □

W.l.o.g. we may assume that $\mathcal{L}(X)$ is non-degenerate, since a degenerate r.v. is a constant.

If the d.f. of $\mathcal{L}(X)$ is not continuous, let ξ denote an exponential $\mathbf{E}(1)$ r.v. independent of $X, X_1, ..., X_n$. Then the d.f. $F_{X-\varepsilon\xi}$ of $\mathcal{L}(X-\varepsilon\xi)$ and the inverse $F^{-1}_{X-\varepsilon\xi}$ are continuous and strictly monotone.

Since the d.f. F of $\mathcal{L}(X)$ is right-continuous, $\mathbb{P}(S_n-\varepsilon\xi\leq x)\to\mathbb{P}(S_n\leq x)$ as $\varepsilon\downarrow 0$ for any $x\in\mathbb{R}$. Therefore, we may assume in the sequel that the d.f. F and the inverse F^{-1} are continuous.

Let $p\in(0;1)$. Then decomposition (10.19) holds with $B=[h_-;h_+]$, where $h_+\geq h_-$,

$$\mathbb{P}(X<h_-)=p/2, \quad \mathbb{P}(X>h_+)=p/2. \tag{10.20}$$

If $h_+=h_-$, then $A=I_{h_+}$, $X=h_++\tau(X''-h_+)$, and we can apply the estimate $d_{TV}(\sum_{i=1}^{\nu_n}(X_i''-h_+);\sum_{i=1}^{\pi_{np}}(X_i''-h_+))\leq d_{TV}(\nu_n;\pi_{np})\leq p$. Hence we may assume that $h_+>h_-$.

Distribution A can be shifted by its mean $m_A=\int x\,A(dx)$. Therefore, we may assume that $m_A=0$. Set

$$h=\max\{|h_-|;h_+\}, \quad h_o=h_+-h_-.$$

Then $h_-\leq 0$, $h_+\geq 0$, and $1-Q_X(h)\geq 1-Q_X(h_o)\geq p/2$.

For the proofs of Theorems 10.2 – 10.4 we need few auxiliary lemmas. The next lemma describes the properties of $\widehat{A}$ in a neighbourhood of 0. Let σ^2 denote the variance of A.

Lemma 10.9 *For any* $t\in\mathbb{R}$

$$|\widehat{A}-1|\leq\sigma^2t^2/2. \tag{10.21}$$

If $|t|\leq c/h$, *where* $c\in(0;\sqrt{2})$, *then*

$$\sigma^2t^2\leq c^2, \quad |\exp(\widehat{A}-1)|\leq e^{-\frac{\sigma^2t^2}{2}(1-\frac{c}{3})}, \quad |\widehat{A}|\leq e^{-\frac{\sigma^2t^2}{2}(1-\frac{c}{3})}. \tag{10.22}$$

Proof of Lemma 10.9. Since $m_A=0$, estimate (10.21) follows from Taylor's formula (D.3).

The first inequality in (10.22) is a trivial consequence of the fact that $\sigma^2\leq h^2$. Clearly,

$$\begin{aligned}\left|\widehat{A}-1+\sigma^2t^2/2\right| &= \left|\int_{|x|\leq h}(e^{itx}-1-itx+t^2x^2/2)\,A(dx)\right| \\ &\leq \tfrac{1}{6}\int_{|x|\leq h}|tx|^3A(dx)\leq\tfrac{c}{6}\int_{|x|\leq h}|tx|^2A(dx)=\tfrac{c}{6}\sigma^2t^2\end{aligned}$$

if $|t|\leq c/h$. Therefore,

$$|\widehat{A}|\leq\left|\widehat{A}-1+\frac{\sigma^2t^2}{2}\right|+\left|1-\frac{\sigma^2t^2}{2}\right|\leq\frac{c}{6}\sigma^2t^2+1-\frac{\sigma^2t^2}{2}=1-\frac{\sigma^2t^2}{2}\left(1-\frac{c}{3}\right)\leq e^{-\frac{\sigma^2t^2}{2}(1-\frac{c}{3})},$$

which proves the last estimate in (10.22). Observe that

$$\left|\exp(\widehat{A}-1)\right|=\exp(Re\,\widehat{A}-1)$$

and

$$\begin{aligned}Re\,\widehat{A}-1 &= -\sigma^2t^2/2+Re\,(\widehat{A}-1+\sigma^2t^2/2) \\ &\leq -\sigma^2t^2/2+\left|\widehat{A}-1+\sigma^2t^2/2\right|\leq-\frac{\sigma^2t^2}{2}\left(1-\frac{c}{3}\right)\end{aligned}$$

which proves the middle estimate in (10.22). □

The following result is of independent interest.

Lemma 10.10 *For any* $n \geq 1$ *and* $p \leq c < 1$

$$d_K(F^{*n}; \exp(nq(A-I)) * (qI+pV)^{*n}) \leq C(np)^{-1/2}. \tag{10.23}$$

Lemma 10.10 can be reformulated in the following way.

Let $X', X'', \tau, \pi_{nq}, \nu_{n,p}$ be independent r.v.s with distributions A, V, $\mathbf{B}(p)$, $\mathbf{\Pi}(nq)$ and $\mathbf{B}(n,p)$, respectively. Recall that

$$X \stackrel{d}{=} (1-\tau)X' + \tau X''.$$

Then (10.23) can be rewritten as

$$d_K\left(S_n; \sum_{i=1}^{\pi_{nq}} X_i' + \sum_{i=1}^{\nu_{n,p}} X_i''\right) \leq C(np)^{-1/2}. \tag{10.23*}$$

Recall that $\sum_{i=1}^{\nu_{n,p}} X_i''$ can be approximated by $\sum_{i=1}^{\pi_{np}} X_i''$, where Poisson $\mathbf{\Pi}(np)$ random variable π_{np} is independent of $\pi_{nq}, X_1', X_2', ..., X_1'', X_2'', ...$ (see Lemma 6.1): $d_{TV}(\sum_{i=1}^{\nu_{n,p}} X_i''; \sum_{i=1}^{\pi_{np}} X_i'') \leq p$. This and (10.23*) entail the following estimate of the accuracy of compound Poisson approximation:

$$d_K\left(S_n; \sum_{i=1}^{\pi_{nq}} X_i' + \sum_{i=1}^{\pi_{np}} X_i''\right) \leq p + C(np)^{-1/2}, \tag{10.23⋆}$$

Estimate $O(n^{-1/3})$ of the rate of the accuracy of compound Poisson approximation follows by letting $p \equiv p(n) = n^{-1/3}$.

Proof of Lemma 10.10. By the properties (E.5)–(E.6) of concentration functions,

$$\max\{Q_{F^{*n}}(h); Q_{\exp(n(F-I))}(h); Q_{(qI+pV)^{*n}}(h)\} \leq C(np)^{-1/2}. \tag{10.24}$$

Applying (D.21), (E.1), and (10.24), we get

$$\begin{aligned} &d_K(F^{*n}; \exp(nq(A-I)) * (qI+pV)^{*n}) \\ &\leq \int_{-1/h}^{1/h} |\widehat{F}^n(t) - \exp(nq(\widehat{A}-1))(q+p\widehat{V})^n| \, |t|^{-1} \mathrm{d}t + C(np)^{-1/2}. \end{aligned} \tag{10.25}$$

We need to evaluate

$$\int_0^{1/h} \left|\widehat{F}^n(t) - \exp(nq(\widehat{A}(t)-1))(q+p\widehat{V}(t))^n\right| t^{-1} \mathrm{d}t.$$

Recall that $\widehat{F} = q\widehat{A} + p\widehat{V}$, where $q = 1-p$. Hence

$$\widehat{F}^n = \sum_{k=0}^n \binom{n}{k} q^k p^{n-k} \widehat{A}^k \widehat{V}^{n-k}, \quad (q+p\widehat{V})^n = \sum_{k=0}^n \binom{n}{k} q^k p^{n-k} \widehat{V}^{n-k}.$$

Therefore,

$$\begin{aligned} &\left|\widehat{F}^n - \exp\left(nq(\widehat{A}-1)\right)(q+p\widehat{V})^n\right| \\ \leq\ & \sum_{k=0}^n \binom{n}{k} q^k p^{n-k} |\widehat{V}^{n-k}| \, |\widehat{A}^k - \mathrm{e}^{nq(\widehat{A}-1)}| \\ \leq\ & p^n |1 - \mathrm{e}^{nq(\widehat{A}-1)}| + \sum_{k=1}^n \binom{n}{k} q^k p^{n-k} \left(|\widehat{A}^k - \mathrm{e}^{k(\widehat{A}-1)}| + \left|\mathrm{e}^{k(\widehat{A}-1)} - \mathrm{e}^{nq(\widehat{A}-1)}\right|\right). \end{aligned}$$

Recall that $\widehat{A}^k$ is the characteristic function of $X_1'+\cdots+X_k'$, where $X', X_1', \ldots$ are i.i.d.r.v.s, $\mathcal{L}(X')=A$, $\mathrm{supp}(A)=[h_-;h_+]\subset[-h;h]$. Clearly, $\sigma\le h$.

Concerning the first term corresponding to $k=0$, it suffices to note that

$$\int_0^{1/h}\left|1-\mathrm{e}^{nq(\widehat{A}(t)-1)}\right|t^{-1}\mathrm{d}t\le\int_0^{1/h}n\sigma^2t\,\mathrm{d}t\le n\sigma^2/2h^2\le n/2,$$

leading to the term np^n on the RHS.

By Taylor's formula,

$$|\widehat{A}-\exp(\widehat{A}-1)|\le|\widehat{A}-1|^2/2.$$

Applying (10.21) and (10.22), we obtain

$$\left|\widehat{A}^k-\exp\Big(k(\widehat{A}-1)\Big)\right|\le k\sigma^4t^4\exp(-k\sigma^2t^2/3)\qquad(k\ge1).$$

Therefore,

$$\int_0^{1/h}\left|\widehat{A}^k-\exp\Big(k(\widehat{A}-1)\Big)\right|t^{-1}\,\mathrm{d}t\le C/k\le 2C/(k+1)\qquad(k\ge1).$$

By (I.10), (10.21), and (10.22),

$$\left|\exp\Big(k(\widehat{A}-1)\Big)-\exp\Big(nq(\widehat{A}-1)\Big)\right|\le|k-nq|\,\sigma^2t^2\exp(-\min\{k;nq\}\sigma^2t^2/3)\,.$$

Hence

$$\int_0^{1/h}\left|\mathrm{e}^{k(\widehat{A}-1)}-\mathrm{e}^{nq(\widehat{A}-1)}\right|t^{-1}\,\mathrm{d}t\le C|k-nq|/\min\{k;nq\}.$$

Thus,

$$\begin{aligned}\int_0^{1/h}&\left|\widehat{F}^n-\exp\Big(nq(\widehat{A}-1)\Big)\,(q+p\widehat{V})^n\right|t^{-1}\mathrm{d}t\\ &\le\; np^n+C\sum_{k=1}^n q^kp^{n-k}\binom{n}{k}\Big(1/(k+1)+|k-nq|/\min\{k;nq\}\Big)\\ &\le\; np^n+C\mathbb{E}(\nu_{n,q}+1)^{-1}+C\mathbb{E}|\nu_{n,q}-nq|\mathbb{I}^{>}/\min\{\nu_{n,q};nq\},\end{aligned}\qquad(10.26)$$

where $\nu_{n,q}$ is a r.v. with the distribution $\mathcal{L}(\nu_{n,q})=\mathbf{B}(n,q)$, $\mathbb{I}^{>}=\mathbb{I}\{\nu_{n,q}>0\}$.

By (I.3),

$$\mathbb{E}(\nu_{n,q}+1)^{-1}\le 1/nq.$$

One can check that

$$\mathbb{E}|\nu_{n,q}-nq|\mathbb{I}^{>}/\min\{\nu_{n,q};nq\}\le C/\sqrt{nq}\,.$$

Substituting the last estimate into (10.26) and taking into account (10.25), we complete the proof of Lemma 10.10. □

The following lemma deals with convolutions of $(A-I)$ and $F=qA+pV$, where $q=1-p$. Recall that $A([-h,h])=1$, $h>0$, and that (10.20) holds.

Lemma 10.11 *Denote* $\tilde{p} = \mathbb{P}(|X|>h)$. *Then for any* $m\in\mathbb{N}$, $\delta>0$,

$$\|(A-I)^{*m} * \mathbf{\Pi}(\delta, F)\|_K \le C(m)(q\delta)^{-(m+1)/2}\tilde{p}^{-1/2}. \tag{10.27}$$

In particular,

$$\|(A-I) * \mathbf{\Pi}(\delta, F)\|_K \le C(q\delta)^{-1}\tilde{p}^{-1/2}.$$

Proof of Lemma 10.11. We apply Lemma D.14, which states that

$$\|W*H\|_K \;\le\; C\int_{-1/h}^{1/h}\Big|\frac{\widehat{W}(t)\widehat{H}(t)}{t}\Big|\,dt + C\|W\|_{TV}\,Q_G(h)$$

if W is a finite measure, $W(\mathbb{R})=0$, $H\in\mathcal{F}$, $|\widehat{H}(t)|\le C\widehat{G}(t)$ as $|t|\le 1/h$ $(\exists G\in\mathcal{F}_+)$, $h>0$.
Let

$$H = \mathbf{\Pi}(\delta/2; F), \quad W = (A-I)^m * \mathbf{\Pi}(\delta/2; F), \quad \widehat{G}(t) = \exp(\delta(ReF-1)).$$

Using Le Cam's inequality (E.6), we get

$$Q_G(h) \le C(m)/\sqrt{\delta\tilde{p}}.$$

Lemma C.1 yields

$$\|W\|_{TV} \le C(m)\delta^{-m/2}.$$

Thus, $\|W\|_{TV}\,Q_G(h) \le C(m)\delta^{-(m+1)/2}\tilde{p}^{-1/2}$.

Note that $|\hat{H}(t)| = |\exp(\delta(\hat{F}(t)-1))| \le \exp(\delta(\mathrm{Re}\hat{F}(t)-1))$ $(\forall t)$. It follows from Lemma 10.9 that $|\widehat{A}-1| \le \sigma^2 t^2/2$. Therefore, for any $h>0$

$$\int_{-1/h}^{1/h}\frac{|\widehat{H}(t)\widehat{W}(t)|}{|t|}\mathrm{d}t \le 2^{-m}\int_{-1/h}^{1/h}\sigma^{2m}|t|^{2m-1}\mathrm{e}^{-\delta q\sigma^2 t^2/3}\mathrm{d}t \le C(m)(q\delta)^{-m}.$$

Collecting these estimates leads to (10.27). □

Proof of Theorem 10.2. The LHS of (10.6) is less than $s2^s n^s$. Therefore, w.l.o.g. we may assume that $n>10(s+1)$ – otherwise we can apply the trivial estimate $s2^s n^s \le s2^s n^s(10(s+1)/n)^{s+1}$.

Recall that $h>0$. Set $p = n^{-1/(2k+3)}$. Then $np\ge 1$. We recall that Θ denotes any measure satisfying $\|\Theta\|_{TV}\le 1$. We are evaluating

$$d_K\Big(F^{*n}; \sum_{j=0}^{k}\Delta_{ja}\Big),$$

where $\Delta_{ja} = \binom{n}{j}\mathbf{\Pi}(n-j, F) * (F-\mathbf{\Pi}(1, F))^{*j}$.

We have

$$d_K\Big(F^{*n}; \sum_{j=0}^{k}\Delta_{ja}\Big) \le \mathrm{d}_1+\mathrm{d}_2+\mathrm{d}_3+\mathrm{d}_4, \tag{10.28}$$

where

$$\begin{aligned}
\mathrm{d}_1 &= d_K(F^{*n}; \mathbf{\Pi}(nq; A) * (qI+pV)^{*n}),\\
\mathrm{d}_2 &= d_K\Big(\mathbf{\Pi}(nq; A) * (qI+pV)^{*n}; \mathbf{\Pi}(nq; A) * \sum_{j=0}^{k} \tilde{\Delta}_{ja}\Big),\\
\mathrm{d}_3 &= d_K\Big(\mathbf{\Pi}(nq; A) * \sum_{j=0}^{k} \tilde{\Delta}_{ja}; \sum_{j=0}^{k} \hat{\Delta}_{ja}\Big),\\
\mathrm{d}_4 &= d_K\Big(\sum_{j=0}^{k} \hat{\Delta}_{ja}; \sum_{j=0}^{k} \Delta_{ja}\Big),\\
\tilde{\Delta}_{ja} &= \binom{n}{j} \mathbf{\Pi}((n-j)p, V) * (qI+pV-\mathbf{\Pi}(p,V))^{*j},\\
\hat{\Delta}_{ja} &= \binom{n}{j} \mathbf{\Pi}(n-j, F) * (qI+pV-\mathbf{\Pi}(p,V))^{*j}.
\end{aligned}$$

Lemma 10.10 can be applied in order to evaluate d_1: $\mathrm{d}_1 \le C(np)^{-1/2}$. For the estimate of d_2 note that

$$qI+pV - \mathbf{\Pi}(p,V) = p^2(V-I)^{*2}\Theta \tag{10.29}$$

by (B.2). It follows from Lemma C.1 that

$$\|(P-I)^{*m} * \mathbf{\Pi}(\lambda, P)\|_{TV} \le C(m)\lambda^{-m/2} \tag{10.30}$$

for any $P \in \mathcal{F}$, $\lambda > 0$ and $m \in \mathbb{N}$. A similar estimate holds for a compound Binomial distribution, see (C.7).

By (A.6), (B.9), and (10.29),

$$\begin{aligned}
\mathrm{d}_2 &= \Big\|\mathbf{\Pi}(nq, A) * \Big((qI+pV)^{*n} - \sum_{j=0}^{k}\binom{n}{j}\mathbf{\Pi}((n-j)p, V) * (qI+pV-\mathbf{\Pi}(p,V))^{*j}\Big)\Big\|_K\\
&\le \Big\|(qI+pV)^{*n} - \sum_{j=0}^{k}\binom{n}{j}\mathbf{\Pi}((n-j)p, V) * (qI+pV-\mathbf{\Pi}(p,V))^{*j}\Big\|_{TV}\\
&\le \sum_{m=k+1}^{n}\binom{m-1}{k}\Big\|(qI+pV)^{*(n-m)} * \mathbf{\Pi}((m-k-1)p, V) * (qI+pV-\mathbf{\Pi}(p,V))^{*(k+1)}\Big\|_{TV}\\
&\le C(k)\sum_{m=k+1}^{n}\binom{m-1}{k}\Big\|(qI+pV)^{*(n-m)} * \mathbf{\Pi}((m-k-1)p, V) * (p(V-I))^{*2(k+1)}\Big\|_{TV}\\
&\le C(k)\Big(\sum_{m\le\lfloor n/2\rfloor} + \sum_{m\ge\lfloor n/2\rfloor+1}\Big) \le C(k)n^{k+1}p^{2(k+1)}\\
&\times \Big(\|(qI+pV)^{*\lfloor n/3\rfloor} * (V-I)^{*2(k+1)}\|_{TV} + \|\exp(\lfloor n/3\rfloor p(V-I)) * (V-I)^{*2(k+1)}\|_{TV}\Big)\\
&\le C(k)n^{k+1}p^{2(k+1)}(np)^{-(k+1)} = C(k)p^{k+1}.
\end{aligned}$$

In order to estimate d_3, note that

$$\mathbf{\Pi}((n-j)p, V) = \mathbf{\Pi}(np/3, V) * \mathbf{\Pi}((n-j-1/3)p, V)$$

and $3n-3j-1 \geq 30(k+1)-3k-1 \geq 9k$. Therefore, $\boldsymbol{\Pi}((3n-3j-1)p/3, V)$ is a compound Poisson distribution, its norm equals 1, and we can write

$$\boldsymbol{\Pi}((n-j)p, V) = \boldsymbol{\Pi}(np/3, V) * \Theta.$$

Similarly,

$$\boldsymbol{\Pi}((n-j)p, A) = \boldsymbol{\Pi}(np/3, A) * \Theta. \tag{10.31}$$

By (B.2) and (10.31),

$$\boldsymbol{\Pi}((n-j)q; A) - \boldsymbol{\Pi}(nq; A) = C(j)\boldsymbol{\Pi}(nq/3; A) * (A-I) * \Theta.$$

From (10.29); (10.27), (A.6), and (10.30) we get

$$\begin{aligned}
\mathrm{d}_3 &\leq \sum_{j=1}^{k} \binom{n}{j} \|(\boldsymbol{\Pi}((n-j)q; A) - \boldsymbol{\Pi}(nq; A)) * \boldsymbol{\Pi}((n-j)p; V) * (qI + pV - \boldsymbol{\Pi}(p; V))^{*j}\|_K \\
&\leq C(k) \sum_{j=1}^{k} \binom{n}{j} p^{2j} \|\boldsymbol{\Pi}(nq/3; A) * (A-I) * \boldsymbol{\Pi}(np/3; V) * (V-I)^{*2j}\|_K \\
&= C(k) \sum_{j=1}^{k} \binom{n}{j} p^{2j} \|\boldsymbol{\Pi}(nq/6; F) * (A-I) * \boldsymbol{\Pi}(nq/6; A) * \boldsymbol{\Pi}(np/6; V) * (V-I)^{*2j}\|_K \\
&\leq C(k) \sum_{j=1}^{k} \binom{n}{j} p^{2j} \|\boldsymbol{\Pi}(n/6; F) * (A-I)\|_K \|\boldsymbol{\Pi}(np/6; V) * (V-I)^{*2j}\|_{TV} \\
&\leq C(k) \sum_{j=1}^{k} p^{2j-1/2} \cdot n^j \cdot n^{-1} \cdot (np)^{-j} \leq C(k) p^{1/2}/n.
\end{aligned}$$

It remains to estimate d_4. To shorten notation, we set

$$M(V) = qI + pV - \boldsymbol{\Pi}(p, V), \quad M(F) = F - \tilde{F} = F - \boldsymbol{\Pi}(1, F).$$

By (B.2),

$$\begin{aligned}
&M(F) = (F-I)^{*2} * \Theta, \\
&M(V) - M(F) = -p^2(V-I)^{*2}/2 + (F-I)^{*2}/2 + p^3(V-I)^{*3}\Theta/3 + (F-I)^{*3}/3 \\
&= q^2(A-I)^{*2}/2 + pq(A-I) * (V-I) + p^3(V-I)^{*3}\Theta/3 + (F-I)^{*3}\Theta/3.
\end{aligned}$$

Taking into account (10.29), we can write

$$\begin{aligned}
M(V)^{*j} - M(F)^{*j} &= \sum_{m=0}^{j} M(V)^{*m} * M(F)^{*(j-m-1)} * \big(q^2(A-I)^{*2}/2 \\
&+ pq(A-I) * (V-I) + p^3(V-I)^{*3}\Theta/3 + (F-I)^{*3}\Theta/3\big) \\
&= \sum_{m=0}^{j} (p^2(V-I)^{*2} * \Theta)^{*m} * ((F-I)^{*2} * \Theta)^{*(j-m-1)} * \big(q^2(A-I)^{*2}/2 \\
&+ pq(A-I) * (V-I) + p^3(V-I)^{*3}\Theta/3 + (F-I)^{*3}\Theta/3\big).
\end{aligned}$$

Therefore, it follows from (10.30) and (10.27) that

$$\begin{aligned}
\mathrm{d}_4 &\le \sum_{j=1}^{k}\binom{n}{j}\|\mathbf{\Pi}(n-j,F)*(M(V)^{*j}-M(F)^{*j})\|_K\\
&\le C(k)\sum_{j=1}^{k}\binom{n}{j}\|\mathbf{\Pi}(n/3,F)*(M(V)^{*j}-M(F)^{*j})\|_K\\
&\le C(k)\sum_{j=1}^{n}n^j\sum_{m=0}^{j}\|p^{2m}(V-I)^{*2m}*\mathbf{\Pi}(np/9;V)\|_{TV}\\
&\times \|(F-I)^{*2(j-m-1)}*\mathbf{\Pi}(n/9;F)\|_{TV}\Big(q^2\|(A-I)^{*2}*\mathbf{\Pi}(n/9;F)\|_K\\
&+ pq\|(A-I)*\mathbf{\Pi}(n/18;F)\|_K\|(V-I)*\mathbf{\Pi}(np/18;V)\|_{TV}\\
&+ \|(F-I)^{*3}*\mathbf{\Pi}(n/9;F)\|_{TV}+\|p^3(V-I)^{*3}*\mathbf{\Pi}(np/9;V)\|_{TV}\Big)\\
&\le C(k)(np)^{-1/2}.
\end{aligned}$$

Substituting estimates of $\mathrm{d}_1,\mathrm{d}_2,\mathrm{d}_3,\mathrm{d}_4$ into (10.28), we get

$$d_K\Big(F^{*n};\sum_{j=0}^{k}\Delta_{ja}\Big)\le C(np)^{-1/2}+C(k)p^{k+1}+C(k)p^{1/2}n^{-1}+C(k)(np)^{-1/2}.$$

Since $p=n^{-1/(2k+3)}$, this leads to $d_K\Big(F^{*n};\sum_{j=0}^{k}\Delta_{ja}\Big)\le C(k)n^{-(k+1)/(2k+3)}$. The proof is complete. □

Proof of Theorem 10.3. Recall that

$$A_{1a}=-n(F_a-I)^{*2}/2,\quad A_{2a}=n^2(F_a-I)^{*4}/8.$$

By (B.2*),

$$\mathbf{\Pi}(1,F)-I=(F-I)*\Theta,$$

where $\|\Theta\|_{TV}\le 1$.

We apply Theorem 10.2 with $k=1$, (B.2) and (C.10):

$$\begin{aligned}
&\|F^{*n}-\mathbf{\Pi}(n,F)*(I-n(F-I)^{*2}/2)\|_K\\
\le\ &\|F^{*n}-\mathbf{\Pi}(n,F)-n\mathbf{\Pi}(n-1,F)*(F-\mathbf{\Pi}(1,F))\|_K\\
+\ &\|n\mathbf{\Pi}(n-1,F)*(F-\mathbf{\Pi}(1,F)+(F-I)^{*2}/2)\|_K\\
+\ &\tfrac{1}{2}n\|\mathbf{\Pi}(n-1,F)*(F-I)^{*2}*(\mathbf{\Pi}(1,F)-I)\|_K\\
\le\ &Cn^{-2/5}+Cn\|(F-I)^{*3}*\mathbf{\Pi}(n-1,F)\|_K\le Cn^{-2/5}.
\end{aligned}$$

The proof of (10.7) is very similar and is therefore omitted. □

Proof of Theorem 10.4. We aim to show that there exist distribution $V=V(F)$ and an absolute constant C such that

$$d_K\Big(F^{*n};\exp(n(F-I)-n(V-I)^{*2}/2)\Big)\le Cn^{-1/2}. \tag{10.32}$$

If (after suitable shifting) $F=(1-p)I+pV$, where $0<c\le p\le C<1$, then (10.32) follows from (7.16); we take into account (A.3). Otherwise we apply decomposition (10.19) with $p=0.1$: $F=qA+pV$, where $q=(1-p)$.

Denote

$$G = \exp\big(n(F-I) - np(V-I)^{*2}/2\big) = \exp\big(nq(A-I) + np(V-I) - np^2(V-I)^{*2}/2\big).$$

By the triangle inequality,

$$\begin{aligned} d_K(F^{*n}; G) &\le d_K(F^{*n}; \exp(nq(A-I)) * (qI+pV)^{*n}) \\ &+ d_{TV}((qI+pV)^{*n}; \exp\big(np(V-I) - np^2(V-I)^{*2}/2\big)). \end{aligned}$$

We apply (10.23) to get

$$d_{TV}((qI+pV)^{*n}; \exp\big(np(V-I) - np^2(V-I)^{*2}/2\big)) \le Cn^{-1/2}.$$

It follows from (10.23) that

$$d_K(F^{*n}; \exp(nq(A-I)) * (qI+pV)^{*n}) \le C(np)^{-1/2} = C_1 n^{-1/2},$$

and (10.32) follows. □

Proof of Theorem 10.5. We need to show that

$$\sup_{U\in\mathcal{F}_s} \sup_{V\in\mathcal{F}} d_K(\mathcal{L}(S_n); \mathcal{L}(\tilde{S}_n)) \le C\big(n^{-1/2} + (p/nq)^{1/2} + p\big).$$

First, we notice that the upper bound in (10.8) holds when $p=0$ (i.e., when $F=U\in\mathcal{F}_s$ in (6.1*):

$$\sup_{F\in\mathcal{F}_s} d_K(S_n; \tilde{S}_n) \le C/\sqrt{n+1}. \tag{10.33}$$

Indeed, according to (C.25),

$$\|U^{*n} - \exp(n(U-I))\|_K \le C/\sqrt{n+1} \qquad (U\in\mathcal{F}_s).$$

We consider now the case where $p>0$. Set $q = 1-p$.

Let M be a finite measure, and let V be a probability distribution. By the properties of norms,

$$\|M * V^{*j}\|_K \le \|M\|_K \, \|V^{*j}\|_{TV} = \|M\|_K$$

for any $j\in\mathbb{N}$. Applying (10.33) and inequality (I.3), we derive

$$\begin{aligned} \big\|(qU+pV)^{*n} - (qe^{U-I}+pV)^{*n}\big\|_K &\le \sum_{j=0}^{n} \binom{n}{j} q^j p^{n-j} \|U^{*j} - \exp(j(U-I))\|_K \\ &\le C\sum_{j=0}^{n} \binom{n}{j} \frac{q^j p^{n-j}}{\sqrt{j+1}} \le Cn^{-1/2}. \end{aligned}$$

According to (C.13),

$$\|e^{a(P-I)} - e^{b(P-I)}\|_K \le C\,|a-b|/\max\{a;b\} \qquad (a>0,\ b>0).$$

Let $\mathcal{L}(\nu_{n,q}) = \mathbf{B}(n,q)$. Then

$$\begin{aligned} &\big\|(qe^{U-I} + pV)^{*n} - e^{nq(U-I)} * (qI+pV)^{*n}\big\|_K \\ \le\ & \sum_{j=0}^{n} \binom{n}{j} q^j p^{n-j} \big\|e^{j(U-I)} - e^{nq(U-I)}\big\|_K \\ \le\ & \sum_{j=0}^{n} \binom{n}{j} q^j p^{n-j} |j-nq|/\max\{j;nq\} \le \sqrt{\operatorname{var}\nu_{n,q}}/nq \le C(p/nq)^{1/2}. \end{aligned}$$

Combining the above estimates, we get

$$\big\|(qU+pV)^{*n} - \mathrm{e}^{nq(U-I)} * (qI+pV)^{*n}\big\|_{K} \le C\big(n^{-1/2} + (p/nq)^{1/2}\big). \tag{10.34}$$

We apply properties of metrics, (6.8) and (7.15) to get

$$\begin{aligned} &\big\|\mathrm{e}^{nq(U-I)} * (qI+pV)^{*n} - \mathrm{e}^{nq(U-I)} * \exp(np(V-I))\big\|_{K} \\ \le\; & \big\|\mathrm{e}^{nq(U-I)}\big\|_{TV} \cdot \big\|(qI+pV)^{*n} - \exp(np(V-I))\big\|_{K} \\ \le\; & \big\|(qI+pI_1)^{*n} - \exp(np(I_1-I))\big\|_{TV} = 2d_{TV}(\mathbf{B}(n,p);\mathbf{\Pi}(np)) \le 2p. \end{aligned}$$

Since $p \le \tilde{C} < 1$, the proof of (10.9) is complete.

We now prove (10.10):

$$\sup_{U\in\mathcal{F}_s}\sup_{V\in\mathcal{F}} d_K(\mathcal{L}(S_n);\mathcal{L}(\tilde{S}_n) * L_1^{*n}) \le C\big(n^{-1/2} + \sqrt{p/n}\big),$$

where $L_1 = \exp\big(-p^2(V-I)^{*2}/2\big)$. Observe that $d_K(\mathcal{L}(S_n);\mathcal{L}(\tilde{S}_n) * L_1^{*n})$ equals

$$\begin{aligned} &\big\|(qU+pV)^{*n} - \mathrm{e}^{nq(U-I)} * \exp\big(np(V-I) - np^2(V-I)^{*2}/2\big)\big\|_{K} \\ \le\; & \big\|(qU+pV)^{*n} - \mathrm{e}^{nq(U-I)} * (qI+pV)^{*n}\big\|_{K} \\ +\; & \big\|(qI+pV)^{*n} - \exp\big(np(V-I) - np^2(V-I)^{*2}\big)\big\|_{K} \big\|\mathrm{e}^{nq(U-I)}\big\|_{TV}. \end{aligned}$$

Estimate (10.10) follows from (10.34), (A.3), and (7.16). □

Proof of Theorem 10.6. Recall that $S_n \stackrel{d}{=} \sum_{i=1}^{n} \tau_i X_i'$, where $\mathcal{L}(\tau_i) = \mathbf{B}(p_i)$. Hence

$$\mathcal{L}(S_n) = \prod_{i=1}^{n} ((1-p_i)I + p_i F),$$

where $F = \mathcal{L}(X')$. We aim to show that

$$d_K(\mathcal{L}(S_n); H_{n,k}) \le C(k)\lambda_{k+1}\lambda^{-k-1} \qquad (k\in\mathbb{N}),$$

where $H_{n,k} = \exp\Big(\sum_{j=1}^{k}(-1)^{j+1}\lambda_j(F-I)^{*j}/j\Big)$.

The proof is very similar to the proof of (7.33), therefore we use the same notation:

$$A(F) = \sum_{j=1}^{k}(-1)^{j+1}\lambda_j(F-I)^{*j}/j, \quad B(F) = \sum_{j=k+1}^{\infty}(-1)^{j+1}\lambda_j(F-I)^{*j}/j.$$

We have

$$\begin{aligned} d_K(\mathcal{L}(S_n); H_{n,k}) &= \big\|\exp(A_k(F)+B_k(F)) - \exp(A_k(F))\big\|_{K} \\ &= \big\|\exp(A_k(F)) * (\exp(B_k(F)-I))\big\|_{K} \\ &= \Big\|\exp(A_k(F)) * B_k(F) * \int_0^1 \exp(\tau B_k(F))\mathrm{d}\tau\Big\|_{K} \\ &\le \int_0^1 \big\|B_k(F) * \exp\big(A_k(F)+\tau B_k(F)\big)\big\|_{K}\mathrm{d}\tau. \end{aligned} \tag{10.35}$$

By the assumptions, $\max_i p_i \le 1/4$. Consequently,

$$\lambda_j = \sum_{i=1}^n p_i^j \le (1/4)^{j-k-1} \sum_{i=1}^n p_i^{k+1} = \lambda_{k+1}/4^{j-k-1} \quad (j > k+1).$$

Therefore,

$$\sum_{j=k+1}^{\infty} \frac{\lambda_j}{j} \|F-I\|_{TV}^{j-k-1} \le \lambda_{k+1} \sum_{j=k+1}^{\infty} \frac{2^{j-k-1}}{j4^{j-k-1}} \le C(k)\lambda_{k+1}.$$

Note that $B_k(F) = (F-I)^{*(k+1)}\lambda_{k+1}C(k)\Theta$ and

$$\begin{aligned} &\left\|B_k(F) * \exp\left(A_k(F)+\tau B_k(F)\right)\right\|_K \\ \le\ & C(k)\lambda_{k+1}\left\|(F-I)^{*(k+1)} * \exp\left(A_k(F)+\tau B_k(F)\right)\right\|_K. \end{aligned} \tag{10.36}$$

Similarly, $\lambda_j \le \lambda/4^{j-1}$ $(j > 1)$ and

$$\left\|\sum_{j=2}^{k} \frac{\lambda_j(-1)^{j+1}(F-I)^{*(j-2)}}{j} + \tau \sum_{j=k+1}^{\infty} \frac{\lambda_j(-1)^{j+1}(F-I)^{*(j-2)}}{j}\right\|_{TV} \le \lambda \sum_{j=2}^{\infty} \frac{2^{j-2}}{4^{j-1}j} = \frac{\lambda}{4}$$

for any $\tau \in [0,1]$. This allows us to write

$$A(F) + \tau B(F) = \lambda(F-I) + (\lambda/4)(F-I)^{*2} * \Theta.$$

By Lemma B.1,

$$\left\|\exp\left(\frac{7}{8}\lambda(F-I) + (\lambda/4)(F-I)^{*2} * \Theta\right)\right\|_{TV} \le C.$$

Therefore,

$$\left\|(F-I)^{*(k+1)} * \exp\left(A_k(F)+\tau B_k(F)\right)\right\|_K \le C\left\|(F-I)^{*(k+1)} * \exp\left(\lambda(F-I)/8\right)\right\|_K. \tag{10.37}$$

It follows from (10.35), (10.36), and (10.37) that

$$\|\mathcal{L}(S_n) - H_{n,k}\|_K \le C(k)\|\lambda_{k+1}(F-I)^{k+1} * \exp(\lambda(F-I)/8)\|_K.$$

An application of (C.24) completes the proof. □

Proof of Theorem 10.7. Estimates (10.14) and (10.16) follow from Theorem 10.8 with $s=0$. Estimate (10.15) follows from the triangle inequality and Lemma C.10, which states that

$$\|(qU+pV)^{*n}*(U-I)^{*k}*(V-I)^{*m}\|_K \le C_{k,m}(nq)^{-k}\min\{1;(np)^{-m/2}\} \quad (k,m\in\mathbb{Z}_+). \tag{10.38}$$

Indeed, we aim to show that

$$\sup_{U\in\mathcal{F}_+}\sup_{V\in\mathcal{F}} d_K(\mathcal{L}(S_n);\mathcal{L}(\tilde{S}_n) * (I-n(\mathcal{L}(X)-I)^{*2}/2)) \le C\left(n^{-1}+p\right)^2.$$

According to Theorem 10.8,

$$\|\mathcal{L}(S_n) - \mathbf{\Pi}(n,X) - n\mathbf{\Pi}(n-1,X) * (\mathcal{L}(X)-\mathbf{\Pi}(1,X))\|_K \le C(n^{-1}+p)^2.$$

Recall that $B_0 = \mathbf{\Pi}(1,X)$. It remains to estimate

$$\begin{aligned} & B_0^{*n} + B_0^{*(n-1)} * (\mathcal{L}(X) - B_0) - B_0^{*n} * (I - n(\mathcal{L}(X) - I)^{*2}/2) \\ = \; & nB_0^{*(n-1)} * (I - B_0) * (\mathcal{L}(X) - B_0) + nB_0^{*n}(\mathcal{L}(X) - B_0 + (\mathcal{L}(X) - I)^{*2}/2) \\ = \; & CnB_0^{*(n-1)} * (\mathcal{L}(X) - I)^{*3} * \Theta + nB_0^{*n} * (\mathcal{L}(X) - I)^{*3} * \Theta. \end{aligned}$$

Hence

$$\|B_0^{*n} + nB_0^{*(n-1)} * (\mathcal{L}(X) - B_0) - B_0^{*n} * (I - n(\mathcal{L}(X) - I)^{*2}/2)\|_{K} \le n^{-1/2}$$

by (10.38). Estimate (10.15) follows from (10.38).

In order to prove of (10.18), it suffices to apply Theorem 10.8 with $s = 2$ and Lemma C.10. □

Proof of Theorem 10.8. Recall that

$$B_0 = \mathbf{\Pi}(1,X) = \exp(\mathcal{L}(X) - I), \quad B_1 = B_0 * \exp\left(-p^2(V - I)^{*2}/2\right),$$

$$\Delta_{00} = \mathbf{\Pi}(n,X) = \mathcal{L}(\tilde{S}_n), \quad \Delta_{01} = n\mathbf{\Pi}(n-1,X) * (\mathcal{L}(X) - \mathbf{\Pi}(1,X)).$$

We will show that

$$\sup_{U\in\mathcal{F}_+}\sup_{V\in\mathcal{F}} d_K\Big(\mathcal{L}(S_n); \sum_{j=0}^{s} \Delta_{0j}\Big) \le C(s)(n^{-1} + p)^{s+1} \qquad (s \in \mathbb{Z}_+),$$

where

$$\sup_{U\in\mathcal{F}_+}\sup_{V\in\mathcal{F}} \|\Delta_{0j}\|_K \le C(j)(n^{-1} + p)^j \quad (j = 1, 2, \ldots, s), \tag{10.39}$$

and that

$$\sup_{U\in\mathcal{F}_+}\sup_{V\in\mathcal{F}} d_K\left(\mathcal{L}(S_n); \sum_{j=0}^{s} \Delta_{1j}\right) \le C(s)(n^{-1} + \sqrt{p/n})^{s+1},$$

where $\sup_{U\in\mathcal{F}_+}\sup_{V\in\mathcal{F}} \|\Delta_{1j}\|_K \le C(j)(n^{-1} + \sqrt{p/n})^j \quad (j = 1, 2, \ldots, s)$.

Observe that

$$\begin{aligned} B_0 - \mathcal{L}(X) &= \sum_{j=0}^{\infty} (\mathcal{L}(X) - I)^{*j}/j! - \mathcal{L}(X) \\ &= (\mathcal{L}(X) - I)^{*2} * \sum_{j=2}^{\infty} (\mathcal{L}(X) - I)^{*(j-2)}/j! \\ &= C(\mathcal{L}(X) - I)^{*2} * \Theta, \end{aligned}$$

since

$$\left\|\sum_{j=2}^{\infty} (\mathcal{L}(X) - I)^{*(j-2)}/j!\right\|_{TV} \le \sum_{j=2}^{\infty} \|\mathcal{L}(X) - I\|_{TV}^{j-2}/j! \le \sum_{j=2}^{\infty} 2^{j-2}/j! < 1.$$

We have $\|\mathcal{L}(X) - I\|_{TV} \le \|\mathcal{L}(X)\|_{TV} + \|I\|_{TV} = 2$. Thus,

$$\begin{aligned} \|\Delta_{0j}\|_K &\le C(j)\binom{n}{j} \exp(j\|\mathcal{L}(X) - I\|_{TV})\|B_0^{*n} * (\mathcal{L}(X) - I)^{*2j}\|_K \\ &\le C(j)\exp(4j)n^j\|B_0^{*n} * (\mathcal{L}(X) - I)^{*2j}\|_K. \end{aligned}$$

Estimate (10.39) now follows from Lemma C.10.

Recall that

$$\Delta_{1j} = \binom{n}{j} B_1^{*(n-j)} * (\mathcal{L}(X) - B_1)^{*j}\,.$$

To prove that

$$\|\Delta_{1j}\|_K \le C(j)(n^{-1} + \sqrt{p/n}\,)^j \qquad (j = 1, 2, \ldots, s), \tag{10.40}$$

we apply Lemma B.2, which states that

$$\exp(np(P-I) - np^2(P-I)^{*2}/2) = \exp(np(1-\tilde{c})(P-I)/2) * \Theta$$

for any $P \in \mathcal{F}$ if $0 \le p \le \tilde{c} < 1$, $n \in \mathbb{N}$. We get

$$\|\Delta_{1j}\|_K \le C\big\|\exp(nq(U-I) + np(1-\tilde{c})(V-I)/2) * (\mathcal{L}(X) - B_1)^{*j}\big\|_K. \tag{10.41}$$

In order to evaluate $d_K\Big(\mathcal{L}(S_n); \sum_{j=0}^{s} \Delta_{1j}\Big)$, we note that

$$\begin{aligned}
B_1 &= \exp(q(U-I)) * \exp\big(p(V-I) - p^2(V-I)^{*2}/2\big) \\
&= \exp(q(U-I)) * \big(I + p(V-I) + Cp^3(V-I)^{*3} * \Theta\big) \\
&= \big(I + q(U-I) + Cq^2(U-I)^{*2}\big) * \big(\big(I + p(V-I)\big) \\
&+ \exp(q(U-I)) * Cp^3(V-I)^{*3} * \Theta \\
&= I + q(U-I) + p(V-I) + q(U-I)p(V-I) \\
&+ Cq^2(U-I)^{*2} * \Theta + Cp^3(V-I)^{*3} * \Theta.
\end{aligned}$$

Therefore,

$$B_1 - \mathcal{L}(X) = q(U-I)p(V-I) + Cq^2(U-I)^{*2} * \Theta + Cp^3(V-I)^{*3} * \Theta. \tag{10.42}$$

Estimate (10.40) now follows from (10.41) and Lemma C.10.

Next we prove that

$$\sup_{U \in \mathcal{F}_+} \sup_{V \in \mathcal{F}} d_K\Big(\mathcal{L}(S_n); \sum_{j=0}^{s} \Delta_{1j}\Big) \le C(s)(n^{-1} + \sqrt{p/n}\,)^{s+1} \qquad (s \in \mathbb{Z}_+). \tag{10.43}$$

We apply Bergström's identity (B.9) and Lemma B.2, which states that

$$\exp(np(P-I) - np^2(P-I)^{*2}/2) = \exp(np(1-\tilde{c})(P-I)/2) * \Theta$$

for any $P \in \mathcal{F}$ if $0 \le p \le \tilde{c} < 1$, $n \in \mathbb{N}$. We get

$$\Big\|\mathcal{L}(X)^{*n} - \sum_{j=0}^{s} \Delta_{1j}\Big\|_K \le \sum_{j=s+1}^{n} f_j\,,$$

where $f_j = \binom{j-1}{s} \|\mathcal{L}(X)^{*(n-j)} * B_1^{*(j-s-1)} * (\mathcal{L}(X) - B_1)^{*(s+1)}\|_K$.

We consider separately $\sum_{s<j\le\lfloor n/2\rfloor} f_j$ and $\sum_{\lfloor n/2\rfloor+1\le j\le n} f_j$. Taking into account Lemma B.2, it is easy to check that $(j \ge s+1)$

$$\|B_1^{*(s+1)}\|_{TV} \le C_s\,, \quad \|B_1^{*(j-s-1)}\|_{TV} \le C_s\Big\|\exp\Big(jp(1-\tilde{C})(V-I)/2\Big)\Big\|_{TV} \le C_s\,.$$

Therefore, taking into account (B.5),

$$\begin{aligned}
\sum_{j\le\lfloor n/2\rfloor} f_j &\le C_s \sum_{j\le\lfloor n/2\rfloor} \binom{j-1}{s} \|\mathcal{L}(X)^{*\lfloor n/2\rfloor} * (\mathcal{L}(X) - B_1)^{*(s+1)}\|_K \\
&\le C(s) n^{s+1} \|\mathcal{L}(X)^{*\lfloor n/2\rfloor} * (\mathcal{L}(X) - B_1)^{*(s+1)}\|_K.
\end{aligned}$$

It follows from (10.42) that

$$\begin{aligned}&\|\mathcal{L}(X)^{*\lfloor n/2\rfloor} * (L(X) - B_1)^{*(s+1)}\|_K \\ \le\ & C(s)\sum_{l=0}^{s+1}\sum_{j=0}^{s+1-l} (1-p)^{l+2j}\|\mathcal{L}(X)^{*\lfloor n/2\rfloor} * (U-I)^{*(l+2j)} * \big(p(V-I)\big)^{*\lfloor 3(s+1)-3j-2l\rfloor}\|_K.\end{aligned}$$

Applying Lemma C.10, we prove that

$$\sum_{j\le\lfloor n/2\rfloor} f_j \le C(s)(n^{-1}+\sqrt{p/n}\,)^{s+1}.$$

By Lemma B.2, $\sum_{j\ge\lfloor n/2\rfloor+1} f_j$ is bounded by

$$C_s n^{s+1}\Big\|\exp\Big(\lfloor n/2\rfloor(1-p)(U-I) + \lfloor n/2\rfloor(1-\tilde{C})(V-I)/2\Big) * (\mathcal{L}(X)-B_1)^{*(s+1)}\Big\|_K.$$

Arguing as above, we prove that

$$\sum_{j\ge\lfloor n/2\rfloor+1} f_j \le C(s)(n^{-1}+\sqrt{p/n}\,)^{s+1},$$

which completes the proof of (10.43).

The proof of

$$\sup_{U\in\mathcal{F}_+}\sup_{V\in\mathcal{F}} d_K\Big(\mathcal{L}(S_n);\sum_{j=0}^{s}\Delta_{1j}\Big) \le C(s)(n^{-1}+p)^{s+1} \qquad (s\in\mathbb{Z}_+)$$

is very similar and, therefore, omitted. □

Exercises

10.1. Let $X, X_1, X_2, \dots$ be symmetric i.i.d.r.v.s. Let ν, $\pi_{n/5}$, $X_1, X_2, \dots$ be independent r.v.s, $\mathcal{L}(\nu) = \mathbf{NB}(n, 1/6)$, $\mathcal{L}(\pi_{n/5}) = \mathbf{\Pi}(n/5)$. Prove that

$$d_K\left(\sum_{i=1}^{\nu} X_i; \sum_{i=1}^{\pi_{n/5}} X_i\right) \le Cn^{-1}.$$

10.2. Let $S_n = \sum_{i=1}^n X_i$, where $X, X_1, \dots, X_n$ are i.i.d.r.v.s with non-negative characteristic functions. We assume that $X = \sum_{j=1}^{\nu}\sum_{m=1}^{\tilde{\nu}} X'_j X''_m$, where

$$\begin{aligned}\mathbb{P}(\nu=1,\tilde{\nu}=0) &= \mathbb{P}(\nu=0,\tilde{\nu}=1) = \mathbb{P}(\nu=1,\tilde{\nu}=1) = \mathbb{P}(\nu=2,\tilde{\nu}=0)\\ &= \mathbb{P}(\nu=0,\tilde{\nu}=2) = 1/5,\end{aligned}$$

$X', X'_1, X'_2, \dots$ are i.i.d.r.v.s with $\mathcal{L}(X') = F \in \mathcal{F}_+$, $X'', X''_1, X''_2, \dots$ are i.i.d.r.v.s with $\mathcal{L}(X'') = G \in \mathcal{F}_+$, the sequence $\{X'_j\}$ is independent of $\{X''_m\}$. Prove that

$$d_K(\mathcal{L}(S_n); \mathbf{\Pi}(4n/5; X') * \mathbf{\Pi}(4n/5; X'')) \le C/n.$$

10.3. Let $S_n = \sum_{i=1}^n \tau_i X_i$, where $\mathcal{L}(\tau_i) = \mathbf{B}(p_i)$, $X, X_1, X_2, \dots, X_n$ are i.i.d.r.v.s such that $\mathbb{E}X = 0$, $\mathbb{E}X^2 < \infty$, satisfying Cramer's condition (11.5). Denote $F = \mathcal{L}(X)$. Prove that

$$d_K(\mathcal{L}(S_n); \hat{H}_{n,k}) \leq C(F,k)\lambda_{k+1}\lambda^{-k-1}$$

if $\max_i p_i \leq 1/4$, where $\hat{H}_{n,k}$, λ, λ_2 are the same as in Theorem 10.6.

Notes

Theorem 10.1 is from [323].
Theorems 10.2–10.4 are from [65].
Theorem 10.5 is a partial case of Theorems 3 and 4 from [63].
Theorem 10.6 is a generalisation of Theorem 2.10 from [89].
Theorems 10.7 and 10.8 are from [61, 70].
Theorem 10.8 is from [70].

11

Compound Poisson approximation for sequences of random variables

Let $X, X_1, X_2, \ldots$ be a sequence of i.i.d.r.v.s, and let $S_n = X_1 + X_2 + \cdots + X_n$. The situation where $\mathcal{L}(X)$ does not depend on n is often referred to as a "scheme of sequences".

In the previous chapters distributions were often allowed to depend on n (i.e., $\{X_1, \ldots, X_n\}_{n \geq 1}$ was usually a triangle array of r.v.s). In this chapter, we present results concerning compound Poisson approximation to $\mathcal{L}(S_n)$ in a scheme of sequences. Typically the estimates contain constants depending on the distribution: C_X, $C(F)$, etc.

11.1 General remarks

Denote by Y_{μ,σ^2} a normal r.v. with the mean μ and variance σ^2. If $\mathbb{E}|X|^3 < \infty$, then

$$d_K(S_n; Y_{\mu,\sigma^2}) \leq C_X n^{-1/2} \qquad (n \geq 1) \tag{11.1}$$

by (H.1), where $\mu = \mathbb{E}S_n$, $\sigma^2 = \text{var}\, S_n$.

A similar bound is applicable to any compound Poisson distribution with a finite third absolute moment that matches the first two moments of $\mathcal{L}(S_n)$. Therefore, the accuracy of compound Poisson approximation to $\mathcal{L}(S_n)$ is at least $C_X n^{-1/2}$. Thus, of interest are only results that in a sense improve (11.1).

Zaitsev [331] has conjectured that for any distribution $\mathcal{L}(X)$ there exists constant C_X such that

$$\begin{aligned} &d_K(\mathcal{L}(S_n); \mathcal{D}) \leq C_X n^{-1}, \\ &\inf_{a \in \mathbb{R}} d_K(\mathcal{L}(S_n + na); \mathbf{\Pi}(n, X + a)) \leq C_X n^{-1/2}. \end{aligned} \tag{11.2}$$

One can expect that in the case of integer-valued r.v.s the rate of the accuracy of approximation in terms of d_o is $n^{-3/2}$:

$$d_o(\mathcal{L}(S_n); \mathcal{D}) \leq C_X n^{-3/2}. \tag{11.3}$$

In this chapter, we present results related to these conjectures.

A feature of CP approximation (comparing to normal one) is that bounds to the accuracy of CP approximation can be derived even if $\mathbb{E}X^2 = \infty$.

As usual, $\tilde{S}_n$ denotes the sum of accompanying r.v.s, see (1.9).

DOI: 10.1201/9781003478164-11

Theorem 11.1 *Let $X, X_1, \dots$ be i.i.d.r.v.s, $\mathbb{E}X = 0$, $\exists \beta \in (1;2]$ such that $\mathbb{E}|X|^\beta < \infty$. Then there exists a constant C_X such that*

$$d_K(S_n; \tilde{S}_n) \le C_X n^{-\alpha}, \tag{11.4}$$

where $\alpha = \min\{1/2; \beta-1\}$. If $\beta \in (3/2; 2)$ and $\mathbb{E}X^2 = \infty$, then

$$d_K(S_n; \tilde{S}_n) = o(n^{-1/2}).$$

In the next two sections, we show that (11.4) can be improved if additional assumptions are made.

11.2 Distributions obeying Cramér's condition

Estimate (11.4) can be improved if Cramér's condition

$$\limsup_{|t|\to\infty} |\widehat{F}(t)| < 1 \tag{11.5}$$

holds. Note that Cramér's condition excludes lattice distributions.

Studnev [289] has shown that

$$d_K(S_n; \tilde{S}_n) = O(n^{-1}) \tag{11.6}$$

if (11.5) holds. Studnev [289] presents also a first-order asymptotic expansion in (11.6) assuming $\mathbb{E}X^4 < \infty$.

Theorem 11.2 *If assumptions of Theorem 11.1 and (11.5) are satisfied, then there exists a constant C_X such that*

$$d_K(S_n; \tilde{S}_n) \le C_X n^{1-\beta}, \quad d_K(\mathcal{L}(S_n); G_\beta) \le C_X n^{2-2\beta},$$

where $G_\beta = \mathcal{L}(\tilde{S}_n) * (I - n(\mathcal{L}(X) - I)^{*2}/2)$.

Measure G_β is a first-order asymptotic expansion of $\mathcal{L}(\tilde{S}_n)$.

If $\mathbb{E}X^2 < \infty$, then Theorem 11.2 entails Studnev's result. Further results on the topic can be found in [77, 137].

If distributions $\{\mathcal{L}(X_i)\}$ are absolutely continuous, then Cramér's condition is satisfied. The next theorem demonstrates that compound Poisson approximation may have further advantages over normal approximation: the following bound holds in terms of the stronger total variation distance, and the rate of the accuracy of approximation is n^{-1}.

Theorem 11.3 *Let $X, X_1, X_2, \dots, X_n$ be i.i.d. absolutely continuous r.v.s with a bounded density. Assume that $\mathbb{E}X = 0$, $\mathbb{E}X^2 < \infty$. Let $\tilde{S}_n$ denote the sum of accompanying random variables. Then for any $n \in \mathbb{N}$*

$$d_{TV}(S_n; \tilde{S}_n) \le C_X n^{-1}. \tag{11.7}$$

11.3 Discrete random variables

Let $X, X_1, X_2, \ldots$ be a sequence of i.i.d. lattice r.v.s with maximum span 1. Assume that $\mathbb{E}|X|^3 < \infty$. W.l.o.g. we may consider r.v.s $\{X_i\}$ integer-valued. Set

$$\mu = \mathbb{E}X, \quad u = \mathrm{var}\, X/\mathbb{E}X^2 .$$

Denote by $Y_{n,u}$ a compound Poisson $\mathbf{\Pi}(nu, X)$ r.v. If $X_i \stackrel{d}{=} X$ $(\forall i)$, then [77] there exists constant C_X (which depends on $\mathcal{L}(X)$) such that

$$d_{TV}(S_n; \lfloor n\mu^3/\mathbb{E}X^2 \rfloor + Y_{n,u}) \leq C_X n^{-1/2} . \tag{11.8}$$

If $\mathbb{E}X^4 < \infty$, then, according to Theorem 7.24, the rate $C_X n^{-1}$ of the accuracy of approximation can be achieved in terms of the total variation distance.

It is natural to consider also estimates in terms of the point distance. We assume that $\mathcal{L}(X)$ is non-degenerate, i.e., there exist numbers $a<b$ such that $\mathbb{P}(X=a)\mathbb{P}(X=b)>0$.

Let $\beta \in (0;1]$ be such a number that $n\mathbb{E}X+\beta \in \mathbb{Z}$, and let G_X be the shifted compound Poisson distribution with the Fourier transform

$$\begin{aligned} \widehat{G}_X(t) &= \exp\Big\{ \mathrm{i}t\frac{\gamma}{n} - \frac{\beta}{n(b-a)}(\mathrm{e}^{\mathrm{i}t(b-a)}-1) + \frac{\beta}{6n(b-a)}(\mathrm{e}^{2\mathrm{i}t(b-a)}-1) \\ &+ \frac{\beta}{3n(b-a)}(\mathrm{e}^{-\mathrm{i}t(b-a)}-1) + \frac{1}{6}\Big(\widehat{F}_X(2t)\widehat{F}_X^2(-t)-1\Big)\Big\}, \end{aligned}$$

where $F_X = \mathcal{L}(X)$, $\gamma = n\mathbb{E}X+\beta$.

Observe that $\widehat{F}_X(2t)\widehat{F}_X^2(-t)$ is the characteristic function of $2X_1-X_2-X_3$. Applying Lemma D.2, it is not difficult to check that the first 3 moments of F_X and G_X match.

Theorem 11.4 *For any* $n \geq 6/F_X(\{a\})F_X^2(\{b\})$

$$d_o(F_X^{*n}; G_X^{*n}) \leq C_X n^{-3/2} . \tag{11.9}$$

Estimate (11.9) supplements conjecture (11.3).

Next we consider discrete non-lattice r.v.s. The following theorem is an example of Arak's technique applied to a discrete non-lattice distribution.

Let $X, X_1, \ldots$ be i.i.d.r.v.s taking values $0, x_1, \ldots, x_N$. Denote by D_n the compound Poisson distribution with the Fourier transform

$$\widehat{D}_n(t) = \exp\Bigg(\sum_{j=1}^{N} \Big(\mathrm{i}tx_j\lambda_j + \varepsilon_j(\mathrm{e}^{\mathrm{i}tx_j}-1)\Big) + n\big(\widehat{F}^2(-t)\widehat{F}(2t)-1\big)/6 \Bigg), \tag{11.10}$$

where $\lambda_j = [nF(\{x_j\})]$, $\varepsilon_j = nF(\{x_j\})-\lambda_j$, $F = \mathcal{L}(X)$.

Theorem 11.5 *For any* $n \in \mathbb{N}$

$$d_K(\mathcal{L}(S_n); D_n) \leq C_X n^{-1} .$$

Open problems

11.1. Prove or disprove (11.2).

11.2. Evaluate constant C_X in (11.7).

11.4 Proofs

Proof of Theorem 11.1. Since $\mathbb{E}X=0$, we have

$$\widehat{F}-1=\int_{-\infty}^{\infty}(\mathrm{e}^{\mathrm{i}tx}-1-\mathrm{i}tx)F(\mathrm{d}x).$$

By (I.10),

$$|\mathrm{e}^{\mathrm{i}t}-1|\le|t|,\quad |\mathrm{e}^{\mathrm{i}t}-1-\mathrm{i}t|\le\min\{2|t|;t^2/2\}.$$

Therefore,

$$\begin{aligned}\left|\int_{|x|\le y}(\mathrm{e}^{\mathrm{i}tx}-1-\mathrm{i}tx)F(\mathrm{d}x)\right| &\le \int_{|x|\le y}t^2x^2F(\mathrm{d}x)\le t^2\mathbb{E}|X|^\beta y^{2-\beta},\\ \left|\int_{|x|>y}(\mathrm{e}^{\mathrm{i}tx}-1-\mathrm{i}tx)F(\mathrm{d}x)\right| &\le 2\int_{|x|>y}|tx|F(\mathrm{d}x)\le 2|t|\mathbb{E}|X|^\beta y^{1-\beta}\end{aligned}$$

for any positive y. Taking $y=|t|^{-1}$, we get

$$|\widehat{F}(t)-1|\le 3|t|^\beta\mathbb{E}|X|^\beta\qquad(t\in\mathbb{R}).\tag{11.11}$$

According to Lemma D.3, there exist $T>0$ such that

$$|\widehat{F}|\le 1-C(F)t^2\le\mathrm{e}^{-C(F)t^2}\qquad(|t|\le T).$$

It follows from (E.3) that $Q_{F^{*n}}(T^{-1})\le C(F)n^{-1/2}$; if $\mathbb{E}X^2=\infty$, then $Q_{F^{*n}}(T^{-1})=o(n^{-1/2})$ by (E.4). Now the statement of the theorem follows from Lemma D.21. □

Proof of Theorem 11.2. Let $v=\beta$–1. Without loss of generality we may assume that $n\ge4$. Note that

$$|\widehat{F}-1|^2=(Re\widehat{F}-1)^2+(Im\widehat{F})^2=(Re\widehat{F})^2-2Re\widehat{F}+1+(Im\widehat{F})^2=|\widehat{F}|^2-1-2(Re\widehat{F}-1).$$

By Lemma D.4, for any $\varepsilon>0$ there exists constant $C(F)>0$ such that

$$|\widehat{F}(t)|\le\mathrm{e}^{-C(F)},\quad |\exp(\widehat{F}(t)-1)|\le\exp(|\widehat{F}(t)|-1)\le\mathrm{e}^{-C(F)}\qquad(|t|\ge\varepsilon),\tag{11.12}$$

and

$$\begin{aligned}|\exp(\widehat{F}-1-(\widehat{F}-1)^2/2)| &\le \exp((Re\widehat{F}-1)+|\widehat{F}-1|^2/2)\\ &= \exp((|\widehat{F}|^2-1)/2)\le\mathrm{e}^{-C(F)}\qquad(|t|\ge\varepsilon).\end{aligned}$$

Recall that F is non-degenerate. It follows from Lemma D.3 that

$$\max\left\{|\widehat{F}|;|\exp(\widehat{F}-1)|;\exp((|\widehat{F}|^2-1)/2)\right\}\le\exp\bigl(-C(F)t^2\bigr)$$

if $|t|\le\exists\varepsilon=\varepsilon(F)$.

Set

$$T=n^4,\quad U_1=\exp(F-I),\quad U_2=U_1^{*n}*(I-n(F-I)^{*2}/2).$$

Using (D.21) and (D.22) with $h=1/T$, for any $\varepsilon\in(0;T)$ we derive

$$\|F^{*n}-U_1^{*n}\|_K \le C(F)\int_0^{\varepsilon}\frac{|\widehat{F}^n-\widehat{U}_1^n|}{t}\mathrm{d}t + C(F)n^{-2}+C(F)Q_{U_1^{*n}}(1/T),$$

$$\|F^{*n}-U_2\|_K \le C(F)\int_0^{\varepsilon}\frac{|\widehat{F}^n-\widehat{U}_2|}{t}\mathrm{d}t + C(F)n^{-2}+C(F)\|I-n(F-I)^{*2}\|_K\, Q_{U_1^{*n}}(1/T).$$

Note that by (11.12)

$$\int_{\varepsilon}^{T}\frac{|\widehat{F}^n-\widehat{U}_1^n|}{t}\mathrm{d}t \le \frac{2|n^4-\varepsilon|}{\varepsilon}\,\mathrm{e}^{-C(F)n} \le C(F)n^{-2}, \qquad \int_{\varepsilon}^{T}\frac{|\widehat{F}^n-\widehat{U}_2^n|}{t}\mathrm{d}t \le C(F)n^{-2}.$$

By (E.7),

$$\begin{aligned} Q_{U^{*n}}(1/T) &\le \frac{C(F)}{T}\int_{-T}^{T}|\widehat{U}_1^n|\mathrm{d}t \le \frac{C(F)}{T}\Big(\int_0^{\varepsilon}|\widehat{U}_1^n|\mathrm{d}t+\int_{\varepsilon}^{T}|\widehat{U}_1^n|\mathrm{d}t\Big)\\ &\le \frac{C(F)}{T}\int_0^{\infty}\mathrm{e}^{-C(F)t^2n}\mathrm{d}t + C(F)T\mathrm{e}^{-nC(F)} \le C(F)n^{-4}. \end{aligned}$$

Taking into account (11.11), we get

$$|\widehat{F}^n-\widehat{U}_1^n| \le n\mathrm{e}^{-(n-1)C(F)t^2}|\widehat{F}-\widehat{U}_1| \le C(F)n\mathrm{e}^{-C(F)nt^2}|\widehat{F}-1|^2 \le C(F)\mathrm{e}^{-C(F)nt^2}n|t|^{2+2v}$$

if $|t|\le\varepsilon$. Applying Bergström's identity (B.9), we prove that

$$\begin{aligned} |\widehat{F}^n-\widehat{U}_2| &\le |\widehat{F}^n-\widehat{U}_1^n-n\widehat{U}_1^{n-1}(\widehat{F}-\widehat{U}_1)| + n|(\widehat{U}_1^{n-1}-\widehat{U}_1^n)(\widehat{F}-\widehat{U}_1)|\\ &+ n|\widehat{U}_1^n(\widehat{F}-\widehat{U}_1)+(\widehat{F}-1)^2/2)| \le C(F)\exp(-C(F)nt^2)\left(n^2|\widehat{F}-1|^4+n|\widehat{F}-1|^3\right)\\ &\le C(F)\exp(-C(F)nt^2)\,|t|^{4v}. \end{aligned}$$

Collecting these estimates, we get the assertions of Theorem 11.2. □

Proof of Theorem 11.3. Since $\mathcal{L}(\tilde{S}_n)=\exp(n(F-I))$, we have

$$d_{TV}(S_n;\tilde{S}_n)=\tfrac{1}{2}\|F^{*n}-\exp(n(F-I))\,\|_{TV},$$

where $F=\mathcal{L}(X)$. Recall that F^{*n} has a bounded density, while $\exp(n(F-I))$ is not absolutely continuous, since it has an atom at zero: $\mathbb{P}(\tilde{S}_n=0)=\mathrm{e}^{-n}$.

W.l.o.g. we may assume that $n\ge 10$. Indeed, if $n<10$, then $d_{TV}(S_n;\tilde{S}_n)\le 1\le 10/n$, and (11.7) follows.

Using (B.3), $F^{*n}-\exp(n(F-I))$ can be presented as a sum of three parts:

$$F^{*n}-\exp(n(F-I)) = (F-\exp(F-I))*\sum_{k=1}^{n}F^{*(k-1)}*\exp((n-k)(F-I)) = J_1+J_2+J_3,$$

where

$$\begin{aligned} J_1 &= (F-\exp(F-I))*(F+\exp(F-I))*\exp((n-2)(F-I)),\\ J_2 &= (F-\exp(F-I))*\sum_{k=3}^{[n/2]+1} F^{*(k-1)}*\exp((n-k)(F-I)),\\ J_3 &= (F-\exp(F-I))*\sum_{k=[n/2]+1}^{n-1} F^{*k}*\exp((n-k-1)(F-I)). \end{aligned}$$

According to (1.6*),

$$\exp((F-I)) - F = \tfrac{1}{2}(F-I)^{*2}*\Theta,$$

where $\|\Theta\|_{TV}\le 1$. It follows from (A.1) and (C.9) that

$$\begin{aligned} \|J_1\|_{TV} &\le \tfrac{1}{2}\|(F-I)^{*2}*(F+\exp(F-I))*\exp((n-2)(F-I))\,\|_{TV} \\ &\le \|(F-I)^{*2}*\exp((n-2)(F-I))\,\|_{TV}\le 3/\mathrm{e}(n-2). \end{aligned} \tag{11.13}$$

However, this argument does not provide sharp estimates of J_2, J_3:

$$\begin{aligned} \|J_2\|_{TV} &\le \max_{k\le[n/2]} \tfrac{n}{4}\|(F-I)^{*2}*F^{*2}*\exp((n-k)(F-I))\,\|_{TV}\le 3/2\mathrm{e},\\ \|J_3\|_{TV} &\le \tfrac{n}{4}\|(F-I)^{*2}*F^{\lceil n/2\rceil}\|_{TV}\le 3/2\mathrm{e}. \end{aligned}$$

We proceed using the method of characteristic functions.

We denote by $\widehat{F}_X$ the characteristic function of X. According to Lemmas D.3 and D.4, there exist constants $\varepsilon>0$ and $C_X>0$ such that

$$\begin{aligned} |\widehat{F}_X(t)| &\le \exp(-C_Xt^2), \quad \left|\exp\left(\widehat{F}_X(t)-1\right)\right|\le\exp(-C_Xt^2) && (|t|\le\varepsilon),\\ |\widehat{F}_X(t)| &\le \exp(-C_X), \quad |\widehat{F}'_X(t)|\le\mathbb{E}|X| && (|t|>\varepsilon). \end{aligned} \tag{11.14}$$

Note that $|\exp(\widehat{F}_X(t)-1)|\le 1$. An analogue of (B.2*) is

$$|\widehat{F}_X(t)-\exp(\widehat{F}_X(t)-1)|\le|\widehat{F}_X(t)-1|^2/2.$$

By the properties (D.3), (D.4) of characteristic functions,

$$|\widehat{F}_X(t)-1|\le C_Xt^2,\quad |\widehat{F}'_X(t)|\le C_X|t| \qquad (\forall t). \tag{11.15}$$

Therefore, as $|t|\le\varepsilon$,

$$|\widehat{J}_2(t)| \le C_Xnt^4\exp(-nC_Xt^2)\le C_X\exp(-nC_Xt^2/2)\,n^{-1}, \tag{11.16}$$

$$|\widehat{J}'_2(t)| \le nC_X\exp(-nC_Xt^2)\,(|t|^3+n|t|^5)\le C_Xn^{-1/2}\exp(-nC_Xt^2/2), \tag{11.17}$$

If $|t|>\varepsilon$, then by (11.14),

$$|\widehat{J}_2(t)| \le C_X|\widehat{F}_X(t)|^2\exp(-nC_X), \tag{11.18}$$

$$|\widehat{J}'_2(t)| \le nC_X|\widehat{F}_X(t)||\widehat{F}_X(t)'|\exp(-nC_X)\le C_X\exp(-nC_X). \tag{11.19}$$

Applying Lemma D.10 with $b=\sqrt{n}$, we derive from (11.16) – (11.19) that $\|J_2\|_{TV}\le C_Xn^{-1}$.

The same argument yields $\|J_3\|_{TV} \le C_X n^{-1}$. Combining these estimates with (11.13), we complete the proof of Theorem 11.3. □

Proof of Theorem 11.4. By ϑ we denote complex numbers satisfying $|\vartheta|\le 1$.

Recall that

$$\widehat{G}_X(t) = \exp\Big\{ \mathrm{i}t\frac{\gamma}{n} - \frac{\beta}{n(b-a)}(\mathrm{e}^{\mathrm{i}t(b-a)}-1) + \frac{\beta}{6n(b-a)}(\mathrm{e}^{2\mathrm{i}t(b-a)}-1)$$
$$+ \frac{\beta}{3n(b-a)}(\mathrm{e}^{-\mathrm{i}t(b-a)}-1) + \frac{1}{6}\Big(\widehat{F}_X(2t)\widehat{F}^2_X(-t)-1\Big)\Big\}.$$

Expanding the characteristic function in powers of $\mathrm{i}t$, we get

$$\widehat{F}_X(2t)\widehat{F}^2_X(-t)-1 = 3(\mathrm{i}t)^2\mathrm{var}\,X + (\mathrm{i}t)^3\mathbb{E}(X-\mathbb{E}X)^3 + 11\vartheta C_X t^4.$$

Applying (I.8) to $\mathrm{e}^{\mathrm{i}t(b-a)}$, we arrive at

$$\widehat{G}_X(t) = \exp\bigg(\mathrm{i}t\mathbb{E}X + \frac{(\mathrm{i}t)^2\mathrm{var}\,X}{2} + \frac{(\mathrm{i}t)^3\mathbb{E}(X-\mathbb{E}X)^3}{6} + \vartheta C_X t^4\bigg).$$

According to Lemma D.5,

$$\max\{|\widehat{F}^n_X(t)|; |\widehat{G}^n_X(t)|\} \le \exp(-nC_X t^2) \qquad (|t|\le\pi).$$

It follows from (D.3) that

$$\widehat{F}_X(t)\mathrm{e}^{-\mathrm{i}t\mathbb{E}X} = 1 + \frac{(\mathrm{i}t)^2\mathrm{var}\,X}{2} + \frac{(\mathrm{i}t)^3\mathbb{E}(X-\mathbb{E}X)^3}{6} + \vartheta C_X t^4.$$

Therefore,

$$|\widehat{F}_X(t) - \widehat{G}_X(t)| = |\widehat{F}_X(t)\mathrm{e}^{-\mathrm{i}t\mu} - \widehat{G}(t)\mathrm{e}^{-\mathrm{i}t\mu}| \le C_X t^4$$

and

$$|\widehat{F}^n_X(t)-\widehat{G}^n_X(t)| \le C_X n\mathrm{e}^{-C_X(n-1)t^2}|\widehat{F}_X(t)-\widehat{G}(t)| \le C_X nt^4\mathrm{e}^{-C_X(n-1)t^2}$$

if $|t|\le\pi$. Applying (D.12), we get

$$d_o(F^{*n}; G^{*n}_X) \le C\int_{|t|\le\pi} nt^4\mathrm{e}^{-C_X nt^2}\,\mathrm{d}t \le C_X n^{-3/2}.$$

The proof is complete. □

Proof of Theorem 11.5. Denote $F = \mathcal{L}(X)$, $F\{x_j\} = F(\{x_j\})$, $p_0 = F(\{0\})$. Symbol $C(F)$ denotes constants depending on F. Symbol ϑ denotes complex numbers such that $|\vartheta|\le 1$.

Note that $\lambda_j \in \{0,1,\dots,N\}$, $0\le\varepsilon_j<1$. Without loss of generality we may assume that $0<x_1<x_2<\dots<x_N$.

Let r.v. Y have compound Poisson distribution D_n with ch.f. (11.10). We show below the first three moments of Y are very close to corresponding moments of S_n.

We apply Lemma D.16 with $m = 26n$, $\mathbf{u} = (0, x_1, x_2, \dots, x_N)$, $h = x_N/\pi$ and

$$\widehat{U}(t) = \exp\Big(\tfrac{n}{10}p_0\big(Re\,\widehat{F}(t)-1\big)\Big).$$

First, we estimate the difference of characteristic functions $|\widehat{F}^n(t)-\widehat{D}_n(t)|$ when t is close to zero. Note that

$$\widehat{F}(t) = p_0 + \sum_{j=1}^N F\{x_j\}\mathrm{e}^{\mathrm{i}tx_j} = p_0 + (1-p_0)\sum_{j=1}^N \frac{F\{x_j\}}{1-p_0}\mathrm{e}^{\mathrm{i}tx_j}.$$

By (D.6),

$$|\widehat{F}(t)| \le \exp\Big(p_0(Re\,\widehat{F}(t)-1)\Big) = \exp\left(p_0\sum_{j=1}^{N} F\{x_j\}(\cos(tx_j)-1)\right). \tag{11.20}$$

Note that $1-\cos(tx_j) = 2\sin^2(tx_j/2) \ge 2t^2x_j^2/\pi^2$ as $|tx_j| < \pi$, cf. (I.11). Hence

$$|\widehat{F}(t)| \le \exp\left(-2p_0\sum_{j=1}^{N} F\{x_j\}\sin^2(tx_j/2)\right) \le \exp\left(-\frac{2p_0t^2}{\pi^2}\sum_{j=1}^{N} F\{x_j\}x_j^2\right) \tag{11.21}$$

as $|t|\le 1/h$.

The absolute value of any characteristic function does not exceed 1. Therefore,

$$|Re\,\widehat{F}^2(-t)\widehat{F}(2t)| \le |\widehat{F}^2(-t)||\widehat{F}(2t)| \le |\widehat{F}^2(-t)| = |\widehat{F}(t)|^2.$$

It follows from the proof of (D.6) that

$$|\widehat{F}(t)|^2 - 1 \le 2p_0(Re\widehat{F}(t) - 1).$$

Therefore,

$$\begin{aligned} |\widehat{D}_n(t)| &\le \exp\left(2\sum_{k=1}^{N}\varepsilon_k\right)\exp\left(\frac{n}{6}\big(Re\,\widehat{F}^2(-t)\widehat{F}(2t)-1\big)\right) \le C(F)\exp\left(\frac{n}{6}\big(|\widehat{F}(t)|^2-1\big)\right) \\ &\le C(F)\exp\left(-\frac{2np_0}{3}\sum_{j=1}^{N} F\{x_j\}\sin^2(tx_j/2)\right) \le C(F)\exp\left(-\frac{2np_0t^2}{3\pi^2}\sum_{j=1}^{N} F\{x_j\}x_j^2\right) \end{aligned}$$

as $|t|\le 1/h$.

By Lemma D.1 and Lemma D.2,

$$\left|it\mathbb{E}X + \big(\widehat{F}^2(-t)\widehat{F}(2t)-1\big)/6 - \big(\widehat{F}(t)-1\big) + \big(\widehat{F}(t)-1\big)^2/2 - \big(\widehat{F}(t)-1\big)^3/3\right| \le C(F)t^4.$$

Consequently,

$$\begin{aligned} &\left|\exp\Big((\widehat{F}(t)-1) - (\widehat{F}(t)-1)^2/2 + (\widehat{F}(t)-1)^3/3\Big) - \widehat{D}^{1/n}(t)\right| \\ &\le C(F)\Big|(\widehat{F}(t)-1) - (\widehat{F}(t)-1)^2/2 + (\widehat{F}(t)-1)^3/3 \\ &- \frac{1}{n}\sum_{j=1}^{N}\Big(\lambda_j itx_j + \varepsilon_j(\exp(itx_j)-1)\Big) - \big(\widehat{F}^2(-t)\widehat{F}(2t)-1\big)/6\Big| \\ &\le C(F)\Big|(\widehat{F}(t)-1) - (\widehat{F}(t)-1)^2/2 + (\widehat{F}(t)-1)^3/3 \\ &- it\mathbb{E}X + \vartheta C(F)t^2/n - \big(\widehat{F}^2(-t)\widehat{F}(2t)-1\big)/6\Big| \\ &\le C(F)(t^2/n + t^4) \end{aligned}$$

and

$$
\begin{aligned}
|\widehat{F}^n(t)-\widehat{D}_n(t)| &\le C(F)n\exp\Big((n-1)p_0\big(Re\,\widehat{F}(t)-1\big)/3\Big)\,|\widehat{F}(t)-\widehat{D}^{1/n}(t)|\\
&\le C(F)ne^{-C(F)nt^2}\Big(\Big|\widehat{F}(t)-\exp\Big((\widehat{F}(t)-1)-(\widehat{F}(t)-1)^2/2+(\widehat{F}(t)-1)^3/3\Big)\Big|\\
&+\Big|\exp\Big((\widehat{F}(t)-1)-(\widehat{F}(t)-1)^2/2+(\widehat{F}(t)-1)^3/3\Big)-\widehat{D}^{1/n}(t)\Big|\Big)\\
&\le C(F)ne^{-C(F)nt^2}(|\widehat{F}(t)-1|^4+C(F)t^2/n+C(F)t^4)\\
&\le C(F)\mathrm{e}^{-C(F)nt^2}(t^2+nt^4) \qquad (11.22)
\end{aligned}
$$

if $|t|<1/h$. Thus,

$$
\int_{|t|\le 1/h}\frac{|\widehat{F}^n(t)-\widehat{D}_n(t)|}{|t|}\,\mathrm{d}t\le C(F)\int_{-\infty}^{\infty}(|t|+n|t|^3)\mathrm{e}^{-C(F)nt^2}\mathrm{d}t\le C(F)n^{-1}. \qquad (11.23)
$$

Next, we estimate the difference $|\widehat{F}^n(t)-\widehat{D}_n(t)|$ for all $t\in\mathbb{R}$.

Functions $\{\mathrm{e}^{\mathrm{i}tx_j}\}$ are $2\pi/x_j$-periodical. Hence we can replace tx_j by t_jx_j, where t_j is determined by the equation

$$
\mathrm{e}^{\mathrm{i}tx_j}=\mathrm{e}^{\mathrm{i}t_jx_j}\qquad (|t_jx_j|\le\pi).
$$

We have

$$
\begin{aligned}
\widehat{F}(t) &= \widehat{F}(\mathbf{t})=p_0+\sum_{j=1}^{N}F\{x_j\}\mathrm{e}^{\mathrm{i}t_jx_j},\\
\widehat{D}_n(t) &= \widehat{D}_n(\mathbf{t})=\exp\left(\sum_{j=1}^{N}\Big(\mathrm{i}t_jx_j\lambda_j+\varepsilon_j(\mathrm{e}^{\mathrm{i}t_jx_j}-1)\Big)+\frac{n}{6}(\widehat{F}^2(-\mathbf{t})\widehat{F}(2\mathbf{t})-1)\right),\\
\widehat{U}(t) &= \widehat{U}(\mathbf{t})=\exp\left(\frac{n}{10}p_0\big(Re\,\widehat{F}(\mathbf{t})-1\big)\right),
\end{aligned}
$$

where $\mathbf{t}=(t_1,t_2,\dots,t_N)$.

By introducing $\mathbf{t}$, we have replaced one-dimensional characteristic functions $\widehat{U}(t)$, $\widehat{F}(t)$, $\widehat{D}_n(t)$ by multivariate characteristic functions $\widehat{U}(\mathbf{t})$, $\widehat{F}(\mathbf{t})$ and $\widehat{D}_n(\mathbf{t})$, respectively. The main benefit of such replacement is that

$$
\sin^2(tx_j/2)=\sin^2(t_jx_j/2)\ge t_j^2x_j^2/\pi^2
$$

by (I.11) if $|t_jx_j|\le\pi$. Since $\sin^2(t_jx_j/2)\le t_j^2x_j^2/4$, similarly to (11.20) we obtain

$$
\begin{aligned}
|\widehat{F}(\mathbf{t})| &\le \exp\Big(p_0(Re\,\widehat{F}(\mathbf{t})-1)\Big)\le\exp\left(-\frac{2p_o}{\pi^2}\sum_{j=1}^{N}F\{x_j\}x_j^2t_j^2\right),\\
|\widehat{D}_n(\mathbf{t})| &\le C(F)\exp\left(-\frac{2np_0}{3\pi^2}\sum_{j=1}^{N}F\{x_j\}x_j^2t_j^2\right),\\
|U(\mathbf{t})| &= \exp\left(-\frac{2np_0}{10}\sum_{j=1}^{N}F\{x_j\}\sin^2(t_jx_j/2)\right)\ge\exp\left(-\frac{np_0}{20}\sum_{j=1}^{N}F\{x_j\}x_j^2t_j^2\right).
\end{aligned}
$$

Applying (I.8) and arguing as in (11.22), we get

$$|\widehat{F}^n(\mathbf{t}) - \widehat{D}_n(\mathbf{t})| \leq C(F)\Big(\sum_{j=1}^N t_j^2 + n\sum_{j=1}^N t_j^4\Big)\exp\Bigg(-\frac{2np_0}{3\pi^2}\sum_{j=1}^N F\{x_j\}x_j^2t_j^2\Bigg).$$

Let $c(F) = 0.01p_0 \min_j F\{x_j\}x_j^2$. Then

$$\begin{aligned}
|\widehat{F}^n(t)-\widehat{D}(t)|/\widehat{U}(t) &= |\widehat{F}^n(\mathbf{t})-\widehat{D}(\mathbf{t})|/\widehat{U}(\mathbf{t})\\
&\leq C(F)\exp\Bigg(-c(F)n\sum_{j=1}^N t_j^2\Bigg)\Bigg(\sum_{j=1}^N t_j^2 + n\sum_{j=1}^N t_j^4\Bigg)\\
&\leq C(F)\sum_{j=1}^N \Big(t_j^2\mathrm{e}^{-c(F)nt_j^2} + nt_j^4\mathrm{e}^{-c(F)nt_j^2}\Big) \leq C(F)n^{-1}. \quad (11.24)
\end{aligned}$$

Using properties (E.7), (E.8) of concentration function, we obtain

$$Q_U(h) \leq Ch\int\limits_{|t|<1/h} \exp(-C(F)nt^2)\,\mathrm{d}t \leq C(F)hn^{-1/2}. \tag{11.25}$$

Note that $\ln m \leq C\ln(n+1)$. Inequality $\ln(n+1) \leq 2\sqrt{n}$ and (11.24) allow us to write

$$\ln(m+1)Q)U(h)\sup_{t\in\mathbb{R}}|\widehat{F}^n(t)-\widehat{D}(t)|/\widehat{U}(t) \leq C(F)/n. \tag{11.26}$$

Next, we estimate $\delta(F^n - D, m, \mathbf{u})$.

Set $F^{(-)}(A) = F(-A)$, $F^{(2)}(A) = F(A/2)$ for any Borel set A. If $n \leq C(F)$, then $d_K(F^{*n}, D) \leq 1 \leq C(F)/n$. Therefore, w.l.o.g. we may assume that $n > C(F)$, where $C(F)$ is sufficiently large to ensure that

$$\operatorname{supp}(F^{(-)})^{*2} * F^{(2)} \subset K_4(\mathbf{u}), \quad \Big\|\prod_{j=1}^{*n} I_{x_j\lambda_j}\Big\|_{TV} = 1,$$

$$\sum_{j=1}^N \|\varepsilon_j(I_{x_j} - I)\|_{TV} + \frac{n}{6}\|(F^{(-)})^{*2} * F^{(2)} - I\|_{TV} \leq 2N + n/3 \leq 3n.$$

Therefore,

$$\delta(F^n - D, m, \mathbf{u}) \leq \mathrm{e}^{-n} \leq C/n \tag{11.27}$$

by Lemma D.17. Combining (11.23), (11.26), (11.27) and applying Lemma D.16, we derive the assertion of Theorem 11.5. □

Exercises

11.1. Let $X, X_1, X_2, \ldots$ be i.i.d. integer-valued r.v.s with maximum span 1, $S_n = X_1 + \ldots + X_n$. Assume that $\mathbb{E}X = 0$, $\mathbb{E}X^2 < \infty$. Let $\tilde{S}_n$ denote the sum of accompanying r.v.s. Prove that

$$d_{\mathrm{o}}(S_n; \tilde{S}_n) \leq C_X n^{-3/2}. \tag{11.28}$$

11.2. Let assumptions of the previous exercise hold. Prove that

$$d_K(S_n; \tilde{S}_n) \le C_X n^{-1}. \tag{11.29}$$

Notes

Theorem 11.1 was proved in [331].
Theorem 11.3 is essentially Theorem 3 in [77].
Theorem 11.4 is Theorem 10 from [77].
Theorem 11.5 is Theorem 2.1 from [81].

12

Multivariate compound Poisson approximation

12.1 Multivariate compound Poisson limit theorem

In this section, we present a multivariate compound Poisson limit theorem.

Let $\{X, X_1, ..., X_n\}$ be a sequence of k-dimensional random vectors $X_i = (X_i^{(1)}, ..., X_i^{(k)})$ that are non-zero with "small" probabilities. Set

$$S_n = X_1 + ... + X_n \,.$$

Below we assume that random vectors $\{X_i\}_{i \geq 1}$ have non-negative integer-valued components.

Example 12.1 Let $\{\xi_i\}$ be a sequence of random variables. Denote

$$N_n[a, b) = \sum_{i=1}^{n} \mathbb{1}\{a \geq \xi_i > b\} \qquad (a > b).$$

Given a set $x_1 > ... > x_k$ of levels, set

$$N_n = (N_n(x_1), N_n[x_1; x_2), ..., N_n[x_{k-1}; x_k)) \,. \tag{12.1}$$

Then $N_n = X_1 + ... + X_n$, where

$$X_i = (\mathbb{1}\{\xi_i > x_1\}, \mathbb{1}\{x_1 \geq \xi_i > x_2\}, ..., \mathbb{1}\{x_{k-1} \geq \xi_i > x_k\}) \,.$$

Random vector N_n plays a role in extreme value theory when one deals with a joint distribution of exceedances of several level (cf. (12.5) and [221], Chapter 6). □

The distribution of the random vector S_n can be approximated by a multivariate compound Poisson distribution. Indeed, Theorem 5.1 holds if $\{X_i\}$ are i.i.d. random vectors.

Consider now a situation where $\{X, X_1, ..., X_n\}$ is a (strictly) stationary sequence of random vectors. The argument behind the proof of Theorem 5.3 remains valid. Hence Theorem 5.3 holds if $\{X_i\}$ are random vectors.

Theorem 5.3* *Assume that*

$$\limsup_{n \to \infty} n\mathbb{P}(X \neq \bar{0}) < \infty, \tag{12.2}$$

DOI: 10.1201/9781003478164-12

and there exists the limit

$$\lim_{n\to\infty} \mathbb{P}(S_n = \bar{0}) := \mathrm{e}^{-\lambda} \qquad (\exists \lambda > 0). \tag{12.3}$$

If $\mathcal{L}(S_r | S_r \neq \bar{0}) \Rightarrow \mathcal{L}(\zeta)$ as $n \to \infty$ for a random vector ζ and a sequence $\{r = r_n\} \in \mathcal{R}$, then

$$\mathcal{L}(S_n) \Rightarrow \mathbf{\Pi}(\lambda, \zeta) \qquad (n \to \infty). \tag{5.12*}$$

The limiting distribution $\mathbf{\Pi}(\lambda, \zeta)$ in (5.12) does not depend on the choice of a sequence $\{r_n\}$.*

If $\mathcal{L}(S_n)$ converges weakly to a random vector S, then there exists $\lambda \geq 0$ and a random vector ζ such that $\mathcal{L}(S) = \mathbf{\Pi}(\lambda, \zeta)$, where $\lambda = -\ln \mathbb{P}(S = \bar{0})$, and (12.3) holds. If $\lambda > 0$, then there exist a sequence $\{r = r_n\} \in \mathcal{R}$ such that

$$\mathcal{L}(S_r | S_r \neq 0) \Rightarrow \mathcal{L}(\zeta) \qquad (n \to \infty). \tag{5.11*}$$

Random vector ζ in Theorem 5.3* may have dependent components. The following theorem presents a necessary and sufficient condition for a weak convergence of $\mathcal{L}(S_n)$ to a vector with *independent* compound Poisson components.

Given $t_0 = 0 < t_1 < \ldots < t_k < \infty$, denote $\bar{t} = (t_1, \ldots, t_k)$. Set

$$S_n^{(j)} = X_1^{(j)} + \ldots + X_n^{(j)}, \quad p_j = (t_j - t_{j-1})/t_k \qquad (1 \leq j \leq k).$$

Condition $(C_{\bar{t}})$. *We say that condition $(C_{\bar{t}})$ holds if there exists a random variable η taking values in $\mathbb{N}$ and a sequence $\{r = r_n\}$ such that $n \gg r \gg 1$,*
(a) for every $1 \leq i \leq m$, $\ell \geq 1$,

$$\mathbb{P}(S_r^{(i)} = \ell) \sim \frac{r}{n} \mathbb{P}(\eta = \ell)(t_i - t_{i-1}) \qquad (n \to \infty),$$

(b) for every $1 \leq i < j \leq k$

$$\mathbb{P}(S_r^{(i)} > 0, S_r^{(j)} > 0) = o(r/n) \qquad (n \to \infty).$$

Condition $(C_{\bar{t}})$ is necessary and sufficient for the weak convergence of $\mathcal{L}(S_n)$ to $\mathbf{\Pi}(\lambda, \eta)$, the distribution of a vector with *independent* compound Poisson components. It stipulates the "regular" way of asymptotic clustering of extremes comparing to a more general situation where components may be dependent.

Note that conditions (a) and (12.2) yield

$$\mathbb{P}(S_r^{(i)} > 0) \sim (t_i - t_{i-1}) r/n \qquad (n \to \infty) \tag{12.4}$$

$(1 \leq i \leq k,\ \ell \geq 1)$. Hence (a) means

$$\mathbb{P}(S_r^{(i)} = \ell | S_r^{(i)} > 0) \sim \mathbb{P}(\eta = \ell) \qquad (n \to \infty). \tag{a^*}$$

If condition (Δ) holds, then (12.4) is equivalent to

$$\lim_{n\to\infty} \mathbb{P}(S_n^{(i)} = 0) = \mathrm{e}^{-t_i - t_{i-1}}.$$

Thus, instead of assuming (12.3), one could have added (12.4) as item (c) of condition $(C_{\bar{t}})$ (cf. [220, 221]).

Condition (b) means components of a random vector ζ_r with the distribution $\mathcal{L}(\zeta_r) = \mathcal{L}(S_r|S_r \neq 0)$ are asymptotically independent.

Condition (C). We say that condition (C) holds if condition $(C_{\bar{t}})$ is valid for every $\bar{t} = (t_1, ..., t_k)$, where $0 < t_1 < ... < t_k < \infty$, $k \in \mathbb{N}$.

Let $\{\pi(s), s \geq 0\}$ be a Poisson process with intensity rate 1, and let $\eta, \eta_1, \eta_2, ...$ be a sequence of i.i.d.r.v.s taking values in $\mathbb{N}$. Denote

$$Q(t) = \sum_{j=1}^{\pi(t)} \eta_j \, .$$

Then $\{Q(t), t \geq 0\}$ is a compound Poisson jump process. Equivalently,

$$\tilde{Q}(B) := \int_B Q(dt)$$

is a compound Poisson point process with the Lebesgue intensity measure and multiplicity distribution $\mathcal{L}(\eta)$.

Denote

$$S_{\bar{t}} = \{Q(t_1), Q(t_2) - Q(t_1), ..., Q(t_k) - Q(t_{k-1})\}.$$

Clearly, $S_{\bar{t}}$ is a random vector with *independent* compound Poisson components.

The ch.f. of $S_{\bar{t}}$ is

$$\mathbb{E}\exp(i\bar{s}S_{\bar{t}}) = \exp\Big(t_m\Big(\sum_{j=1}^{m} p_j \varphi_\eta(s_j) - 1\Big)\Big) \qquad (\forall \bar{s} = (s_1, ..., s_m) \in \mathbb{R}^k),$$

where φ_η is a ch.f. of r.v. η.

In Theorem 12.6, we assume conditions (12.2) and (Δ), see p. 261.

Theorem 12.6 *Assume conditions (12.2) and* (Δ). *Given a vector* $\bar{t} = (t_1, ..., t_m)$, *where* $t_0 = 0 < t_1 < ... < t_k < \infty$, *suppose that there exist the limits*

$$\lim_{n\to\infty} \mathbb{P}(S_n^{(j)} = 0) = \mathrm{e}^{-t_j + t_{j-1}} \qquad (\forall j \in \{1, ..., k\}).$$

Weak convergence

$$S_n \Rightarrow S_{\bar{t}} \qquad (n \to \infty)$$

holds if and only if condition $(C_{\bar{t}})$ *holds.*

Example 12.2 deals with sample extremes. We rewrite the sample $X_1, \ldots, X_n$ in the non-increasing order:

$$X_{1,n} \geq ... \geq X_{n,n} \, .$$

Then $X_{1,n}, ..., X_{n,n}$ are called the "order statistics". In particular, $M_n = X_{1,n}$ is the sample maximum, $X_{l,n}$ is the lth sample maximum.

Denote by $N_n(x) = \sum_{i=1}^n \mathbb{I}\{X_i > x\}$ the number of exceedances over threshold x. It is easy to see that

$$\{X_{l,n} \leq x\} = \{N_n(x) < l\} \qquad (1 \leq l \leq n). \tag{12.5}$$

Let $\{u_n(\cdot)\}$ be a non-decreasing normalising sequence:

$$\limsup_{n\to\infty} n\mathbb{P}(X > u_n(t)) < \infty, \ \lim_{n\to\infty} \mathbb{P}(M_n \leq u_n(t)) = \mathrm{e}^{-t} \qquad (\forall t > 0). \tag{12.6}$$

Assume condition (Δ). The following result can be deduced from Theorem 12.6 for the joint limiting distribution of $X_{1,n}$ and $X_{l,n}$: if $0<s<t$, then

$$\lim_{n\to\infty} \mathbb{P}(X_{1,n}\le u_n(s),\ X_{l,n}\le u_n(t)) \tag{12.7}$$
$$= e^{-t}\left\{1+\sum_{j=1}^{l-1}(t-s)^j\,\mathbb{P}\left(\sum_{i=1}^{j}\eta_i<k\right)\Big/j!\right\} \qquad (l\ge 2).$$

In particular,

$$\lim_{n\to\infty} \mathbb{P}(X_{1,n}\le u_n(s), X_{2,n}\le u_n(t)) = e^{-t}\left(1+(t-s)\mathbb{P}(\eta=1)\right).$$

Similarly, if $0<q<s<t$, then Theorem 12.6 yields

$$\lim_{n\to\infty} \mathbb{P}(X_{1,n}\le u_n(q), X_{2,n}\le u_n(s), X_{3,n}\le u_n(t)) \tag{12.8}$$
$$= e^{-t}\Big(1+(t-q)\mathbb{P}(\eta=1) + (t-s)^2\mathbb{P}^2(\eta=1)/2$$
$$+ (t-s)(s-q)\mathbb{P}^2(\eta=1) + (t-s)\mathbb{P}(\eta=2)\Big).$$

Formulas (12.7), (12.8) demonstrate the impact of the asymptotic clustering of extremes on the limiting distribution of upper order statistics.

Denote

$$\bar N_n(\bar t) = \{N_n(u_n(t_1)), N_n[u_n(t_1);u_n(t_2)), ..., N_n[u_n(t_{m-1});u_n(t_m))\}.$$

According to Theorem 12.6,

$$\bar N_n(\bar t) \Rightarrow S_{\bar t} \tag{12.9}$$

if and only if $(C_{\bar t})$ holds; instead of (Δ) we may assume here $(\Delta(\{u_n(\bar t)\}))$. □

Condition $(C_{\bar t})$ stipulates the "regular" way of asymptotic clustering of extremes. Waiving it makes the situation more complicated (cf. formula (6.10) in [221]).

Example 12.3 Let $\{X_i, i\ge 1\}$ be a strictly stationary α–mixing sequence. Hsing [154] has shown that $\lim_{n\to\infty}\mathbb{P}(X_{1,n}\le u_n(s), X_{l,n}\le u_n(t))$, if exists, is necessarily expressed via a compound Poisson distribution, cf. (12.10).

The necessary and sufficient conditions for the convergence of $\mathbb{P}(X_{1,n}\le u_n(s), X_{l,n}\le u_n(t))$ as $n\to\infty$: if (12.6) holds, then the probability

$$\mathbb{P}(X_{1,n}\le u_n(s), X_{l,n}\le u_n(t)) = \mathbb{P}(N_n(u_n(s))=0, N_n(u_n(t))<l)$$

converges for every $t>s>0$ if and only if there exist functions $f_i(\cdot)$ and a sequence $\{r=r_n\}$ such that $n\gg r\gg 1$,

$$\lim_{n\to\infty} \mathbb{P}(N_r(u_n(s))=0,\ N_r(u_n(t))=i\,|\,N_r(u_n(t))>0) = f_i(s/t)$$

for each $t>s>0$ and $i\in\{1,\dots,l-1\}$.

The limit is expressed via a compound Poisson distribution:

$$\lim_{n\to\infty} \mathbb{P}\left(N_n(u_n(s))=0,\ N_n(u_n(t))<l\right) = \mathbb{P}\left(\sum_{i=1}^{\pi(t)}\zeta_i^*<l\right), \tag{12.10}$$

where $\{\zeta_i^*\}$ are i.i.d.r.v.s, $\mathbb{P}(\zeta^*=i) = f_i(s/t)$ (Novak [220], see also [221], p. 107). In particular, the limiting cluster size distribution depends on the ratio s/t. □

12.2 Accuracy of multivariate CP approximation: rare events

This section presents results concerning the accuracy of multivariate compound Poisson approximation to the distribution of the sum $S_n = X_1 + ... + X_n$ of random vectors $X_1, ..., X_n$.

Let $\{X_i\}$ be independent random vectors that are non-zero with small probabilities. Recall that $\tilde{S}_n$ denotes the sum of accompanying random vectors, see (1.9). Set

$$p_i = \mathbb{P}(X_i \neq \bar{0}), \ \lambda = p_1 + ... + p_n \quad (i \geq 1).$$

Khintchine's formula (1.12*) holds for random vectors:

$$X_i \stackrel{d}{=} \tau_i X_i',$$

where τ_i and X_i' are independent r.v.s,

$$\mathcal{L}(X_i') = \mathcal{L}(X_i | X_i \neq 0), \ \ \mathcal{L}(\tau_i) = \mathbf{B}(p_i).$$

Hence (5.3) holds:

$$S_n \stackrel{d}{=} \tau_1 X_1' + ... + \tau_n X_n'.$$

If $\{X_i'\}$ are identically distributed, then

$$d_{TV}(S_n; Y) \equiv d_{TV}\Big(\sum_{i=1}^{\nu_n} X_i'; \sum_{i=1}^{\pi_\lambda} X_i'\Big) \leq d_{TV}(\nu_n; \pi_\lambda) \tag{5.6*}$$

by (5.6). In particular, (6.10) entails

$$d_{TV}(S_n; \tilde{S}_n) \leq \sum_{i=1}^{n} p_i^2.$$

A number of univariate results have been generalised to the multidimensional case.

Inequality (6.13) has been generalised to the case of independent random vectors taking values in $\mathbb{R}^k$ that are non-zero with small probabilities by Zaitsev [327]: there exists an absolute constant C_k such that

$$d_K(S_n; \tilde{S}_n) \leq C_k \max_{1 \leq i \leq n} p_i. \tag{12.11}$$

The following estimate is due to Roos [265]. Denote

$$\bar{P}(\cdot) = \frac{1}{\lambda} \sum_{j=1}^{n} p_i \mathbb{P}(X_i' \in \cdot).$$

Assume that $\min_{1 \leq i \leq n} p_i > 0$. Then $\mathcal{L}(X_i')$ is absolutely continuous with respect to $\bar{P}$ for every i. Let f_i denote the density of $\mathcal{L}(X_i')$ with respect to $\bar{P}$. Set $\rho_i = \int f_i^2 d\bar{P}$. Then

$$d_{TV}(S_n; \tilde{S}_n) \leq 8.8 \sum_{i=1}^{n} p_i^2 \min\{1; \rho_i/\lambda\}. \tag{12.12}$$

Estimates (6.24), (6.25) hold in the multivariate case (cf. [89], p. 28). Generalisation of (6.24) was proved by Roos [265]. Multivariate versions of (7.15) for independent random vectors have been given by Roos [261, 263]. In [139] was possibility to use stronger than Kolmogorov metric was considered.

Independent 0-1 random vectors.

Let $\{X_i\}$ be independent k-dimensional random vectors. Denote

$$\bar{0} = (0,\dots,0),\ \bar{e}_r = (0,\dots,1,\dots,0) \quad (1\le r\le k),$$

where vector $\bar{e}_r$ has the rth coordinate equal to 1, the other coordinates equal to 0. Set

$$p_{i,r} = \mathbb{P}(X_i=\bar{e}_r),\quad p_i=\sum_{r=1}^{k} p_{i,r},\quad \lambda_r = \sum_{j=1}^{n} p_{j,r}>0,\quad \tilde{p}_0 = \sum_{r=1}^{k} \max_{1\le j\le n} p_{j,r}\,.$$

As before, I denotes the distribution concentrated at $\bar{0}$, $I_{\bar{e}_j}$ is the distribution concentrated at $\bar{e}_j$. Denote

$$A_j = \sum_{r=1}^{k} p_{j,r}(I_{\bar{e}_r} - I),\quad H_{n,s}^{(k)} = \exp\left(\sum_{j=1}^{n}\sum_{m=1}^{s}\frac{(-1)^{m+1}}{m}A_j^{*m}\right) \qquad (s\ge 1).$$

Then

$$\mathcal{L}(S_n) = \prod_{j=1}^{*n}(I+A_j).$$

It follows from Theorem 1 in Roos [261] that there exists constant C_s such that

$$d_{TV}(\mathcal{L}(S_n); H_{n,s}^{(k)}) \le C_s \sum_{j=1}^{n}\left(\min\left\{\sum_{r=1}^{k}\frac{p_{j,r}^2}{\lambda_r}; p_j^2\right\}\right)^{(s+1)/2} \tag{12.13}$$

if $\tilde{p}_0\le 1/4$.

If k,s are fixed and $p_{j,r}\asymp C$, then the rate of accuracy in (12.13) is $O(n^{-(s-1)/2})$. Further SCP approximations to the sum of 0-1 random vectors can be found in [263]. A one-dimensional version of (12.13) is given in (7.18). See [266] for a further discussions.

Dependent 0-1 random vectors.

We consider now the case of weakly dependent 0-1 random vectors.

Let $X, X_1,\dots,X_n$ be a strictly stationary sequence of k-dimensional 0-1 random vectors such that not more than one coordinate of a vector may equal 1. Set $X_i = (X_i^{(1)},...,X_i^{(k)})$,

$$S_n = X_1 +...+ X_n\,.$$

Denote $\bar{e}_j = (0,\dots,1,\dots,0)$, where vector $\bar{e}_j$ has the jth coordinate equal 1, the other coordinates equal zero. Assume that

$$\mathbb{P}(X=\bar{0}) = 1-p,\ \mathbb{P}(X=\bar{e}_j) = p_j \qquad (1\le j\le k), \tag{12.14}$$

where $p := p_1 + ... + p_k = \mathbb{P}(X\ne 0)$.

If $\{X_i\}$ are independent, then $\mathcal{L}(S_n)$ is multinomial $\mathbf{B}(n,p_1,...,p_k)$ distribution. Estimates of the accuracy of Poisson approximation to $\mathbf{B}(n,p_1,...,p_k)$ can be found in Section 3.2, see also [224].

Recall that symbol $\mathbf{\Pi}(\lambda,\zeta)$ denotes the compound Poisson distribution with intensity λ and multiplicity distribution $\mathcal{L}(\zeta)$:

$$\mathbf{\Pi}(\lambda,\zeta) = \mathcal{L}\left(\sum\nolimits_{i=0}^{\pi_\lambda}\zeta_i\right),$$

where $\pi(\lambda),\zeta,\zeta_1,...$ are independent, $\mathcal{L}(\pi_\lambda) = \mathbf{\Pi}(\lambda)$, $\zeta_i \stackrel{d}{=} \zeta\ (i\ge 1)$, $\zeta_0 = \bar{0}$.

Given $r \in \{1,...,n\}$, let $\zeta, \zeta_1, \zeta_2, \ldots$ be independent random vectors with the common distribution

$$\mathcal{L}(\zeta) = \mathcal{L}(S_r | S_r \neq \bar{0}).$$

Denote

$$q = \mathbb{P}(S_r \neq 0), \ k = [n/r], \ r' = n - rk, \ \lambda = kq.$$

We approximate $\mathcal{L}(S_n)$ by the multivariate compound Poisson distribution $\mathbf{\Pi}(kq, \zeta)$. The following theorem is cited from [221], Theorem 6.8.

Theorem 12.7 *If* $n > r > l \geq 0$ *and* $\mathcal{L}(Y) = \mathbf{\Pi}(kq, \zeta)$, *then*

$$d_{TV}(S_n; Y) \leq C_{n,r} rp + (2nr^{-1}l + r')p + nr^{-1} \min\{\beta(l); \kappa(l)\}, \tag{12.15}$$

where $C_{n,r} = \min\{1 - e^{-np}; 3/4e + (1 - e^{-np})rp\}$ *and* $\kappa(l) = 2(1+2/k)\left(2^{k-1}k^2\alpha^2(l)\right)^{1/(2+k)}$ *if* $k2^{(k-1)/2}\alpha(l) \leq 1$, *otherwise* $\kappa(l) = 1$.

If random vectors $\{X_i\}$ are independent, then (12.15) with $l = 0$ and $r = 1$ becomes an estimate of the accuracy of multivariate Poisson approximation to $\mathcal{L}(S_n)$ with a correct constant $3/4e$ at the leading term:

$$d_{TV}(\mathcal{L}(S_n); \mathbf{\Pi}(n, X')) \leq 3p/4e + (1 - e^{-np})p^2,$$

where $\mathcal{L}(X') = \mathcal{L}(X | X \neq 0)$.

The next corollary applies (12.15) to the case of m-dependent random vectors.

Corollary 12.8 *If vectors* $\{X_i\}$ *are* m*-dependent and* $m < r < n$, *then*

$$d_{TV}(S_n; Y) \leq C_{n,r} rp + (r + 2nm/r)p. \tag{12.16}$$

The right-hand side of (12.16) is $O(p\sqrt{n})$ if $r \asymp \sqrt{n}$.

12.3 Symmetric integer-valued random vectors

Let $\{X_j\}$ be independent symmetric random vectors concentrated on coordinate axes of $\mathbb{R}^k$:

$$\sum_{r=1}^{k} \sum_{m=-\infty}^{\infty} \mathbb{P}(X_j = m\bar{e}_r) = \sum_{r=1}^{k} \mathbb{P}(X_j \in \bar{e}_r \mathbb{Z}) = 1.$$

Set $p_j = \mathbb{P}(X_j \neq \bar{0})$. We assume that $p_j > 0$ $(\forall j)$.

Obviously, Khintchine's decomposition (6.1) holds:

$$\mathcal{L}(X_j) = (1 - p_j)I + p_j \mathcal{L}(X_j'),$$

where $\mathcal{L}(X_j') = \mathcal{L}(X_j | X_j \neq \bar{0})$.

In this section, we investigate the case where $\mathcal{L}(X_j')$ can be presented as a mixture of distributions concentrated on coordinate axes. More precisely, assume that

$$\mathcal{L}(X_j') = \sum_{r=1}^{k} p_{j,r} F_r / p_j,$$

for some $0 \le p_{j,r} \le 1$ such that $\sum_{r=1}^k p_{j,r} = p_j > 0$, where F_r is a symmetric distribution concentrated on the r'th coordinate axis except zero: $\operatorname{supp} F_r = \bar{e}_r \mathbb{Z}$. Then

$$\mathcal{L}(S_n) = \prod_{j=1}^{*n} \left(I + \sum_{r=1}^{k} p_{j,r}(F_r - I) \right), \quad \mathcal{L}(\tilde{S}_n) = \exp\left(\sum_{j=1}^{n} \sum_{r=1}^{k} p_{j,r}(F_r - I) \right).$$

Let σ_r^2 denote the variance of F_r. Set

$$g(z) = 2\mathrm{e}^z(\mathrm{e}^{-z} - 1 - z)/z^2 = 2\sum_{s=2}^{\infty} \frac{z^{s-2}}{s!}(s-1), \quad \lambda_{n,r} = \sum_{j=1}^{n} p_{j,r},$$

$$\alpha_0 = \sum_{i=1}^{n} g(2p_i) \min\left\{ 2^{-3/2} \sum_{r=1}^{k} p_{i,r}^2 / \lambda_{n,r}; p_i^2 \right\}.$$

Theorem 12.9 *Assume that F_r is a symmetric distribution with variance $\sigma_r^2 < \infty$ $(r = 1, \dots, k)$. If $2\alpha_0 \mathrm{e} < 1$, then*

$$d_{TV}(S_n; \tilde{S}_n) \le \frac{8}{(1 - 2\alpha_0 \mathrm{e})^{3/2}} \sum_{i=1}^{k} (\sigma_i + 1) \sum_{r=1}^{k} \lambda_{nr}^{-2} \sum_{j=1}^{n} p_{j,r}^2. \tag{12.17}$$

Bound (12.17) is a (slightly weaker) multidimensional analogue of (7.30) that states that

$$d_{TV}(\mathcal{L}(S_n); \mathbf{\Pi}(\lambda, F)) \le \frac{C\sqrt{\sigma' + 1}}{(1 - c_0)^{5/2}} \frac{\lambda_2}{\lambda^2},$$

where $\sigma'^2 = \operatorname{var} X'$. If $\{\sigma_r, p_{j,r}\}$ are bounded away from 0, then RHS of (12.17) is $O(n^{-1})$.

It is natural to consider local estimates. Let

$$a_r^* = \sum_{j=1}^{n} (1 - p_j) p_{j,r}.$$

Theorem 12.10 *Assume that F_r is a symmetric distribution $(r = 1, 2, \dots, k)$. Then*

$$\sup_{\mathbf{m} \in \mathbb{Z}^k} |\mathbb{P}(S_n = \mathbf{m}) - \mathbb{P}(\tilde{S}_n = \mathbf{m})| \le 3.12 \left(\prod_{l=1}^{k} \sqrt{\frac{2}{a_l^* \mathrm{e}}} \right) \sum_{j=1}^{n} \left(\sum_{r=1}^{k} \frac{p_{j,r}}{a_r^*} \right)^2. \tag{12.18}$$

If $k = 1$, the rate of the accuracy in (12.18) is the same as in (7.29). If $p_{j,r} \asymp C$, then the rate of accuracy in Theorem 12.10 is $O(n^{-(1+k/2)})$.

Local estimate (12.18) hold under much weaker assumption $0 \le p_{j,r} \le 1$ than assumption $2\alpha_0 \mathrm{e} < 1$ required in (12.17). Note also that the accuracy of estimate (12.18) directly depends on the dimension k of the space – for larger k the accuracy is better.

12.4 Accuracy of multivariate CP approximation: general case

Let $X, X_1, \dots, X_n$ be independent and identically distributed random vectors taking values in $\mathbb{R}^k$, $S_n = X_1 + \cdots + X_n$. Given $a \in \mathbb{R}$, denote by $\tilde{X}_{1,a}, ..., \tilde{X}_{n,a}$ accompanying $X_1 + a, ..., X_n + a$ independent random vectors (see (1.7)), and let

$$\tilde{S}_{n,a} = \tilde{X}_{1,a} + ... + \tilde{X}_{n,a}, \quad \tilde{S}_n = \tilde{S}_{n,0}.$$

Recall that $\tilde{S}_{n,a}$ is a compound Poisson random vector.

Throughout this chapter C_k denotes (possibly different) absolute positive constants that depends only on space dimension k.

Estimate (10.5) of the accuracy of compound Poisson approximation has been generalised to the multivariate case by Presman [236]: there exists absolute constant C_k such that

$$\sup_{\mathcal{L}(X)} \inf_a d_K(S_n; \tilde{S}_{n,a} - na) \leq C_k n^{-1/3}. \tag{12.19}$$

Relation (12.19) is still the best available estimate of the accuracy of multivariate compound Poisson approximation to $\mathcal{L}(S_n)$ without additional assumptions on $\mathcal{L}(X)$.

Bentkus et al. [39] state that the rate of approximation to $\mathcal{L}(S_n)$ by $\mathcal{L}(\tilde{S}_n)$ is $O(n^{-1})$ if $\mathbb{E}|X|^{8/3} < \infty$. Namely, assume that $\mathcal{L}(X)$ is not concentrated on a hyperspace in $\mathbb{R}^k$, $\mathbb{E}X = 0$, $\mathbb{E}\|X\|^{8/3} < \infty$. Then for any $a \in \mathbb{R}^k$, as $n \to \infty$,

$$d_K(\|S_n - a\|_k; \|\tilde{S}_n - a\|_k) = O((1 + \|a\|^4) n^{-1}), \tag{12.20}$$

where $\|\cdot\|_k$ denotes the Euclidean norm in $\mathbb{R}^k$. Note that estimate (12.20) is not uniform in a.

Results concerning asymptotic expansions can be found in [236, 67].

Symmetric random vectors.

Multivariate analogues of (10.8) and (10.13) for independent random vectors have been established by Zaitsev [326, 328]. More precisely, let $X, X_1, X_2, \ldots$ be i.i.d. symmetric random vectors taking values in $\mathbb{R}^k$. Then

$$d_K(S_n, \tilde{S}_n) \leq C_k n^{-1/2}. \tag{12.21}$$

If, in addition, there exists $\alpha \in (0; 2)$ such that

$$\mathbb{E}\exp(itX) \geq -1 + \alpha \qquad (\forall t \in \mathbb{R}^k)$$

then

$$d_K(S_n, \tilde{S}_n) \leq C_k \Big(n^{-1} + \exp(-n\alpha + ck \ln^3 n) \Big) = O(n^{-1}). \tag{12.22}$$

Mixtures of distributions with a dominant symmetric part.

Let $X, X_1, X_2, \ldots$ be i.i.d. symmetric random vectors taking values in $\mathbb{R}^k$. Any random vector X_i admits representation

$$X_i \stackrel{d}{=} (1-\tau)X'' + \tau X',$$

where $\mathcal{L}(\tau) = \mathbf{B}(p)$, $p \in [0; 1]$, random vectors X_i', X_i'' and r.v. τ are independent.

Assume that

$$X_i'' = \sum_{j=1}^{\nu} Z_j,$$

where $\nu, Z, Z_1, \ldots$ are independent, $Z, Z_1, \ldots$ are i.i.d. symmetric random vectors with non-negative characteristic functions, r.v. ν takes values in $\{1, 2, \ldots, s\}$. Denote

$$q_j = \mathbb{P}(\nu = j) \quad (j = 1, 2, \ldots, s), \quad P = \mathcal{L}(Z), \quad V = \mathcal{L}(X').$$

Assume that P is a symmetric distribution with a non-negative ch.f., V is a distribution of an arbitrary k-dimensional random vector. Then

$$\mathcal{L}(X_i) = (1-p)\sum_{j=1}^{s} q_j P^{*j} + pV, \quad \mathcal{L}(S_n) = \left((1-p)\sum_{j=1}^{s} q_j P^{*j} + pV\right)^{*n},$$

and $q_1 + \cdots + q_s = 1$. Denote $a_s = \mathbb{E}\nu = \sum_{i=1}^{s} iq_i$,

$$\tilde{D}_n = \exp(na_s(1-p)(P-I) + np(V-I)),$$
$$D_n = \exp\big(na_s(1-p)(P-I) + np(V-I) - np^2(V-I)^{*2}/2\big).$$

$\tilde{D}_n, D_n$ can be written as a convolution of CP and SCP measures:

$$\tilde{D}_n = \mathbf{\Pi}(na_s(1-p); P) * \mathbf{\Pi}(np; V), \quad D_n = \mathbf{\Pi}(na_s; P) * \mathbf{\Pi}(np(1+p); V) * \mathbf{\Pi}(-np^2/2; V^{*2}).$$

Proposition 12.11 *If* $p \in [0, 1/2]$, *then*

$$d_K(\mathcal{L}(S_n); \tilde{D}_n) \le C(k, s)\big(1/na_s + \sqrt{p/n} + p\big), \tag{12.23}$$
$$d_K(\mathcal{L}(S_n); D_n) \le C(k, s)\big(1/na_s + \sqrt{p/n}\big), \tag{12.24}$$

where constant $C(k, s)$ *depends only on* k *and* s.

Example 12.4 Let P be symmetric k-dimensional distribution with a non-negative ch.f.;

$$F = 0.2P + 0.4P^{*2} + 0.4P^{*5}, \quad D = \exp(3(P-I)).$$

Obviously, F also has non-negative ch.f. Therefore from (12.22) and (12.23) it follows that

$$d_K(F^{*n}; \exp(n(F-I))) \le C_k n^{-1}, \quad d_K(F^{*n}; D^{*n}) \le C_k n^{-1},$$

i.e. the order of accuracy is the same for accompanying distribution and for D^{*n}. However

$$\exp((F-I)) = \exp(0.2(P-I) + 0.4(P^{*2}-I) + 0.4(P^{*5}-I))$$

is structurally much more complicated than D.

The accuracy of approximation in (12.23)–(12.24) depends on p, s and $\{q_j\}$. If $s = 1$, then RHS of (12.23) is $O(p^{1/2}n^{-1/2})$. Note also that $a_s = \mathbb{E}\nu$. Note also, that for $s = 1$, (12.23) and (12.24) are multidimensional analogues of (10.16) and (10.14), respectively.

Infinite-dimensional spaces.

Very few results on the accuracy of compound Poisson approximation are known for sums of random elements taking values in a general measurable space.

Bakštys & Paulauskas [12, 13] dealt with random elements $X_1, ..., X_n$ taking values in a separable Banach space. They proved that the sum $S_n = X_1 + ... + X_n$ is in a sense "close" to the sum $\tilde{S}_n = \tilde{X}_1 + \cdots + \tilde{X}_n$ of accompanying compound Poisson random elements.

Let $X, X_1, X_2, \ldots$ be a sequence of i.i.d. random elements taking values in a separable Banach space $\mathcal{B}$. Denote by $\tilde{X}_a, \tilde{X}_{1,a}, ..., \tilde{X}_{n,a}$ accompanying $X+a, X_1+a, ..., X_n+a$ independent random elements, and let $\tilde{S}_{n,a} = \tilde{X}_{1,a} + ... + \tilde{X}_{n,a}$.

Let $\mathcal{U}$ denote the set of all convex Borel subsets of $\mathcal{B}$. Suppose that for any $\varepsilon > 0$ there exists a finite-dimensional subspace $\mathcal{V}_\varepsilon$ such that $\mathbb{P}(X \in \mathcal{V}_\varepsilon) \le \varepsilon$. Then [13]

$$\lim_{n\to\infty} \inf_{a\in\mathcal{B}} \sup_{A\in\mathcal{U}} |\mathbb{P}(S_n \in A) - \mathbb{P}(\tilde{S}_{n,a} - na \in A)| = 0. \tag{12.25}$$

Bakštys [14] proved that if X is a symmetric random vector taking values in a Hilbert space, then

$$\lim_{n\to\infty} \sup_{A\in\mathcal{V}} |\mathbb{P}(S_n \in A) - \mathbb{P}(\tilde{S}_n \in A)| = 0, \tag{12.26}$$

where $\mathcal{V}$ denotes a set of all open balls.

Nagaev [216] has derived estimates of the accuracy of approximation to S_n by a sum $\tilde{S}_n$ of accompanying random vectors. We present his result below.

Let $X, X', X_1, \dots, X_n$ be a sequence of i.i.d.r.v.s taking values in a real separable Hilbert space with scalar product $(\cdot,\cdot)$ and norm $\|\cdot\|_H$. Given a constant $C>0$, set

$$B^2(x;C) = \mathbb{E}(X-X',x)^2 \mathbb{1}\{\|X\|_H \wedge \|X'\|_H \le C\},$$

$B^2(a) = B^2(a,\infty) = \mathbb{E}(X-X',x)^2$. Let

$$\sigma_1^2(C) \ge \sigma_2^2(C) \ge \dots$$

be the eigenvalues of the quadratic form $B(\cdot;C) \times B(\cdot;C)$. Denote by $V_r(a) = \{x \in H : \|x-a\| \le r\}$ the neighbourhood of a, and let

$$\sigma^2(C) = \prod_{j=1}^{\infty} \sigma_j^2(C), \ \Lambda_l(C) = \prod_{j=1}^{l} \sigma_j^2(L), \ \sigma^2 = \mathbb{E}\|X\|_H^2 .$$

Theorem 12.12 [216] If $\mathbb{E}X=0$ and $\sigma^2<\infty$, then for any a and $C \in (0;\infty)$

$$\sup_r |\mathbb{P}(S_n \in V_r(a)) - \mathbb{P}(Y_n \in V_r(a))| \le \frac{\sigma^4 + B^2(a)/n}{\Lambda_5^{2/5}(C)\sqrt{n}} + \frac{C\sigma(C)}{\Lambda_3^{1/3}(C)\sqrt{n}} . \tag{12.27}$$

Further results on SCP approximation to the sum of 0-1 random vectors can be found in [261, 263, 266]. Approximation in ℓ_p-metric for a sequence of 1-dependent identically distributed vectors has been suggested in [92].

Open problems

12.1. The term $(2nr^{-1}l+r')p$ in (12.15) appears because of the method (Bernstein's blocks approach). An open question is if it can be removed.
12.2. Establish multivariate analogues of (10.8) and Theorem 10.8.
12.3. What is the rate of growth of C_k as a function of k in (12.19), (12.22)?

12.5 Proofs

Proof of Theorem 5.3*. The argument of the proof of Theorem 5.3 remains valid. Hence Theorem 5.3 holds if $\{X_i\}$ are random vectors. □

Proof of Theorem 12.7 follows the lines of the proof of Theorem 6.8 (the blocks method, Lindeberg's device, Berbee's lemma and Lemma 6.1) and is therefore omitted. □

Proof of Theorem 12.9 will employ multidimensional analogues of (C.8), (C.9), and (C.10) that are presented below.

Lemma 12.13 [257] *For any* $t_r>0$

$$\Big\|\sum_{r=1}^k p_{jr}(I_r-I)*\exp\Big(\sum_{r=1}^k t_r(I_r-I)\Big)\Big\|_{TV} \le \min\Big\{\Big(\sum_{r=1}^k p_{jr}^2/t_r\Big)^{1/2};2\sum_{r=1}^k p_{jr}\Big\},$$

$$\Big\|\Big(\sum_{r=1}^k p_{jr}(I_r-I)\Big)^{*2}*\exp\Big(\sum_{r=1}^k t_r(I_r-I)\Big)\Big\|_{TV} \le \min\Big\{\sqrt{2}\sum_{r=1}^k\frac{p_{jr}^2}{t_r};\Big(2\sum_{r=1}^k p_{jr}\Big)^2\Big\},$$

$$\Big\|\Big(\sum_{r=1}^k p_{jr}(I_r-I)\Big)^{*3}*\exp\Big(\sum_{r=1}^k t_r(I_r-I)\Big)\Big\|_{TV} \le \min\Big\{\frac{3}{2^{1/3}}\sum_{r=1}^k\frac{p_{jr}^2}{t_r};\Big(2\sum_{r=1}^k p_{jr}\Big)^2\Big\}^{3/2}.$$

Lemma 12.13 is a consequence of relations (18), (19), (26) in [257].

Denote

$$M_j=\sum_{r=1}^k p_{j,r}(F_r-I),\quad M=\sum_{j=1}^n M_j=\sum_{r=1}^k\Big(\sum_{j=1}^n p_{j,r}\Big)(F_r-I).$$

Thanks to (A.3) and (C.8)

$$\begin{aligned}\big\|\mathcal{L}(S_n)-\mathcal{L}(\tilde S_n)\big\|_{TV} &\le \sum_{j=1}^n\sum_{1\le i_1<\cdots<i_j\le n}\prod_{m=1}^j\Big\|L_{i_m}*\exp(M/j)\Big\|_{TV}\\ &\le \sum_{j=1}^n\Big(\sum_{i=1}^n\big\|L_i*\mathrm{e}^{M/j}\big\|_{TV}\Big)^j/j!,\end{aligned}$$

where

$$L_i=-M_i^{*2}\sum_{m=2}^\infty(-M_i)^{*(m-2)}(m-1)/m!$$

Since $\|F_r-I\|_{TV}\le 2$, we have $\|M_i\|_{TV}\le 2p_i$. Therefore,

$$2\big\|L_i*\mathrm{e}^{M/j}\big\|_{TV}\le g(2p_i)\big\|M_i^{*2}*\mathrm{e}^{M/j}\big\|_{TV}\le g(2p_i)\big\|M_i*\mathrm{e}^{M/2j}\big\|_{TV}^2.$$

Applying Lemma 12.13 with $t_r=\sum_{j=1}^n p_{j,r}/j$, we get

$$\big\|M_i^{*2}\mathrm{e}^{M/j}\big\|_{TV}\le 4\min\Big\{2^{-3/2}j\sum_{r=1}^k p_{i,r}^2/\lambda_{n,r};\Big(\sum_{r=1}^k p_{i,r}\Big)^2\Big\}.$$

Therefore, $2\big\|\mathcal{L}(S_n)-\mathcal{L}(\tilde S_n)\big\|_{TV}$ is bounded by

$$\begin{aligned}2\sum_{j=1}^n\Big(\sum_{i=1}^n\big\|L_i*\mathrm{e}^{M/j}\big\|_{TV}\Big)^j/j! &\le \sum_{j=1}^n\Big(\sum_{i=1}^n\frac{g(2p_i)}{2j!}\big\|M_i^{*2}*\mathrm{e}^{M/j}\big\|_{TV}\Big)^{j-1}\sum_{i=1}^n g(2p_i)\big\|M_i*\mathrm{e}^{M/2j}\big\|_{TV}^2\\ &\le \sum_{j=1}^n\frac{(2\alpha_0 j)^{j-1}}{j!}\sum_{i=1}^n g(2p_i)\big\|M_i*\mathrm{e}^{M/2j}\big\|_{TV}^2.\end{aligned}\qquad(12.28)$$

Taking into account the fact that the norm of any distribution equals 1 and applying Lemma 12.13, we obtain

$$
\begin{aligned}
\big\|M_i * \mathrm{e}^{M/2j}\big\|^2_{TV} &\le \left(\sum_{r=1}^{k} p_{i,r}\left\|(F_r-I)*\exp\left(\frac{1}{2j}\sum_{j=1}^{n} p_{j,r}(F_r-I)\right)\right\|_{TV}\right)^2 \\
&\le \left(\frac{7.2j}{\mathrm{e}}\sum_{r=1}^{k} p_{i,r}\sqrt{1+\sigma_r}\frac{1}{\lambda_{n,r}}\right)^2 \\
&\le \left(\frac{7.2}{\mathrm{e}}\right)^2 j^2\sum_{s=1}^{k}(1+\sigma_s)\sum_{r=1}^{k}\frac{p^2_{i,r}}{\lambda^2_{n,r}}\,.
\end{aligned}
$$

By Stirling's formula and the definition of $g(x)$,

$$
\frac{j^j}{j!} \le \frac{\mathrm{e}^j}{\sqrt{2\pi j}}\,, \quad g(2p_i) \le g(2) \le 4.2.
$$

Recalling that $2\alpha_0\mathrm{e} < 1$ by assumption, we get

$$
\begin{aligned}
\sum_{j=1}^{\infty}(2\alpha_0 j)^{j-1}j^2/j! &\le \frac{1}{\alpha_0\sqrt{8\pi}}\sum_{j=1}^{\infty}\sqrt{j}(2\alpha_0\mathrm{e})^j \\
&\le \frac{1}{\alpha_0\sqrt{8\pi}}\sqrt{\sum_{j=1}^{\infty}(2\alpha_0\mathrm{e})^j\sum_{j=1}^{\infty}j(2\alpha_0\mathrm{e})^j} \le \frac{\mathrm{e}}{\sqrt{2\pi}(1-2\alpha_0\mathrm{e})^{3/2}}\,.
\end{aligned}
$$

It remains to apply (12.28). □

Proof of Theorem 12.10. We keep notation M_j, M from the previous proof.

By a multivariate version of the inversion formula (D.11),

$$
\sup_{\mathbf{m}\in\mathbb{Z}^k}|W(\{\mathbf{m}\})| \le \frac{1}{(2\pi)^k}\int_{-\pi}^{\pi}\cdots\int_{-\pi}^{\pi}|\widehat{W}(\mathbf{t})|\mathrm{d}t_1\ldots\mathrm{d}t_k \tag{12.29}
$$

for any finite measure W concentrated on $\mathbb{Z}^k$.

Estimate (12.18) is trivial if $a^*_r=0$. Therefore, we may assume that $a^*_r>0$ $(\forall r)$.

Applying (D.6), we get

$$
\left|1+\widehat{M}_j\right| = \left|1-p_j + p_j\sum_{r=1}^{k}\frac{p_{j,r}}{p_j}\widehat{F}_r(t_r)\right| \le \exp\left(\sum_{r=1}^{k}(1-p_j)p_{j,r}(\widehat{F}_r(t_r)-1)\right).
$$

Taylor's formula (I.9) entails

$$
\left|1+\widehat{M}_j - \exp(\widehat{M}_j)\right| \le |\widehat{M}_j|^2/2\,.
$$

Therefore,

$$
\begin{aligned}
\left|\prod_{j=1}^{n}(1+\widehat{M}_j) - \mathrm{e}^{\widehat{M}}\right| &\le \sum_{j=1}^{n}\left|1+\widehat{M}_j-\mathrm{e}^{\widehat{M}_j}\right|\prod_{l=1}^{j-1}\left|1+\widehat{M}_l\right|\prod_{l=j+1}^{n}\mathrm{e}^{\widehat{M}_l} \\
&\le \sum_{j=1}^{n}\left|\widehat{M}_j\right|^2\exp\left(\sum_{l\ne j}^{n}\sum_{r=1}^{k}(1-p_l)p_{l,r}(\widehat{F}_r-1)\right) \\
&\le \sqrt{\mathrm{e}}\,\exp\left(\sum_{r=1}^{k}a_r(\widehat{F}_r-1)\right)\sum_{j=1}^{n}\left|\widehat{M}_j\right|^2.
\end{aligned} \tag{12.30}
$$

Note that

$$|\widehat{M}_j|^2 \le \Big(\sum_{r=1}^k p_{j,r}(1-\widehat{F}_r)\Big)^2 = \sum_{r=1}^k p_{j,r}^2(1-\widehat{F}_r(t_r)) + \sum_{l\ne r} p_{j,l}(1-\widehat{F}_l(t_l))p_{j,r}(1-\widehat{F}_r(t_r)).$$

Substituting this expression into (12.30) and applying (12.29) and Lemma I.1 completes the proof of Theorem 12.10. □

Proof of Proposition 12.11. The beginning of the proof practically repeats the proof of (10.9). Denote $A=\sum_{j=1}^s q_j P^{*j}$, $q=1-p$. Recall that

$$d_K(S_n,\tilde{S}_n) \equiv \|A^{*n} - \exp(n(A-I))\|_K \le C_k n^{-1}$$

by (12.22) with $\alpha=1$. Applying (12.32), (I.3), and (I.4), we get

$$\begin{aligned}\|(qA+pV)^{*n} - (qe^{A-I}+pV)^{*n}\|_K &\le \sum_{j=0}^n \binom{n}{j} q^j p^{n-j} \|A^{*j} - \exp(j(A-I))\|_K \\ &\le \sum_{j=0}^n \binom{n}{j} \frac{q^j p^{n-j}}{j+1} \le C_k/n.\end{aligned}$$

By (12.32),

$$d_K(\exp(a(G-I)); \exp(b(G-I))) \le C_k|a-b|/\max\{1;a;b\}$$

for any symmetric distribution G. Therefore,

$$\begin{aligned}&\|(qe^{A-I}+pV)^{*n} - e^{nq(A-I)} * (qI+pV)^{*n}\|_K \\ &\le \sum_{j=0}^n \binom{n}{j} q^j p^{n-j} \|e^{j(A-I)} - e^{nq(A-I)}\|_K \\ &\le C_k \sum_{j=0}^n \binom{n}{j} q^j p^{n-j} |j-nq|/n \max\{q;j\} \le C_k\sqrt{p/qn}\,.\end{aligned}$$

By the properties of norms, $\|(qI+pV)^{*n}\|_{TV}=1$. Therefore,

$$\|(e^{nq(A-I)} - e^{nqa_s(P-I)}) * (qI+pV)^{*n}\|_K \le \|(e^{nq(A-I)} - e^{nqa_s(P-I)})\|_K\,.$$

Since $A=\sum_{j=1}^s q_j P^{*j}$, $a_s=\sum_{i=1}^s iq_i$, we have

$$\begin{aligned}A-I &= \sum_{j=1}^s q^j(P^{*j}-I) = a_s\sum_{j=1}^s \frac{q^j}{a_s}(P^{*j}-I) \\ &= a_s\sum_{j=1}^s \frac{jq^j}{ja_s}(P^{*j}-I) = a_s\sum_{j=1}^s \frac{jq^j}{a_s}\frac{(P^{*j}-I)}{j}.\end{aligned}$$

Therefore, $\exp(nq(A-I)) = \exp\Big(nqa_s\sum_{j=1}^s \frac{jq_j}{a_s}\frac{(P^{*j}-I)}{j}\Big)$.

Denote

$$b=nqa_s\,,\quad \beta_1 = \min_{j=1,2,\ldots,s}\{jq_j/a_s\},\quad \beta_2=\min_{j=2,\ldots,s}\{jq_j/a_s\},\ldots,\quad \beta_s = sq_s/a_s\,.$$

Then $\beta_1 \le \beta_2 \le \cdots \le \beta_s$. Set

$$h_0 = 1, \quad m_i = \lfloor b(\beta_1 + \ldots + \beta_s) \rfloor + 1.$$

By (12.33),

$$\begin{aligned}
&\left\| \exp\left(b \sum_{j=1}^{s} \beta_j (P - I) \right) - \exp\left(b \sum_{j=1}^{s} \beta_j (P^{*h_j} - I)/h_j \right) \right\|_K \\
&\le \left\| \exp\left(b \sum_{j=1}^{s} \beta_j (P - I) \right) - \exp\left(b \sum_{j=1}^{s} \beta_j (P^{*h_1} - I)/h_1 \right) \right\|_K \\
&+ \left\| \exp\big(b\beta_1 (P^{*h_1} - I)/h_1 \big) \right\|_K \\
&\quad \left\| \exp\left(b \sum_{j=2}^{s} \beta_j (P^{*h_1} - I)/h_1 \right) - \exp\left(b \sum_{j=2}^{s} \beta_j (P^{*h_j} - I)/h_j \right) \right\|_K \le \ldots \\
&\le \sum_{i=1}^{s} \left\| \exp\left(b \sum_{j=i}^{s} \beta_j (P^{*h_{i-1}} - I)/h_{i-1} \right) - \exp\left(b \sum_{j=i}^{s} \beta_j (P^{*h_i} - I)/h_i \right) \right\|_K \\
&\le \sum_{i=1}^{s} \left\| \exp\left(b \sum_{j=i}^{s} \beta_j (P^{*h_{i-1}} - I)/h_{i-1} \right) - P^{*m_i} \right\|_K \\
&+ \sum_{i=1}^{s} \left\| P^{*m_i} - \exp\left(b \sum_{j=i}^{s} \beta_j (P^{*h_i} - I)/h_i \right) \right\|_K \le C(k,s)/na_s .
\end{aligned}$$

The last estimate is due to the fact that $h_i \le s$, $\beta_s \ge 1/s$. Collecting these estimates, we can write:

$$\left\| (qA + pV)^{*n} - \exp(na_s q(P - I)) * (qI + pV)^{*n} \right\|_K \le C(k,s)(1/na_s + \sqrt{p/n}).$$

Recall that

$$\tilde{D}_n = \mathbf{\Pi}(nqa_s; P) * \mathbf{\Pi}(np; V) = \exp(na_s q(P - I)) * \exp(np(V - I)).$$

Using (5.6*) and (3.15), we get

$$\begin{aligned}
&\| \exp(na_s q(P - I)) * (qI + pV)^{*n} - \tilde{D}_n \|_K \\
&\quad \le \| (qI + pV)^{*n} - \exp(np(V - I)) \|_K \\
&\quad \le \| (qI + pI_1)^{*n} - \exp(np(I_1 - I)) \|_{TV} = 2d_{TV}(\mathbf{B}(n,p); \mathbf{\Pi}(np)) \le 2p.
\end{aligned}$$

Therefore, (12.23) follows from the triangle inequality. Estimate (12.24) can be proved similarly with an application of (5.6*) and the upper bound in (7.16). □

Exercises

12.1. Zaitsev [326] has proved that if P is a k-dimensional distribution with a non-negative characteristic function, then

$$d_K(P^{*n}; P^{*(n+1)}) \le C_k/n. \qquad (12.31)$$

Using (12.31), prove that that for any $a>0, b>0$ and any symmetric distribution G

$$d_K(\exp(a(G-I))\,;\exp(b(G-I))) \le C_k|a-b|/\max\{1;a;b\}. \qquad (12.32)$$

12.2. Let F be a k-dimensional distribution with a non-negative characteristic function. Prove that

$$\|F^{*\lfloor a\rfloor+1} - \exp\big(a(F^{*j}-I)/j\big)\|_K \le C_k j/a \qquad (a>0, j\in\mathbb{N}). \qquad (12.33)$$

In particular, (12.33) with $j=1$ entails

$$\|F^{*n+1} - \exp(n(F-I))\|_K \le C_k/n \qquad (\forall n\in\mathbb{N}).$$

12.3 Let $X, X_1, \ldots, X_n$ be i.i.d. two-dimensional random vectors, $\mathbb{P}(X=(0,0)) = 1/2$,

$$\mathbb{P}(X=(1,0)) = \mathbb{P}(X=(-1,0)) = \mathbb{P}(X=(0,1)) = \mathbb{P}(X=(0,-1)) = 1/8.$$

Let $S_n = X_1+\cdots+X_n$, and let $\tilde{S}_n$ denote the sum of accompanying random vectors. Prove that

$$\sum_{m,j=-\infty}^{\infty} \Big(\mathbb{P}(S_n=(m,j)) - \mathbb{P}(\tilde{S}_n=(m,j))\Big)^2 \le 100n^{-3}. \qquad (12.34)$$

This is an estimate in terms of the ℓ_2-metric.

Notes

Theorem 12.6 is essentially Theorem 6.3 from [221].
Theorem 12.7 is Theorem 6.8 from [221].
Theorems 12.9 and 12.10 are Theorems 3.1 and 4.1 in [187].
Estimates (12.24) and (12.23) are proved in [63].

13

Compound Poisson process approximation

Let r.v.s $X_1, X_2, ...$ represent rare events (i.e., $\{X_j\}$ are non-zero with "small" probability). Poisson process approximation to corresponding empirical point processes of exceedances has been studied by many authors (see, e.g., [221, 224], and references therein). However, Poisson process approximation holds only if the limiting cluster size distribution is degenerate.

If the limiting cluster size distribution is not degenerate, then the limiting distribution of an empirical point processes of exceedances is typically compound Poisson (cf. [221], Chapter 7).

This chapter presents results concerning compound Poisson process approximation.

Many general facts concerning Poisson and compound Poisson processes can be found in existing literature (see, e.g., [168, 188, 205, 96]). The main attention in this chapter is given to processes of exceedances.

13.1 Empirical processes

Compound Poisson process is a process with independent compound Poisson increments. Namely, a point process $S(\cdot)$ is called a compound Poisson process with intensity measure Q and compounding (multiplicity) distribution $\mathcal{L}(\zeta)$ if it has independent increments (i.e., for arbitrary disjoint measurable sets $A_1, ..., A_k$ r.v.s $S(A_1), ..., S(A_k)$ are independent) and for any measurable set A random variable $S(A)$ is compound Poisson $\mathbf{\Pi}(Q(A), \mathcal{L}(\zeta))$.

If $Q = \lambda m$, where $\lambda > 0$ is a constant and m is a Lebesgue measure, then we say that $S(\cdot)$ is a compound Poisson process with intensity rate λ and multiplicity distribution $\mathcal{L}(\zeta)$.

Let r.v.s $X, X_1, ..., X_n$ represent rare events (i.e., they are non-zero with "small" probability). Let the point process $S_n(\cdot)$ on $(0;1]$ be defined as

$$S_n(\cdot) = \sum_{i=1}^{n} X_i \mathbb{I}\{i/n \in \cdot\}. \tag{13.1}$$

If $\{X_i\}$ are Bernoulli r.v.s, then point process (13.1) counts *locations* of rare events.

Example 13.1 of a process of exceedances of a "high" threshold. Let $\{\xi_i, i \geq 1\}$ be a stationary sequence of random variables, and let $\{u_n\}$ be a sequence of levels. Set $X_i = \mathbb{I}\{\xi_i > u_n\}$. Denote

$$N_n(B, u_n) = \sum_{i=1}^{n} X_i \mathbb{I}\{i/n \in B\} \tag{13.2}$$

DOI: 10.1201/9781003478164-13

for any measurable set $B \subset (0;1]$. Process $S_n(\cdot) = N_n(\cdot, u_n)$ counts locations of exceedances of level u_n. The distribution of process $N_n(\cdot, u_n)$ can often be approximated by the distribution of a compound Poisson process. □

Given a sequence $X_1, ..., X_n$ of r.v.s, we may consider the jump process

$$S_{n,t} \equiv S_{n,t}^X = \sum_{i=1}^{[nt]} X_i \qquad (0<t\leq 1)$$

known also as a random broken line. Note that $S_{n,t} = S_n((0;t])$.

Example 13.2 Let $X_1, ..., X_n$ be i.i.d. Bernoulli $\mathbf{B}(p)$ r.v.s, and let $Y_1, ..., Y_n$ be i.i.d. Poisson $\mathbf{\Pi}(p)$ r.v.s. Process $S_{n,\cdot}^X$ can be called a "Bernoulli process". Clearly, $S_{n,t}^Y = \sum_{i=1}^{[nt]} Y_i$ $(0<t\leq 1)$ is a Poisson jump process.

Note that

$$\sup_{0<t\leq 1} |S_{n,t}^X - S_{n,t}^Y| = \max_{1\leq k\leq n} |S_k^X - S_k^Y|.$$

Inequalities (3.5), (3.6) provide estimates of the accuracy of Poisson process approximation to the "Bernoulli process" $S_{n,\cdot}^X$ on $(0;1]$. □

Often a natural approximation to $S_n(\cdot)$ is a compound Poisson process.

Let $\{X, X_1, ..., X_n\} \equiv \{X_{n,0}, X_{n,1}, ..., X_{n,n}\}$, where $n \geq 1$, be a triangle array of dependent r.v.s, strictly stationary in each row. In applications r.v.s $\{X_i\}$ are typically non-negative; they usually represent rare events. Therefore, in this section we assume that $X_i \geq 0$ $(\forall i)$.

Set $S_n = X_{n,1} + ... + X_{n,n}$. If a sequence $\{r = r_n\}$ of natural numbers obeys $n \gg r_n \geq 1$, we denote by $\zeta_r \equiv \zeta_{r,n}$ a r.v. with distribution (2.9):

$$\mathcal{L}(\zeta_r) = \mathcal{L}(S_r \,|\, S_r > 0).$$

The next theorem presents necessary and sufficient conditions for the weak convergence of $S_n(\cdot)$ to a compound Poisson point process.

Theorem 13.1 *Assume mixing condition* (Δ), *and suppose that (5.10) holds. If, as* $n \to \infty$,

$$\mathbb{P}(S_n = 0) \to e^{-\lambda} \qquad (\exists \lambda > 0), \tag{2.12*}$$

$$\mathcal{L}(S_r | S_r \neq 0) \Rightarrow \mathcal{L}(\zeta) \tag{5.11*}$$

for a sequence $\{r = r_n\}$ *obeying (2.8), then*

$$S_n(\cdot) \Rightarrow S(\cdot), \tag{13.3}$$

where $S(\cdot)$ *is a compound Poisson point process with intensity rate* λ *and multiplicity distribution* $\mathcal{L}(\zeta)$.

If $S_n(\cdot)$ *converges weakly to a point process* $S(\cdot)$, *then* $S(\cdot)$ *is a compound Poisson process on* $(0;1]$ *with intensity rate* λ *given by (2.12*). If* $\lambda > 0$, *then (5.11*) is valid for some r.v.* ζ *and sequence* $\{r_n\}$ *obeying (2.8).*

Condition (5.10) can be relaxed to allow for $np \to \infty$ as $n \to \infty$ at a certain "slow" rate, where $p \equiv p(n) = \mathbb{P}(X \neq 0)$.

Example 13.3 Let $\{X_i, i \geq 0\}$ be a *regenerative* process, i.e., there exist integer-valued r.v.s $0 < S_0 < S_1 < \ldots$ such that the "cycles"

$$\{X_i,\, 0 \leq i < S_0\},\ \{X_i,\, S_0 \leq i < S_1\}, \ldots$$

are i.i.d. We define r.v.s $Y, Y_1, Y_2, \ldots$ as follows:

$$Y = \max_{0 \leq i < S_0} X_i\,,\ Y_1 = \max_{S_0 \leq i < S_1} X_i\,, \ldots$$

Let

$$T_r = \sum_{i=S_{r-1}}^{S_r - 1} \mathbb{1}\{X_i > u_n\} \qquad (r \in \mathbb{N}),$$

where $\{u_n\}$ is a sequence of levels.

Suppose that S_0 is aperiodic, $\mu := \mathbb{E}S_0 < \infty$ and $\mathbb{P}(Y > \max_{1 \leq j \leq k} Y_i) \to 0$ as $k \to \infty$.

Process $N_n(\cdot, u_n)$ converges weakly to a non-degenerate point process N if and only if there exist $\lambda > 0$ and a distribution P such that

$$n\mathbb{P}(Y > u_n)/\mu \to \lambda \text{ and } \mathcal{L}(T_1 | T_1 > 0) \Rightarrow P$$

as $n \to \infty$; necessarily N a compound Poisson point process with intensity rate λ and multiplicity distribution P (Rootzén [267]). □

13.2 Excess process

Let $X, X_1, X_2, \ldots, X_n$ be a stationary sequence of r.v.s. When one is interested in the joint distribution of exceedances of several levels among $X_1, \ldots, X_n$, a natural tool is the excess process $N_n^\varepsilon(\cdot)$. This section presents necessary and sufficient conditions for the weak convergence of the excess process to a compound Poisson process.

Given a sequence $\{u_n(\cdot), n \geq 1\}$ of functions on $[0; \infty)$, denote

$$N_n^\varepsilon(t) = \sum_{i=1}^{n} \mathbb{1}\{X_i > u_n(t)\} \qquad (t > 0).$$

Let $T > 0$. We call $\{N_n^\varepsilon(t), t \in [0; T]\}$ the *excess process*.

Process $N_n^\varepsilon(\cdot)$ describes variability in the *heights* of the observations $X_1, X_2, \ldots, X_n$.

Note that $N_n^\varepsilon(\cdot)$ is the "tail empirical process" (13.4) for $Y_{n,1}, \ldots, Y_{n,n}$, where $Y_{n,i} = u_n^{-1}(X_i)$:

$$N_n^\varepsilon(t) = \sum_{i=1}^{n} \mathbb{1}\{Y_{n,i} < t\}. \tag{13.4}$$

There is a considerable amount of research on the topic of tail empirical processes (see, e.g., [95, 202] and references therein).

We present necessary and sufficient conditions for the weak convergence of the excess process to a compound Poisson process in Theorem 13.2.

Suppose that function $u_n(\cdot)$ is strictly decreasing for all large enough n, $u_n(0) = \infty$,

$$\limsup_{n \to \infty} n\mathbb{P}(X > u_n(t)) < \infty \qquad (0 < t < \infty), \tag{13.5}$$

$$\lim_{n \to \infty} \mathbb{P}(N_n(u_n(t)) = 0) = \mathrm{e}^{-t} \qquad (t > 0). \tag{13.6}$$

Condition (13.6) means $u_n(\cdot)$ is a "proper" normalising sequence for the sample maximum $M_n = \max\{X_1, ..., X_n\}$.

Let $\{\pi_s, s \geq 0\}$ be a Poisson process with intensity rate 1, and let $\zeta_1, \zeta_2, ...$ be a sequence of i.i.d. copies of ζ. Denote

$$Q_\zeta(t) = \sum_{j=1}^{\pi_t} \zeta_j \,. \tag{13.7}$$

Then $\{Q_\zeta(t), t \geq 0\}$ is a compound Poisson jump process. Equivalently, $\tilde{Q}_\zeta(B) = \int_B Q_\zeta(dt)$ is a compound Poisson point process with the Lebesgue intensity measure and multiplicity distribution $\mathcal{L}(\zeta)$. We do not distinguish between Q_ζ and $\tilde{Q}_\zeta$ in the sequel.

Recall that condition (C) was introduced on p. 205.

Theorem 13.2 *Assume mixing condition* (Δ), *(13.5), (13.6), and let* $\pi_\zeta(\cdot)$ *denote a compound Poisson process with intensity rate 1 and multiplicity distribution* $\mathcal{L}(\zeta)$. *Then*

$$N_n^\varepsilon(\cdot) \Rightarrow Q_\zeta(\cdot)$$

as $n \to \infty$ *if and only if condition* (C) *holds.*

General situation.

Excess process $N_n^\varepsilon(\cdot)$ may converge weakly to a process of a more complex structure:

$$\{N_n^\varepsilon(t),\, t \leq T\} \Rightarrow \left\{ \sum_{j=1}^{\pi_T} \gamma_j(t/T),\, t \leq T \right\} \tag{13.8}$$

as $n \to \infty$, where π_T is a Poisson r.v., $\{\gamma_j(\cdot)\}$ are independent jump processes.

Process $\left\{ \sum_{j=1}^{\pi_T} \gamma_j(\cdot) \right\}$ can be called *Poisson cluster process* or *compound Poisson process of the second order* (regarding the standard compound Poisson process as a "compound Poisson process of the first order").

Results concerning approximation (13.8) can be found in [221], Chapter 8. In particular, the accuracy of approximation to the distribution of an excess process can be evaluated in terms of the total variation distance (cf. [221], Theorem 8.3).

13.3 General point processes of exceedances

Both (13.2) and (13.4) are one–dimensional processes of exceedances. In this section, we deal with a general point process of exceedances N_n^*, which counts locations of extremes (rare events) as well as their heights.

For any Borel set $A \subset (0; 1] \times [0; \infty)$ denote

$$N_n^*(A) = \sum_{i=1}^{n} \mathbb{1}\{\, (i/n, u_n^{-1}(X_i)) \in A\,\}.$$

Two–dimensional point process N_n^* counts locations of exceedances of "high" levels as well as their "heights".

If $\{X_i\}$ are i.d.d.r.v.s, or if $\{X_i, i \geq 1\}$ is a strictly stationary sequence obeying a certain mixing condition, then $N_n^*(\cdot)$ converges weakly to a pure Poisson point process (Adler [1], see also [224]).

The following theorem presents necessary and sufficient conditions for the weak convergence of point process $N_n^*(\cdot)$ to a compound Poisson point process.

Denote by $N^*(\cdot)$ a compound Poisson point process on $(0;1]\times[0;\infty)$ with the Lebesgue intensity measure and multiplicity distribution $\mathcal{L}(\zeta)$. Note that

$$Q_\zeta(t) \stackrel{d}{=} N^*((0;1] \times [0;t)).$$

Recall that condition (C) was introduced on p. 205.

Theorem 13.3 *Assume conditions* (Δ), *(13.5), (13.6). Then*

$$N_n^* \Rightarrow N^* \tag{13.9}$$

if and only if condition (C) *holds.*

Results concerning weak convergence of point process N_n^* to a Poisson cluster process can be found in [221], Chapter 8.

Example 13.4 Let $\{\xi_i\}, \{\alpha_i\}$ be independent sequences of i.i.d.r.v.s, $\mathbb{P}(\xi \leq x) = F(x)$, $\mathcal{L}(\alpha) = \mathbf{B}(\theta)$, where $\theta \in (0;1)$. Put $X_1 = \xi_1$, and let

$$X_i = \alpha_i \xi_i + (1-\alpha_i)X_{i-1} \qquad (i \geq 2).$$

Then $\{X_i, i \geq 1\}$ is a stationary sequence of r.v.s with the marginal d.f. F, the cluster sizes have the geometric $\mathbf{\Gamma}(1-\theta)$ distribution.

Notice that sequence $\{X_i, i \geq 1\}$ is φ–mixing, and

$$\varphi(k) \leq (1-\theta)^k \qquad (k \geq 1).$$

Indeed, suppose that $A \in \sigma\{X_1, ..., X_m\}$, $B \in \sigma\{X_{m+k+1}, ...\}$, and let Z be the length of a 0–run starting at α_{m+1} (we set $Z = 0$ if $\alpha_{m+1} = 1$). Then event $B \cap \{Z \leq k\}$ is independent of $\sigma\{X_1, ..., X_m\}$. Hence $\mathbb{P}(B, Z \leq k|A) = \mathbb{P}(B, Z \leq k)$ and

$$\mathbb{P}(B, Z > k|A) \leq \mathbb{P}(Z > k|A) = \mathbb{P}(\alpha_1 = ... = \alpha_{k+1} = 0) = (1-\theta)^{k+1}.$$

Furthermore,

$$\mathbb{P}(M_n \leq u) = F(u)\mathbb{E}(1-p)^N = F(u)(1-\theta p)^{n-1},$$

where $N = \sum_{i=2}^n \alpha_i$ is Binomial $\mathbf{B}(n-1, \theta)$ r.v.

Denote $K^* = \sup\{x\colon F(x) < 1\}$, and assume that

$$\mathbb{P}(X \geq x)/\mathbb{P}(X > x) \to 1 \tag{13.10}$$

as $x \to K^*$ (Gnedenko's condition [133]). Then there exists a sequence $\{u_n\}$ such that $n\mathbb{P}(X > u_n) \to 1$ (cf. Theorem 1.7.13 in [189]). Let

$$u_n(t) = u_{[\theta n/t]}.$$

Then

$$\mathbb{P}(X > u_n(s)) \sim s/n\theta, \ \mathbb{P}(N_r(u_n(s)) > 0) \sim sr/n,$$

and $\{u_n(\cdot)\}$ obeys (13.6).

We now check condition (C). Let $s<t<v$. Condition (b) follows from the estimate

$$\begin{aligned}&\mathbb{P}(N_r[u_n(t);u_n(v))>0, N_r[u_n(s);u_n(t))>0)\\ &\le r^2\mathbb{P}(u_n(v)<\xi\le u_n(t))\mathbb{P}(u_n(t)<\xi\le u_n(s)) = O((r/n)^2)\end{aligned}$$

($N_r[x;y)$ has been introduced on p. 203). Random variables $\{X_i,...,X_{i+m-1}\}$ form a cluster of 0's of size m if $\alpha_{i-1}=1$, $\alpha_i= ... =\alpha_{i+m-1}=0, \alpha_{i+m}=1$. Denote

$$W_r = \mathbb{I}_1 + \sum_{i=2}^{r} \alpha_i \mathbb{I}_i\,,$$

where $\mathbb{I}_i = \mathbb{I}\{\xi_i \in (u_n(t);u_n(s)]\}$. Asymptotically, only one cluster among $X_1,...,X_r$ may hit $(u_n(t);u_n(s)]$. Therefore,

$$\begin{aligned}&\mathbb{P}(N_r[u_n(s);u_n(t)) = j) \sim \mathbb{P}(N_r[u_n(s);u_n(t))=j, N_r(u_n(s))=0)\\ &= \mathbb{P}(N_r[u_n(s);u_n(t))=j, N_r(u_n(s))=0, W_r=1) + O((r/n)^2)\\ &\sim r\theta^2(1-\theta)^{j-1}\mathbb{P}(u_n(t)<\xi\le u_n(s)) \sim (t-s)\mathbb{P}(\zeta=j)r\theta/n, \qquad (13.11)\end{aligned}$$

where $\mathcal{L}(\zeta) = \mathbf{\Gamma}(1-\theta)$. Thus, condition (a) holds, and Theorem 13.3 entails

$$N_n^* \Rightarrow N^*,$$

where N^* is the compound Poisson point process with the Lebesgue intensity measure and multiplicity distribution $\mathbf{\Gamma}(1-\theta)$. □

An estimate of the accuracy of approximation to $\mathcal{L}(N_n^*(\cdot))$ in terms of a d_G-type distance has been established in [30].

Open problem

13.1. Improve the estimate of the accuracy of approximation $N_n^* \approx N^*$ from [30].

13.4 Proofs

Proofs of theorems (sketch) involve Bernstein's blocks method, Lindeberg's device and results concerning defining a random vector X and an independent copy of X on a common probability space (cf. the proof of Theorem 6.8).

Lindeberg's device means replacing $f(X_1,...,X_n)$ with $f(Y_1,...,Y_n)$ step by step, where f is an arbitrary function:

$$f(X_1,...,X_n)-f(Y_1,...,Y_n) = \sum_{i=1}^{n}(f(Y_1,...,Y_{i-1},X_i,...,X_n)-f(Y_1,...,Y_i,...,X_n)).$$

The well-known relation $a^n - b^n = (a-b)\sum_{i=1}^n b^{i-1}a^{n-i}$ is an example of this approach.

Bernstein's blocks method suggests splitting the sample $X_1,...,X_n$ of dependent r.v.s into blocks $(X_1,...,X_r),...$ of length $r\in\mathbb{N}$ separated by subblocks $(X_{r+1},...,X_{r+l}),...$ of length $l\ll r$. The contribution of the subblocks is "small" if so is $lp[n/r]$, where p is the probability of an exceedance/rare event. Hence the subblocks can be omitted if the term $o(1)$ has been added.

The remaining (reduced) blocks $(X_1, ..., X_r), (X_{r+l+1}, ..., X_{2r+l}), ...$ are "almost independent" (the corresponding σ-fields are separated by l steps), and can be replaced with their independent copies thanks to Berbee's lemma G.1 and Lemma G.2.

We can demonstrate this approach by proving that condition (C) entails (13.9).

Suppose that condition (C) holds. We need to show that

$$\{N_n^*(A_1), \ldots, N_n^*(A_k)\} \Rightarrow \{N^*(A_1), \ldots, N^*(A_k)\} \tag{13.12}$$

for any array $\{A_1, \ldots, A_k\}$ of Borel sets on $(0;1] \times [0;\infty)$.

Since Borel sets can be approximated by finite unions of rectangles, it suffices proving (13.12) for finite unions of rectangles $\{A_i\}$. Splitting unions of rectangles in a proper way, we observe that it suffices proving (13.12) in the case

$$A_i = (a_i; b_i] \times \bigcup_{j=1}^{m_i} [c_{ij}; d_{ij}),$$

where intervals $(a_i; b_i]$ are disjoint, and for each i intervals $[c_{ij}; d_{ij})$ are disjoint too.

By the argument mentioned above, random variables $\{N_n^*(A_i)\}$ are asymptotically independent. Thus, it remains to show that

$$N_n^*(A) \Rightarrow N^*(A) \qquad (n \to \infty) \tag{13.13}$$

where $A = (a; b] \times \bigcup_{j=1}^{m} [c_j; d_j)$. But (13.13) is given by (12.9) – though (12.9) has been proven for the case $(a; b] = (0; 1]$, the argument is evidently valid for an arbitrary interval $(a; b] \subset (0; 1]$. The proof of (13.9) is complete. □

Exercises

13.1. Let $\Pi_{\lambda_1}^{\zeta}$ and $\Pi_{\lambda_2}^{\zeta}$ be independent compound Poisson point processes on [0;1] with intensity rates λ_1, λ_2 and multiplicity distribution $\mathcal{L}(\zeta)$. Show that $\Pi_{\lambda_1}^{\zeta} + \Pi_{\lambda_2}^{\zeta} \stackrel{d}{=} \Pi_{\lambda_1+\lambda_2}^{\zeta}$ and

$$d_{TV}\left(\Pi_{\lambda_1}^{\zeta}; \Pi_{\lambda_2}^{\zeta}\right) \leq 1 - \exp\left(-|\lambda_1 - \lambda_2|\right).$$

13.2. Show that compound Poisson process is infinitely divisible.

13.3. Let $\pi, \xi, \xi_1, ..., \zeta, \zeta_1, ...$ be independent r.v.s, $\mathcal{L}(\pi) = \mathbf{\Pi}(1)$, $\xi_i \stackrel{d}{=} \xi \in \mathbf{U}(0;1)$, $\zeta_i \stackrel{d}{=} \zeta$, ζ takes values in $\mathbb{Z}_+$. Define the jump process

$$N(t) = \sum_{j=1}^{\pi} \zeta_j \mathbb{1}\{\xi_j \leq t\}.$$

Prove that $\{N(t), t \in [0; 1]\}$ is a compound Poisson process.

Notes

The theory of compound Poisson processes goes back to Lundberg [200] and Erlang [114].

Theorem 13.1 in the case of process $N_n(\cdot, u_n) = \sum_{i=1}^n \mathbb{I}\{\xi_i > u_n\}\mathbb{I}\{i/n \in \cdot\}$ is due to Hsing et al. [153].

Theorem 13.2 is Theorem 7.2 from [221]. Theorem 13.3 is Theorem 7.4 from [221].

14

Appendix

This chapter collects various auxiliary results about metrics, characteristic functions, inversion lemmas, Stein's method, properties of convolutions, and concentration functions.

A Norms and metrics

Measure M on $\mathbb{R}$ (equipped with a particular sigma-algebra $\mathcal{A}$) is called *finite* if

$$-\infty < \inf_{A\in\mathcal{A}} M(A) \le \sup_{A\in\mathcal{A}} M(A) < \infty$$

(in general, there may exist a measurable set A such that $M(A)<0$).

Given a finite (signed) measure M, let

$$M = M^+ - M^-$$

denote the Hahn-Jordan decomposition of M, where M^+ and M^- are non-negative measures.

Given a set $A\in\mathcal{A}$, let A^c denote a complement to A. There exists set $A_+\in\mathcal{A}$ such that

$$M^+(\cdot) = M(A_+\cap\cdot), \quad M^-(\cdot) = M(A_+^c\cap\cdot).$$

Clearly, $A_+ = \text{supp } M^+$, $A_+^c = \text{supp } M^-$. The sum

$$\|M\|_{TV} = M^+(\mathbb{R})+M^-(\mathbb{R}) = M^+(A_+)+M^-(A_+^c)$$

is called the *total variation norm* of M.

If M is a finite measure, then its total variation norm $\|M\|_{TV}<\infty$.

Recall that the class of all probability distributions is denoted by $\mathcal{F}$. The total variation norm of any probability distribution equals 1.

An important property of the total variation norm is that

$$\|P*Q\|_{TV} \le \|P\|_{TV}\,\|Q\|_{TV} \tag{A.1}$$

for any (signed) measures P and Q, where $*$ denotes convolution of measures.

If $F_1,\dots,F_n,G_1,\dots,G_n$ are probability distributions, then (A.1) and the triangle inequality entail

$$\|F_1*\cdots*F_n - G_1*\cdots*G_n\|_{TV} \le \sum_{j=1}^n \|F_j - G_j\|_{TV}\,. \tag{A.2}$$

Given $F\in\mathcal{F}$ and a sequence $a=\{a_i\}_{i\ge1}$ of numbers such that $\sum_{i=0}^\infty |a_i| < \infty$, we call

$$\varphi_a(F) = \sum_{i=0}^\infty a_i F^{*i}$$

DOI: 10.1201/9781003478164-A

a compound measure. Clearly,

$$\|\varphi_a(F)\|_{TV} \le \sum_{j=0}^{\infty} |a_j| = \|\varphi(I_1)\|_{TV} \,. \tag{A.3}$$

Bound (A.3) allows one to evaluate the total variation distance between two compound Poisson r.v.s with the same compounding distribution, cf. [89], p. 28.

If $0 \le a_j \le 1$ $(\forall j)$, $\sum_{j=0}^{\infty} a_j = 1$, then

$$\varphi_a(F) = \mathcal{L}\left(\sum_{i=1}^{\nu} \zeta_i\right),$$

where $\nu, \zeta, \zeta_1, \ldots$ are independent r.v.s, $\mathcal{L}(\zeta) = F$, $\mathbb{P}(\nu = k) = a_k$ $(k \in \mathbb{Z})$.

The following lemma relates the total variation norm and the total variation distance.

Lemma A.1 *Let M be a measure of finite variation (i.e., a finite measure) defined on $(\mathbb{R}, \mathcal{A})$. If $M(\mathbb{R}) = 0$, then*

$$\sup_{A \in \mathcal{A}} |M(A)| = \|M\|_{TV}/2. \tag{A.4}$$

Proof of Lemma A.1. We apply the Hahn-Jordan decomposition $M = M^+ + M^-$, where M^+, M^- are non-negative measures. Denote $B = \text{supp}\, M^+$. Then $B^c = \text{supp}\, M^-$, $B \cap B^c = \emptyset$,

$$\|M\|_{TV} = M^+(\mathbb{R}) + M^-(\mathbb{R}) = M^+(B) + M^-(B^c).$$

By assumption, $M^+(\mathbb{R}) = M^-(\mathbb{R})$. Hence

$$\|M\|_{TV} = 2M^+(\mathbb{R}) = 2M^+(B) = 2M^-(B^c).$$

For any $A \in \mathcal{A}$

$$|M(A)| = |M^+(A) - M^-(A)| \le M^+(A) \vee M^-(A) \le M^+(B) \vee M^-(B^c) = \|M\|_{TV}/2.$$

The supremum is achieved on B: $M(B) = M^+(B) = \|M\|_{TV}/2$. □

Recall that the *total variation distance* $d_{TV}(X;Y)$ between the distributions of r.v.s X and Y is

$$d_{TV}(X;Y) \equiv d_{TV}(\mathcal{L}(X);\mathcal{L}(Y)) = \sup_{A \in \mathcal{A}} |\mathbb{P}(X \in A) - \mathbb{P}(Y \in A)|\,,$$

where $\mathcal{A}$ is a Borel σ-field. Lemma A.1 states that

$$d_{TV}(X;Y) = \tfrac{1}{2}\|\mathcal{L}(X) - \mathcal{L}(Y)\|_{TV} \,.$$

According to Dobrushin's theorem (see [109, 48]),

$$d_{TV}(X;Y) = \inf_{X',Y'} \mathbb{P}(X' \ne Y'),$$

where the infimum is taken over all random pairs (X', Y') such that $\mathcal{L}(X') = \mathcal{L}(X)$, $\mathcal{L}(Y') = \mathcal{L}(Y)$. The total variation distance can also be expressed as

$$d_{TV}(X;Y) = \sup_f |\mathbb{E}f(X) - \mathbb{E}f(Y)|,$$

where the supremum is over all measurable functions f taking values in $[0;1]$. If X and Y are integer-valued r.v.s, then

$$d_{TV}(X,Y) = \frac{1}{2}\sum_{j\in\mathbb{Z}} |\mathbb{P}(X=j) - \mathbb{P}(Y=j)|.$$

If X and Y are absolutely continuous r.v.s with densities $p_X(\cdot), p_Y(\cdot)$, then

$$d_{TV}(X;Y) = \frac{1}{2}\int_{-\infty}^{\infty} |p_X(x) - p_Y(x)|\mathrm{d}x. \tag{A.5}$$

Historically the accuracy of approximation to $\mathcal{L}(S_n)$ was first studied in terms of the *uniform* distance (sometimes called the *Kolmogorov* distance).

The uniform distance $d_K(X;Y) \equiv d_K(F_X;F_Y)$ between the distributions of random variables X and Y with distribution functions (d.f.s) F_X and F_Y is defined as

$$d_K(X;Y) \equiv d_K(\mathcal{L}(X);\mathcal{L}(Y)) = \sup_x |F_X(x) - F_Y(x)|$$

(in the multidimensional case F_X , F_Y denote multivariate distribution functions). Evidently, $d_K(X;Y) \le d_{TV}(X;Y)$.

Given a measure M, we denote by

$$\|M\|_K = \sup_{x\in\mathbb{R}} |M((-\infty;x])|$$

the norm corresponding to Kolmogorov's distance. Obviously, $\|M\|_K \le \|M\|_{TV}$.

For any finite measures M_1, M_2

$$\|M_1 * M_2\|_K \le \|M_1\|_K \|M_2\|_{TV} \wedge \|M_1\|_{TV} \|M_2\|_K. \tag{A.6}$$

Let X, Y be integer-valued r.v.s. The *point metric* (*local metric*) is defined as

$$d_{\mathrm{o}}(X;Y) = d_{\mathrm{o}}(\mathcal{L}(X);\mathcal{L}(Y)) = \sup_{k\in\mathbb{Z}} |\mathbb{P}(X=k) - \mathbb{P}(Y=k)|.$$

If measure the support of M is concentrated on the set $\mathbb{Z}$ of integer numbers, the *local norm* is defined as

$$\|M\|_{\mathrm{o}} = \sup_{k\in\mathbb{Z}} |M(\{k\})|.$$

In the sequel we may use notation $M\{k\} := M(\{k\})$.

The following relations between metrics hold true:

$$\|M\|_{\mathrm{o}} \le \|M\|_{TV}, \quad \|M\|_{\mathrm{o}} = \|(I_1 - I) * M\|_K \le \|(I_1 - I) * M\|_{TV}, \tag{A.7}$$

cf. [89], p. 63. Observe that $\|(I_1 - I) * M\|_{TV} = \sum_k |\Delta M(\{k\})|$, where

$$\Delta M(\{k\}) = M(\{k+1\}) - M(\{k\}).$$

The Gini–Kantorovich distance $d_G(X;Y) \equiv d_G(\mathcal{L}(X);\mathcal{L}(Y))$ between the distributions of r.v.s X and Y with finite first moments (known also as the Kantorovich-Wasserstein distance) is given by (1.11). Note that

$$d_G(X;Y) = \inf_{X',Y'} \mathbb{E}|X' - Y'|,$$

where the infimum is taken over all random pairs (X',Y') such that $\mathcal{L}(X')=\mathcal{L}(X)$, $\mathcal{L}(Y')=\mathcal{L}(Y)$ [242, 303]. If X and Y take values in $\mathbb{Z}_+$, then [272, 101]

$$d_G(X;Y)=\sum_{i\geq 1}|\mathbb{P}(X\geq i)-\mathbb{P}(Y\geq i)|.$$

Distance d_G was apparently introduced by Gini [131]; Kantorovich [170] has introduced a class of metrics that includes d_G.

Let M be a measure concentrated on the set of integer numbers. Assume that $M(\mathbb{Z})=0$. Denote by

$$\|M\|_G=\sum_{k=-\infty}^{\infty}|M((-\infty;k])|=\sum_{k=-\infty}^{\infty}|M((k;\infty))| \tag{A.8}$$

the norm corresponding to the Gini-Kantorovich distance. Observe that

$$\|M\|_{TV}=\|(I_1-I)*M\|_G. \tag{A.9}$$

Indeed, $M*(I_1-I)((-\infty,k])=M((-\infty,k-1])-M((-\infty,k])=-M(\{k\})$.

Relations (A.7), (A.9) may hint why the rate of approximation in terms of $\|\cdot\|_{\circ}$ is typically sharper than in terms of $\|\cdot\|_{TV}$, and the rate of approximation in terms of d_{TV} is typically sharper than in terms of d_G.

A generalisation of d_G is the Kantorovich-Rubinstein distance

$$d_t(\mathbb{P}_X;\mathbb{P}_Y)=\inf_{X,Y}(\mathbb{E}|X-Y|^t)^{1/t} \qquad (t\geq 1),$$

where the infimum is taken over all random pairs (X,Y) with the marginal distributions $\mathbb{P}_X$ and $\mathbb{P}_Y$ respectively. Here one can put $X=F_X^{-1}(\xi)$, $Y=F_Y^{-1}(\xi)$, where F_X and F_Y are the d.f.s of $\mathcal{L}(X)$ and $\mathcal{L}(Y)$, ξ is a uniform $\mathbf{U}[0;1]$ r.v. It is known that

$$d_t(\mathbb{P}_X;\mathbb{P}_Y)=\sup\int|g(F_X(y))-g(F_Y(y))|\mathrm{d}y,$$

where the supremum is taken over all smooth functions g such that $\int_0^1|g'(z)|^{t/(t-1)}\mathrm{d}z\leq 1$.

If distributions P_1 and P_2 have densities f_1 and f_2 with respect to a measure μ, set

$$d_H^2(P_1;P_2)=\frac{1}{2}\int\left(f_1^{1/2}-f_2^{1/2}\right)^2\mathrm{d}\mu=1-\int\sqrt{f_1f_2}\,\mathrm{d}\mu.$$

Then d_H denotes the *Hellinger* distance. It is known that

$$d_H^2\leq d_{TV}\leq d_H\sqrt{2-d_H^2}\,.$$

Denote

$$\chi^2(P_1;P_2)=\int_{\mathrm{supp}P_2}(\mathrm{d}P_1/\mathrm{d}P_2-1)^2\,\mathrm{d}P_2\,.$$

By the Cauchy-Bunyakovski inequality,

$$2d_{TV}(P_1;P_2)\leq\chi(P_1;P_2).$$

Let

$$d_{KL}^2(P_1;P_2)=\int_{\mathrm{supp}P_2}\ln(\mathrm{d}P_1/\mathrm{d}P_2)\,\mathrm{d}P_1$$

denote the *Kullback–Leibler* divergence. According to a Pinsker-type inequality,

$$d_{TV} \leq d_{KL}/\sqrt{2}\,.$$

Though d^2_{KL} is not a metric, it plays a role in statistics (cf. [148]) and in the theory of large deviations (cf. [221], formula (14.40), and ex. 41 on p. 324).

Lévy's metric is defined as

$$d_L(X;Y) = \inf\{\varepsilon>0 : \mathbb{P}(X<x-\varepsilon) - \varepsilon \leq \mathbb{P}(Y<x) \leq \mathbb{P}(X<x+\varepsilon) + \varepsilon \quad (\forall x\in\mathbb{R})\}.$$

It is weaker than the total variation distance: $d_L(X;Y) \leq d_{TV}(X;Y)$. Convergence in d_L entails weak convergence of distributions. Unlike Kolmogorov's or the total variation distance, Lévy's metric is not invariant under the scale transformation.

Certain other distances can be found in [198, 221, 247, 248, 260]. For the relations between metrics, see, e.g., [130, 292, 221].

Exercises

A.1 Let $P_t = \mathcal{N}(t;1)$, $t>0$. Show that

$$d_{TV}(P_0;P_t) = 2P_0([0;t/2]).$$

Hence $d_{TV}(P_0;P_t) \sim t/\sqrt{2\pi}$ as $t\to 0$.

B Convolutions of measures

The distribution of a sum of independent r.v.s is a convolution of measures. The commutative property of convolution and the exponential structure of CP distributions can be used to establish sharp bounds of the accuracy of compound Poisson approximation. This section contains a collection of such bounds.

This approach can be called the method if convolutions, cf. [89]. The method of convolutions is relatively simple to apply, cf. the proofs of Theorems 6.6 and 10.8. The main drawback of the method is that it cannot be used for sums of dependent r.v.s.

Exponent of a measure.

Recall that distribution P is infinitely divisible if for any $n\in\mathbb{N}$ there exists distribution P_n such that $P_n^{*n}=P$.

Any compound Poisson distribution is infinitely divisible:

$$\exp(\lambda(P-I)) = (\exp(\lambda(P-I)/n))^{*n}.$$

By the same argument any SCP measure is infinitely divisible.

The following easily verifiable identities were repetitiously used in the proofs:

$$\mathrm{e}^M - \mathrm{e}^V = (M-V) * \int_0^1 \exp(tM+(1-t)V)\,\mathrm{d}t, \tag{B.1}$$

$$\mathrm{e}^P = I + P + P^{*2}/2 + \ldots + \frac{P^{*k}}{k!} + \frac{P^{*(k+1)}}{k!} * \int_0^1 \mathrm{e}^{tP}(1-t)^k\mathrm{d}t$$

for any finite measures M, V. Indeed, using (1.5), we get

$$\begin{aligned} e^M - e^V &= e^V * (e^{M-V} - I) \\ &= e^V * \int_0^1 (M-V) * e^{(M-V)t} dt = (M-V) * \int_0^1 \exp(tM+(1-t)V)\, dt. \end{aligned}$$

The second equation is equation (1.6).

Applied to CP distribution $\mathbf{\Pi}(\lambda, P) = \exp(\lambda(P-I))$, where $\lambda>0$, (1.6) yields

$$\exp(\lambda(P-I)) = I + \sum_{i=1}^{k} \frac{\lambda^i (P-I)^{*i}}{i!} + \frac{\lambda^{k+1}}{(k+1)!}(P-I)^{*(k+1)} * \Theta \qquad (k \in \mathbb{Z}_+). \tag{B.2}$$

In particular,

$$\exp(\lambda(P-I)) - I = \lambda(P-I) * \Theta \stackrel{d}{=} \lambda(P-I) + \frac{\lambda^2}{2}(P-I)^{*2} * \Theta. \tag{B.2*}$$

If $\lambda=1$, then this yields

$$\exp(P-I) - P = \tfrac{1}{2}(P-I)^{*2} * \Theta. \tag{B.2*}$$

The next lemma is essentially Lemma 2.5 from [89]. It appears in SCP approximations. Recall that Θ denotes a measure satisfying $\|\Theta\|_{TV} \le 1$.

Lemma B.1 *Let $a>0$, $0<b<a/3$. There exists absolute constant C such that for any $P \in \mathcal{F}$*

$$\left\| \exp\big(a(P-I) + b(P-I)^{*2} * \Theta\big) \right\|_{TV} \le C.$$

Lemma B.2 [70] *Let $0 \le p \le \tilde{c} < 1$, $n \in \mathbb{N}$. Then for any $P \in \mathcal{F}$*

$$\exp\big(np(P-I) - np^2(P-I)^{*2}/2\big) = \exp(np(1-\tilde{c})(P-I)/2) * \Theta_1,$$

where Θ_1 is a measure such that $\|\Theta_1\|_{TV} \le C$.

Note that

$$\exp\big(np(P-I) - np^2(P-I)^{*2}/2\big) = \exp\big(np(1+p)(P-I) - np^2(P^{*2}-I)/2\big).$$

Therefore, the statement of Lemma B.2 can be rewritten as

$$\mathbf{\Pi}(np(1+p); P) * \mathbf{\Pi}(-np^2/2; P^{*2}) = \mathbf{\Pi}(np(1-\tilde{c})/2; P) * \Theta_1,$$

where

$$\Theta_1 = \mathbf{\Pi}(np(p+(1+\tilde{c})/2; P) * \mathbf{\Pi}(-np^2/2; P^{*2}).$$

The same statement can be written in terms of r.v.s. Let $X_1, X_2, \ldots,\ X', X_1', X_2', \ldots,$ $Y, Y_1, Y_2, \ldots$ be independent r.v.s, $\mathcal{L}(X_i) = \mathcal{L}(X_i') = P$, $L(Y_i) = \mathcal{L}(X_i + X_i') = P^{*2}$ $(\forall i)$. Let independent generalised r.v.s $\pi_{-p^2/2}$, $\pi_{np(1+p)}$, $\pi_{np(1-\tilde{c})/2}$, η be independent of other r.v.s, where $\|\mathcal{L}(\eta)\|_{TV} \le 1$. Then Lemma B.2 states that

$$\sum_{i=0}^{\pi_{np(1+p)}} X_i + \sum_{k=0}^{\pi_{-np^2/2}} Y_k \stackrel{d}{=} \sum_{i=1}^{\pi_{np(1-\tilde{c})/2}} X_i' + \eta.$$

Bergström's identity.
Given if $a\in\mathbb{R}$, $b\in\mathbb{R}$, recall that

$$a^n - b^n = (a-b)\sum_{i=1}^{n} b^{i-1}a^{n-i}\,, \quad a^n = (a-b+b)^n = b^n + \sum_{m=1}^{n}\binom{n}{m}(a-b)^m b^{n-m}\,. \qquad \text{(B.3)}$$

Thus,

$$a^n - b^n = \sum_{m=1}^{k}\binom{n}{m}(a-b)^m b^{n-m} + (a-b)^{k+1}\sum_{m=k+1}^{n}\binom{n}{m}(a-b)^{m-k-1}b^{n-m}\,.$$

Bergström's identity [41] states that for any $k\in\{0,...,n-1\}$

$$a^n - b^n = \sum_{m=1}^{k}\binom{n}{m}(a-b)^m b^{n-m} + (a-b)^{k+1}\sum_{m=k}^{n-1}\binom{m}{k}a^{n-1-m}b^{m-k}\,. \qquad \text{(B.4)}$$

For instance, (B.4) with $k=0$ is the first equation in (B.3); (B.4) with $k=1$ means

$$a^n - b^n = n(a-b)b^{n-1} + (a-b)^2\sum_{m=1}^{n-1} m a^{n-1-m}b^{m-1}\,. \qquad \text{(B.4*)}$$

Note that

$$\sum_{m=k}^{n-1}\binom{m}{k} = \binom{n}{k+1}. \qquad \text{(B.5)}$$

Proof of (B.4) involves mathematical induction. (B.4) holds if $k=0$:

$$a^n - b^n = (a-b)\sum_{m=0}^{n-1} a^{n-m-1}b^m\,. \qquad \text{(B.6)}$$

Next we prove (B.4*). By (B.6),

$$a^{n-1-m} = b^{n-1-m} + (a-b)\sum_{i=1}^{n-1-m} b^{i-1}a^{n-1-m-i} \qquad (1\le m\le n-1).$$

Therefore,

$$\begin{aligned} a^n - b^n &= (a-b)\sum_{m=0}^{n-1} b^m\Big(b^{n-1-m} + (a-b)\sum_{i=1}^{n-1-m} b^{i-1}a^{n-1-m-i}\Big) \\ &= n(a-b)b^{n-1} + (a-b)^2\sum_{m=0}^{n-1}\sum_{i=1}^{n-m-1} b^{m+i-1}a^{n-1-m-i}\,. \qquad \text{(B.7)} \end{aligned}$$

Set $j=m+i$. Then

$$\sum_{m=0}^{n-1}\sum_{i=1}^{n-m-1} b^{m+i-1}a^{n-1-m-i} = \sum_{j=1}^{n-1} b^{j-1}a^{n-1-j}\sum_{m=0}^{n-1}\mathbb{1}\{1\le j-m\le n-1-m\}$$

$$= \sum_{j=1}^{n-1} b^{j-1}a^{n-1-j}\sum_{m=0}^{j-1} 1 = \sum_{j=1}^{n-1} j b^{j-1}a^{n-1-j}\,, \qquad \text{(B.8)}$$

and (B.4*) follows.

Assume that $a>b$ and (B.4) holds for a particular $k\geq 1$. Since

$$a^{n-1-m}=b^{n-1-m}+(a-b)\sum_{i=1}^{n-m-1}a^{n-1-m-i}b^{i-1}$$

by (B.6), the reminder term in (B.4) is

$$(a-b)^{k+1}\sum_{m=k}^{n-1}\binom{m}{k}a^{n-1-m}b^{m-k}=(a-b)^{k+1}$$

$$\sum_{m=k}^{n-1}\binom{m}{k}\Big(b^{n-1-m}+(a-b)\sum_{i=1}^{n-m-1}a^{n-1-m-i}b^{i-1}\Big)b^{m-k}$$

$$=(a-b)^{k+1}\sum_{m=k}^{n-1}\binom{m}{k}b^{n-1-m}+(a-b)^{k+2}\sum_{m=k}^{n-1}\binom{m}{k}\sum_{i=1}^{n-1-m}a^{n-1-m-i}b^{m-k+i-1}.$$

Note that $\sum_{i=1}^{n-1-m}(b/a)^i=\frac{b}{a-b}(1-(b/a)^{n-1-m})$. Recall that $\sum_{m=k}^{n-1}\binom{m}{k}=\binom{n}{k+1}$. Therefore,

$$\sum_{m=k}^{n-1}\binom{m}{k}\sum_{i=1}^{n-1-m}a^{n-1-m-i}b^{m-k+i-1}$$

$$=\sum_{m=k}^{n-1}\binom{m}{k}a^{n-1-m}b^{m-k-1}\frac{b}{a-b}(1-(b/a)^{n-1-m})$$

$$=\sum_{m=k}^{n-1}\binom{m}{k}(a^{n-1-m}b^{m-k}-b^{n-1-k})/(a-b)$$

$$=\sum_{m=k}^{n-2}\binom{m}{k}(a^{n-1-m}b^{m-k}-b^{n-1-k})/(a-b)$$

$$=\sum_{m=k}^{n-2}\binom{m}{k}a^{n-1}b^{-k-1}\sum_{j=m+1}^{n-1}(b/a)^j$$

$$=\sum_{m=k}^{n-2}\binom{m}{k}\sum_{j=m+1}^{n-1}a^{n-j-1}b^{j-k-1}=\sum_{j=k+1}^{n-1}a^{n-j-1}b^{j-k-1}\sum_{m=k}^{j-1}\binom{m}{k}$$

$$=\sum_{j=k+1}^{n-1}a^{n-1-j}b^{j-k-1}\binom{j}{k+1},$$

where we have applied (B.5). Thus, (B.4) holds also for $k+1$. □

These identities can be rewritten for convolutions of finite measures P and H, see [41]:

$$P^{*n}-H^{*n}=\sum_{m=1}^{n}\binom{n}{m}(P-H)^{*m}*H^{*(n-m)},$$

$$P^{*n}-H^{*n}=\sum_{m=1}^{k}\binom{n}{m}(P-H)^{*m}*H^{*(n-m)} \tag{B.9}$$

$$+\ (P-H)^{*(k+1)}*\sum_{m=k}^{n-1}\binom{m}{k}P^{*(n-1-m)}*H^{*(m-k)}\qquad(k\in\{1,\dots,n-1\}).$$

If $k=0$, then (B.9) becomes

$$P^{*n}-H^{*n}=(P-H)*\sum_{m=0}^{n-1}P^{*(n-1-m)}*H^{*(m)}.$$

Bergström identity can be used when constructing asymptotic expansions as H may have beneficial properties (for example, H may be infinitely divisible).

The following example presents an application of (B.9).

Example B.1 Let X be an non-negative integer-valued random variable with distribution $\mathbb{P}_X=\mathcal{L}(X)$ such that $\mathbb{E}X^{k+1}<\infty$, where $k\in\mathbb{Z}_+$. Then

$$\mathbb{P}_X=I+\sum_{j=1}^{k}\frac{\mathbb{E}X^{(j)}}{j!}(I_1-I)^{*j}+\frac{\mathbb{E}X^{(k+1)}}{(k+1)!}(I_1-I)^{*(k+1)}*\Theta, \tag{B.10}$$

where $\|\Theta\|_{TV}\le 1$. Indeed,

$$\mathbb{P}_X=\sum_{m\ge 0}\mathbb{P}_X(m)I_m=I+\sum_{m=1}^{\infty}\mathbb{P}_X(m)(I_m-I).$$

Note that $I_m=I_1^{*m}$ $(m\in\mathbb{Z}_+)$. By (B.9),

$$I_m-I=\sum_{j=1}^{k}\binom{m}{j}(I_1-I)^{*j}+(I_1-I)^{*(k+1)}*\sum_{j=k}^{m-1}\binom{j}{k}I_1^{*(m-1-j)} \qquad (m\in\mathbb{N}).$$

Hence

$$\begin{aligned}\mathbb{P}_X &= I+\sum_{j=1}^{k}(I_1-I)^{*j}\sum_{m=1}^{\infty}\binom{m}{j}\mathbb{P}_X(m)\\ &+ (I_1-I)^{*(k+1)}*\sum_{m=1}^{\infty}\mathbb{P}_X(m)\sum_{j=k}^{m-1}\binom{j}{k}I_1^{*(m-1-j)}.\end{aligned}$$

Clearly, $\sum_{m=1}^{\infty}\binom{m}{j}\mathbb{P}_X(m)=\mathbb{E}X^{(j)}/j!$ Besides,

$$\begin{aligned}\Big\|\sum_{m=1}^{\infty}\mathbb{P}_X(m)\sum_{j=k}^{m-1}\binom{j}{k}I_1^{*(m-1-j)}\Big\|_{TV} &\le \sum_{m=1}^{\infty}\mathbb{P}_X(m)\sum_{j=k}^{m-1}\binom{j}{k}\\ &= \sum_{m=1}^{\infty}\mathbb{P}_X(m)\binom{m}{k+1}=\mathbb{E}X^{(k+1)}/(k+1)!\end{aligned}$$

This completes the proof of (B.10). □

An extension of (B.9) to a convolution of non-identical distributions has been given in [74]. Assume that $\sum_{m=j}^{k}\equiv 0$, $\prod_{m=j}^{*k}\equiv I$ if $j>k$.

Lemma B.3 [74] *Let $F_1,\dots,F_n,H_1,\dots,H_n$ be measures of finite variation. Then*

$$\prod_{j=1}^{*n}F_j-\prod_{j=1}^{*n}H_j=\sum_{l=1}^{s}\sum_{m_1,\dots m_n}^{*}\prod_{j=1}^{*n}(F_j-H_j)^{*m_j}*H_j^{*(1-m_j)}+R_n(s) \tag{B.11}$$

for any $s\in\{0,1,...,n\}$, *where*

$$R_n(s) = \sum_{l=s+1}^{n} (F_l-H_l) * \prod_{j=l+1}^{*n} F_j * \sum_{m_1,...m_n}^{*} \prod_{j=1}^{*(l-1)} (F_j-H_j)^{*m_j} * H_j^{*(1-m_j)},$$

$\sum^*_{m_1,...m_n}$ *denotes summation over all* $m_1\in\{0,1\},...m_n\in\{0,1\}$ *such that* $m_1+...+m_n=l$.

In particular, (B.11) with $s=0$ yields

$$\prod_{j=1}^{*n} F_j - \prod_{j=1}^{*n} H_j = \sum_{j=1}^{n} H_1 * ... * H_{j-1} * (F_j-H_j) * F_{j+1} * ... * F_n . \tag{B.12}$$

Similarly, (B.11) with $s=1$ yields

$$\prod_{j=1}^{*n} F_j - \prod_{j=1}^{*n} H_j = \sum_{j=1}^{n}(F_j-H_j) * \prod_{l=1,l\neq j}^{*n} H_l + \sum_{j=2}^{n}(F_j-H_j) * \prod_{l=j+1}^{*n} F_l * \sum_{m=1}^{j-1}(F_m-H_m) * \prod_{l=1,l\neq m}^{*(j-1)} H_l. \tag{B.13}$$

C Smoothing inequalities

Smoothing usually means that one switches from S_n to $S_n+\xi$, where r.v.s S_n, ξ are independent, $\mathcal{L}(\xi)$ has special properties (e.g., ch.f. of $\mathcal{L}(\xi)$ may be non-negative).

In the case of integer-valued random variables one often needs to switch from S_n+1 to S_n, from $S_{\pi_\lambda+1}$ to S_{π_λ}, where π_λ is a Poisson $\mathbf{\Pi}(\lambda)$ r.v., etc. Proofs of results concerning Poisson and compound Poisson approximation to the distribution of a sum S_n of integer-valued random variables $X_1,...,X_n$ often require estimates of

$$d_{TV}(S_n;S_n+1), \quad d_{TV}(\pi_\lambda;\pi_\lambda+1), \quad d_{TV}(S_{\pi_\lambda};S_{\pi_\lambda+1}) .$$

This section presents a number of such results.

We start with smoothing inequalities for the Poisson distribution.
If $\lambda>0$, then

$$\|(I_1-I) * e^{\lambda(I_1-I)}\|_{TV} \leq \sqrt{2/e\lambda}, \tag{C.1}$$
$$\|(I_1-I)^{*2} * e^{\lambda(I_1-I)}\|_{TV} \leq 3/e\lambda. \tag{C.2}$$

Inequality (C.1) is Proposition A.2.7 in [21]. It states that

$$d_{TV}(\pi_\lambda;\pi_\lambda+1) \leq 1/\sqrt{2e\lambda} .$$

More precisely,

$$d_{TV}(\pi_\lambda;\pi_\lambda+1) \leq \min\left\{1/\sqrt{2e\lambda};1/\sqrt{2\pi\lfloor\lambda\rfloor}\right\}. \tag{C.1*}$$

Indeed, since $\mathcal{L}(\pi_\lambda)$ is unimodal,

$$d_{TV}(\pi_\lambda;\pi_\lambda+1) = \tfrac{1}{2}\|\Delta\mathbb{P}(\pi_\lambda=\cdot)\|_{TV} = \max_k \mathbb{P}(\pi_\lambda=k) = \mathbb{P}(\pi_\lambda=\lfloor\lambda\rfloor),$$

where

$$\Delta \mathbb{P}_{\pi_\lambda}(j) = \mathbb{P}_{\pi_\lambda}(j+1) - \mathbb{P}_{\pi_\lambda}(j), \quad \Delta^k = \Delta(\Delta^{k-1}).$$

It is known that $\mathbb{P}(\pi_\lambda = \lfloor\lambda\rfloor) \le 1/\sqrt{2\pi\,\lfloor\lambda\rfloor}$ (see, e.g., formula (4.58) in [221]). Hence (C.1*) holds.

Estimate (C.2) was proved by Roos ([260], Lemma 3). Bound (C.2) states that

$$\|\Delta^2 \mathbb{P}_{\pi_\lambda}\|_{TV} \le 3/\mathrm{e}\lambda. \tag{C.2*}$$

It follows from (A.1) and (C.1) that

$$\|(I_1 - I)^{*k} * \mathrm{e}^{\lambda(I_1 - I)}\|_{TV} \le \left(2k/\mathrm{e}\lambda\right)^{k/2}. \tag{C.3}$$

A similar to (C.3) bound is valid for the Binomial distribution $\mathbf{B}(n,p)$, see (C.7).

Shorgin ([279], Lemma 6) has shown that

$$\|\Delta^k \mathbb{P}_{\pi_\lambda}(\cdot)\|_{\mathrm{o}} \le \tfrac{1}{2}\left(1+\sqrt{\pi/2}\right)\sqrt{\mathrm{e}}\,(k/\mathrm{e}\lambda)^{(k+1)/2} \qquad (k \in \mathbb{N}). \tag{C.4}$$

In particular,

$$\|\Delta \mathbb{P}_{\pi_\lambda}(\cdot)\|_{\mathrm{o}} \le \left(1+\sqrt{\pi/2}\right)/2\sqrt{\mathrm{e}}\,\lambda,$$

where $\left(1+\sqrt{\pi/2}\right)/2\sqrt{\mathrm{e}} \approx 0.68$.

Estimates in terms of the point metric follow from bounds in terms of the total variation thanks to (C.3) and (A.7): for any measure M on $\mathbb{Z}$

$$\|M\|_{\mathrm{o}} = \|(I_1 - I) * M\|_K \le \|(I_1 - I) * M\|_{TV}.$$

For example,

$$\|\Delta \mathbb{P}_{\pi_\lambda}(\cdot)\|_{\mathrm{o}} \equiv \|(I_1 - I) * \mathrm{e}^{\lambda(I_1 - I)}\|_{\mathrm{o}} \le 3/\mathrm{e}\lambda\,, \tag{C.5}$$

$$\|\Delta^k \mathbb{P}_{\pi_\lambda}(\cdot)\|_{\mathrm{o}} \equiv \|(I_1 - I)^{*k} * \mathrm{e}^{\lambda(I_1 - I)}\|_{\mathrm{o}} \le \left(2(k+1)/\mathrm{e}\lambda\right)^{(k+1)/2}. \tag{C.6}$$

Similar bounds are valid for the Binomial distribution $\mathbf{B}(n,p)$. Let $q = 1-p$. If $0<p<1$, $j, n \in \mathbb{N}$, then (see Lemma 4 in [258] or Lemma 3.1 in [83])

$$\begin{aligned}
\|(I_1 - I)^{*j} * (qI + pI_1)^{*n}\|_{TV} &\le \binom{n+j}{j}^{-1/2} (pq)^{-j/2} \\
&\le \sqrt{\mathrm{e}}\, j^{1/4} \left(\frac{n}{n+j}\right)^{n/2} \left(\frac{j}{(n+j)pq}\right)^{j/2}, \qquad \text{(C.7)} \\
\|(I_1 - I)^{*j} * (qI + pI_1)^{*n}\|_{\mathrm{o}} &\le \frac{\sqrt{\mathrm{e}}}{2}\left(1+\sqrt{\frac{\pi}{2j}}\right)\left(\frac{n}{n+j+1}\right)^{(n+j+1)/2}\left(\frac{j}{npq}\right)^{(j+1)/2}.
\end{aligned}$$

If $j = 1$, then these bounds mean

$$2d_{TV}(S_n; S_n+1) \le 1/\sqrt{(n+1)pq}\,, \quad d_{\mathrm{o}}(S_n; S_n+1) \le \frac{1}{2\sqrt{\mathrm{e}}}\left(1+\sqrt{\frac{\pi}{2}}\right)\Big/ npq.$$

Inequalities (C.1)–(C.3) can be generalised as follows.

Lemma C.1 *If* $\lambda>0$ *and* $k\geq 3$, *then*

$$\sup_{P\in\mathcal{F}} \|(P-I)*\mathrm{e}^{\lambda(P-I)}\|_{TV} \leq \|(I_1-I)*\mathrm{e}^{\lambda(I_1-I)}\|_{TV} \leq \sqrt{2/\mathrm{e}\lambda}, \tag{C.8}$$

$$\sup_{P\in\mathcal{F}} \|(P-I)^{*2}*\mathrm{e}^{\lambda(P-I)}\|_{TV} \leq \|(I_1-I)^{*2}*\mathrm{e}^{\lambda(I_1-I)}\|_{TV} \leq 3/\mathrm{e}\lambda, \tag{C.9}$$

$$\sup_{P\in\mathcal{F}} \|(P-I)^{*k}*\mathrm{e}^{\lambda(P-I)}\|_{TV} \leq \|(I_1-I)^{*k}*\mathrm{e}^{\lambda(I_1-I)}\|_{TV} \leq \left(2k/\mathrm{e}\lambda\right)^{k/2}. \tag{C.10}$$

These bounds are instrumental for the proofs of a number of theorems. Inequality (C.8) is a consequence of (5.6), see also (A.3). It means that

$$d_{TV}\left(\sum_{i=1}^{\pi_\lambda} X_i; \sum_{i=1}^{\pi_\lambda+1} X_i\right) \leq d_{TV}(\pi_\lambda; \pi_\lambda+1) \leq 1/\sqrt{2\mathrm{e}\lambda}.$$

A similar bound in terms of d_G is trivial: due to (6.7),

$$d_G\left(\sum_{i=1}^{\pi_\lambda} X_i; \sum_{i=1}^{\pi_\lambda+1} X_i\right) \leq \mathbb{E}|X|. \tag{C.11}$$

Inequality (C.9) follows from (C.8) and (A.3); for $P=I_1$ it was proved in [260], Lemma 3. Estimate (C.10) follows from (A.1) and (C.8).

Note also the following consequence of (3.83*): if $a>0$, $b>0$, then

$$\begin{aligned} \tfrac{1}{2}\|\mathrm{e}^{a(P-I)}-\mathrm{e}^{b(P-I)}\|_{TV} &= d_{TV}(\mathbf{\Pi}(a;X);\mathbf{\Pi}(b;X)) \leq d_{TV}(\pi_a;\pi_b) \\ &\leq |a-b|\min\{1;\sqrt{2}/\sqrt{\mathrm{e}}(\sqrt{a}+\sqrt{b})\} \end{aligned} \tag{C.12}$$

for any r.v. X, where $P=\mathcal{L}(X)$.

Proposition C.2 *If* $P\in\mathcal{F}_s$, *then*

$$\|\mathrm{e}^{a(P-I)}-\mathrm{e}^{b(P-I)}\|_K \leq C|a-b|/\max\{a;b\}. \tag{C.13}$$

A multivariate analogue of (C.13) is (12.32).

Proof of (C.13). Without loss of generality we may assume that $a\leq b$.

If $a\leq b\leq 2a$, then it follows from (B.1), (A.6), and (C.24*) that

$$\begin{aligned} \left\|\mathrm{e}^{a(P-I)}-\mathrm{e}^{b(P-I)}\right\|_K &= \left\|\mathrm{e}^{a(P-I)}*(P-I)*\int_0^1 (b-a)\mathrm{e}^{(b-a)t(P-I)}\mathrm{d}t\right\|_K \\ &\leq \left\|(P-I)*\mathrm{e}^{a(P-I)}\right\|_K |b-a| \int_0^1 \left\|\mathrm{e}^{(b-a)t(P-I)}\right\|_{TV}\mathrm{d}t \\ &\leq C|b-a|/a \leq 2C|b-a|/b \end{aligned}$$

since $\mathrm{e}^{(b-a)t(P-I)}$ is a CP distribution and its total variation norm equals 1.

If $b>2a$, then

$$\left\|\mathrm{e}^{a(P-I)}-\mathrm{e}^{b(P-I)}\right\|_K \leq 1 \leq 2(b-a)/b.$$

This proves (C.13). □

Recall that

$$\begin{aligned}\|(I_1-I)*\mathrm{e}^{\lambda(I_1-I)}\|_{TV} &= \|\Delta\mathbb{P}_{\pi_\lambda}\|_{TV} = 2d_{TV}(\pi_\lambda;\pi_\lambda+1),\\ \|(I_1-I)^{*k}*\mathrm{e}^{\lambda(I_1-I)}\|_{TV} &= \sum_{j=-\infty}^{\infty}|\Delta^k\mathbb{P}_{\pi_\lambda}(j)|.\end{aligned}$$

Let

$$\|f\|_1=\int_{\mathbb{R}}|f(x)|\,\mathrm{d}x$$

denote the L_1-norm of function f. Set

$$\varphi_k(x)=\frac{1}{\sqrt{2\pi}}\frac{\mathrm{d}^k}{\mathrm{d}x^k}\,\mathrm{e}^{-x^2/2}.$$

Note that

$$\|\varphi_1\|_1=\sqrt{2/\pi},\quad \|\varphi_2\|_1=4/\sqrt{2\pi\mathrm{e}},\quad \|\varphi_3\|_1=\sqrt{2/\pi}\left(1+4\mathrm{e}^{-3/2}\right).$$

Lemma C.3 [263] *Let $\lambda>0$. For any $k\geq1$ there exists an absolute constant C_k such that*

$$\begin{aligned}\left|\|(I_1-I)^{*k}*\mathrm{e}^{\lambda(I_1-I)}\|_{TV}-\|\varphi_k\|_1\lambda^{-k/2}\right| &\leq C_k\lambda^{-(k+1)/2},\\ \left|\|(I_1-I)^{*k}*\mathrm{e}^{\lambda(I_1-I)}\|_{G}-\|\varphi_{k-1}\|_1\lambda^{-(k-1)/2}\right| &\leq C_k\lambda^{-k/2}.\end{aligned}$$

The Kanter-Barbour-Xia-Mattner-Roos inequality.
An estimate similar to (C.1) holds for any sum of independent integer-valued r.v.s. Let $\{X_j\}$ be independent integer-valued random variables,

$$S_n=X_1+\cdots+X_n,\quad S_{n,i}=S_n-X_i.$$

Recall that

$$2d_{TV}(S_n;S_n+1)=\sum_k|\Delta\mathbb{P}_{S_n}(k)|=\|(I_1-I)*\mathcal{L}(S_n)\|_{TV}.$$

Similarly,

$$\sum_{-\infty}^{\infty}|\Delta^2\mathbb{P}_{S_n}(k)|=\|(I_1-I)^{*2}*\mathcal{L}(S_n)\|_{TV}.$$

Lemma C.4 *Let $U=\sum_{i=1}^n(1-d_{TV}(X_i;X_i+1))$. Then*

$$d_{TV}(S_n;S_n+1)\leq\sqrt{2/\pi(1/4+U)}. \tag{C.14}$$

Lemma C.4 is often applied to $S_{n,i}=S_n-X_i$:

$$d_{TV}(S_{n,i};S_{n,i}+1)\leq\sqrt{2/\pi\max\{1/4;U-3/4\}}. \tag{C.15}$$

Estimate (C.15) follows from (C.14) and the obvious inequality $0\leq d_{TV}(X_i;X_i+1)\leq1$.

Corollary C.5 *If $U>0$, $n>1$, then*

$$\sum_{-\infty}^{\infty}|\Delta^2\mathbb{P}_{S_n}(k)|\leq7/U,\quad \sum_{-\infty}^{\infty}|\Delta^2\mathbb{P}_{S_{n,i}}(k)|\leq17.1/U. \tag{C.16}$$

Inequality (C.14) is based on the observation by Barbour & Xia ([25], Proposition 4.6) that the "Mineka coupling" (Lindvall [196], p. 44) yields

$$d_{TV}(S_n; S_n+1) \le \mathbb{P}(S_n^Y \in \{0,1\}),$$

where $S_n^Y = Y_1 + ... + Y_n$, $Y_1, Y_2, ...$ are independent r.v.s with the distribution

$$\mathbb{P}(Y_i=0)=d(X_i; X_i+1), \ \ \mathbb{P}(Y_i=1) = \mathbb{P}(Y_i=-1) = (1-d_{TV}(X_i; X_i+1))/2.$$

Denote

$$G(\lambda) = \mathbb{P}(\pi_{\lambda/2}-\pi'_{\lambda/2} \in \{0,1\}) \qquad (\lambda>0).$$

Kanter ([169], formula (4.3)) has shown that

$$\mathbb{P}(S_n^Y \in \{0,1\}) \le G(U).$$

Estimate (C.14) is Corollary 1.6 in Mattner & Roos [204] who have proved that

$$G(\lambda) \le \sqrt{2/\pi(1/4+\lambda)} \qquad (\lambda>0).$$

Proof of Corollary C.5. If $U \le 7/4$, then $\|(I_1-I)^{*2} * \mathcal{L}(S_n)\|_{TV} \le \|(I_1-I)^{*2}\|_{TV} \le 4 \le 7/U$ by the properties of total variation norm.

Assume that $U>7/4$. Denote

$$v_i = 1-d_{TV}(X_i; X_i+1), \quad V_i = v_1+\cdots+v_i \quad (1 \le i \le n).$$

Without loss of generality we may assume that $0 \le v_1 \le v_2 \le \ldots \le v_n$.

Let $\nu = \max\{j\colon V_j \le U/2\}$. Then

$$V_\nu \le U/2, \quad V_\nu + v_{j+1} > U/2, \quad U-V_\nu = v_{\nu+1}+\cdots+v_n \ge U/2.$$

Denote $U_1 = V_\nu$, $U_2 = U-V_\nu$,

$$S_1 = X_1+\cdots+X_\nu, \quad S_2 = X_{\nu+1}+\cdots+X_n.$$

Then $S_n = S_1 + S_2$. Hence

$$\|(I_1-I)^{*2} * \mathcal{L}(S_n)\|_{TV} \le \|(I_1-I) * \mathcal{L}(S_1)\|_{TV} \, \|(I_1-I) * \mathcal{L}(S_2)\|_{TV}.$$

Since $V_\nu \ge U/2-1$, it follows from (C.14) that

$$\begin{aligned}
\|(I_1-I)^{*2} * \mathcal{L}(S_n)\|_{TV} &\le \frac{2}{\pi}(1/4+U_1)^{-1/2}(1/4+U_2)^{-1/2} \\
&\le \frac{2}{\pi}(U/2-3/4)^{-1/2}(1/4+U/2)^{-1/2} \\
&\le 4/\pi(U-1) < 7/U.
\end{aligned}$$

The second estimate in (C.16) can be proved similarly – one replaces U with $U-v_i \ge U-1$. □

Posfai [235] gives the following analogue of (C.14).

Proposition C.6 *Let $X_1, X_2, \ldots, X_n$ be independent integer-valued random variables. Then for any even natural number m*

$$d_{TV}(S_n; S_n+1) \le \frac{4}{m\sqrt{nmp}} + \frac{8\, d_{TV}(X_n; X_n+1)}{nmp} \qquad (n \ge 2),$$

where $p = \min\{\mathbb{P}(X_i=k)\colon k = 1, \ldots, m, \ j = 1, \ldots, n\}$.

Let $X, X_1, X_2, \dots$ be a sequence of i.i.d. integer-valued random variables with maximum span 1, $S_n = X_1 + ... + X_n$. Then [122] $d_{TV}(S_n; S_n+1) \to 0$ as $n \to \infty$; if X is bounded, then

$$d_{TV}(S_n; S_n+1) = 1/\sigma\sqrt{2\pi n} + O(1/n) \qquad (n \to \infty), \tag{C.17}$$

where $\sigma^2 = \operatorname{var} X$.

If S_n is an integer-valued random variable and $\mathcal{L}(S_n)$ is unimodal, then

$$d_{TV}(S_n; S_n+1) = \max_k \mathbb{P}(S_n = k).$$

Bounds (E.2), (E.5) with $h=0$ yield estimates of $\max_k \mathbb{P}(S_n = k)$.

The following bound is due to Gamkrelidze [122]. Let $X, X_1, X_2, ...$ be a sequence of independent integer-valued unimodal r.v.s. Denote $p_j = \max_k \mathbb{P}(X_j = k)$. Then

$$\max_{k \in \mathbb{Z}} \mathbb{P}(S_n = k) \le \sqrt{\frac{6}{\pi}} \left(\sum_{i=1}^{n} (p_i^{-2} - 1) \right)^{-1/2} (1 + 2/\sqrt{n}). \tag{C.18}$$

Example C.2 Let $X, X_1, X_2, ...$ be i.i.d. Bernoulli $\mathbf{B}(1/2)$ r.v.s. Then

$$d_{TV}(S_n; S_n+1) = \max_{k \in \mathbb{Z}_+} \mathbb{P}(S_n = k) < \sqrt{2/\pi n}$$

([299], p. 57). Inequality (C.14) yields

$$d_{TV}(S_n; S_n+1) \le \sqrt{2/\pi(1/4+U)} \sim 2/\sqrt{\pi n} \quad (n \to \infty)$$

while (C.18) entails $d_{TV}(S_n; S_n+1) \le \sqrt{2/\pi n}(1 + 2/\sqrt{n})$. □

Concerning $d_K(S_n; S_{n+1})$, if $F \in \mathcal{F}$, then there exists an absolute constant C such that

$$\inf_{a \in \mathbb{R}} \|(I_a * F - I) * (I_a * F)^{*n}\|_K \le C/\sqrt{n}. \tag{C.19}$$

Bound (C.19) follows from Lemma 10.10, (A.1), (A.3), (C.7), and the triangle inequality.

If $X, X_1, X_2, ...$ are i.i.d. Bernoulli $\mathbf{B}(p)$ r.v.s, then evidently

$$d_{TV}(S_n; S_{n+1}) \le p d_{TV}(S_n; S_n+1).$$

Symmetric distributions.

Recall that $\mathcal{F}_s$ is class of symmetric distributions, $\mathcal{F}_+$ is the class of symmetric distributions with non-negative characteristic functions. Below constants depending on m are denoted by $C(m)$, $C(k,m)$ denotes a positive constant that depends on k and m only.

Lemma C.7 *Let $F \in \mathcal{F}_+$, $n, k, m \in \mathbb{N}$. Then*

$$\|F^{*n} * (F-I)^{*k}\|_K \le C(k) n^{-k}, \tag{C.20}$$

$$\|F^{*n} - F^{*m}\|_K \le C|m-n|/\max\{m; n\}. \tag{C.21}$$

In particular,

$$\|(F-I) * F^{*n}\|_K \le C/n. \tag{C.21*}$$

Lemma C.8 *If $G \in \mathcal{F}_s$, $\lambda \in (0, \infty)$, $n, m \in \mathbb{N}$, $p = 1-q \in (0,1)$, then*

$$\|(G-I)^{*m} * (qI+pG)^{*n}\|_K \le C(m) q^{-1} (npq)^{-m}, \tag{C.22}$$

$$\|G^{*n} - G^{*m}\|_K \le C/\sqrt{n+1} + C|m-n|/\max\{m; n\}, \tag{C.23}$$

$$\|(G-I)^{*m} * e^{\lambda(G-I)}\|_K \le C(m) \lambda^{-m}. \tag{C.24}$$

In particular,

$$\|(G-I)*\mathrm{e}^{\lambda(G-I)}\|_K \le C/\lambda, \tag{C.24*}$$

$$\|(G-I)*(qI+pG)^{*n}\|_K \le C/npq^2\,. \tag{C.22*}$$

It follows from (C.23) that

$$\|U^{*n}-\exp(n(U-I))\,\|_K \le C/\sqrt{n+1} \qquad (n\in\mathbb{Z}_+,\, U\in\mathcal{F}_s). \tag{C.25}$$

In other words, if $X, X_1, X_2, \ldots$ are i.i.d.r.v.s with $\mathcal{L}(X)=U$, then

$$d_K(S_n;\tilde{S}_n) \le C/\sqrt{n+1}\,.$$

Indeed, let π_n denote a Poisson $\mathbf{\Pi}(n)$ r.v. Then

$$\begin{aligned}
\|U^{*n}-\exp(n(U-I))\,\|_K &= \|U^{*n}-\mathrm{e}^{-n}\sum_{j=0}^{\infty}U^{*j}n^j/j!\|_K \le \Big\|\sum_{j=0}^{\infty}(U^{*n}-U^{*j})\Big\|_K \mathrm{e}^{-n}n^j/j! \\
&\le \sum_{j=0}^{\infty}\Big(C/\sqrt{n+1}+C|j-n|/n\Big)\mathrm{e}^{-n}n^j/j! \\
&= C/\sqrt{n+1}+C\mathbb{E}|\pi_n-n|/n \le C/\sqrt{n+1}\,.
\end{aligned}$$

Lemma C.9 *If* $F\in\mathcal{F}_+$, $G,H\in\mathcal{F}_s$, $a,b\in(0,\infty)$, $n,k,m\in\mathbb{N}$, *then*

$$\|(F-I)^{*k}*F^{*n}*(G-I)^{*m}*\mathrm{e}^{a(G-I)}\|_K \le C(k,m)n^{-k}a^{-m}, \tag{C.26}$$

$$\|(G-I)^{*k}*\mathrm{e}^{a(G-I)}*(H-I)^{*m}*\mathrm{e}^{b(H-I)}\|_K \le C(k,m)a^{-k}b^{-m}. \tag{C.27}$$

Estimate (C.20) is a corollary of Theorem 5.5.1 in [8]. Estimate (C.21) is (2.35) from [89]; a version with $\min\{m;n\}$ instead of $\max\{m;n\}$ is Theorem 5.5.1 in [8]. Estimate (C.19) is Theorem 5.4.2 from [8]. Estimate (C.23) with $\min\{m;n\}$ instead of $\max\{m;n\}$ is from Theorem 5.5.2 in [8], the present version is Theorem 2.6 in [89]. Estimates (C.24), (C.26), (C.27) were proved in [65]. Estimate (C.22) is from [83], Proposition 3.1.

Combining (B.2) with (C.24), we get

$$\Big\|\mathrm{e}^{a(G-I)}*\Big(\mathrm{e}^{b(G-I)}-\sum_{j=0}^{m}b^j(G-I)^{*j}/j!\Big)\Big\|_K \le C(m)b^{m+1}\min\{1;a^{-(m+1)}\} \tag{C.28}$$

for any $a,b>0$, $m\in\mathbb{Z}_+$, $G\in\mathcal{F}_s$.

Recall that by Khintchine's formula (6.1) any distribution can be presented as

$$\mathcal{L}(X) \stackrel{d}{=} qU+pV,$$

where $p\in(0;1)$, $q=1-p$.

Lemma C.10 [70] *Assume that* $\mathcal{L}(X) \stackrel{d}{=} qU+pV$, *where* $U\in\mathcal{F}_+$, $V\in\mathcal{F}$, $0<p\le c_1<1$, $q=1-p$. *If* $n\in\mathbb{N}$, $k,m\in\mathbb{Z}_+$, $c_2>0$, *then there exists an absolute constant* $C_{k,m}$ *such that*

$$\|(qU+pV)^{*n}*(U-I)^{*k}*(V-I)^{*m}\|_K \le C_{k,m}(nq)^{-k}\min\{1;(np)^{-m/2}\},$$

$$\|\exp(nq(U-I)+c_2np(V-I))*(U-E)^{*k}(V-I)^{*m}\|_K \le C_{k,m}(nq)^{-k}\min\{1;(np)^{-m/2}\}.$$

Lemma C.10 is crucial for the proof of Theorem 10.7.

In the case of integer-valued r.v.s one would prefer estimates in terms of the total variation norm.

Let P be a *symmetric* distribution concentrated on $\mathbb{Z}\backslash\{0\}$. Assume that P has a finite variance σ^2. It is stated in [84], Lemma 4.6, that for any $\lambda>0$ and $j\in\mathbb{N}$

$$\begin{aligned} \|(P-I)^{*j} * \exp(\lambda(P-I))\|_{TV} &\leq 3.6\sqrt{1+\sigma}\, j^{j+1/4}\lambda^{-j}\mathrm{e}^{-j}, \qquad (C.29)\\ \|(P-I)^{*j} * \exp(\lambda(P-I))\|_{\circ} &\leq 2\left(\frac{j+1/2}{\lambda\mathrm{e}}\right)^{j+1/2}. \end{aligned}$$

In particular,

$$d_{TV}\left(\sum_{i=1}^{\pi_\lambda} X_i; \sum_{i=1}^{\pi_\lambda+1} X_i\right) \leq 1.8\sqrt{1+\sigma}\,/\lambda\mathrm{e}, \quad d_{\mathrm{o}}\left(\sum_{i=1}^{\pi_\lambda} X_i; \sum_{i=1}^{\pi_\lambda+1} X_i\right) \leq 2(1.5/\lambda\mathrm{e})^{3/2}. \qquad (C.29^*)$$

The argument used to prove (C.13) can be applied to deduce from (C.29) that

$$\|\mathrm{e}^{a(P-I)} - \mathrm{e}^{b(P-I)}\|_{TV} \leq \frac{|a-b|}{\max\{a;b\}} \max\{2; 1.33\sqrt{\sigma+1}\} \qquad (a,b>0) \qquad (C.30)$$

if P is a *symmetric* distribution concentrated on $\mathbb{Z}\backslash\{0\}$.

Note that (C.29) does not hold if $j=0$.

An analogue of (C.29*) holds in terms of the Gini-Kantorovich distance if the r.v.s are symmetric, compare with (C.11).

Lemma C.11 *Let $X, X_1, X_2, \ldots$ be i.i.d. symmetric r.v.s taking values in $\mathbb{Z}\backslash\{0\}$. Assume that $\mathbb{E}X^2 = \sigma^2 < \infty$, $\lambda>0$. Then there exists an absolute constant C such that*

$$d_G\left(\sum_{i=0}^{\pi_\lambda} X_i; \sum_{i=0}^{\pi_\lambda+1} X_i\right) \leq C(\sigma+1)^{3/2}\lambda^{-1/2}. \qquad (C.31)$$

Proof of lemma C.11. The bound trivially follows from (C.11) if $\lambda\leq\sigma+1$. Therefore, we may assume that $\lambda>\sigma+1$.

Denote $F = \mathcal{L}(X)$. Similarly to the proof of (7.31), we note that

$$\begin{aligned} F-I &= \sum_{k=1}^{\infty} \mathbb{P}(X=k)(I_k+I_{-k}-2I) = \sum_{k=1}^{\infty} \mathbb{P}(X=k) * (I_k-I)(I-I_{-k})\\ &= (I_1-I) * \sum_{k=1}^{\infty} \mathbb{P}(X=k)(I_{k-1}+I_{k-2}+\cdots+I) * (I-I_{-k}). \end{aligned}$$

Taking into account (A.9), $\|(F-I) * \exp(\lambda(F-I))\|_G = \|M\|_{TV}$, where

$$M = \sum_{k=1}^{\infty} \mathbb{P}(X=k)(I_{k-1}+I_{k-2}+\cdots+I) * (I-I_{-k}) * \exp(\lambda(F-I)).$$

Hence

$$\widehat{M}(t) = \sum_{k=1}^{\infty} \mathbb{P}(X=k)(\mathrm{e}^{\mathrm{i}t(k-1)}+\ldots+1)(1-\mathrm{e}^{-\mathrm{i}tk}) \exp\left(\lambda(\widehat{F}(t)-1)\right).$$

Recall that

$$1-\widehat{F}(t) = 2\sum_{j=1}^{\infty}(1-\cos(tj))\mathbb{P}(X=j) = 4\sum_{j=1}^{\infty} \sin^2(tj/2)\mathbb{P}(X=j)$$

due to symmetry. Since $|1-\mathrm{e}^{-\mathrm{i}tk}| = 2|\sin(tk/2)|$, we have

$$|\widehat{M}(t)| \le 2\exp\Big(\lambda(\widehat{F}(t)-1)\Big)\sum_{k=1}^{\infty}\mathbb{P}(X=k)k|\sin(tk/2)| \le \sigma\exp\Big(\lambda(\widehat{F}(t)-1)\Big)\,|1-\widehat{F}(t)|^{1/2}$$

by the Bunyakovskiy-Cauchy-Schwartz inequality. Note that

$$|1-\widehat{F}(t)|^{1/2}\,\exp\Big(\lambda(\widehat{F}(t)-1)/2\Big) \le \sup_{x>0} x\mathrm{e}^{-\lambda x^2/2} \le 1/\sqrt{\mathrm{e}\lambda}\,.$$

Therefore,

$$|\widehat{M}(t)| \le C\sigma\lambda^{-1/2}\exp\Big(\lambda(\widehat{F}(t)-1)/2\Big)\,.$$

Next,

$$\begin{aligned}
|\widehat{M}'(t)| &\le \sum_{k=1}^{\infty}\mathbb{P}(X=k)(k-1+k-2+\cdots+1)|\mathrm{e}^{\mathrm{i}t}-1|\exp\Big(\lambda(\widehat{F}-1)\Big)\\
&+\ \sum_{k=1}^{\infty}\mathbb{P}(X=k)k^2\exp\Big(\lambda(\widehat{F}-1)\Big) + 2\sum_{k=1}^{\infty}\mathbb{P}(X=k)k|\sin(tk/2)|\lambda|\widehat{F}'|\exp\Big(\lambda(\widehat{F}-1)\Big)\\
&\le\ C\sigma^2\exp\Big(\lambda(\widehat{F}-1)\Big) + C\sigma^2\Big(\lambda|\widehat{F}-1|\exp\Big(\lambda(\widehat{F}-1)/2\Big)\Big)\exp\Big(\lambda(\widehat{F}-1)/2\Big)\\
&\le\ C\sigma^2\exp\Big(\lambda(\widehat{F}-1)/2\Big)\,.
\end{aligned}$$

Note that

$$\int_{-\pi}^{\pi}\exp\Big(\lambda(\widehat{F}(t)-1)/2\Big)\,dt = 2\pi\mathbb{P}(Y=0) \le C/\sqrt{\lambda\mathbb{P}(|X|\ne 0)}$$

by (E.6), where Y denotes a r.v. with distribution $\mathcal{L}(Y) = \mathbf{\Pi}(\lambda, X)$. Hence

$$\int_{-\pi}^{\pi}|\widehat{M}(t)|^2dt \le C\sigma^2\lambda^{-3/2}\,, \qquad \int_{-\pi}^{\pi}|\widehat{M}'(t)|^2dt \le C\sigma^4\lambda^{-1/2}\,.$$

We apply (D.10) with $b=\lambda^{1/2}\max\{1;\sigma\}$, $v=0$:

$$\begin{aligned}
\|M\|^2_{TV} &\le \frac{1+b\pi}{2\pi}\int_{-\pi}^{\pi}\left(|\widehat{M}(t)|^2 + b^{-2}\left|\Big(\widehat{M}(t)\Big)'\right|^2\right)dt\\
&\le \int_{-\pi}^{\pi}\left((1+b)|\widehat{M}(t)|^2 + (b^{-1}+b^{-2})\left|\Big(\widehat{M}(t)\Big)'\right|^2\right)\mathrm{d}t.
\end{aligned}$$

Hence

$$\|M\|^2_{TV} \le C\max\{1;\sigma^3\}\lambda^{-1}\,,$$

leading to (C.31). □

Similar estimates hold for compound Binomial distribution. If distribution F is *symmetric* concentrated on $\mathbb{Z}\backslash\{0\}$ with finite variance σ^2, then

$$\|(F-I)^{*j} * (I+p(F-I))^{*n}\|_{TV} \ \le\ 6.73\sqrt{\sigma}\frac{j}{q^{1/4}}\left(\frac{j}{enpq}\right)^j, \tag{C.32}$$

$$\|(F-I)^{*j} * (I+p(F-I))^{*n}\|_{\circ} \ \le\ 2\sqrt{eq}\left(\frac{j+1/2}{enpq}\right)^{j+1/2}$$

for any $j\in\mathbb{N}$, $p\in(0,1)$, where $q=1-p$ (see [83], Proposition 3.2).

Bound (C.32) with $j=1$ means

$$d_{TV}\left(\sum_{i=1}^{\nu} X_i; \sum_{i=1}^{\nu+1} X_i\right) \leq 3.37\sqrt{\sigma}/nepq^{5/4},$$

where $\nu, X, X_1, X_2, \ldots$ are independent r.v.s, $\mathcal{L}(\nu) = \mathbf{B}(n,p)$, $X, X_1, X_2, \ldots$ are i.i.d., $\mathcal{L}(X)$ is symmetric.

Shifted distributions.

It is shown in [83], Corollary 3.1, that for any $F \in \mathcal{F}$, $\lambda > 0$, $j \in \mathbb{N}$,

$$\inf_{\alpha\in\mathbb{R}} \|(F * I_\alpha - I)^{*j} * \exp(\lambda(F * I_\alpha - I))\|_K \leq 17.6\lambda^{-j/2-j/(2j+2)} (j/\mathrm{e})^{(j+1)/2}. \tag{C.33}$$

If $j=1$, then this means

$$\inf_{\alpha\in\mathbb{R}} \|(F * I_\alpha - I) * \exp(\lambda(F * I_\alpha - I))\|_K \leq 17.6\lambda^{-3/4}\mathrm{e}^{-1} \leq 6.5\lambda^{-3/4}. \tag{C.34}$$

This can be rewritten as follows: for any r.v. X

$$\inf_{\alpha\in\mathbb{R}} d_K\left(\sum_{i=1}^{\pi_\lambda}(X_i+\alpha); \sum_{i=1}^{\pi_\lambda+1}(X_i+\alpha)\right) \leq 6.5\lambda^{-3/4}.$$

Recall that

$$d_K(\mathbf{\Pi}(a,P); \mathbf{\Pi}(b,P)) \equiv \|\mathrm{e}^{a(P-I)} - \mathrm{e}^{b(P-I)}\|_K \leq C|a-b|/\max\{a;b\} \qquad (a>0,\ b>0)$$

by (C.13). An analogue of (C.13) for shifted distributions holds as well.

Lemma C.12 *For any $a>0$, $b>0$, and any r.v. X*

$$\inf_{c\in\mathbb{R}} d_K(\mathbf{\Pi}(a, X+c); \mathbf{\Pi}(b, X+c)) \leq 9|a-b|/\max\{a;b\}^{3/4}. \tag{C.35}$$

Proof of Lemma C.12. Without loss of generality we may assume that $a \leq b$.

Let $F_c = \mathcal{L}(X+c)$. Then

$$\begin{aligned} d_K(\mathbf{\Pi}(a, X+c); \mathbf{\Pi}(b, X+c)) &= \left\|\mathrm{e}^{a(F_c-I)} - \mathrm{e}^{b(F_c-I)}\right\|_K \\ &\leq |a-b|\|F_c - I\|_{TV} \int_0^1 \left\|\mathrm{e}^{(a\tau+b(1-\tau))(F_c-I)}\right\|_{TV} \mathrm{d}\tau \\ &\leq 2(b-a), \end{aligned} \tag{C.36}$$

which completes proof in the case $a \leq b \leq 1$.

Next, consider the case $a<1\leq b\leq 2$. From (C.36) we deduce

$$\left\|\mathrm{e}^{a(F_c-I)} - \mathrm{e}^{b(F_c-I)}\right\|_K \leq 2(b-a) \leq 4(b-a)b^{-1} \leq 4(b-a)b^{-3/4}.$$

Now consider the case $a<1<2<b$. Then

$$\left\|\mathrm{e}^{a(F_c-I)} - \mathrm{e}^{b(F_c-I)}\right\|_K \leq 1 \leq 2(b-1)b^{-1} \leq 2(b-a)b^{-1} \leq 2(b-a)b^{-3/4}.$$

Let now $1\leq a\leq b\leq 1.5a$. Repeating the proof of (C.13) and applying (C.34), we get

$$\begin{aligned} \inf_{c\in\mathbb{R}} \left\|\mathrm{e}^{a(F_c-I)} - \mathrm{e}^{b(F_c-I)}\right\|_K &\leq (b-a) \inf_{c\in\mathbb{R}} \|(F_c - I) * \exp(a(F_c - I))\|_K \\ &\leq 6.5(b-a)a^{-3/4} \leq 9(b-a)b^{-3/4}. \end{aligned}$$

Finally, let $1\le a<1.5a<b$. Then

$$\left\|e^{a(F_c-I)}-e^{b(F_c-I)}\right\|_K \le 2 \le 6(b-a)b^{-1} \le 6(b-a)b^{-3/4}.$$

The proof is complete. □

Smoothing under moment and Cramér's conditions.

Let X be a r.v. with distribution F and characteristic function $\widehat{F}_X$. Assume that $\mathbb{E}X=0$, $\mathbb{E}|X|^{1+\delta}<\infty$ $(\exists 0<\delta\le 1)$, and let Cramér's condition (11.5) be satisfied:

$$\limsup_{|t|\to\infty} |\widehat{F}_X(t)| < 1$$

Then there exists constant $C(X,k,\delta)$ such that

$$\|(F-I)^{*k} * \exp(a(F-I))\,\|_K \le C(X,k,\delta)a^{-k(1+\delta)/2} \tag{C.37}$$

for any $a>0$.

For $a\in\mathbb{N}$ bound (C.37) is given in [89], Theorem 9.3. Bound (C.37) holds also if $0<a\notin\mathbb{N}$. Indeed,

$$\begin{aligned}\|(F-I)^{*k}\exp(a(F-I))\,\|_K &\le \|(F-I)^{*k} * \exp(\lceil a\rceil(F-I))\,\|_K \|\exp(\delta(I-F))\,\|_{TV}\\ &\le e^2\|(F-I)^{*k} * \exp(\lceil a\rceil(F-I))\,\|_K \le C(X,k)\lceil a\rceil^{-k(1+\delta)/2} \le C(X,k)a^{-k(1+\delta)/2}.\end{aligned}$$

Exercises

C.1 Let S be an integer-valued r.v. Assume that $\mathcal{L}(S)$ has a finite variance $\sigma^2 := \operatorname{var} S \in (1;\infty)$. Prove that there exists an absolute constant $c>0$ such that

$$d_{TV}(S;S+1) \ge c\sigma^{-1}, \quad d_{o}(S;S+1) \ge c\sigma^{-2}. \tag{C.38}$$

Estimate (C.38) complements Lemma C.4.

C.2 Let π_λ denote a Poisson $\mathbf{\Pi}(\lambda)$ r.v. Check that

$$d_K(\pi_\lambda;\pi_\lambda+1) \ge (1+o(1))/\sqrt{2\pi\lambda} \qquad (\lambda\to\infty).$$

Let r.v. τ be independent of π_λ, $\mathbb{P}(\tau=-1)=\mathbb{P}(\tau=1)=1/2$. Show that

$$d_K(\pi_\lambda;\pi_\lambda+\tau) = O(1/\lambda) \qquad (\lambda\to\infty).$$

C.3 Prove (C.19).

D Characteristic functions

Basic properties of characteristic functions can be found in many books (see, e.g., Chapter 1 in [8], Section 1.3 in [231], or Section 1.12 in [280]). We present below basic properties of characteristic functions that are relevant to the subject of this book.

General properties of characteristic functions.

Let X be a r.v. with distribution function F_X and characteristic function

$$\widehat{F}_X(t) = \mathbb{E}e^{itX} = \mathbb{E}\cos(tX) + i\mathbb{E}\sin(tX) \equiv Re\widehat{F}_X(t) + iIm\widehat{F}_X(t),$$

where $Re\widehat{F}_X$ and $Im\widehat{F}_X$ denote real and imaginary parts of $\widehat{F}_X$. Obviously,

- $\widehat{F}_X(0) = 1, \ |\widehat{F}_X(t)| \le 1 \quad (t \in \mathbb{R})$.
- For any $a, b \in \mathbb{R}$, $\widehat{F}_{aX+b}(t) = e^{itb}\widehat{F}_X(at)$.
- $\widehat{F}_{-X}(t) = \widehat{F}_X(-t) = Re\widehat{F}_X(t) - iIm\widehat{F}_X(t)$.

We can mention the following properties of characteristic functions:

- If r.v.s X and Y are independent, then $\widehat{F}_{X+Y}(t) = \widehat{F}_X(t)\widehat{F}_Y(t)$.
- If X, X_1, X_2 are i.i.d.r.v.s, then

$$\widehat{F}_{X_1-X_2}(t) = \widehat{F}_X(t)\widehat{F}_X(-t) = |\widehat{F}_X(t)|^2 .$$

- For all $t, \lambda \in \mathbb{R}$

$$\left|\exp\Big(\lambda(\widehat{F}_X(t) - 1)\Big)\right| = \exp\Big(\lambda(Re\widehat{F}_X(t) - 1)\Big).$$

- $\widehat{F}_X \in \mathbb{R} \Leftrightarrow \mathcal{L}(X)$ is symmetric.
- The distribution of a random variable X is lattice if and only if there exists $t_0 \neq 0$ such that $|\widehat{F}_X(t_0)| = 1$.
- The inversion formula: if x, y are the continuity points of d.f. F_X, then

$$F(y)_X - F_X(x) = \frac{1}{2\pi} \lim_{T\to\infty} \int_{-T}^{T} \frac{e^{-itx} - e^{-ity}}{it} \widehat{F}_X(t)dt. \tag{D.1}$$

If characteristic function $\widehat{F}_X$ is integrable on $\mathbb{R}$, then $\mathcal{L}(X)$ has continuous density

$$f_X(x) = \frac{1}{2\pi} \int_{-\infty}^{\infty} e^{-itx}\widehat{F}_X(t)dt. \tag{D.2}$$

- If $\mathbb{E}|X|^k < \infty$ for some $k \in \mathbb{N}$, then $\widehat{F}_X^{(j)}(0) = i^j\mathbb{E}X^j \quad (j \le k)$.
- If $\mathbb{E}|X|^k < \infty$ for some $k \in \mathbb{N}$, then the following expansions hold ($|\vartheta| \le 1$):

$$\widehat{F}_X(t) = 1 + (it)\,\mathbb{E}X + \frac{(it)^2}{2}\,\mathbb{E}X^2 + \ldots + \frac{(it)^{k-1}}{(k-1)!}\,\mathbb{E}X^{k-1} + \vartheta\frac{|t|^k}{k!}\,\mathbb{E}|X|^k, \tag{D.3}$$

$$\widehat{F}'_X(t) = i\left\{\mathbb{E}X + (it)\,\mathbb{E}X^2 + \frac{(it)^2}{2!}\,\mathbb{E}X^3 + \ldots + \frac{(it)^{k-2}}{(k-2)!}\,\mathbb{E}X^{k-1}\right\} + \frac{\vartheta|t|^{k-1}}{(k-1)!}\,\mathbb{E}|X|^k. \tag{D.4}$$

- For all $t \in \mathbb{R}$

$$|1 - \widehat{F}_X(t)|^2 \le 2(1 - Re\widehat{F}_X(t)). \tag{D.5}$$

- Let $0 \le p \le 1$. Then for any $t \in \mathbb{R}$

$$|(1-p) + p\widehat{F}_X(t)| \le \exp\Big(p(1-p)(Re\widehat{F}_X(t) - 1)\Big). \tag{D.6}$$

- If $\mathcal{L}(X)$ is concentrated on $\mathbb{Z}_+$ and $\mu := \mathbb{E}X - \mathbb{E}^2X - \mathbb{E}X(X-1) > 0$, then

$$|\widehat{F}_X(t)| \leq \exp\big(-2\mu\sin^2(t/2)\big) \qquad (\forall t \in \mathbb{R}). \tag{D.7}$$

These properties are well-known (see, for example, [280], p. 335 or [231], Section 1.3). For instance,

$$\begin{aligned} |\widehat{F}_X(t)-1|^2 &= |\mathbb{E}(e^{itX}-1)|^2 \leq \mathbb{E}(\cos(tX)-1)^2 + \mathbb{E}\sin^2(tX) \\ &= 2\big(1-\mathbb{E}\cos(tX)\big) = 2(1-Re\widehat{F}_X(t)). \end{aligned}$$

To prove (D.6), observe that $(Re\widehat{F}_X(t))^2 + (Im\widehat{F}_X(t))^2 = |\widehat{F}_X(t)|^2 \leq 1$. Therefore,

$$\begin{aligned} |1-p+p\widehat{F}_X(t)|^2 &= (1-p+pRe\widehat{F}_X(t))^2 + p^2(Im\widehat{F}_X(t))^2 \\ &= (1-p)^2 + 2p(1-p)Re\widehat{F}_X(t) + p^2(Re\widehat{F}_X(t))^2 + p^2(Im\widehat{F}_X(t))^2 \\ &\leq (1-p)^2 + 2p(1-p)Re\widehat{F}_X(t) + p^2 = 1 + 2p(1-p)(Re\widehat{F}_X(t)-1) \\ &\leq \exp\Big(2p(1-p)(Re\widehat{F}_X(t)-1)\Big). \end{aligned}$$

Property (D.6) was proved by Franken [120].

Next lemma presents moment expansions in terms of powers of $e^{it}-1$ instead of powers of t. It is a particular case of Lemma 3 in Šiaulys & Čekanavičius [281], see also (1.12), (1.13) in [89]. The proof of Lemma D.1 is almost identical to the proof of (B.10).

Lemma D.1 *Let X be a r.v. taking values in $\mathbb{Z}_+$. If $\mathbb{E}X^{k+1} < \infty$ $(\exists k \in \mathbb{N})$, then*

$$\Big|\widehat{F}_X(t) - \sum_{j=0}^{k} \mathbb{E}X^{(j)}(e^{it}-1)^j/j!\Big| \leq \mathbb{E}X^{(k+1)}|e^{it}-1|^{k+1}/(k+1)!$$

$$\Big|\widehat{F}'_X(t) - ie^{it}\sum_{j=1}^{k} \mathbb{E}X^{(j)}(e^{it}-1)^{j-1}/(j-1)!\Big| \leq \mathbb{E}X^{(k+1)}|e^{it}-1|^{k}/k!$$

for any $t \in \mathbb{R}$.

The following lemma is instrumental to the work on Kolmogorov's problem.

Lemma D.2 *Suppose that $\mathbb{E}X^4 < \infty$. Then for any $t \in \mathbb{R}$*

$$\begin{aligned} \sum_{k=1}^{3}(-1)^{k-1}(\widehat{F}_X(t)-1)^k/k &= \sum_{k=1}^{3}\kappa_k(it)^k/k! + \tfrac{9}{4}\vartheta t^4\mathbb{E}X^4, \\ \widehat{F}_X(2t)\widehat{F}_X^2(-t) &= 1 + 6\sum_{k=2}^{3}\kappa_k\frac{(it)^k}{k!} + \tfrac{32}{3}\vartheta t^4\mathbb{E}X^4, \end{aligned}$$

where $\kappa_1 = \mathbb{E}X$, $\kappa_2 = \mathrm{var}\,X$, $\kappa_3 = \mathbb{E}(X-\mathbb{E}X)^3$ are the first three cumulants of $\mathcal{L}(X)$, $\widehat{F}_X$ is the characteristic function of $\mathcal{L}(X)$, ϑ is a complex number such that $|\vartheta| \leq 1$.

The proof of Lemma D.2 can be found in [8], p. 161. Note that $\widehat{F}_X(2t)\widehat{F}_X^2(-t)$ is the characteristic function of $Y = 2X - X_1 - X_2$, where X, X_1, X_2 are i.i.d.r.v.s, $\mathbb{E}Y = 0$, $\mathrm{var}\,Y = 6\mathrm{var}\,X$, $\mathbb{E}(Y-\mathbb{E}Y)^3 = 6\mathbb{E}(X-\mathbb{E}X)^3$.

The following two lemmas can be found, e.g., in Petrov [231], p. 14. We denote by C_X constants depending on $\mathcal{L}(X)$.

Lemma D.3 *Let X be non-degenerate r.v. Then there exist $\varepsilon = \varepsilon(\mathcal{L}(X)) > 0$ and constant C_X such that*

$$|\widehat{F}_X(t)| \leq 1 - C_X t^2 \leq \exp(-C_X t^2) \qquad (|t| \leq \varepsilon).$$

Lemma D.4 *Let $\widehat{F}_X(t)$ satisfy Cramér's condition (11.5). Then for every $\varepsilon > 0$ there exists constant $C_{X,\varepsilon} < 1$ such that*

$$|\widehat{F}_X(t)| \leq C_{X,\varepsilon} \qquad (|t| \geq \varepsilon).$$

A more specific bound holds if $\mathcal{L}(X)$ has a bounded density f and $\operatorname{var} X < \infty$. Denote $C_0 = \sup_x f(x)$, $\sigma^2 = \operatorname{var} X$. Then inequality $|\widehat{F}_X(t)| \leq C_X$ $(|t| \geq \varepsilon)$ follows from the estimate

$$|\widehat{F}_X(t)| \leq \exp\left(-\frac{t^2}{96(2\sigma|t| + \pi)^2 C_0^2} \right) \qquad (t \in \mathbb{R}) \tag{D.8}$$

(see Statulevičius [287], formula (4.2)).

Next lemma due to Gamkrelidze [121] is an analogue of Lemma D.3 for integer-valued distributions with maximum span 1.

Lemma D.5 *If $\mathcal{L}(X)$ is a non-degenerate integer-valued distribution with maximum span 1, then there exists constant $c_X > 0$ such that for any $t \in [-\pi; \pi]$*

$$|\mathbb{E}e^{itX}| \leq \exp(-c_X t^2).$$

Next lemma is an analogue of Lemma D.4 for integer-valued distributions (see, for example, [135], p. 60).

Lemma D.6 *Let $\mathcal{L}(X)$ be concentrated on $\mathbb{Z}$ with the maximum span equal to 1. Then for any $\varepsilon > 0$ there exists constant $C_{X,\varepsilon} \in (0; 1)$ such that*

$$|\widehat{F}_X(t)| \leq C_{X,\varepsilon} \qquad (\varepsilon < |t| \leq \pi).$$

The method of characteristic functions.

The closeness of distributions can be estimated via the closeness of their characteristic functions – this is the essence the method of characteristic functions.

The method of characteristic functions is arguably the most popular approach when proving limit theorems, see [135, 162, 231]. It goes back to Liapounoff [194, 195]. Kolmogorov's uniform problem was solved by the offspring of the method of characteristic function – the so-called method of triangle functions, see [8].

The method of characteristic functions is especially effective when dealing with sums of independent r.v.s, since it allows one to switch to products of characteristic functions.

The drawback of the method is that it is not easily applicable to dependent r.v.s.

In the next four Lemmas we assume that measure M is concentrated on $\mathbb{Z}$ with the maximal span equal to 1. The following two results are due to Presman [237, 238].

Lemma D.7 *Let $P_X := \mathcal{L}(X)$ and $P_Y := \mathcal{L}(Y)$ be two unit measures on $\mathbb{Z}$ with finite second moments. Let $a \in \mathbb{R}$ be an arbitrary number. Denote $\varphi(t) = (\hat{P}_X(t) - \hat{P}_Y(t))e^{-ita}$. Then for any $b > 0$*

$$\|P_X - P_Y\|_{TV} \leq \frac{1 + \pi b}{2\pi} \int_{|t| \leq \pi} |\varphi(t) - b^{-2}\varphi''(t)| dt. \tag{D.9}$$

Bound (D.9) follows from (I.17) and the fact that

$$k^2 P_X(k) = -\frac{1}{2\pi}\int_{|t|\le\pi} \mathrm{e}^{-itk}\hat{P}_X''(t)\mathrm{d}t \qquad (k\in\mathbb{Z}).$$

Close to Lemma D.7 is the following

Lemma D.8 *Let M be a measure of finite variation on $\mathbb{Z}$ with a finite absolute first moment. Then for any $v\in\mathbb{R}$, $b>0$,*

$$\|M\|_{TV}^2 \le \frac{1+b\pi}{2\pi}\int_{-\pi}^{\pi}\left(|\mathrm{e}^{-itv}\widehat{M}(t)|^2 + b^{-2}\left|\left(\mathrm{e}^{-itv}\widehat{M}(t)\right)'\right|^2\right)\mathrm{d}t. \tag{D.10}$$

The proof of Lemma D.8 can be found in Presman [238], formula (14), or [89], Lemma 5.1.

The inversion formula for finite measures on $\mathbb{Z}$:

$$M(\{k\}) = \frac{1}{2\pi}\int_{-\pi}^{\pi}\mathrm{e}^{-itk}\widehat{M}(t)\,\mathrm{d}t \qquad (k\in\mathbb{Z}). \tag{D.11}$$

Hence

$$\|M\|_{\circ} \le \frac{1}{2\pi}\int_{-\pi}^{\pi}|\widehat{M}(t)|\,\mathrm{d}t. \tag{D.12}$$

By the Parseval identity,

$$\sum_{k\in\mathbb{Z}} M^2(\{k\}) = \frac{1}{2\pi}\int_{-\pi}^{\pi}|\widehat{M}(t)|^2\mathrm{d}t. \tag{D.13}$$

Lemma D.9 *Let M be a measure of finite variation concentrated on $\mathbb{Z}$. Then*

$$\|M\|_{K} \le \frac{1}{4\pi}\int_{-\pi}^{\pi}\frac{|\widehat{M}(t)|}{|\sin\frac{t}{2}|}\,\mathrm{d}t \le \frac{1}{4}\int_{-\pi}^{\pi}\frac{|\widehat{M}(t)|}{|t|}\,\mathrm{d}t. \tag{D.14}$$

If $\sum_k k^2|M(\{k\})| \le \infty$, then for any $a\in\mathbb{R}$

$$(k-a)^2\,|M(\{k\})| \;\le\; \frac{1}{2\pi}\int_{-\pi}^{\pi}\left|\left(\widehat{M}(t)\mathrm{e}^{-ita}\right)''\right|\mathrm{d}t, \tag{D.15}$$

$$(k-a)^2\,|M((-\infty,k))| \;\le\; \frac{1}{2\pi}\int_{-\pi}^{\pi}\left|\left(\frac{\widehat{M}(t)\mathrm{e}^{-ita}}{\mathrm{e}^{-it}-1}\right)''\right|\mathrm{d}t. \tag{D.16}$$

Note that (D.11) is a particular case of the Parseval identity: if μ,ν are finite measures on $\mathbb{Z}_+$ and $\hat{\mu},\hat{\nu}$ are their Fourier transforms, then

$$\sum_k \mu(k)\nu(k) = \frac{1}{2\pi}\int_{-\pi}^{\pi}\hat{\mu}(t)\hat{\nu}(-t)\mathrm{d}t.$$

More generally, if functions $|f|, f^2, |g|, g^2$ are integrable on $\mathbb{R}$ and φ, ψ are the Fourier transforms of f, g, then

$$\int_{-\infty}^{\infty} f(x)\bar{g}(x)\mathrm{d}x = \frac{1}{2\pi}\int_{-\infty}^{\infty} \varphi(t)\bar{\psi}(t)\mathrm{d}t.$$

If f is continuous and functions f, φ are integrable on $\mathbb{R}$, then for any distribution F

$$\int_{-\infty}^{\infty} f(x)F(\mathrm{d}x) = \frac{1}{2\pi}\int_{-\infty}^{\infty} \varphi(t)\widehat{F}(-t)\mathrm{d}t \tag{D.17}$$

(see [8], Theorem 1.12).

If $M = \mathcal{L}(X) - \mathcal{L}(Y)$, where X, Y are integer-valued r.v.s, then (D.9)–(D.16) yield bounds to distances between distributions.

The next lemma is essentially Lemma 1.5.1 from Ibragimov & Linnik [162], see also Lemma 8.1 in [89].

Lemma D.10 *Let $\mathcal{L}(X), \mathcal{L}(Y)$ have bounded densities, $\mathbb{E}|X|<\infty$, $\mathbb{E}|Y|<\infty$. Then for any $b>0$*

$$d^2_{TV}(X;Y) \le \frac{1}{2}\int_{-\infty}^{\infty} (b|\widehat{M}(t)|^2 + b^{-1}|\widehat{M}'(t)|^2)\mathrm{d}t, \tag{D.18}$$

where $\widehat{M}(t) = \widehat{P}_X(t) - \widehat{P}_Y(t)$.

The following analogue of (D.14) holds for arbitrary distributions (Esseen [115], p. 33; see also (8.10) in [89]).

Lemma D.11 *Let F, G be two distributions with characteristic functions $\widehat{F}(t), \widehat{G}(t)$. Then*

$$d_K(F;G) \le \frac{1}{2\pi}\int_{-\infty}^{\infty} \frac{|\widehat{F}(t)-\widehat{G}(t)|}{|t|}\mathrm{d}t. \tag{D.19}$$

Lemma D.11 is a consequence of the inversion formula: integrating by parts, one gets

$$(\widehat{F}(t)-\widehat{G}(t))/\mathrm{i}t = \int_{-\infty}^{\infty} \mathrm{e}^{\mathrm{i}tx}(G(x)-F(x))\mathrm{d}x,$$

i.e., function $(\widehat{F}(t)-\widehat{G}(t))/it$ is a Fourier transform of $G-F$. Hence

$$G(x)-F(x) = \frac{1}{2\pi}\lim_{T\to\infty}\int_{-T}^{T} \mathrm{e}^{-\mathrm{i}tx}\frac{\widehat{F}(t)-\widehat{G}(t)}{\mathrm{i}t}\mathrm{d}t$$

if function $(\widehat{F}(t)-\widehat{G}(t))/it$ is integrable (the bound is trivial if it is not).

The following lemma is known in the literature on the method of characteristic functions (cf. Theorem 2a in [115], Proposition 3 in [191], and Theorem 2.1.2 in [8], Chapter III).

Let $F\in\mathcal{F}$, $G\in\mathcal{F}$, $H\in\mathcal{F}$, and let ζ denote a r.v. with distribution $\mathcal{L}(\zeta)=H$.

Lemma D.12 *If $\tau>0$ and $\mathbb{P}(|\zeta|\le\tau/2)>1/2$, then*

$$d_K(F;G) \le C_\tau d_K(F*H;G*H) + C^*_\tau \min\{Q_F(\tau);Q_G(\tau)\}, \tag{D.20}$$

where $C_\tau = 1/(2\mathbb{P}(|\zeta|\le\tau/2)-1)$, $C^*_\tau = \mathbb{P}(|\zeta|\le\tau/2)/(2\mathbb{P}(|\zeta|\le\tau/2)-1)$.

If symmetric r.v. $\zeta = h\xi_o$, where $h>0$, d.f. F_{ξ_o} of $\mathcal{L}(\xi_o)$ is continuous strictly monotone and ch.f. φ_{ξ_o} of $\mathcal{L}(\xi_o)$ vanishes outside of $[-1;1]$, then one can choose $\tau = ch$, where $c = 2F_{\xi_o}^{-1}(7/8)$. Hence $\mathbb{P}(|\zeta| \le \tau/2) = 3/4$, and (D.20) becomes

$$d_K(F;G) \le 2d_K(F*H;G*H) + \tfrac{3}{2}\min\{Q_F(\tau);Q_G(\tau)\}. \tag{D.20*}$$

Since φ_{ξ_o} vanishes outside of $[-1;1]$, (D.19) entails $(\tau>0)$

$$d_K(F;G) \le \frac{1}{\pi}\int_{-c/\tau}^{c/\tau}\left|\frac{\widehat{F}(t)-\widehat{G}(t)}{t}\right|dt + \tfrac{3}{2}\min\{Q_F(\tau);Q_G(\tau)\}. \tag{D.21}$$

Using (E.1), this can be rewritten as $(h>0)$

$$d_K(F;G) \le \frac{1}{\pi}\int_{-1/h}^{1/h}\left|\frac{\widehat{F}(t)-\widehat{G}(t)}{t}\right|dt + \tfrac{3}{2}\lceil c\rceil\min\{Q_F(h);Q_G(h)\}. \tag{D.21*}$$

Lemma D.12 is a step forward comparing to Lemma D.11: one may have a sharp estimate of $|\hat{F}(t)-\hat{G}(t)|$ as $t\in[-1/h;1/h]$ but not as $t\notin[-1/h;1/h]$ yet $\min\{Q_F(h);Q_G(h)\}$ may be "small". Switching from the task of evaluating $d_K(F;G)$ to the task of evaluating $d_K(F*H;G*H)$ is an example of "smoothing".

The following corollary of Lemma D.12 can be useful for asymptotic expansions.

Corollary D.13 *Let $F\in\mathcal{F}$, $G\in\mathcal{F}$, $h>0$, and let M be a measure of finite variation such that $M(\mathbb{R})=0$. Then*

$$d_K(F;G*(I+M)) \le \int_{-1/h}^{1/h}\left|\frac{\widehat{F}(t)-\widehat{G}(t)(1+\widehat{M}(t))}{t}\right|dt + \tfrac{3}{2}(1+\|M\|_{TV})Q_G(h). \tag{D.22}$$

Proof of Corollary D.13. Let $M = M^+ - M^-$ be the Jordan-Hahn decomposition of M, where M^+ and M^- are non-negative measures. It follows from $M(\mathbb{R})=0$ that $M^+(\mathbb{R}) = M^-(\mathbb{R}) = \|M\|_{TV}/2$. Set

$$H_1 = \frac{F+G*M^-}{1+\|M\|_{TV}/2}, \quad H_2 = \frac{I+M^+}{1+\|M\|_{TV}/2}.$$

Then $H_1, H_2\in\mathcal{F}$ and $(1+\|M\|_{TV}/2)(H_1 - G*H_2) = F - G*(I+M)$. By (D.21) and (A.6),

$$\begin{aligned}\|F-G*(I+M)\|_K &= (1+\|M\|_{TV}/2)\|H_1 - G*H_2\|_K\\ &\le (1+\|M\|_{TV}/2)\int_{-1/h}^{1/h}\left|\frac{\widehat{H}_1(t)-\widehat{G}(t)\widehat{H}_2(t)}{t}\right|dt + C(1+\|M\|_{TV}/2)Q_{G*H_2}(h)\\ &\le \int_{-1/h}^{1/h}\left|\frac{\widehat{F}(t)-\widehat{G}(t)(1+\widehat{M}(t))}{t}\right|dt + C(1+\|M\|_{TV})Q_G(h).\end{aligned}$$

Corollary D.13 is proved. □

The following version of Lemma D.12 was proved in [87]. It can be useful when dealing with finite signed measures concentrated on a lattice.

Lemma D.14 *Given $h>0$, let W be a measure of finite variation, $W(\mathbb{R})=0$, $H\in\mathcal{F}$. If $|\widehat{H}(t)| \le C\widehat{G}(t)$ as $|t|\le 1/h$ $(\exists G\in\mathcal{F}_+)$, then*

$$\begin{aligned}\|W*H\|_K &\le C\int_{-1/h}^{1/h}\left|\frac{\widehat{W}(t)\widehat{H}(t)}{t}\right|dt + C\|W\|_{TV}\,Q_G(h)\\ &\le Ch^{-1}\sup_{|t|\le 1/h}|\widehat{W}(t)|/|t| + C\|W\|_{TV}\,Q_G(h).\end{aligned}$$

The following lemma can be used to prove lower bounds.

Lemma D.15 *Let M be a measure of finite variation concentrated on $\mathbb{Z}$, $a\in\mathbb{R}$, $b>1$. There exists an absolute constant $c>0$ such that*

$$\|M\|_{TV} \geq \|M\|_{K} \ \geq\ c\left|\int_{-\infty}^{\infty} \mathrm{e}^{-t^2/2}\widehat{M}(t/b)\mathrm{e}^{-\mathrm{i}ta}\mathrm{d}t\right|, \tag{D.23}$$

$$\|M\|_{\circ} \ \geq\ \frac{c}{b}\left|\int_{-\infty}^{\infty} \mathrm{e}^{-t^2/2}\widehat{M}\Big(t/b\Big)\mathrm{e}^{-\mathrm{i}ta}\mathrm{d}t\right|. \tag{D.24}$$

Estimates (D.23), (D.24) remain valid if $\mathrm{e}^{-t^2/2}$ is replaced by $t\mathrm{e}^{-t^2/2}$.

The proof of Lemma D.15 can be found in [281] and [89], Lemma 10.4.

Some inequalities are related to the structure of the support of a distribution. The following inequality is due to Arak. It is instrumental to the work on Kolmogorov's problem.

Given numbers $m\in\mathbb{N}$, $N\in\mathbb{N}$, $\mathbf{u}=(u_1,\ldots,u_N)\in\mathbb{R}^N$, we denote

$$K_m(\mathbf{u}) = \Big\{ j_1u_1 + ... + j_Nu_N : j_i\in\{-m,-m+1,\ldots,m\} \ \ (i=1,\ldots,N)\Big\}. \tag{D.25}$$

If $X_1, X_2, \ldots, X_n$ are independent random variables taking values in the set $\{0, u_1, u_2, \ldots, u_N\}$, then $S_n = X_1 + \cdots + X_n$ takes values in the set

$$\{j_1u_1 + ... + j_Nu_N\,, \quad \text{where } j_i\in\{0,1,\ldots,n\},\ j_1+\cdots+j_N\leq n\}$$

so that $\mathbb{P}(S_n\in K_m(\mathbf{u}))=1$. If we take m sufficiently large, then the probability for the approximating CP distribution not to take values from $K_m(\mathbf{u})$ will be "small". In a sense we deal with $X_i = u_1\xi_1 + \cdots + u_N\xi_N$, where $(\xi_1, \xi_2, \ldots, \xi_N)$ is an N-dimensional random vector taking values in $\mathbb{Z}^n$.

Due to (E.2), in general (D.21) provides the accuracy no better than $O(n^{-1/2})$. The following lemma due to Arak to some extent allows one to overcome this shortage.

For any finite signed measure M let $M=M^+-M^-$ denote the Hahn-Jordan decomposition of M. For instance, if F_X, F_Y are two distributions, then one can chose $M=F_X-F_Y$. Set

$$\delta(M,m,\mathbf{u}) = M^+\{\mathbb{R}\backslash K_m(\mathbf{u})\} + M^-\{\mathbb{R}\backslash K_m(\mathbf{u})\} \qquad (m\in\mathbb{N}).$$

Lemma D.16 [6] *Let M be a measure of finite variation such that $M(\mathbb{R})=0$, and let $U\in\mathcal{F}_+$. There exists an absolute constant C such that for any $m\in\mathbb{N}$, $N\in\mathbb{N}$, $\mathbf{u}\in\mathbb{R}^N$, $h>0$,*

$$\|M\|_{K} \leq C\int\limits_{|t|\leq 1/h}\left|\frac{\widehat{M}(t)}{t}\right|\mathrm{d}t + C\ln(m+1)Q_U(h)\sup_{t\in\mathbb{R}}\frac{|\widehat{M}(t)|}{\widehat{U}(t)} + \delta(M,m,\mathbf{u}). \tag{D.26}$$

Lemma D.16 contains an unspecified non-negative characteristic function $\widehat{U}$. In applications $\widehat{U}$ usually corresponds to a symmetric CP distribution.

For an effective application of (D.26) we need estimates of $\delta(M,m,\mathbf{u})$. The following lemma presents such estimates in a particular situation where $M=W*\exp(V)$, V,W are measures of finite variation.

Lemma D.17 [82] *Let V,W be measures of finite variation. If $N\in\mathbb{N}$, $a\in\mathbb{N}$, $s\in\mathbb{N}$, $\mathbf{u}\in\mathbb{R}^N$, $\operatorname{supp}W\subset K_s(\mathbf{u})$, $\operatorname{supp}V\subset K_a(\mathbf{u})$, then for any $y\in\mathbb{N}$*

$$\delta(W*\exp(V), s+ay, \mathbf{u}) \leq \|W\|_{TV}\exp(3\|V\|_{TV}-y)\,,\quad \delta(\exp(V), ay, \mathbf{u}) \leq \exp(3\|V\|_{TV}-y)\,.$$

The choice of y depends on the supports of the distributions. An example of an application of Lemmas D.16 and D.17 is Theorem 7.8.

Exercises

D.1 Let $\mathcal{L}(X)$ have bounded density f: $\sup_x f(x) \le C_0$. Show that

$$\frac{1}{2\pi}\int_{-\infty}^{\infty} |\widehat{F}_X(t)|^2 \le C_0 ,$$

where $\widehat{F}_X = \mathbb{E}e^{itX}$.

D.2 Let $\mathcal{L}(X)$ be a distribution with density

$$f_X(x) = (1-|x|)\mathbb{I}\{|x| \le 1\}.$$

Let φ_X denote the ch.f. of $\mathcal{L}(X)$. Check that

$$\varphi_X(2t) = (\sin t)^2/t^2 .$$

D.3 Inversion formula (D.1) can be derived from the following fact:

$$\lim_{T\to\infty}\int_{-T}^{T}\frac{\sin t}{t}dt = \pi \tag{D.27}$$

Show that (D.1) entails (D.27). Thus, (D.1) and (D.27) are equivalent.

D.4 Check that

$$\lim_{T\to\infty}\int_{-T}^{T}\frac{\sin t}{t}dt = \int_{-\infty}^{\infty}\frac{\sin^2 t}{t^2}dt = \pi.$$

D.5 Let $h > 0$. Denote

$$J_s = \frac{1}{\pi}\lim_{T\to\infty}\int_{-T}^{T}\frac{\sin t}{t}\cos(st)dt.$$

Show that $J_s = 1$ if $|s| < 1$, $J_s = 0$ if $|s| > 1$, $J_s = 1/2$ if $|s| = 1$.

E Concentration functions

Concentration functions are important tools when one applies smoothing lemmas, e.g., Lemma D.12. Properties of concentration functions can be found, e.g., in [8], Chapter II, [231], Chapter III, [89], Section 1.5.

Let $S_n = X_1 + ... + X_n$. Recall that $Q_Y(h) \equiv Q_{\mathcal{L}(Y)}(h) = \sup_x \mathbb{P}(x \le Y \le x+h)$ is the concentration function of $\mathcal{L}(Y)$.

- If $F, G \in \mathcal{F}$, $h > 0$, $c \in (0; h]$, then

$$Q_{F*G}(h) \le \min\{Q_F(h); Q_G(h)\}, \quad Q_F(h) \le \lceil h/c \rceil Q_F(c). \tag{E.1}$$

- The Kolmogorov-Rogozin inequality [251]. Let $X_1, \ldots, X_n$ be independent r.v.s. There exists an absolute constant C such that

$$Q_{S_n}(h) \leq C \left(\sum\nolimits_{i=1}^{n} (1 - Q_{X_i}(h)) \right)^{-1/2}. \tag{E.2}$$

- In particular, if $X, X_1, X_2, \ldots, X_n$ are i.i.d.r.v.s and $\mathcal{L}(X)$ is not degenerate, then

$$Q_{S_n}(h) \leq C_X(h+1)/\sqrt{n} \tag{E.3}$$

for all $h \geq 0$ and $n \in \mathbb{N}$ (cf. Rogozin [251], Petrov [230]).

- If $\mathbb{E}X^2 = \infty$, then

$$Q_{S_n}(h) = o(n^{-1/2}) \quad \text{as } n \to \infty \tag{E.4}$$

(Esseen [117], Theorem 4.1).

- The Kolmogorov-Rogozin inequality (E.2) has been improved by Kanter [169] and Mattner & Roos ([204], Corollary 1.5):

$$Q_{S_n}(h) \leq \sqrt{2/\pi} \Big/ \left(1/4 + \sum_{i=1}^{n} (1 - Q_{X_i}(2h)) \right)^{1/2}. \tag{E.5}$$

Namely, Kanter ([169], Theorem 4.2) has shown that

$$Q_{S_n}(h) \leq G\left(\sum_{i=1}^{n} (1 - Q_{X_i}(2h)) \right),$$

where

$$G(\lambda) = \mathbb{P}(\pi_{\lambda/2} - \pi'_{\lambda/2} \in \{0, 1\}) \quad (\lambda > 0).$$

Mattner & Roos [204] have proved that $G(\lambda) \leq \sqrt{2/\pi(1/4+\lambda)}$.

- Le Cam's inequality:

$$Q_{\mathbf{\Pi}(\lambda,X)}(h) \leq 1/\sqrt{2e\lambda p(h)} \leq 0.61/\sqrt{\lambda \mathbb{P}(|X| > h)}, \tag{E.6}$$

where $p(h) = \max\{\mathbb{P}(X < -h); \mathbb{P}(X > h)\}$ (cf. Proposition 3 and Remark 5 in [264]).

- Esseen's inequality ([117], see also [8] Lemma 4.1.4):

$$Q_F(h) \leq \left(\frac{96}{95} \right)^2 h \int\limits_{|t|<1/h} |\widehat{F}(t)| \, \mathrm{d}t \qquad (F \in \mathcal{F}). \tag{E.7}$$

- If $F \in \mathcal{F}_+$, then ([8], Lemma 1.5)

$$h \int\limits_{|t|<1/h} \widehat{F}(t) \, \mathrm{d}t \leq 13 Q_F(h). \tag{E.8}$$

A multidimensional version of the Kolmogorov-Rogozin inequality has been given by Esseen [117]. More results concerning concentration functions can be found in [215, 119, 118, 136, 138].

Exercises

E.1 Let $X, X_1, X_2, \ldots$ be independent integer-valued r.v.s. Denote $p_j^* = \sup_{k\in\mathbb{Z}} \mathbb{P}(X_j = k)$. Verify that there exists an absolute constant $C > 0$ such that

$$Q_{S_n}(0) \equiv \sup_{k\in\mathbb{Z}} \mathbb{P}(S_n = k) \le C\left(\sum_{j=1}^{n}(1-p_j^*)\right)^{-1/2}.$$

E.2 Show that

$$\mathbb{P}(\pi_{\lambda/2} - \pi'_{\lambda/2} = 0) \wedge \mathbb{P}(\pi_{\lambda/2} - \pi'_{\lambda/2} = 1) \ge 1/\sqrt{2\pi\lambda}(1+o(1)) \qquad (\lambda \to \infty).$$

E.3 Let X be an arbitrary r.v., and let standard normal r.v. Y be independent of X. Show that

$$d_K(X; X+cY) \le Q_X(c).$$

Thus, the accuracy of "smoothing" $\mathcal{L}(X)$ can be evaluated with the help of the estimates of the concentration function.

F Stein's method

Stein's method has been presented in many books and papers (see, for example, [21, 27, 25, 221, 58]).

The main advantage of Stein's method over other approximation methods is its applicability to dependent r.v.s. Constants derived via Stein's method are usually "small".

A drawback of Stein's method in the case of integer-valued r.v.s is that it does not apparently provide any advantage if the distributions are symmetric.

We summarise below basic points of the method.

Suppose one wants to approximate the distribution $\mathcal{L}(S)$ of a random variable S by distribution $\mathcal{L}(Y)$ of a random variable Y. One may be interested in evaluating $\mathbb{E}f(S) - \mathbb{E}f(Y)$ for a particular function f or

$$\sup_{f\in\mathbb{F}} |\mathbb{E}f(S) - \mathbb{E}f(Y)|$$

for a particular class $\mathbb{F}$ of functions.

According to Stein's method, there exist function $g(\cdot) \equiv g(\cdot, f, \mathcal{L}(Y))$ and operator $\mathcal{A} = \mathcal{A}(\mathcal{L}(Y))$ such that

$$f(\cdot) - \mathbb{E}f(Y) = \mathcal{A}g(\cdot), \qquad \text{(F.1)}$$

cf. (12.25), (12.29) in [221]. Hence $f(S) - \mathbb{E}f(Y) = \mathcal{A}g(S)$, and one only needs to evaluate $\mathbb{E}\mathcal{A}g(S)$.

We present further details in the case where r.v.s S and Y take values in $\mathbb{Z}_+$.

Given a function h on $\mathbb{Z}_+$, denote

$$\Delta h(j) = h(j+1) - h(j), \quad \Delta^2 = \Delta\Delta, \quad \|\Delta h\| = \sup_j |\Delta h(j)|.$$

Set
$$p(k) = \mathbb{P}(Y{=}k), \quad L_Y(k) = p^{-1}(k)\Delta p(k{-}1) \qquad (k \in \operatorname{supp}\mathcal{L}(Y)).$$
Then function $g \equiv g(\cdot, f, \mathcal{L}(Y))$ can be defined as follows: $g(0){=}0$,
$$g(n) \;=\; [\mathbb{E}f(Y)\mathbb{1}\{Y{<}n\} - \mathbb{E}f(Y)\mathbb{P}(Y{<}n)]/p(n{-}1) \qquad (n \in \operatorname{supp}\mathcal{L}(Y)) \tag{F.2}$$
(cf. [221], Section 12.2). Function g is a solution of equation
$$\Delta g(k) + g(k)L_Y(k) = f(k) - \mathbb{E}f(Y) \qquad (k{\ge}1), \tag{F.3}$$
i.e.,
$$\mathcal{A}g(\cdot) = \Delta g(\cdot) + g(\cdot)L_Y(\cdot).$$
We call (F.3) "the Stein equation". Thus,
$$f(\cdot) - \mathbb{E}f(Y) = \Delta g(\cdot) + g(\cdot)L_Y(\cdot) = \Delta g(\cdot) + g(\cdot)p^{-1}(\cdot)\Delta p(\cdot{-}1),$$
so that
$$(f(k) - \mathbb{E}f(Y))p(k) = p(k)\Delta g(k) + g(k)\Delta p(k{-}1) = \Delta(p(k{-}1)g(k)).$$
Typically $\lim_{k\to\infty} p(k)g(k{-}1){=}0$, hence $\mathcal{L}(Y)$ obeys (F.3). One can expect that $\mathcal{L}(S)$ is "close" to $\mathcal{L}(Y)$ if $\mathbb{E}\mathcal{A}g(S)$ is "small".

Note that it is possible to write down the Stein equation in a different manner (cf. [221], Remark 4.1). The way shown in (F.2) is in line with the general approach to characterisation of discrete distributions (see [221], Chapter 12).

Estimates of $\mathbb{E}f(S){-}\mathbb{E}f(Y)$ derived by evaluating $\mathbb{E}\mathcal{A}g(S)$ usually take the form
$$|\mathbb{E}f(S) - \mathbb{E}f(Y)| \le \|g\|\delta_1 + \|\Delta g\|\delta_2 + \|\Delta^2 g\|\delta_3 \tag{F.4}$$
for particular quantities $\delta_1, \delta_2, \delta_3$ (cf. the proof of Theorem 3.6).

A number of metrics (e.g., the uniform distance, the total variation distance, the Gini-Kantorovich distance, etc.) can be defined as
$$d_{\mathbb{F}_*}(X;Y) \equiv d_{\mathbb{F}_*}(\mathcal{L}(X);\mathcal{L}(Y)) = \sup_{f\in\mathbb{F}_*} |\mathbb{E}f(X){-}\mathbb{E}f(Y)|,$$
where $\mathbb{F}_*$ is a particular class of functions.

For instance, recall that the total variation distance between the distributions of r.v.s X and Y is
$$d_{TV}(\mathcal{L}(X);\mathcal{L}(Y)) = \sup_{f\in\mathcal{S}} \big(\mathbb{E}f(X){-}\mathbb{E}f(Y)\big),$$
where $\mathcal{S}$ denotes the class of measurable functions taking values in $[0;1]$.

In view of (F.4), the problem of evaluating $d_{\mathbb{F}_*}(S;Y)$ reduces to the problem of evaluating $\sup_{f\in\mathbb{F}_*}\|g\|$, $\sup_{f\in\mathbb{F}_*}\|\Delta g\|$, $\sup_{f\in\mathbb{F}_*}\|\Delta^2 g\|$, cf. (3.91), (3.92).

Since one needs evaluating $\mathbb{E}\Delta^k g(X)$, note that
$$|\mathbb{E}\Delta^k g(X)| \le 2\|\Delta^{k-1}g\| d_{TV}(X;X{+}1) \qquad (k{\ge}1) \tag{F.5}$$
for any integer-valued r.v. X and bounded function $g(\cdot)$. Indeed,
$$\begin{aligned}|\mathbb{E}\Delta^k g(X)| \;&=\; \Big|\sum_{j=-\infty}^{\infty} \Delta^k g(j)\mathbb{P}(X{=}j)\Big| = \Big|\sum_{j=-\infty}^{\infty} \Delta^{k-1} g(j)\Delta\mathbb{P}_X(j{-}1)\Big| \\ &\le\; \|\Delta^{k-1}g\| \sum_{j=-\infty}^{\infty} |\Delta\mathbb{P}_X(j{-}1)| = 2\|\Delta^{k-1}g\| d_{TV}(X;X{+}1).\end{aligned}$$

Similarly ($k\geq 2$),

$$|\mathbb{E}\Delta^k g(X)| = \Big|\sum_{j=-\infty}^{\infty} \Delta^{k-2}g(j)\Delta^2\mathbb{P}_X(j-2)\Big| \leq \|\Delta^{k-2}g\|\,\|\mathcal{L}(X)*(I_1-I)^{*2}\|_{TV}. \qquad (F.6)$$

Example F.1 *Poisson approximation.* If $Y=\pi_\lambda$ is a Poisson $\mathbf{\Pi}(\lambda)$ random variable, then the Stein equation (F.3) is

$$g(n+1) - \lambda^{-1}ng(n) = f(n) - \mathbb{E}f(\pi_\lambda) \qquad (n\in\mathbb{Z}_+). \qquad (F.7)$$

Clearly, $\mathbb{E}\pi_\lambda g(\pi_\lambda) = \lambda\mathbb{E}g(\pi_\lambda+1)$ by (3.56), hence $\mathbb{E}(g(\pi_\lambda+1) - \lambda^{-1}\pi_\lambda g(\pi_\lambda)) = 0$ in line with intuition.

One aims at evaluating $\mathbb{E}g(S_n+1) - \lambda^{-1}\mathbb{E}S_n g(S_n)$. The solution of (F.7) is

$$g(n) = \left(\mathbb{E}f(\pi_\lambda)\mathbb{I}\{\pi_\lambda<n\} - \mathbb{E}f(\pi_\lambda)\mathbb{P}(\pi_\lambda<n)\right)/\mathbb{P}(\pi_\lambda=n-1) \qquad (n\geq 1) \qquad (F.8)$$

(the value of $g(0)$ is irrelevant).

If $f(\cdot)\equiv f_A(\cdot) = \mathbb{I}\{\cdot\in A\}$, $A\subset\mathbb{Z}_+$, then (3.91) holds:

$$\|g\| \leq \min\{\lambda;\sqrt{2\lambda/\mathrm{e}}\}, \quad \|\Delta g\| \leq 1-\mathrm{e}^{-\lambda}$$

(see, e.g., [21], Remark 1.1.2). This bound is typically sufficient for deriving estimates of the accuracy of Poisson approximation in terms of the total variation distance.

If $f\in\mathcal{L}_1$, where $\mathcal{L}_1 = \{h\colon |h(x)-h(y)| \leq |x-y|\}$ is the set of Lipschitz functions, then (3.92) holds:

$$\|g\| \leq \lambda, \quad \|\Delta g\| \leq \lambda\wedge\tfrac{4}{3}\sqrt{2\lambda/\mathrm{e}}, \quad \|\Delta^2 g\| \leq \min\{4\lambda/3;2\}.$$

This bound allows one deriving estimates of the accuracy of Poisson approximation in terms of the Gini-Kantorovich distance. Note that $\frac{4}{3}\sqrt{2/\mathrm{e}} \leq 1.1437$.

If $f(\cdot)=\mathbb{I}\{\cdot=m\}$ $(m\in\mathbb{Z})$, estimates of $\|g\|$, $\|\Delta g\|$, $\|\Delta^2 g\|$ can be found in [226].

The following is a version of Stein's equation (F.7):

$$\lambda\tilde{g}(j+1) - j\tilde{g}(j) = f(j) - \mathbb{E}f(\pi_\lambda) \qquad (j\geq 0). \qquad (F.7^*)$$

If $f(j) = \mathbb{I}\{j\in A\}$, where $A\in\mathbb{Z}_+$, then there exist a solution $\tilde{g}$ to this equation satisfying

$$\|\tilde{g}\| \leq \min\{1;\sqrt{2/\lambda\mathrm{e}}\}, \quad \|\Delta\tilde{g}\| \leq (1-\mathrm{e}^{-\lambda})/\lambda$$

(cf. [21]), leading to estimates of the accuracy of Poisson approximation in terms of d_{TV}.

If $f\in\mathcal{L}_1$, we can chose $\tilde{g}(0) = \tilde{g}(1)$. Then (3.92) yields

$$\|\tilde{g}\| \leq 1, \quad \|\Delta\tilde{g}\| \leq \min\{1;1.1437/\sqrt{\lambda}\}, \quad \|\Delta^2\tilde{g}\| \leq \min\{4/3;2/\lambda\}$$

leading to estimates of the accuracy of Poisson approximation in terms of the Gini-Kantorovich distance. □

Example F.2 *Shifted Poisson approximation* $S_n \approx a+\pi_\mu$. Suppose that one aims at evaluating $|\mathbb{E}h(S_n) - \mathbb{E}h(\pi_\mu+a)|$, where $a\in\mathbb{N}, \mu>0,\ h\colon\mathbb{Z}\to[0;1]$.

Let $h_A(\cdot) := \mathbb{I}\{\cdot\in A\}$ $(A\subset\mathbb{Z}_+)$. The basic equation is

$$f(k+1) - \mu^{-1}(k-a)f(k) = h_A(k) - \mathbb{E}h_A(\pi_\mu+a) \qquad (k\geq a).$$

The solution $f \equiv f_{h_A}$ is

$$f(k)=0 \quad (k \le a), \quad f(k) = g_{\tilde{h}}(k-a) \quad (k \ge a),$$

where $\tilde{h}(m) = h_A(m+a)$ $(m \ge 0)$ and $g_{\tilde{h}}$ is given by (F.8) with λ replaced with μ. Hence

$$\mathbb{E}h_A(S_n) - \mathbb{E}h_A(\pi_\mu + a) = \mathbb{E}f(S_n+1) - \mu^{-1}\mathbb{E}(S_n-a)f(S_n).$$

One proceeds by evaluating $\mathbb{E}f(S_n+1) - \mu^{-1}\mathbb{E}(S_n-a)f(S_n)$. □

Example F.3 *Negative Binomial approximation.* Let $\mathcal{L}(Y)$ be a Negative Binomial distribution $\mathbf{NB}(r,p)$ with parameters $p \in (0;1)$ and $r > 0$, see (1.4). The advantage of the Negative Binomial approximation is that it involves two parameters while Poisson distribution $\mathbf{\Pi}(\lambda)$ has only one.

Equation (F.3) for the Negative Binomial distribution $\mathbf{NB}(r,p)$ is

$$g(j+1) - jg(j)/(j+r-1)p = \mathbb{I}\{j \in A\} - \mathbf{NB}(r,p)(A) \qquad (j \ge 1, A \subset \mathbb{Z}_+). \tag{F.9}$$

Let $S_{r,p}$ be a Negative Binomial $\mathbf{NB}(r,p)$ r.v. One can check that

$$\mathbb{E}S_{r,p}g(S_{r,p})/(S_{r,p}+r-1)p = \mathbb{E}g(S_{r,p}+1)$$

in line with intuition.

Another version of Stein's equation for NB distribution is given in Phillips & Brown [54]:

$$(r+j)pg(j+1) - jg(j) = \mathbb{I}\{j \in A\} - \mathbf{NB}(r,p)(A) \qquad (j \ge 1, A \subset \mathbb{Z}_+), \tag{F.10}$$

where (cf. [55])

$$\|g\| \le \min\left\{\frac{1}{1-p}; \frac{1.75}{\sqrt{rp(1-p)}}\right\}, \quad \|\Delta g\| \le \frac{1}{rp}, \quad |\Delta g(j)| \le \min\left\{\frac{1}{p(r+j)}; \frac{1}{j}\right\}. \tag{F.11}$$

Phillips & Brown [54] effectively set

$$\mathcal{A}g(j) = \frac{(j+1)p_{j+1}}{p_j}g(j+1) - jg(j) = (r+j)pg(j+1) - jg(j),$$

where $p_j = \mathbb{P}(S_{r,p}=j)$ $(j \in \mathbb{Z}_+)$. Note that $p\mathbb{E}(r+S_{r,p})g(S_{r,p}+1) = \mathbb{E}S_{r,p}g(S_{r,p})$. □

Example F.4 *Compound Poisson approximation.* Given a set $\lambda_1, \lambda_2, ...$ of positive numbers such that $\lambda := \sum_{j \ge 1} \lambda_j < \infty$, let

$$Y = \sum_{j=1}^{\infty} j\pi_{\lambda_j},$$

where $\{\pi_{\lambda_j}\}$ are independent Poisson $\mathbf{\Pi}(\lambda_j)$ random variables. Then Y is a compound Poisson r.v. (cf. (1.2)).

Using (3.57), one can check that

$$\mathbb{E}Yg(Y) = \sum_{j=1}^{\infty} j\lambda_j \mathbb{E}g(j+Y)$$

if $\mathbb{E}|Yg(Y)| < \infty$. This suggest the following "Stein equation" [22]:

$$\sum_{j\geq 1} j\lambda_j g(j+k) - kg(k) = f(k) - \mathbb{E}f(Y) \qquad (k\in\mathbb{Z}_+). \tag{F.12}$$

If $\|f\| \leq 1$, then

$$\|\Delta g\| \leq \mathcal{M}(\lambda),$$

where $\mathcal{M}(\lambda) = \min\left\{1; \lambda_{n,1}^{-1}\right\} e^{\lambda}$ (Barbour et al. [22]); factor $\mathcal{M}(\lambda)$ obeys (6.16)– (6.17). Estimates of $\mathcal{M}(\lambda)$ can be found in [100]. □

Example F.5 *Shifted compound Poisson approximation.* Let $X_1, ..., X_n$ be independent integer-valued random variables, $S_n = X_1 + \cdots + X_n$. Given $\gamma\in\mathbb{Z}$, $K\in\mathbb{N}$, $\lambda_{-t}, \ldots, \lambda_t \in \mathbb{R}$, let $\mathbf{\Pi}_{\gamma,\lambda}$ denote the distribution $\mathcal{L}(Y)$ of r.v.

$$Y = \gamma + \sum_{1\leq|k|\leq K} k\pi_{\lambda_k},$$

where $\{\pi_{\lambda_k}\}$ are independent Poisson (generalised) r.v.s.

The "Stein equation" (F.1) in this case is

$$f(\cdot) - \mathbb{E}f(Y) = \mathcal{A}g(\cdot),$$

where f is a bounded function,

$$(\mathcal{A}g)(j) = \sum_{1\leq|k|\leq K} k\lambda_k g(j+k) - (j-\gamma)g(j).$$

Set

$$\lambda = \sum_{1\leq|k|\leq K} k\lambda_k\,, \quad \tilde{\theta} = \sum_{|k|\leq K} k(k-1)\lambda_k/\lambda.$$

Assume that $\tilde{\theta} < 1/2$, $\lambda > 0$. One aims to show that

$$|\mathbb{E}(\mathcal{A}g)(S_n)| \leq \delta_1\|g\| + \delta_2\|\Delta g\|$$

for particular δ_1, δ_2. It is known [29] that

$$\|g\| \leq \frac{2\|f\|}{1-2\tilde{\theta}}(1\wedge\lambda^{-1/2}), \quad \|\Delta g\| \leq \frac{2\|f\|}{1-2\tilde{\theta}}(1\wedge\lambda^{-1}).$$

Then for any $\gamma\in\mathbb{Z}$

$$d_{TV}(\mathcal{L}(S_n); \Pi_{\gamma,\lambda}) \leq \frac{1}{1-2\tilde{\theta}}\left((1\wedge\lambda^{-1/2})\delta_1 + (1\wedge\lambda^{-1})\delta_2 + \varepsilon_\lambda + (1+\varepsilon_\lambda)\mathbb{P}(S_n < \gamma)\right), \tag{F.13}$$

where $\varepsilon_\lambda = \sum_{j<0} |\Pi_{0,\lambda}(\{j\})|$ (see [29], Corollary 2.3). If all $\lambda_i \geq 0$, then

$$\varepsilon_\lambda = \mathbb{P}(Y < 0) \leq \exp(-\lambda/8K), \tag{F.14}$$

where $\mathcal{L}(Y) = \mathbf{\Pi}_{\gamma,\lambda}$ (cf. [29], Lemma 6.6). □

A method of deriving Stein's equations via probability generating functions is described in [297]. A "perturbations approach" to the work on solutions of Stein's equations is given in [34].

Exercises

F.1 Show that the solution of (F.7) is

$$g(n) = (\mathbb{E}f(\pi_\lambda)\mathbb{I}\{\pi_\lambda < n\} - \mathbb{E}f(\pi_\lambda)\mathbb{P}(\pi_\lambda < n)) / \mathbb{P}(\pi_\lambda = n-1) \qquad (n\in\mathbb{N}).$$

G Mixing (weak dependence) conditions

Given a sequence $\{X_{n,i}\}_{i\geq 1}$ of r.v.s, let $\mathcal{F}_{l,m}$ denote the σ–field generated by $\{X_{n,l},...,X_{n,m}\}$. Random variables $\{X_{n,i}\}_{i\geq 1}$ are called m-dependent if $\mathcal{F}_{1,\ell}$ and $\mathcal{F}_{\ell+m+1,\infty}$ are independent for all $\ell\geq 1$.

Set

$$\begin{aligned}\alpha_n(l) &= \sup|\mathbb{P}(AB)-\mathbb{P}(A)\mathbb{P}(B)|, \quad \varphi_n(l) = \sup|\mathbb{P}(B|A) - \mathbb{P}(B)|,\\ \beta_n(l) &= \sup\mathbb{E}\sup_B|\mathbb{P}(B|\mathcal{F}_{1,m})-\mathbb{P}(B)|,\end{aligned}$$

where the supremum is taken over $m\geq 1$, $A\in\mathcal{F}_{1,m}$, $B\in\mathcal{F}_{m+l+1,n}$ such that $\mathbb{P}(A)>0$. Then $\alpha_n(\cdot)$, $\beta_n(\cdot)$, $\varphi_n(\cdot)$ are called mixing (weak dependence) coefficients.

Sequence $\{X_{n,i}\}_{i\geq 1}$ is called α-mixing if $\alpha_n(l_n)\to 0$ for some sequence $\{l=l_n\}$ of natural numbers such that $0\leq l_n \ll n$ as $n\to\infty$; we say for brevity that *condition* $(\Delta\{l_n\})$ holds.

Sequence $\{X_{n,i}\}_{i\geq 1}$ is called φ-mixing if $\varphi_n(l_n)\to 0$ for some sequence $\{l=l_n\}$ of natural numbers such that $0\leq l_n\ll n$ as $n\to\infty$.

Let $\{u_n(\cdot), n\geq 1\}$ be a sequence of functions from $[0;\infty)$ to $\mathbb{R}$ such that $u_n(\cdot)$ is strictly decreasing for all large enough n. Given $\bar{t}=(t_1,...,t_k)$, denote $u_n(\bar{t})=(u_n(t_1),...,u_n(t_k))$, and let $\mathcal{F}_{l,m}(\bar{t})$ be the σ–field generated by the events $\{X_i>u_n(t_j)\}$, $l\leq i\leq m, 1\leq j\leq k$.

We define mixing coefficients $\alpha_n(l,u_n(\bar{t}))$, $\beta_n(l,u_n(\bar{t}))$, $\varphi_n(l,u_n(\bar{t}))$ similarly to $\alpha_n(l)$, $\beta_n(l)$, $\varphi_n(l)$ with σ–fields $\mathcal{F}_{1,m}$, $\mathcal{F}_{m+l+1,n}$ replaced with σ–fields $\mathcal{F}_{1,m}(\bar{t})$, $\mathcal{F}_{m+l+1,n}(\bar{t})$.

Condition $(\Delta\{u_n(\bar{t})\})$ is defined similarly to condition $(\Delta\{l_n\})$ with σ–fields $\mathcal{F}_{l,m}(\bar{t})$ instead of $\mathcal{F}_{l,m}$: condition $(\Delta\{u_n(\bar{t})\})$ holds if $\alpha(l_n,u_n(\bar{t}))\to 0$ for some sequence $\{l_n\}$ (perhaps, dependent on $\{u_n(\bar{t})\}$) such that $l_n\to\infty$, $l_n/n\to 0$.

Condition (Δ) holds if $(\Delta\{u_n(\bar{t})\})$ is in force for all $\bar{t}=(t_1,...,t_k): 0<t_1<\ldots<t_k<\infty$, $k\geq 1$.

Class $\mathcal{R}(\bar{t})$. If condition $(\Delta\{l_n\})$ holds, then there exists a sequence $\{r_n\}$ of natural numbers such that

$$n\gg r_n\gg l_n\geq 0, \; nr_n^{-1}\alpha_n^{2/3}(l_n)\to 0 \qquad (n\to\infty) \tag{2.8*}$$

(for instance, one can take $r_n=\left[\sqrt{n\max\{l_n;n\alpha_n(l_n)\}}\right]$).

We denote by $\mathcal{R}$ the class of all such sequences $\{r_n\}$.

Similarly, if condition $(\Delta\{u_n(\bar{t})\})$ holds, then there exists sequences $\{l_n\},\{r_n\}$ of natural numbers such that (2.8*) holds.

Results concerning relations between different mixing (weak dependence) coefficients can be found, e.g., in [162], [221], Chapter 14.

When dealing with sums of dependent r.v.s one would like to replace some of those r.v.s with their "independent" copies. The following two lemmas allow one to do that.

Let (X,Y) be a random vector taking values in $\mathbb{R}^l\times\mathbb{R}^m$, and let α, β, φ and ρ be the mixing coefficients corresponding to the σ–fields $\sigma(X)$ and $\sigma(Y)$.

For any $v\in\mathbb{R}^m$ we put $|v|=\max_{i\leq m}|v_i|$.

Lemma G.1 *One can construct* X,Y,Y^* *on a common probability space in such a way that* $Y^*\stackrel{d}{=}Y$, Y^* *is independent of* X *and*

$$\mathbb{P}(Y^*\neq Y)=\beta.$$

Lemma G.2 *One can define random vectors X, Y and $\hat{Y}$ on a common probability space in such a way that $\hat{Y}$ is independent of X, $\hat{Y} \stackrel{d}{=} Y$ and $(y>0, K \in \mathbb{N})$*

$$\mathbb{P}\left(|\hat{Y}-Y|>y\right) \leq 2^{(m+3)/2} K^{m/2} \alpha + 2\mathbb{P}(|Y|>Ky). \tag{G.1}$$

H Normal approximation

Let Y_{μ,σ^2} denote normal $\mathcal{N}(\mu;\sigma^2)$ r.v. with mean μ and variance σ^2. Let

$$\Phi_{\mu;\sigma^2}(x) = \frac{1}{\sqrt{2\pi\sigma^2}} \int_{-\infty}^{x} \exp\left(-(t-\mu)^2/2\sigma^2\right) \mathrm{d}t$$

denote the d.f. of $\mathcal{N}(\mu;\sigma^2)$. We write Φ_{σ^2} for $\Phi_{0;\sigma^2}$. Note that ch.f.

$$\widehat{\Phi}_{\mu;\sigma^2}(t) = \mathrm{e}^{it\mu - \sigma^2 t^2/2}.$$

Let $X_1, X_2, \ldots, X_n$ be independent r.v.s. with finite third absolute moments, $S_n = X_1 + \ldots + X_n$. Denote $\mu = \mathbb{E}S_n$, $\sigma^2 = \operatorname{var} S_n$. The Berry-Esseen inequality states that

$$d_K(S_n; Y_{\mu,\sigma^2}) \leq C \sum_{i=1}^{n} \mathbb{E}|X_i - \mathbb{E}X_i|^3/\sigma^3, \tag{H.1}$$

where $C \leq 0.56$ ($C \leq 0.4693$ in the case of i.i.d.r.v.s) [278, 293].

Exercises

H.1 Prove the following estimate from [180] (see also [162], Lemma 15.3.1):

$$d_K(\mathcal{N}(0;\sigma_1^2); \mathcal{N}(0;\sigma_2^2)) \leq |\sigma_1^2 - \sigma_2^2|/\sigma_2^2 \qquad (\sigma_1 \geq \sigma_2 > 0). \tag{H.2}$$

Show that (H.2) holds if d_K is replaced with d_{TV}.

I Useful identities and inequalities

A list of useful identities and inequalities can be found, e.g., in [221], Chapter 14 or [89], Chapter 1. We add below few more useful identities and inequalities relevant to this book.

Let $f_k(t)$, $g_k(t)$, $(k=1,\ldots,n)$ be two sets of twice differentiable functions. Then (omitting dependence on t and assuming $\prod_{j=s}^{m} = 1$ if $s>m$, we can write

$$\prod_{k=1}^{n} f_k - \prod_{k=1}^{n} g_k = \sum_{k=1}^{n} (f_k - g_k) \prod_{j=1}^{k-1} f_j \prod_{j=k+1}^{n} g_j \tag{I.1}$$

(recall that $\prod_{j=1}^{0} f_j = \prod_{j=n+1}^{n} g_j = 1$). Similarly,

$$\Big(\prod_{k=1}^{b} f_k - \prod_{k=1}^{n} g_k\Big)' = \sum_{k=1}^{n} f_k'\Big(\prod_{j\neq k}^{n} f_j - \prod_{j\neq k}^{n} g_j\Big) + \sum_{k=1}^{n}(f_k'-g_k')\prod_{j\neq k}^{n} g_j\,,$$

$$\begin{aligned}\Big(\prod_{k=1}^{b} f_k - \prod_{k=1}^{n} g_k\Big)'' &= \sum_{k=1}^{n} f_k''\Big(\prod_{j\neq k}^{n} f_j - \prod_{j\neq k}^{n} g_j\Big) + \sum_{k=1}^{n}(f_k''-g_k'')\prod_{j\neq k}^{n} g_j \\ &+ \sum_{k=1}^{n} f_k'\sum_{j\neq k}^{n}(f_j'-g_j')\prod_{m\neq k.j}^{n} g_m + \sum_{k=1}^{n}(f_k'-g_k')\sum_{j\neq k}^{n} g_j'\prod_{m\neq k,j}^{n} g_m \\ &+ \sum_{k=1}^{n} f_k'\sum_{j\neq k}^{n} f_j'\Big(\prod_{m\neq j,k}^{n} f_m - \prod_{m\neq j,k}^{n} g_m\Big). \end{aligned} \tag{I.2}$$

Binomial coefficients.

The following simple inequalities deal with binomial coefficients. Let $0<p<1$. Denote $q=1-p$. Let $S_{n,p}$ denote a Binomial $\mathbf{B}(n,p)$ r.v. Then

$$\mathbb{E}(S_{n,p}+1)^{-c} \equiv \sum_{k=0}^{n}\binom{n}{k}p^k q^{n-k}/(k+1)^c \le (p(n+1))^{-c} \qquad (c=1/2,1,2), \tag{I.3}$$

$$\mathbb{E}|S_{n,p}-np| \equiv \sum_{k=0}^{n}\binom{n}{k}p^k q^{n-k}|k-np| \le \sqrt{npq}\,. \tag{I.4}$$

A discrete analogue of Taylor's formula.

Let g be a function on $\mathbb{Z}$. Then

$$\begin{aligned} g(k) &= g(0) + \sum_{i=0}^{k-1}\Delta g(i) && (k\in\mathbb{N}), \\ g(k) &= g(0) - \sum_{j=k+1}^{0}\Delta g(j-1) && (k\in\mathbb{Z}, k<0). \end{aligned}$$

These equations can be reiterated: since $\Delta g(i) = \Delta g(0) + \sum_{j=0}^{i-1}\Delta^2 g(j)$ $(i\in\mathbb{N})$, we get

$$g(k) = g(0) + k\Delta g(0) + R_1(g,k) \qquad (k\ge 2),$$

where $R_1(g,k) = \sum_{i=0}^{k-1}\sum_{j=0}^{i-1}\Delta^2 g(j) = \sum_{j=1}^{k-1}(k-j)\Delta^2 g(j-1)$.

The above expansions can be extended in the following way: one can check that

$$g(\omega+k) = g(\omega) + k\Delta g(\omega) + R_1(g,\omega,k), \tag{I.5}$$

where $\omega\in\mathbb{Z}$,

$$R_1(g,\omega,k) = \begin{cases} \sum_{i=1}^{k-1}(k-i)\Delta^2 g(\omega+i-1) & (k\ge 2), \\ 0 & (k=0,1), \\ -\sum_{j=1}^{-k}(k+j-1)\Delta^2 g(\omega-j) & (k<0). \end{cases}$$

This can be rewritten as

$$g(\omega+l) = g(\omega+1) + (l-1)\Delta g(\omega+1) + R(g,\omega,l), \tag{I.5*}$$

where

$$R(g,\omega,l) = \begin{cases} \sum_{i=1}^{l-2}(l-i-1)\Delta^2 g(\omega+i) & (l\geq 3), \\ 0 & (l=1,2), \\ -\sum_{j=0}^{-l}(l+j-1)\Delta^2 g(\omega-j) & (l\leq 0). \end{cases}$$

Similarly, for any function g

$$g(\omega+k) = g(\omega) + k\Delta g(\omega) + \binom{k}{2}\Delta^2 g(\omega) + R_2(g,\omega,k), \tag{I.6}$$

where $\omega\in\mathbb{Z}$,

$$R_2(g,\omega,k) = \begin{cases} \sum_{i=1}^{k-2}\binom{k-i}{2}\Delta^3 g(\omega+i-1) & (k\geq 3), \\ 0 & (0\leq k\leq 2), \\ -\sum_{j=0}^{-k-1}\binom{1-k-j}{2}\Delta^3 g(\omega-j-1) & (k<0). \end{cases}$$

If $k\geq 0$, then ([270], Lemma 1)

$$f(k) = \sum_{i=0}^{k}\binom{k}{i}\Delta^i f(0) = f(0) + k\Delta f(0) + \frac{1}{2}\Delta^2 f(0) + ... + \Delta^k f(0). \tag{I.7}$$

Taylor series.
Let $t,x\in\mathbb{R}$, $k\in\mathbb{Z}_+$. By Taylor's formula,

$$\mathrm{e}^{\mathrm{i}tx} = 1 + \mathrm{i}tx + \frac{(\mathrm{i}tx)^2}{2!} + \cdots + \frac{(\mathrm{i}tx)^k}{k!} + \theta\frac{|tx|^{k+1}}{(k+1)!}, \tag{I.8}$$

where $|\theta|\leq 1$. More generally, let a be a real or complex number. Then

$$\mathrm{e}^a = 1 + a + \frac{a^2}{2!} + \ldots + \frac{a^k}{k!} + \frac{a^{k+1}}{k!}\int_0^1 \mathrm{e}^{ta}(1-t)^k\,\mathrm{d}t. \tag{I.9}$$

Consequently,

$$\left|\mathrm{e}^a - 1 - \sum_{j=1}^{k}\frac{a^j}{j!}\right| \leq \frac{|a|^{k+1}}{(k+1)!}\exp(|a|\mathbb{1}\{Re\, a>0\}). \tag{I.10}$$

A probabilistic version of Taylor's formula can be found in [221], Section 14.10.

Relation

$$\mathrm{e}^{\mathrm{i}t} = \cos t + \mathrm{i}\sin t \qquad (t\in\mathbb{R})$$

is known as Euler's formula. Clearly, $|\mathrm{e}^{\mathrm{i}t}|=1$.
Let $k\in\mathbb{Z}_+$. Then

$$|\mathrm{e}^{\mathrm{i}t}-1| = 2|\sin(t/2)|, \quad |\sin(kt/2)| \leq k\,|\sin(t/2)|,$$
$$(\mathrm{e}^{\mathrm{i}t}-1) + (\mathrm{e}^{-\mathrm{i}t}-1) = -(\mathrm{e}^{\mathrm{i}t}-1)(\mathrm{e}^{-\mathrm{i}t}-1),$$
$$|(\mathrm{e}^{-\mathrm{i}t}-1)^k - (-1)^k(\mathrm{e}^{\mathrm{i}t}-1)^k| \leq k|\mathrm{e}^{\mathrm{i}t}-1|^{k+1}.$$

Note that

$$|\sin(t/2)| \geq |t|/\pi \qquad (|t|\leq\pi). \tag{I.11}$$

The following two technical lemmas appear helpful when deriving (C.14)-type estimates.

Lemma I.1 [279] *Let* $\lambda>0,\ m\geq 1$. *Then*

$$\int_{-\pi/2}^{\pi/2} \exp\big(-2\lambda\sin^2 t\big)\,|\sin t|^m \mathrm{d}t \leq \pi(1+\sqrt{\pi/2m}\,)2^{-m-1}\mathrm{e}^{-m/2}(m/\lambda)^{-(m+1)/2}. \tag{I.12}$$

If $m=1$, then (I.12) yields

$$\int_0^{\pi/2} \exp\big(-2\lambda\sin^2 t\big)\sin t\mathrm{d}t \leq \frac{\pi}{8\sqrt{\mathrm{e}}}(1+\sqrt{\pi/2}\,)/\lambda. \tag{I.13}$$

Roos ([259], Lemma 6) has shown that

$$\int_0^{\pi/2} \exp\big(-2\lambda\sin^2 t\big)\,\mathrm{d}t \leq \frac{\pi}{4}\big(1+\sqrt{\pi/2}\,\big)/\sqrt{3\lambda}. \tag{I.12*}$$

Lemma I.2 [169] *If* $\beta\in[0;1],\ \gamma\geq 0,\ n\in\mathbb{N},\ r\in\mathbb{N}$, *then*

$$\begin{aligned}&\int_0^{\pi}(\cos t)^{2r}(1+\beta(\cos t-1))^{2n}\exp(\gamma(\cos t-1))(1+\cos t)\mathrm{d}t\\ \leq &\int_0^{\pi}\exp\big((2r+2n\beta+\gamma)(\cos t-1)\big)(1+\cos t)\mathrm{d}t.\end{aligned}$$

Moment inequalities.

- Chebyshev's inequality: if $x>0$, $\sigma^2 = \operatorname{var}\xi < \infty$, then

$$\mathbb{P}(|\xi-\mathbb{E}\xi| \geq x) \leq \sigma^2/x^2.$$

An application of this bound to $x+c$ with a proper choice of c leads to

$$\mathbb{P}(\xi-\mathbb{E}\xi\geq x) \leq \sigma^2/(x^2+\sigma^2).$$

- Lyapunov's inequality [195]: if $1<s\leq t$, then

$$\mathbb{E}|\xi| \leq \big(\mathbb{E}|\xi|^s\big)^{1/s} \leq \big(\mathbb{E}|\xi|^t\big)^{1/t}. \tag{I.14}$$

- Hölder's inequality: if $p,q\in[0;1]$, $p+q=1$, then

$$\mathbb{E}|\xi\eta| \leq \big(\mathbb{E}|\xi|^{1/p}\big)^p\big(\mathbb{E}|\eta|^{1/q}\big)^q.$$

- Minkowski's inequality: if $s\geq 1$, then

$$\big(\mathbb{E}|\xi+\eta|^s\big)^{1/s} \leq \big(\mathbb{E}|\xi|^s\big)^{1/s} + \big(\mathbb{E}|\eta|^s\big)^{1/s}.$$

- Suppose that $X_1, X_2, ..., X_n$ are independent r.v.s, $\mathbb{E}X_i = 0$, $\mathbb{E}|X_i|^t < \infty$ $(t > 0)$. Denote $S_n = X_1 + ... + X_n$,

$$\sigma_n^2 = \operatorname{var} S_n\,, \quad A_n(t) = \sum_{i=1}^{n} \mathbb{E}|X_i|^t\,.$$

  If $t/2 \in \mathbb{N}$, then (Ibragimov & Sharakhmedov [164])

$$\mathbb{E}\,|S_n|^t \le \mathbb{E}(\pi_1 - 1)^t \max\{A_n(t); \sigma_n^t\},$$

  where $\mathcal{L}(\pi_1) = \mathbf{\Pi}(1)$; the constant $\mathbb{E}(\pi_1 - 1)^t$ cannot be improved. By Jensen's inequality,

$$\mathbb{E}\,|S_n|^t \ge \sigma_n^t \qquad (t \ge 2).$$

- Let $X_1, X_2, ..., X_n$ be independent r.v.s with zero means and finite 3rd moments. If $\operatorname{var} S_n = 1$, then (Tyurin [294])

$$|\mathbb{E}|S_n|^3 - 2\sqrt{2/\pi}\,| \le A_n(3). \tag{I.15}$$

Other useful identities and inequalities.

- One can check that

$$1 + x \le \mathrm{e}^x \quad (x \in \mathbb{R}), \qquad 1 + x \ge \mathrm{e}^{x(1-x)} \quad (x > 0). \tag{I.16}$$

- If $a \in \mathbb{R}$ and $b > 0$, then (see, e.g., [238] or [8], p. 83)

$$\sum_{k \in \mathbb{Z}} \left(1 + (k-a)^2/b^2\right)^{-1} \le 1 + b\pi. \tag{I.17}$$

- If $m \in \mathbb{Z}_+$, then

$$(a+b)^{(m)} = \sum_{i=0}^{m} \binom{m}{i} a^{(i)} b^{(m-i)}\,. \tag{I.18}$$

  Recall that $x^{(k)} = x(x-1) \cdots (x-k+1)$ for any $x \in \mathbb{R}, k \in \mathbb{N}$.

- Let $p \in (0;1)$, $q = 1 - p$. Then

$$\sum_{k \ge j} \binom{k}{j} p^j q^{k-j} = 1/p \qquad (j \in \mathbb{Z}_+). \tag{I.19}$$

Exercises

I.1 Prove (I.3)–(I.4).

I.2 Check (I.18).

I.3 Prove (I.19).

Notes

Estimate (C.13) with $\min\{a; b\}$ instead of $\max\{a; b\}$ is Corollary 5.5.2 in [8], the present version is (2.37) in [89].

Lemma C.3 is Proposition 4 in Roos [263]. Lemma C.10 is Theorem 1.2 from [70].

Bound (D.14) is due to Tsaregradskii [291]. Proofs of estimates (D.15) and (D.16) can be found in [89], p. 93–95.

Lemma D.7 is from [237]. Lemma D.8 is from [238].

Esseen [117] was probably the first to evaluate a concentration function via an integral of a characteristic function. He proved (E.7) with an abstract constant. Estimate (E.8) is the Arak's [5] modification of a similar Esseen's [117] result.

Lemma G.1 is from Berbee [40], Chapter 4. Lemma G.2 is Lemma 14.13 in [221].

Lemma I.1 is from Shorgin [279], see also Lemma 6 in [259].

Hints and solutions to selected exercises

> Dr Watson has developed taste for Maths. As a birthday present Sherlock Holmes sent him a book where Fermat famously wrote on a margin:
> – *I know how to prove this CP approximation theorem but not enough space here ...*
> Watson spent many months desperately trying to prove that theorem—no luck. He sent Holmes a postcard:
> – *Sherlock, you can solve any puzzle, how to prove this theorem??*
> Holmes replied by a postcard:
> – *It's elementary, Watson! but I don't have enough space here ...*
>
> Dr Watson, *Adventures of S. Holmes*

Chapter 1

1.1 It is easy to check that

$$\mathbf{\Pi}(-\lambda) * \mathbf{\Pi}(\lambda) = I.$$

By the property (A.1) of the total variation norm,

$$1 = \|I\|_{TV} \leq \|\mathbf{\Pi}(-\lambda)\|_{TV} \, \|\mathbf{\Pi}(\lambda)\|_{TV} = \|\mathbf{\Pi}(-\lambda)\|_{TV}.$$

For an arbitrary r.v. X denote $P_X = \mathcal{L}(X)$. Since P_X^{*j} is a distribution, we have $\|P_X^{*j}\|_{TV} = 1$. Therefore,

$$\|\mathbf{\Pi}(-\lambda, X)\|_{TV} = \mathrm{e}^{\lambda} \| \exp(-\lambda P_X)\|_{TV} \leq \mathrm{e}^{\lambda} \sum_{j=0}^{\infty} \frac{\lambda^j \|P_X^{*j}\|_{TV}}{j!} = \mathrm{e}^{\lambda} \sum_{j=0}^{\infty} \frac{\lambda^j}{j!} = \mathrm{e}^{2\lambda}.$$

1.2 It follows from (D.8) that $\sup_t |\widehat{G}(t)| = \sup_t |1/\widehat{F}(t)| = \infty$. Since

$$\sup_t |\widehat{G}(t)| \leq \int_{-\infty}^{\infty} |M(\mathrm{d}x)| = \|G\|_{TV},$$

the result follows.

1.3 Using the identity $\mathrm{e}^{\mathrm{i}t} = \cos t + \mathrm{i}\sin t$, it is easy to check that

$$|\mathrm{e}^{\mathrm{i}t}-1|^2 = (\cos t-1)^2 + \sin^2 t = 4\sin^2(t/2).$$

Clearly, $|\varphi(t)-1| = |(\mathbb{E}\mathrm{e}^{\mathrm{i}tX}-1)\mathbb{I}\{X\neq 0\}| \leq 2\mathbb{P}(X\neq 0)$. On the other hand,

$$|\mathrm{e}^{\mathrm{i}tk}-1| = |(\mathrm{e}^{\mathrm{i}t}-1)(\mathrm{e}^{\mathrm{i}t(k-1)}+\cdots+1)| \leq k|\mathrm{e}^{\mathrm{i}t}-1|.$$

Hence

$$|\varphi(t)-1| = \left|\sum_{k=1}^{\infty}\mathbb{P}(X=k)(\mathrm{e}^{\mathrm{i}tk}-1)\right| \leq |\mathrm{e}^{\mathrm{i}t}-1|\sum_{k=1}^{\infty}k\mathbb{P}(X=k) = |\mathrm{e}^{\mathrm{i}t}-1|\mathbb{E}X.$$

1.4 First equality follows from corresponding equality of Fourier transforms. Proof of the second relation is straightforward

$$P*\int_0^1 \mathrm{e}^{tP}\mathrm{d}t = \sum_{j=0}^{\infty}\frac{P^{*(j+1)}}{j!}\int_0^1 t^j\mathrm{d}t = \sum_{j=0}^{\infty}\frac{P^{*(j+1)}}{(j+1)!} = \mathrm{e}^P - I.$$

1.5 We have

$$\sum_{k=-\infty}^{\infty}(\mathbb{P}(X=k)-\mathbb{P}(Y=k)) = 0.$$

Therefore,

$$\sum_{k=-\infty}^{\infty}(\mathbb{P}(X=k)-\mathbb{P}(Y=k))^+ = \sum_{k=-\infty}^{\infty}(\mathbb{P}(X=k)-\mathbb{P}(Y=k))^-$$

and

$$\begin{aligned}\sum_{k=-\infty}^{\infty}|\mathbb{P}(X=k)-\mathbb{P}(Y=k)| &= \sum_{k=-\infty}^{\infty}(\mathbb{P}(X=k)-\mathbb{P}(Y=k))^+ \\ &+ \sum_{k=-\infty}^{\infty}(\mathbb{P}(X=k)-\mathbb{P}(Y=k))^- = 2\sum_{k=-\infty}^{\infty}(\mathbb{P}(X=k)-\mathbb{P}(Y=k))^+ \\ &= 2\sum_{k=-\infty}^{\infty}(\mathbb{P}(X=k)-\min\{\mathbb{P}(X=k),\mathbb{P}(Y=k)\}) \\ &= 2-2\sum_{k=-\infty}^{\infty}\min\{\mathbb{P}(X=k)-\mathbb{P}(Y=k)\}.\end{aligned}$$

Chapter 2

2.1 By the inversion formula (D.11),

$$d_{\mathrm{o}}(F_n;G) \leq \frac{1}{2\pi}\int_{-\pi}^{\pi}|\widehat{F}_n(t)-\widehat{G}(t)|\mathrm{d}t \to 0 \quad (n\to\infty).$$

2.2 We need to show that $\sum_{k=-\infty}^{\infty} |\mathbb{P}(X=k) - \mathbb{P}(X_n=k)| \to 0$ as $n\to\infty$.
Note that the assumption entails $\mathbb{P}(X_n = k) \to \mathbb{P}(X = k)$ $(|k| \le K)$ and $\mathbb{P}(|X_n| \le K) \to \mathbb{P}(|X| \le K)$ as $n\to\infty$ for any fixed K. Therefore, $\mathbb{P}(|X_n| > K) \to \mathbb{P}(|X| > K)$ as $n\to\infty$. Observe that

$$\begin{aligned}&\sum_{k=-\infty}^{\infty} |\mathbb{P}(X=k) - \mathbb{P}(X_n=k)| \\ &\le \sum_{k:|k|\le K} |\mathbb{P}(X=k) - \mathbb{P}(X_n=k)| + \mathbb{P}(|X_n|>K) + \mathbb{P}(|X|>K),\end{aligned}$$

and $\mathbb{P}(|X|>K)$ can be made arbitrarily small by the choice of K.

2.3 Define π_{λ_n}, π_λ on a common probability space. W.l.g. we may assume that $\lambda_n > \lambda$. Then $\pi_{\lambda_n} \stackrel{d}{=} \pi_\lambda + \pi_{\varepsilon_n}$, where $\varepsilon_n = \lambda_n - \lambda$. Therefore,

$$d_G(\pi_{\lambda_n};\pi_\lambda) = \mathbb{E}|\pi_{\lambda_n} - \pi_\lambda| = \mathbb{E}\pi_{\varepsilon_n} \to 0 \qquad (n\to\infty).$$

2.4 Observe that $S_{n,p}$ is the sum of n independent geometric r.v.s.

$$\begin{aligned} d_{TV}(S_{n,p};\pi_\lambda) &\le n\, d_{TV}(S_{1,p};\pi_{\lambda/n}) = \frac{n}{2}\sum_{j=0}^{\infty} |(1-p)p^j - (\lambda/n)^j \mathrm{e}^{-\lambda/n}/j!| \\ &\le \frac{n}{2}|1-p-\mathrm{e}^{-\lambda/n}| + \frac{np}{2}|1-p-\mathrm{e}^{-\lambda/n}| + \frac{\mathrm{e}^{-\lambda/n}}{2}|np-\lambda| \\ &+ \frac{n}{2}\sum_{j=2}^{\infty}(1-p)p^j + \frac{n}{2}\sum_{j=2}^{\infty}\frac{(\lambda/n)^j\mathrm{e}^{-\lambda/n}}{j!}. \end{aligned}$$

Obviously,

$$n\sum_{j=2}^{\infty}\frac{(\lambda/n)^j\mathrm{e}^{-\lambda/n}}{j!} = \frac{\lambda^2}{n}\sum_{j=2}^{\infty}\frac{(\lambda/n)^{j-2}\mathrm{e}^{-\lambda/n}}{j!} \le \frac{\lambda^2}{n}$$

and

$$n|1-p-\mathrm{e}^{-\lambda/n}| \le |np-\lambda| + n\sum_{j=2}^{\infty}\frac{(\lambda/n)^j\mathrm{e}^{-\lambda/n}}{j!} \le |np-\lambda| + \lambda^2/n.$$

Therefore,

$$d_{TV}(S_{n,p};\pi_\lambda) \le \frac{|np-\lambda|}{2}(1+p+\mathrm{e}^{-\lambda/n}) + \frac{\lambda^2(2+p)}{2n} + \frac{np^2}{2} \to 0.$$

Chapter 3

3.3 Let $S_{n,i} = S_n - X_i$. Then $\mathbb{E}g(S_n+1) = \mathbb{E}g(S_{n,i}+X_i+1)$. We apply Stein's method:

$$\begin{aligned} \mathbb{E}g(S_n+1) &= (1-p_i-p_i^2/2)\,\mathbb{E}g(S_{n,i}+1) + (p_i-p_i^2)\mathbb{E}g(S_{n,i}+2) + \frac{p_i^2}{2}\mathbb{E}g(S_{n,i}+3) \\ &= \mathbb{E}g(S_{n,i}+1) + p_i\mathbb{E}\Delta g(S_{n,i}+1) - \frac{p_i^2}{2}\mathbb{E}\Delta^2 g(S_{n,i}+1), \\ \mathbb{E}X_i g(S_n) &= p_i\mathbb{E}\Delta g(S_{n,i}+1) + p_i^2\mathbb{E}\Delta^2 g(S_{n,i}+1). \end{aligned}$$

Using equation (F.7*), we get

$$|\mathbb{E}f(S_n) - \mathbb{E}f(\pi_\lambda)| = \Big|\sum_{i=1}^{n}(p_i\mathbb{E}g(S_n) - \mathbb{E}X_i g(S_n))\Big| \le \frac{1}{2}\sum_{i=1}^{n} p_i^3|\mathbb{E}\Delta^2 g(S_{n,i}+1)|.$$

Estimate (3.100) now follows from (3.92).

Observe that $1-d_{TV}(X_j, X_j+1) = p_j(1-p_j/2) \ge 3p_j/4$ and

$$\frac{1}{4} + \sum_{j\ne i}^{n}(1-d_{TV}(X_j, X_j+1)) \ge \frac{1}{4} + \frac{3\lambda}{4} \ge 0.75\lambda.$$

Therefore, $d_{TV}(S_{n,i}, S_{n,i}+1) \le \sqrt{2/0.75\pi\lambda}$. Hence

$$|\mathbb{E}\Delta^2 g(S_{n,i}+1)| \le 2\|\Delta g\| d_{TV}(S_{n,i}, S_{n,i}+1) \le \frac{2}{\lambda\sqrt{\lambda}}\sqrt{\frac{2}{0.75\pi}} \le \frac{2}{\lambda\sqrt{\lambda}}$$

by (C.14), and (3.101) follows.

3.4 We use Θ to denote various measures satisfying $\|\Theta\|_{TV} \le 1$. Let $V = \sum_{k=2}^{\infty}\varkappa_k(I_1-I)^{*(k-2)}/k!$ so that $\mathcal{L}(X) = \exp\big(\lambda(I_1-I) + (I_1-I)^{*2} * V\big)$. Then

$$\|V\|_{TV} \le \sum_{k=2}^{\infty}\frac{|\varkappa_k|2^{k-2}}{k!} \le \lambda\sum_{k=2}^{\infty}\frac{2^{k-2}}{T^{k-1}} \le \frac{7\lambda}{5T} \le \frac{\lambda}{5}.$$

Therefore, $V = (7\lambda/5T)\Theta$ and $V = (\lambda/5)\Theta$. Using (B.1) and Lemma B.1, we get

$$\begin{aligned}\mathcal{L}(X) - \mathbf{\Pi}(\lambda) &= V * (I_1 - I)^{*2} * \int_0^1 \exp\big(\lambda(I_1 - I) + \tau(I_1 - I)^{*2} * V\big)\,\mathrm{d}\tau \\ &= \frac{7\lambda}{5T}(I_1 - I)^{*2} * \exp(0.2\lambda(I_1 - I)) * \Theta.\end{aligned}$$

Applying (C.2), we derive

$$d_{TV}(X;\pi_\lambda) = \frac{1}{2}\|\mathcal{L}(X) - \mathbf{\Pi}(\lambda)\|_{TV} \le \frac{7\lambda}{10T}\|(I_1 - I)^{*2} * \exp(0.2\lambda(I_1 - I))\,\|_{TV} \le \frac{21}{2\mathrm{e}T}.$$

Similarly, from (A.9) and (C.8),

$$\begin{aligned}d_G(X;\pi_\lambda) &= \|\mathcal{L}(X) - \mathbf{\Pi}(\lambda)\|_G \le \frac{7\lambda}{5T}\|(I_1 - I)^{*2} * \exp(0.2\lambda(I_1 - I))\,\|_G \\ &= \frac{7\lambda}{5T}\|(I_1 - I) * \exp(0.2\lambda(I_1 - I))\,\|_{TV} \le 7\sqrt{\frac{2}{5\mathrm{e}}}\frac{\sqrt{\lambda}}{T}.\end{aligned}$$

Recall that $\|\cdot\|_G$ is defined by (A.8).

Chapter 4

4.1 We have

$$\mathbb{P}(S=0) = \mathbb{P}(S-\mathbb{E}S=-\mathbb{E}S) \le \mathbb{P}(|S-\mathbb{E}S|\ge\mathbb{E}S) \le (\operatorname{var} S)/(\mathbb{E}S)^2,$$

where we use Chebyshev's inequality for the last estimate.

Chapter 5

5.1 Clearly, $\mathcal{L}(\tau_i X_i) = (1+p)I - pF$, $\mathcal{L}(\hat{S}_n) = ((1+p)I - pF)^{*n}$, and

$$\|(1+p)I - pF\|_{TV} \le 1+p+p\|F\|_{TV} = 1+2p \le \mathrm{e}^{2p}.$$

Similarly, $\|\mathbf{\Pi}(-\lambda/n, X)\|_{TV} \le \mathrm{e}^{2\lambda/n}$ (cf. Exercise 1.1). Recall that

$$\prod_{j=1}^{*n} F_j - \prod_{j=1}^{*n} H_j = \sum_{j=1}^{n}(F_j - H_j) * \prod_{l=j+1}^{*n} F_l * \prod_{l=1}^{*(j-1)} H_l$$

by (B.12). Therefore,

$$\begin{aligned}
\|\mathcal{L}(\hat{S}_n) - \mathbf{\Pi}(-\lambda; \mathbf{X})\|_{TV} &= \|(I - p(F-I))^{*n} - \exp(-\lambda(F-I))\|_{TV} \\
&\le \|I - p(F-I) - \exp(-\lambda(F-I)/n)\|_{TV} \sum_{k=1}^{n} \mathrm{e}^{(k-1)2p+(n-k)\lambda/n} \\
&\le n\mathrm{e}^{2np+2\lambda}|p - \lambda/n|\|F-I\|_{TV} + \sum_{k=2}^{\infty} \frac{(\lambda/n)^k}{k!}\|F-I\|_{TV}^k \\
&\le \mathrm{e}^{2np+2\lambda}\left(|\lambda - np| + \frac{4\lambda^2}{n}\mathrm{e}^{2\lambda/n}\right) \to 0 \quad (n\to\infty).
\end{aligned}$$

Note that $\|F-I\|_{TV} \le \|F\|_{TV} + \|I\|_{TV} = 2$.

5.2 By the properties of characteristics functions there exists t_0 such that $|\widehat{F}(t_0)| < 1$. Note that for any finite measure M we have

$$\sup_t |\widehat{M}(t)| = \sup_t \left| \int_{-\infty}^{\infty} \mathrm{e}^{itx} M(\mathrm{d}x) \right| \le \int_{-\infty}^{\infty} |M(\mathrm{d}x)| = \|M\|_{TV}.$$

Hence

$$\|(1+p)I - pF\|_{TV} \ge \sup_t |(1+p) - p\widehat{F}(t)| \ge 1+p-p|\widehat{F}(t_0)| > 1.$$

5.3 W.l.o.g. we can assume that $\sum_{i=1}^n p_i > 0$. First of all we prove that $d_{TV}(S_n; \tilde{S}_n) \to 0$, where $\tilde{S}_n$ is the sum of accompanying r.v.s. The proof almost coincides with the proof from Exercise 5.1. Indeed, it follows from (B.9) that

$$\begin{aligned}
d_{TV}(S_n; \tilde{S}_n) &= \frac{1}{2}\left\| \prod_{i=1}^{*n}(I + p_i(F_i - I)) - \prod_{i=1}^{*n} \mathrm{e}^{p_i(F_i - I)} \right\|_{TV} \\
&\le \frac{1}{2}\sum_{i=1}^{n} \|I + p_i(F_i - I) - \mathrm{e}^{p_i(F_i-I)}\|_{TV} \\
&\le 2\sum_{i=1}^{n} p_i^2 \sum_{k=2}^{\infty} p_i^{k-2} 2^{k-2}/k! \le 2\mathrm{e}^2 \sum_{i=1}^{n} p_i^2 \to 0 \quad (n\to\infty).
\end{aligned}$$

For the previous estimate only assumption (a) was needed. Denote $F=\mathcal{L}(Y)$. It follows from (B.1) that

$$\begin{aligned}
\Big\|\prod_{i=1}^{*n} \mathrm{e}^{p_i(F_i-I)} - \mathrm{e}^{\lambda(F-I)}\Big\|_{TV} &\le \Big\|\sum_{i=1}^{n} p_i(F_i-I)-\lambda(F-I)\Big\|_{TV}\\
&\le \Big|\lambda-\sum_{i=1}^{n}p_i\Big| + \Big\|\sum_{i=1}^{n}p_iF_i-\sum_{i=1}^{n}p_iF\Big\|_{TV}\\
&+ \Big|\sum_{i=1}^{n}p_i-\lambda\Big|\|F\|_{TV}\\
&\le 2\Big|\sum_{i=1}^{n}p_i-\lambda\Big| + \|M-F\|_{TV}\sum_{i=1}^{n}p_i,
\end{aligned}$$

where

$$M(k)=\frac{\sum_{i=1}^n p_iF_i(k)}{\sum_{i=1}^n p_i}=\frac{\sum_{i=1}^n p_i\mathbb{P}(X_i'=k)}{\sum_{i=1}^n p_i}\qquad (k\in\mathbb{Z}).$$

It remains to prove that $\|M-F\|_{TV}\to 0$. Observe that $M(k)$ is a distribution of an integer-valued r.v. Moreover, $\lim_{n\to\infty}M(k)=F(k)$ by (a) and (b). The required result follows from the statement of Exercise 2.2 in Chapter 2.

5.4 For the sake of brevity we write p_i instead of $p_{i,n}$. Let $q_i=1-p_i$, $S_{r,i}=S_r-X_i$. We apply Stein's method.

Equation (F.10) states that

$$p(r+j)g(j+1)-jg(j)=\mathbb{I}\{j\in A\}-\mathbf{NB}(r,p)(A)\qquad (j\ge 1, A\subset\mathbb{Z}_+).\qquad (\mathrm{F}.10^*)$$

According to (F.10*), it suffices to estimate

$$|\mathbb{E}p(r+S_r)g(S_r+1)-\mathbb{E}S_rg(S_r)|.$$

Since $\mathbb{P}(X_i=k)=q_ip_i^k$, we have

$$\mathbb{E}g(S_r+1)=\sum_{k=0}^{\infty}q_ip_i^k\mathbb{E}g(S_r^i+1+k);\quad \mathbb{E}X_ig(S_{r,i}+1)=\sum_{k=1}^{\infty}kq_ip_i^k\mathbb{E}g(S_{r,i}+k+1).$$

Similarly,

$$\begin{aligned}
\mathbb{E}X_ig(S_r) &= \sum_{k=1}^{\infty}kq_ip_i^k\mathbb{E}g(S_{r,i}+k)=\sum_{k=0}^{\infty}(k+1)q_ip_i^{k+1}\mathbb{E}g(S_{r,i}+k+1)\\
&= \sum_{k=1}^{\infty}kq_ip_i^{k+1}\mathbb{E}g(S_{r,i}+k+1)+\sum_{k=0}^{\infty}q_ip_i^{k+1}\mathbb{E}g(S_{r,i}+k+1).
\end{aligned}$$

Since $\|g\|\le 1/q$ by (F.11),

$$\begin{aligned}
&|\mathbb{E}p(r+S_r)g(S_r+1)-\mathbb{E}S_rg(S_r)|\\
\le&\ \sum_{i=1}^{r}|p\mathbb{E}g(S_r+1)+p\mathbb{E}X_ig(S_r+1)-\mathbb{E}X_ig(S_r)|\\
\le&\ \sum_{i=1}^{r}\Big(\sum_{k=0}^{\infty}q_ip_i^k|p-p_i||\mathbb{E}g(S_{r,i}+k+1)|+\sum_{k=1}^{\infty}kq_ip_i^k|p-p_i||\mathbb{E}g(S_r^i+k+1)|\Big)\\
\le&\ \|g\|\max_{1\le j\le r}|p_j-p|\sum_{i=1}^{r}\sum_{k=0}^{\infty}(k+1)q_ip_i^k=\|g\|\max_{1\le j\le r}|p_j-p|\sum_{i=1}^{r}p_i\to 0.
\end{aligned}$$

Chapter 6

6.1 We have

$$\Big\|\mathcal{L}\Big(\sum_{j=0}^{\nu_n}\zeta_j\Big)-\mathcal{L}\Big(\sum_{j=0}^{\nu}\zeta_j\Big)\Big\|_{TV}=\Big\|\sum_{j=0}^{\infty}\mathcal{L}^{*j}(\zeta)\mathbb{P}(\nu_n=j)-\sum_{j=0}^{\infty}\mathcal{L}^{*j}(\zeta)\mathbb{P}(\nu=j)\Big\|_{TV}$$

$$\le\sum_{j=0}^{\infty}\|\mathcal{L}^{*j}(\zeta)\|_{TV}|\mathbb{P}(\nu_n=j)-\mathbb{P}(\nu=j)|=\sum_{j=0}^{\infty}|\mathbb{P}(\nu_n=j)-\mathbb{P}(\nu=j)|=2d_{TV}(\nu_n;\nu).$$

6.2 Let

$$\tilde{W}_{n,s}=\sum_{i=1}^{n}\sum_{j=s+1}^{\infty}p_i^j(-1)^{j+1}(\mathcal{L}(X_i')-I)^{*j}/j.$$

Repeating the proof of (6.25), we arrive at

$$2d_{TV}(S_n;H_{n,s})\le\exp\Big(\|\tilde{W}_{s,n}\|_{TV}\Big)-1.$$

It remains to check that

$$\|\tilde{W}_{s,n}\|_{TV}\le\sum_{i=1}^{n}\sum_{j=s+1}^{\infty}\frac{(2p_i)^j}{j}\le\frac{1}{s+1}\sum_{i=1}^{n}\frac{(2p_i)^{s+1}}{1-2p_i}.$$

6.3 Let $f_i(t)$ denote the characteristic function of $\mathcal{L}(X_i')$. Set

$$A(t)=\sum_{i=1}^{n}\sum_{j=1}^{s}\frac{p_i^j(-1)^{j+1}}{j}(f_i(t)-1)^{*j}\quad B(t)=\sum_{i=1}^{n}\sum_{j=s+1}^{\infty}\frac{p_i^j(-1)^{j+1}}{j}(f_i(t)-1)^{*j},$$

so that the characteristic function of $\mathcal{L}(S_n)$ is equal to $\exp(A(t)+B(t))$ and $\widehat{H}_{n,s}(t)=\exp(A(t))$. Observe that

$$\begin{aligned}p_i(Ref_i(t)-1)\ &+\ \sum_{j=2}^{\infty}\frac{p_i^j}{j}|f_i(t)-1|^j\le p_i(Ref_i(t)-1)+p_i^2(1-Ref_i(t))\sum_{j=2}^{\infty}(2p_i)^{j-2}\\ &\le\ p_i(Ref_i(t)-1)+\frac{p_i^2}{1-2p_i}(1-Ref_i(t)\le\frac{p_i}{2}(Ref_i(t)-1).\end{aligned}$$

by (D.5). Therefore, $|\exp(A(t)+\tau B(t))|\le\widehat{\boldsymbol{\Pi}}'(t)$ for any $\tau\in[0,1]$. Similarly, taking into account (1.14),

$$|B(t)|\le\sum_{i=1}^{n}|f_i(t)-1|\sum_{j=s+1}^{\infty}\frac{2^{j-1}p_i^j}{j}\le\frac{1}{s+1}|\sin(t/2)|\sum_{i=1}^{n}\mathbb{E}X_i(2p_i)^{s+1}.$$

Applying estimates from (B.1), we get

$$\Big|e^{A(t)+B(t)}-e^{A(t)}\Big|\le|B(t)|\,\widehat{\boldsymbol{\Pi}}'(t)\le\frac{1}{s+1}|\sin(t/2)|\sum_{i=1}^{n}\mathbb{E}X_i(2p_i)^{s+1}\widehat{\boldsymbol{\Pi}}'(t).$$

It remains to apply (D.14) and (E.8).

6.4 Using (B.12), we get

$$d(S_n;\widehat{Z}_n)=\frac{1}{2}\Big\|\prod_{i=1}^{*n}\mathcal{L}(X_i)-\prod_{i=1}^{*n}\mathcal{L}(Y_i)\Big\|_{TV}\leq\frac{1}{2}\sum_{i=1}^{n}\|\mathcal{L}(X_i)-\mathcal{L}(Y_i)\|_{TV}.$$

Taking into account that $\max\{\|F_{1i}-I\|_{TV};\|F_{2i}-I\|_{TV}\}\leq 2$ and applying (B.2), we derive

$$\mathcal{L}(X_i)=I+\frac{\alpha_i}{2}(F_{1i}-I)+\frac{\beta_i}{2}(F_{2i}-I)+(\alpha_i^2+\beta_i^2)\Theta_1,$$

where $\|\Theta_1\|_{TV}\leq 1$. Similarly,

$$\mathcal{L}(Y_i)=I+\frac{\alpha_i}{2}(F_{1i}-I)+\frac{\beta_i}{2}(F_{2i}-I)+\frac{(\alpha_i+\beta_i)^2}{2}\Theta_2,$$

where $\|\Theta_2\|_{TV}\leq 1$. The statement easily follows since $(\alpha_i+\beta_i)^2\leq 2(\alpha_i^2+\beta_i^2)$.

Chapter 7

7.1 We have

$$\mathcal{L}(S_{2n})=(0.5I+0.5I_1)^{*2n}=(0.5I+0.5I_1)^{*n}*(0.5I+0.5I_{-1})^{*n}*I_n.$$

Therefore, $S_{2n}=n+S_n^{\star}$, where $S_n^{\star}=(X_1-X_1')+\cdots+(X_n-X_n')$, and $X_1,X_1',\ldots,X_n,X_n'$ are independent copies of Bernoulli $\mathbf{B}(0.5)$ variable. It remains to apply (7.12).

7.2 Apply Theorem 7.5.

7.3 Let $\widehat{F}_X(t)$ denote the characteristic function of $\mathcal{L}(X)$. Then $|\widehat{F}_X(t)|\leq\exp\big(-2a\sin^2(t/2)\big)$ by (D.7). Let $\widehat{G}_Z(t)=\exp\big(u(\mathrm{e}^{\mathrm{i}t}-1)+v(\mathrm{e}^{2\mathrm{i}t}-1)\big)$. It can be verified that

$$\widehat{G}_Z(t)=\Big|\exp\Big(\mathbb{E}X(\mathrm{e}^{\mathrm{i}t}-1)+(\mathbb{E}X^{(2)}-\mathbb{E}^2X)(\mathrm{e}^{\mathrm{i}t}-1)^2/2\Big)\Big|\leq\exp\big(-2a\sin^2(t/2)\big).$$

By Lemma D.1,

$$\widehat{F}_X(t)=\mathbb{E}X(\mathrm{e}^{\mathrm{i}t}-1)+\mathbb{E}X^{(2)}(\mathrm{e}^{\mathrm{i}t}-1)^2/2+\vartheta C\mathbb{E}X^{(3)}|\sin(t/2)|^3\qquad(|\vartheta|\leq 1).$$

Therefore, it follows from (I.8) that

$$|\widehat{F}_X(t)-\widehat{G}_Z(t)|\leq C(\mathbb{E}X^{(3)}+\mathbb{E}^3X)|\sin(t/2)|^3.$$

Observe that

$$\begin{aligned}|\mathbb{E}\mathrm{e}^{\mathrm{i}tS_n}-\mathbb{E}\mathrm{e}^{\mathrm{i}tZ_n}| &\leq \sum_{j=1}^{n}|\widehat{F}_X(t)-\widehat{G}_Z(t)||\widehat{F}_X(t)|^{j-1}|\widehat{G}_Z(t)|^{n-j}\\ &\leq C(\mathbb{E}X^{(3)}+\mathbb{E}^3X)|\sin(t/2)|^3\exp\big(-2a(n-1)\sin^2(t/2)\big)\\ &\leq C(\mathbb{E}X^{(3)}+\mathbb{E}^3X)|\sin(t/2)|^3\exp\big(-2an\sin^2(t/2)\big)\end{aligned}$$

by (I.1) since $\exp\big(2a\sin^2(t/2)\big)\leq\mathrm{e}^2$. The result follows by applying (D.14) and (I.12).

7.4 In view of (7.33) it suffices to estimate $d_{TV}(H_{n,2}; G_{n,p})$. It follows from Lemma B.1 that

$$\Big\| \exp\big((2np/3)(F-I) - \tau np^2(F-I)^{*2}/2\big) \Big\|_{TV} \le C$$

for any $\tau \in [0,1]$, since $p \le 1/3$. By (1.5),

$$\begin{aligned}\|H_{n,2} - G_{n,p}\|_{TV} &= \|\exp(np(F-I)) * (\exp\big(-np^2(F-I)^{*2}/2\big) - I + np^2(F-I)^{*2}/2)\|_{TV} \\ &\le Cn^2p^4 \int_0^1 \|(F-I)^{*4} * \exp\big(np(F-I) - \tau np^2(F-I)^{*2}/2\big)\|_{TV}\,\mathrm{d}\tau \\ &\le Cn^2p^4 \|(F-I)^{*4} * \exp(np(F-I)/3)\|_{TV}.\end{aligned}$$

It remains to apply (C.29).

7.5 It is easy to check that the characteristic function of S_n is equal to $\widehat{F}^n(t)$,

$$|\widehat{F}(t) - \widehat{D}(t)| \le C|\widehat{F} - 1|^3; \quad |\widehat{D}(t)| \le \exp\Big(\widehat{F}(t) - 1\Big)$$

and $|\widehat{F}(t)| \le \exp\Big(2(\widehat{F}(t) - 1)/9\Big)$ by (D.6), so that

$$|\widehat{F}^n(t) - \widehat{D}^n(t)| \le Cn|\widehat{F}(t) - 1|^3 \exp\Big(2(n-1)(\widehat{F}(t) - 1)/9\Big) \le Cn^{-2} \exp\Big(n(\widehat{F}(t) - 1)/9\Big)$$

by (I.1) and (I.16). We apply Lemma D.16 with

$$M = F^{*n} - D^{*n}, \quad h = \sqrt{2}/\pi, \quad \widehat{U}(t) = \exp\Big(n(\widehat{F}(t) - 1)/9\Big), \quad m = 26n, \quad \mathbf{u} = (0, 1, \sqrt{2}).$$

Observe that $\mathrm{supp}F^{*n} \subset K_n(\mathbf{u})$, $\mathrm{supp}((F-I) - (F-I)^{*2}/2) \subset K_2(\mathbf{u})$, $\|n(F-I) - n(F-I)^{*2}/2\|_{TV} \le 4n$. Therefore, $\delta(M, m, \mathbf{u}) \le \mathrm{e}^{-n}$ by Lemma D.17.
By (E.8), $Q_U(h) \le Cn^{-1/2}$, $\ln(m+1) \le C\ln(\sqrt{n+1})$, and

$$Q_U(h)\ln(m+1)\sup_{t\in\mathbb{R}} \frac{|\widehat{M}(t)|}{\widehat{U}(t)} \le Cn^{-2}.$$

It remain to estimate $\widehat{M}(t)$ in a neighbourhood of zero.
If $|t| \le \pi/\sqrt{2}$, then $|\sin(t/2)| \ge |t|/\pi$, $|\sin(t/\sqrt{2})| \ge \sqrt{2}|t|/\pi$ and, consequently,

$$\exp\Big(2n(\widehat{F}(t) - 1)/9\Big) = \exp\Big(-8n(\sin^2(t/2) + \sin^2(t/\sqrt{2}))/45\Big) \le \exp\big(-8nt^2/15\pi^2\big),$$

$$|\widehat{F}^n(t) - \widehat{D}^n(t)| \le C|t|^3 \exp\big(-8nt^2/15\pi^2\big).$$

Substituting these estimates into (D.26), we complete the proof.

Chapter 8

8.1 We will use mathematical induction and the fact that $\mathbb{E}Z_lZ_{l+1}\cdots Z_k = 0$. By the induction assumption we have

$$\begin{aligned}\widehat{\mathbb{E}}(Z_1,\dots,Z_{k+1}) &= \mathbb{E}Z_1\cdots Z_{k+1} - \sum_{j=1}^k \widehat{\mathbb{E}}(Z_1,\dots,Z_j)\mathbb{E}Z_{j+1}\cdots Z_{k+1} \\ &= -\widehat{\mathbb{E}}Z_1\cdots Z_k\mathbb{E}Z_{k+1} = (-1)(-1)^{k+1}a^kz^kaz = (-1)^{k+2}a^{k+1}z^{k+1}.\end{aligned}$$

8.2 Note that $|z| = 2|\sin(t/2)| \leq 2$. Arguing as in the proof of Lemma 8.11 and applying mathematical induction, we get

$$\begin{aligned} |\varphi_{k+1}-1| &\leq a|z| + \sum_{j=1}^{k} \frac{|a|^{k-j+2}2^{k-j+2}}{|\varphi_j \cdots \varphi_k|} \leq 2a + 2a\sum_{j=1}^{k}\left(\frac{8a}{3}\right)^{k+1-j} \\ &\leq 2a\Big(1+\sum_{j=1}^{\infty}(3/4)^j\Big) \leq 1/4, \end{aligned}$$

where $a \leq 9/100$.

8.3 For $k>1$ we have

$$\begin{aligned} |\varphi_k - 1 - az| &\leq \sum_{j=1}^{k-1} |z|^{k-j+1}a^{k-j+1}\left(\frac{4}{3}\right)^{k-j} \leq \frac{4}{3}|z|^2a^2\sum_{j=1}^{k-1}2^{k-j-1}(0.91)^{k-j-1}\left(\frac{4}{3}\right)^{k-j-1} \\ &\leq \frac{12}{19}a\sin^2(t/2). \end{aligned}$$

Observe that

$$|1+az|^2 = 1-4a(1-a)\sin^2(t/2) \leq 1-3.64a\sin^2(t/2).$$

Therefore, for $k=1$

$$|\varphi_1| = |1+az| \leq \sqrt{1-3.64\sin^2(t/2)} \leq 1-1.82\sin^2(t/2) \leq \exp\big(-a\sin^2(t/2)\big)\,.$$

Similarly, for $k>1$

$$\begin{aligned} |\varphi_k| &\leq |1+az| + \frac{12}{19}a\sin^2(t/2) \leq 1-1.82a\sin^2(t/2) + 0.64a\sin^2(t/2) \\ &\leq 1-a\sin^2(t/2) \leq \exp\big(-a\sin^2(t/2)\big)\,. \end{aligned}$$

8.4 Just like in Exercise 9.3 we prove that

$$|\varphi_k - 1 - az + a^2z^2/\varphi_{k-1}| \leq C|a||z|^3\,.$$

By Exercise 9.1,

$$\left|\frac{1}{\varphi_{k-1}} - 1\right| \leq \frac{|1-\varphi_{k-1}|}{|\varphi_{k-1}|} \leq \frac{4\cdot 31a\sin^2(t/2)}{3} \leq 21a|z|.$$

It remains to combine both estimates.

8.5 Let

$$\psi_1 = \exp\big(az - a^2z^2/2\big)\,, \quad \psi_k = \exp\big(az - 3a^2z^2/2\big) \qquad (k = 2, 3, \ldots, n).$$

Direct verification shows that

$$|\psi_k| \leq \exp\big(-a\sin^2(t/2)\big)\,, \quad |\varphi_k - \psi_k| \leq C|a|^3|z|^3\,.$$

Therefore,

$$\begin{aligned} |\mathbb{E}e^{itS_n} - \mathbb{E}e^{itY_n}| &= \left|\prod_{k=1}^{n}\varphi_k - \prod_{k=1}^{n}\psi_k\right| \leq \sum_{k=1}^{n}|\varphi_k - \psi_k|\exp\big(-(n-1)a\sin^2(t/2)\big) \\ &\leq Cn|a|^3|z|^3\exp\big(-na\sin^2(t/2)\big) \end{aligned}$$

by (I.1). We can apply Lemmas D.9 and I.1.

Chapter 9

9.1 Observe that

$$\begin{aligned} d_{TV}(A_0;\mathbf{\Pi}(n\gamma_1;K_\beta)) &\leq \|\exp(n\gamma_1(K_\beta-I))*(H-I)*\exp(-p_0\gamma_1(K_\beta-I))\|_{TV}\\ &+ \|\exp(n\gamma_1(K_\beta-I))*(\exp(-p_0\gamma_1(K_\beta-I)))\|_{TV}\\ &\leq e^{2p_0\gamma_1}C|\varkappa_2|\|(K_\beta-I)*\exp(n\gamma_1(K_\beta-I))\|_{TV}\\ &+ e^{2p_0\gamma_1}Cp_0\|(K_\beta-I)*\exp(n\gamma_1(K_\beta-I))\|_{TV}\\ &\leq Cp_0\|(K_\beta-I)*\exp(n\gamma_1(K_\beta-I))\|_{TV} \end{aligned}$$

by (1.6). On one hand,

$$\|(K_\beta-I)*\exp(n\gamma_1(K_\beta-I))\|_{TV}\leq\|(K_\beta-I)\|_{TV}\|\exp(n\gamma_1(K_\beta-I))\|_{TV}\leq 2.$$

On the other hand, it is possible to use (C.8). An application of (9.2) and the triangle inequality complete the proof.

9.2 Let $U_\beta=K_\beta-I$. By (1.6), (9.28), and (C.10) we get

$$\begin{aligned}\|A_3-A_{21}\|_{TV} &= \|A_2*(\exp(n\gamma_3U_\beta^{*3})-I-n\gamma_3U_\beta^{*3})\|_{TV}\leq C\|n^2\gamma_3^2U_\beta^{*6}\exp(n\gamma_1U_\beta/30)\|_{TV}\\ &\leq C\gamma_3^2\gamma_1^{-3}n^{-1}.\end{aligned}$$

Observe that $\gamma_1\geq\alpha/2$, $\gamma_3\leq C\alpha^2(\alpha+\beta)$. Therefore, (9.41) follows from the triangle inequality and Theorem 9.4.

9.3 Let $z=e^{it}-1$. By (D.11),

$$\begin{aligned}\mathbb{P}(\eta_{m,n}=m) &= \frac{1}{2\pi}\int_{-\pi}^{\pi}\exp\Big(\frac{\alpha z}{1-\beta+\alpha}\Big)\exp\big(m\beta z-m\beta^2z^2/2\big)\\ &\times\ \exp\big((n-m-1)\alpha z-(n-m-1)\alpha^2z^2/2\big)\,dt.\end{aligned}$$

Set

$$\begin{aligned} f_1 &= 1+\frac{\alpha z}{1-\beta+\alpha},\quad f_2=1+\beta z,\quad f_3=1+\alpha z,\quad g_1=\exp\Big(\frac{\alpha z}{1-\beta+\alpha}\Big),\\ g_2 &= \exp\big(\beta z-\beta^2z^2/2\big),\quad g_3=\exp\big(\alpha z-\alpha^2z^2/2\big).\end{aligned}$$

Direct estimation and (D.6) allows us to check that

$$|f_1|,|g_1||f_3|,|g_3|\leq\exp\big(-0.5\alpha\sin^2(t/2)\big),\quad |f_2|,|g_2|\leq\exp\big(-0.5\beta\sin^2(t/2)\big).$$

Similarly, from (I.9) we get

$$|f_1-g_1|\leq C\alpha^2|z|^2,\quad |f_2-g_2|\leq C\beta^3|z|^3;\quad |f_3-g_3|\leq C\alpha^3|z|^3.$$

Employing (I.1) and Lemma I.1, we obtain

$$\begin{aligned}
&|\mathbb{P}(S_n=m)-(1-\beta+\alpha)\mathbb{P}(\eta_{m,n}=m)|\\
&\le \frac{(1-\beta+\alpha)}{2\pi}\int_{-\pi}^{\pi}\Big(|f_1-g_1||f_2^m f_3^{n-m-1}|\\
&+ |g_1||f_2^m-g_2^m||g_3^{n-m-1}|+|g_1g_2^m|f_3^{n-m-1}-g_3^{n-m-1}\Big)\mathrm{d}t\\
&\le C\int_{-\pi}^{\pi}(2\alpha^2|z|+m\beta^3|z|^3+(n-m-1)\alpha^3|z|^3)\\
&\times \exp\big(-0.5m\beta\sin^2(t/2)+(n-m)\alpha\sin^2(t/2)\big)\,\mathrm{d}t\\
&\le \frac{C\alpha^2}{m\beta+(n-m)\alpha}+\frac{C(m\beta^3+(n-m-1)\alpha^3)}{(m\beta+(n-m)\alpha)^{3/2}}\le\frac{C(\alpha^2\vee\beta^2)}{n(\alpha\wedge\beta)}.
\end{aligned}$$

Chapter 10

10.1 Denote $F=\mathcal{L}(X)$. Then

$$\mathcal{L}(Y)=\mathbf{\Pi}(n/5;X)*\exp(nA(F)),\quad \text{where } A(F)=\sum_{j=2}^{\infty}(F-I)^{*j}/5^j j.$$

We recall that Θ is used for all measures, satisfying $\|\Theta\|_{TV}\le 1$. Then

$$A(F)=\sum_{j=2}^{\infty}(F-I)^{*j}/5^j j=\frac{1}{40}(F-I)^{*2}*\Theta.$$

By (B.1) and Lemma B.1

$$\begin{aligned}
\mathcal{L}(Y)-\mathbf{\Pi}(n/5;X) &= \mathbf{\Pi}(n/5;X)*(nA(F))\int_0^1 \mathrm{e}^{nA(F)\tau}\mathrm{d}\tau\\
&= \exp\Big(\frac{9n}{80}(F-I)\Big)*(nA(F))*\int_0^1\exp\Big(\frac{7n}{80}(F-I)+\frac{1}{40}(F-I)^{*2}\Big)\,\mathrm{d}\tau\\
&= n(F-I)^{*2}*\exp(9n(F-I)/80)*\Theta.
\end{aligned}$$

It remains to apply (C.24).

10.2 Let

$$\Psi=\mathcal{L}(X)=\frac{1}{5}F+\frac{1}{5}G+\frac{1}{5}F^{*2}+\frac{1}{5}G^{*2}+\frac{1}{5}F*G.$$

Then $\mathcal{L}(S_n)=\Psi^{*n}$. By (10.12),

$$\|\Psi^{*n}-\exp(n(\Psi-I))\|_K\le Cn^{-1}.$$

We have

$$\mathbf{\Pi}(4n/5;X')*\mathbf{\Pi}(4n/5;X'')=\exp(4n(F-I)/5+4n(G-I)/5).$$

For each $\tau\in[0,1]$

$$\begin{aligned}
&\exp(\tau n(\Psi-I)+(1-\tau)(4n(F-I)/5+4n(G-I)/5))\\
&=\exp(n(F-I)/5+n(G-I)/5)*\Theta,
\end{aligned}$$

where

$$\|\Theta\|_{TV} = \Big\| \exp\Big(n\tau(F-I+G-I) + \frac{n(1-\tau)}{5}(F^{*2}-I+G^{*2}-I+F*G-I)\Big) \Big\|_{TV} = 1,$$

since the total variation norm of any distribution equals 1. Observe also that

$$\Psi - I = \frac{4(F-I)}{5} + \frac{4(G-I)}{5} + \frac{(F-I)^{*2}}{5} + \frac{(G-I)^{*2}}{5} + \frac{(F-I)*(G-I)}{5}.$$

By (B.1) and (A.6),

$$\begin{aligned}
&\| \exp(n(\Psi-I)) - \mathbf{\Pi}(4n/5; X') * \mathbf{\Pi}(4n/5; X'')\|_K \\
&\quad \leq \; Cn\|(\Psi - I - 4(F-I)/5 - 4(G-I)) * \exp(n(F-I)/5 + n(G-I))\,\|_K \\
&\quad \leq \; Cn\|(F-I)^{*2} * \exp(n(F-I)/5)\,\|_K + Cn\|(G-I)^{*2} * \exp(n(G-I)/5)\,\|_K \\
&\quad + \; Cn\|(F-I) * (G-I) * \exp(n(F-I)/5 + n(G-I)/5)\,\|_K.
\end{aligned}$$

Application of (C.24), (C.27) and the triangle inequality complete the proof.

10.3 The solution coincides with the proof of (10.11) up to the last step, where (C.37) is used.

Chapter 11

11.1 Arguing as in the proof of Theorem 11.4, we get from Lemma D.3 that there exists $\varepsilon \in (0;\pi)$ such that

$$|\widehat{F}_X(t)| \leq \exp(-C_X t^2)\,, \quad \Big|\exp\Big(\widehat{F}_X(t)-1\Big)\Big| \leq \exp(-C_X t^2)$$

if $|t| \leq \varepsilon$. It follows from Lemma D.6 that

$$|\widehat{F}^n_X(t)| \leq \exp(-nC_X)\,, \quad \Big|\exp\Big(n(\widehat{F}_X(t)-1)\Big)\Big| \leq \exp(-nC_X)$$

if $\varepsilon < |t| \leq \pi$. By (I.1), (I.10) and (D.3),

$$\begin{aligned}
|\widehat{F}^n_X(t) - \exp\Big(n(\widehat{F}^n_X(t)-1)\Big)| &\leq n\Big|\widehat{F}^n_X(t) - \exp\Big(n(\widehat{F}_X(t)-1)\Big)\Big| \exp(-(n-1)C_X t^2) \\
&\leq nC_X(F^n_X(t)-1)^2 \exp(-nC_X t^2) \leq C_X n t^4 \exp(-nC_X t^2)
\end{aligned}$$

if $|t| \leq \varepsilon$. Applying (D.12), we get

$$d_o(F^{*n}; G^{*n}_X) \leq C_X \int_{|t|\leq\varepsilon} nt^4 \mathrm{e}^{-C_X nt^2}\mathrm{d}t + \frac{2}{\pi}\int_\varepsilon^\pi \mathrm{e}^{-C_X n}\mathrm{d}t \leq C_X n^{-3/2}.$$

11.2 This exercise can be solved by the same argument as that in Exercise 13.1 by applying Tsaregradskii's inequality (D.14) instead of (D.12).

Chapter 12

12.1 Denote $J = \| \exp(a(G-I)) - \exp(b(G-I)) \|_K$.

First, we consider the case where $\max\{1; a; b\} = 1$. It follows from (3.83*) that

$$J \le d_{TV}(\pi_a; \pi_b) \le |a-b|.$$

Let now $\max\{1; a; b\} = b$. If $b \ge 3a/2$, then $(b-a)/b > 1/3$ and

$$J \le \| \exp(a(G-I))\|_{TV} + \| \exp(b(G-I))\|_{TV} = 2 \le 6(b-a)/b.$$

If $b < 3a/2$, then $a/(b-a) > 2$. Denote by $m = \lfloor a/(b-a) \rfloor$ the integer part of $a/(b-a)$, and set $\delta = a/(b-a) - m$. Then

$$\begin{aligned} J &= \| \exp((b-a)(m+\delta)(G-I)) - \exp((b-a)(m+1+\delta)(G-I)) \|_K \\ &\le \| \exp((b-a)m(G-I)) - \exp((b-a)(m+1)(G-I)) \|_K \\ &= \left\| (\exp((b-a)(G-I)))^{*m} - (\exp((b-a)(G-I)))^{*(m+1)} \right\|_K . \end{aligned}$$

Note that $\exp((b-a)(G-I))$ has a non-negative characteristic function. By (12.31),

$$J \le C_k/m \le C_k|a-b|/a \le 3C_k|a-b|/2b.$$

12.2 Let $m = \lfloor a \rfloor + 1$. Then $m = j(\lfloor m/j \rfloor + \delta)$, where $0 \le \delta < 1$. By (12.31),

$$\|F^{*m} - F^{*j(\lfloor m/j \rfloor+1)}\|_K \le |j(\lfloor m/j \rfloor + 1) - m| \|F^{*m} * (F-I)\|_{K^*} \le C_k j m^{-1} \le C j a^{-1} .$$

By (12.22) with $\alpha = -1$,

$$\|(F^{*j})^{*(\lfloor m/j \rfloor+1)} - \exp((\lfloor m/j \rfloor + 1)(F^{*j}-I)) \|_K \le C_k j m^{-1} \le C j a^{-1} .$$

Finally, by (12.32),

$$\| \exp\big((\lfloor m/j \rfloor + 1)(F^{*j}-I)\big) - \exp\big(a(F^{*j}-I)/j\big) \|_K \le C_k j a^{-1} .$$

An application of the triangle inequality completes the proof.

12.3 Let $\widehat{F}_X(t_1, t_2) = 1 - 0.5\sin^2(t_1/2) - 0.5\sin^2(t_2/2)$ denote the characteristic function of $\mathcal{L}(X)$. Then the characteristic functions of $\mathcal{L}(S_n)$ and $\mathcal{L}(\tilde{S}_n)$ are equal to $\widehat{F}_X^n$ and $\exp(n(\widehat{F}_X - 1))$, respectively. Observe that $\widehat{F}_X \ge 0$. Therefore,

$$\begin{aligned} \left|\widehat{F}_X^n - \exp\Big(n(\widehat{F}_X-1)\Big)\right| &\le n\left|\widehat{F}_X - \exp\Big(\widehat{F}_X-1\Big)\right| \exp\Big((n-1)(\widehat{F}_X-1)\Big) \\ &\le en(\widehat{F}_X-1)^2 \exp(-(n/2)\sin^2(t_1/2) - (n/2)\sin^2(t_2/2)) \end{aligned}$$

by (I.10). Note that

$$(\widehat{F}_X-1)^4 = (0.25\sin^2(t_1/2) + 0.25\sin^2(t_2/2))^4 \le \sin^8(t_1/2) + \sin^8(t_2/2).$$

Therefore,

$$\begin{aligned} &\left(\widehat{F}_X^n - \exp\Big(n(\widehat{F}_X-1)\Big)\right)^2 \\ &\le \frac{e^2n^2}{2}(\sin^8(t_1/2) + \sin^8(t_2/2)) \exp\big(-n\sin^2(t_1/2) - n\sin^2(t_2/2)\big) . \end{aligned}$$

The multivariate Parseval identity for integer-valued vectors states (see, e.g., [92]):

$$\sum_{\bar{m}} \Big(\mathbb{P}(S_n = \bar{m}) - \mathbb{P}(\tilde{S}_n = \bar{m})\Big)^2 = \frac{1}{(2\pi)^2}\int_{-\pi}^{\pi}\int_{-\pi}^{\pi}\Big(\widehat{F}_X^n - \exp\Big(n(\widehat{F}_X - 1)\Big)\Big)^2 \mathrm{d}t_1\mathrm{d}t_2 .$$

Combining the last two equations and applying Lemma I.1, we prove (12.34).

Appendix

A.1 Let $X \sim \mathcal{N}(0,1)$. By (A.5)

$$\begin{aligned} d_{TV}(P_0;P_t) &= \frac{1}{2\sqrt{2\pi}}\int_{-\infty}^{\infty}\left|\mathrm{e}^{-(x-t)^2/2} - \mathrm{e}^{-x^2/2}\right|\mathrm{d}x = \frac{1}{2\sqrt{2\pi}}\int_{-\infty}^{t/2}\left(\mathrm{e}^{-x^2/2} - \mathrm{e}^{-(x-t)^2/2}\right)\mathrm{d}x \\ &\quad + \frac{1}{2\sqrt{2\pi}}\int_{t/2}^{\infty}\left(\mathrm{e}^{-(x-t)^2/2} - \mathrm{e}^{-x^2/2}\right)\mathrm{d}x = \frac{1}{2}(\mathbb{P}(X \le t/2) - \mathbb{P}(X+t \le t/2) \\ &\quad + \mathbb{P}(X+t \ge t/2) - \mathbb{P}(X \ge t/2)) = \mathbb{P}(X \le t/2) - \mathbb{P}(X+t \le t/2) \\ &= \mathbb{P}(X \le t/2) - \mathbb{P}(X \le -t/2) = \mathbb{P}(|X| \le t/2). \end{aligned}$$

C.1 Let $h>1$ be an absolute constant (it will be chosen later). By (I.8),

$$|\exp(\mathrm{i}t/h\sigma) - 1 - \mathrm{i}t/h\sigma| \le t^2/2h^2\sigma^2 \qquad (|\vartheta| \le 1).$$

By (D.3),

$$|\mathbb{E}\mathrm{e}^{\mathrm{i}t(S-\mathbb{E}S)/h\sigma} - 1| \le t^2/2h^2 .$$

We apply Lemma D.15 with $b = h\sigma$, $a = \mathbb{E}S$. Then

$$d_{TV}(S;S+1) \ \ge\ c\left|\int_{-\infty}^{\infty} t\mathrm{e}^{-t^2/2}(\mathrm{e}^{\mathrm{i}t/h\sigma} - 1)\mathbb{E}\mathrm{e}^{\mathrm{i}t(S-\mathbb{E}S)/h\sigma}\mathrm{d}t\right| .$$

Therefore,

$$\begin{aligned} d_{TV}(S;S+1) &\ge \frac{c}{h\sigma}\left|\int_{-\infty}^{\infty}\mathrm{e}^{-t^2/2}t^2\mathbb{E}\mathrm{e}^{\mathrm{i}t(S-\mathbb{E}S)/h\sigma}\mathrm{d}t\right| - C\int_{-\infty}^{\infty}|t|\mathrm{e}^{-t^2}|\mathrm{e}^{\mathrm{i}t/h\sigma} - 1 - \mathrm{i}t/h\sigma|\mathrm{d}t \\ &\ge \left|\int_{-\infty}^{\infty} t^2\mathrm{e}^{-t^2/2}\mathrm{d}t\right| - \frac{C}{h\sigma}\int_{-\infty}^{\infty} t^2\mathrm{e}^{-t^2/2}|\mathbb{E}\mathrm{e}^{\mathrm{i}t(S-\mathbb{E}S)/h\sigma} - 1|\mathrm{d}t - \frac{C}{h^2\sigma^2} \\ &\ge \frac{c}{h\sigma}\left|\int_{-\infty}^{\infty}\mathrm{e}^{-t^2/2}t^2\mathrm{d}t\right| - \frac{C}{h^3\sigma} - \frac{C}{h^2\sigma^2} \\ &\ge c(1 - C/h)/h\sigma \ge c/\sigma \end{aligned}$$

for some absolute constant C if we choose $h = 2C$. This completes the proof for the total variation distance. The local estimate follows directly from (D.24).

C.2 By (D.11),

$$\begin{aligned} \mathbb{P}(\pi_{\lambda/2} - \pi'_{\lambda/2} = 1) &= \frac{1}{2\pi}\int_{-\pi}^{\pi}\cos t\mathrm{e}^{-\lambda(1-\cos t)}\mathrm{d}t \\ &= \mathbb{P}(\pi_{\lambda/2} - \pi'_{\lambda/2} = 0) + \frac{1}{2\pi}\int_{\pi}^{\pi}(\cos t - 1)\mathrm{e}^{-\lambda(1-\cos t)}\mathrm{d}t \\ &= \mathbb{P}(\pi_{\lambda/2} - \pi'_{\lambda/2} = 0) + O(\lambda^{-1}). \end{aligned}$$

since $\sup_{x>0} xe^{-\lambda x} \le C/\lambda$. Note that $\cos t - 1 \ge -t^2/2$. Therefore,

$$\begin{aligned}\mathbb{P}(\pi_{\lambda/2} - \pi'_{\lambda/2} = 0) &= \frac{1}{2\pi}\int_{-\pi}^{\pi} \mathrm{e}^{-\lambda(1-\cos t)}\mathrm{d}t \\ &\ge \frac{1}{2\pi}\int_{-\pi}^{\pi} \mathrm{e}^{-\lambda t^2/2}\mathrm{d}t \ge \frac{1}{\sqrt{2\pi\lambda}} + O(\lambda^{-1}) \qquad (\lambda\to\infty).\end{aligned}$$

C.3 Let $a \in \mathbb{R}$ be chosen in such a way that $I_a * F = qA + pV$ with $p = 0.1$. Apply Lemma 10.10,

$$\|(I_a * F)^{*n} - \exp(nq(A-I)) * (qI+pV)^{*n}\|_K \le C/\sqrt{n}. \tag{I.20}$$

The total variation of any distribution equals 1. Therefore, from (A.1) and (C.12) we get

$$\begin{aligned}&\|\exp(nq(A-I)) * (qI+pV)^{*n} - \exp((n+1)q(A-I)) * (qI+pV)^{*n}\|_K \\ &\qquad \le \|\exp(nq(A-I)) - \exp((n+1)q(A-I))\|_{TV} \|(qI+pV)^{*n}\|_{TV} \\ &\qquad = \|\exp(nq(A-I)) - \exp((n+1)q(A-I))\|_{TV} \le C/\sqrt{n}. \end{aligned}\tag{I.21}$$

Next by (A.1), (A.3) and by (C.7):

$$\begin{aligned}&\|\exp((n+1)q(A-I)) * (qI+pV)^{*n} - \exp((n+1)q(A-I)) * (qI+pV)^{*(n+1)}\|_K \\ &\qquad \le \|\exp((n+1)q(A-I))\|_{TV} \|(qI+pV)^{*n} - (qI+pV)^{*(n+1)}\|_{TV} \\ &\qquad = p\|(V-I) * (qI+pV)^{*n}\|_{TV} \\ &\qquad \le p\|(I_1 - I) * (qI+pI_1)^{*n}\|_{TV} \le 1/\sqrt{n}. \end{aligned}\tag{I.22}$$

By Lemma 10.10,

$$\|\exp((n+1)q(A-I)) * (qI+pV)^{*(n+1)} - (I_a * F)^{*(n+1)}\|_K \le C/\sqrt{n}. \tag{I.23}$$

It remains to apply the triangle inequality.

D.1 The estimate follows from Plancherel's identity

$$\int_{-\infty}^{\infty} |\widehat{F}_X(t)|^2 \mathrm{d}t = 2\pi \int_{-\infty}^{\infty} f^2(x)\mathrm{d}x \le 2\pi C_0 .$$

D.3 Express $Fx) - F(-x)$ via inversion formula (D.1).

D.4 Applying (D.2) to the distribution from Exercise D.2 and taking $x = 0$, we get

$$1 = \frac{1}{2\pi}\int_{-\infty}^{\infty} \frac{\sin^2(t/2)}{(t/2)^2}\mathrm{d}t = \frac{1}{\pi}\int_{-\infty}^{\infty} \frac{\sin^2 t}{t^2}\mathrm{d}t.$$

In order to show that $\lim_{T\to\infty} \int_{-T}^{T} \frac{\sin t}{t} dt = \pi$, express $F(1) - F(-1)$ via inversion formula (D.1).

D.5 Use the fact that $\pi J_s = \lim_{T\to\infty} \int_{-T}^{T} \frac{\sin t}{t} \mathrm{e}^{\mathrm{i}st} dt$.

E.1 Apply the Kolmogorov-Rogozin inequality (E.2).

E.2 By (D.11),

$$\begin{aligned}\mathbb{P}(\pi_{\lambda/2}-\pi'_{\lambda/2}=1) &= \frac{1}{2\pi}\int_{-\pi}^{\pi}\cos t\mathrm{e}^{-\lambda(1-\cos t)}\mathrm{d}t\\ &= \mathbb{P}(\pi_{\lambda/2}-\pi'_{\lambda/2}=0)+\frac{1}{2\pi}\int_{\pi}^{\pi}(\cos t-1)\mathrm{e}^{-\lambda(1-\cos t)}\mathrm{d}t\\ &= \mathbb{P}(\pi_{\lambda/2}-\pi'_{\lambda/2}=0)+O(\lambda^{-1}).\end{aligned}$$

since $\sup_{x>0} xe^{-\lambda x} \leq C/\lambda$. Note that $\cos t-1\geq -t^2/2$. Therefore,

$$\begin{aligned}\mathbb{P}(\pi_{\lambda/2}-\pi'_{\lambda/2}=0) &= \frac{1}{2\pi}\int_{-\pi}^{\pi}\mathrm{e}^{-\lambda(1-\cos t)}\mathrm{d}t\\ &\geq \frac{1}{2\pi}\int_{-\pi}^{\pi}\mathrm{e}^{-\lambda t^2/2}\mathrm{d}t \geq \frac{1}{\sqrt{2\pi\lambda}}+O(\lambda^{-1}) \qquad (\lambda\to\infty).\end{aligned}$$

E.3 Apply (E.1).

H.1 By (B.1)

$$\mathrm{e}^{-\sigma_2^2t^2}-\mathrm{e}^{-\sigma_1^2t^2/2}=\frac{(\sigma_1^2-\sigma_2^2)\mathrm{e}^{-\sigma_2^2t^2/2}}{2}\int_0^1\mathrm{e}^{-(\sigma_1^2-\sigma_2^2)t^2\tau/2}\mathrm{d}\tau\leq\frac{(\sigma_1^2-\sigma_2^2)}{2}\mathrm{e}^{-\sigma_2^2t^2/2}.$$

It follows from (D.19) that

$$\begin{aligned}d_K(\mathcal{N}(0;\sigma_1^2);\mathcal{N}(0;\sigma_2^2)) &\leq \frac{1}{2\pi}\int_{-\infty}^{\infty}\frac{|\mathrm{e}^{-\sigma_2^2t^2/2}-\mathrm{e}^{-\sigma_1^2t^2/2}|}{|t|}\mathrm{d}t\\ &\leq \frac{\sigma_1^2-\sigma_2^2}{4\pi}\int_{-\infty}^{\infty}|t|\mathrm{e}^{-\sigma_2^2t^2/2}\mathrm{d}t=\frac{\sigma_1^2-\sigma_2^2}{2\pi\sigma_2^2}.\end{aligned}$$

In order to prove the estimate in terms of the total variation distance, use (A.5), see also Lemma 15.3.1 in [162].

I.1 Inequality (I.4) follows from (I.14), since it can be written as $\mathbb{E}|X-\mathbb{E}X|\leq(\operatorname{var}X)^{1/2}$, where $\mathcal{L}(X)=\mathbf{B}(n,p)$. If $c=1$, then (I.3) follows from

$$\binom{n}{k}=\frac{n!}{k!(n-k)!}=\frac{k+1}{n+1}\binom{n+1}{k+1}.$$

An application of Hölder's inequality allows us to derive (I.3) in the case $c=1/2$. Similarly,

$$\begin{aligned}&\sum_{k=0}^{n}\binom{n}{k}p^kq^{n-k}(k+1)^{-2}\leq 2\sum_{k=0}^{n}\binom{n}{k}p^kq^{n-k}/(k+1)(k+2)\\ &=\frac{2}{(n+2)(n+1)p^2}\sum_{k=0}^{n}\binom{n+2}{k+2}p^{k+2}q^{n-k}\leq\frac{2}{(n+2)(n+1)p^2}.\end{aligned}$$

I.2 Let f,g be two m-times differentiable functions. It is known that

$$\frac{\mathrm{d}^m}{\mathrm{d}x^m}(f(x)g(x))=\sum_{i=0}^{m}\binom{m}{i}\left(\frac{\mathrm{d}^i}{\mathrm{d}x^i}f(x)\right)\left(\frac{\mathrm{d}^{m-i}}{\mathrm{d}x^{m-i}}g(x)\right).$$

It suffices to apply this relation to functions $f(x)=x^a$, $g(x)=x^b$ and let $x=1$.

I.3 Observe that

$$\sum_{k\geq j}\binom{k}{j}p^j q^{k-j}=\frac{p^j}{j!}\sum_{k\geq j}k(k-1)(k-2)\cdots(k-j+1)q^{k-j}.$$

Let $x\in(0,1)$. Then

$$\begin{aligned}\sum_{k\geq j}k(k-1)\cdots(k-j+1)x^{k-j} &= \sum_{k\geq j}(x^k)^{(j)}=\left(\sum_{k\geq j}x^k\right)^{(j)}=\left(\sum_{k=0}^{\infty}x^k\right)^{(j)}\\ &= \left((1-x)^{-1}\right)^{(j)}=j!(1-x)^{-j-1}.\end{aligned}$$

Therefore,

$$\sum_{k\geq j}k(k-1)(k-2)\cdots(k-j+1)q^{k-j}=j!p^{-j-1}$$

and

$$\sum_{k\geq j}\binom{k}{j}p^j q^{k-j}=p^{-1}.$$

Bibliography

[1] Adler R.J. (1978) Weak convergence results for extremal processes generated by dependent random variables. — *Ann. Probab.*, **6**(4), 660–667.

[2] Aleškevičienė A. and Statulevičius V. (1995) Asymptotic expansions in the approximation by the Poisson law. — *Lith. Math. J.*, **35**(4), 309–327.

[3] Aleškevičienė A. and Statulevičius V. (2001) Large deviations in the approximation by the Poisson law. — *Teor. Veroyatnost. Primen.*, **46**(4), 625–639 (Russian). Transl.: *Theory Probab. Appl.*, **46**(4), 589–602, 2002.

[4] Arak T.V. (1980) On the approximation of n-fold convolutions of distributions having nonnegative characteristic functions by infinitely divisible distributions. — *Trudy Tallin. Politekhn. Inst.*, **482**, 81–85 (Russian).

[5] Arak T.V. (1980) On the approximation by the accompanying laws of n-fold convolutions of distributions with non-negative characteristic functions. — *Teor. Veroyatn. Primen.*, **25**(2), 225–246 (Russian). Transl.: *Theory Probab. Appl.*, **25**(2), 221–243, 1981.

[6] Arak T.V. (1981) On the convergence rate in Kolmogorov's uniform limit theorem I, II. — *Teor. Veroyatn. Primen.*, **26**(2), 225–245; (3), 449–463 (Russian). Transl.: *Theory Probab. Appl.*, **26**(2), 219–239; (3), 437–451, 1982.

[7] Arak T.V. (1982) An improvement of the lower bound for the rate of convergence in Kolmogorov's uniform limit theorem. — *Teor. Veroyatn. Primen.*, **27**(4), 767–772 (Russian). Transl.: *Theory Probab. Appl.*, **27**(4), 826–832.

[8] Arak T.V. and Zaitsev A.Yu. (1986) Uniform limit theorems for sums of independent random variables. *Trudy Mat. Inst. Steklov (Monograph)*, **174** (Russian). Transl.: *Proc. Steklov Inst. Math.*, **174**, 3–214.

[9] Arenbaev N.K. (1977) Asymptotic behavior of the multinomial distribution. — *Teor. Veroyatn. Primen.*, **21**(4), 826–831 (Russian). Transl.: *Theory Probab. Appl.*, **21**(4), 805–810, 1977.

[10] Arratia R., Goldstein L. and Gordon L. (1989) Two moments suffice for Poisson approximation. — *Ann. Probab.*, **17**(1), 9–25.

[11] Arratia R., Gordon L. and Waterman M.S. (1990) The Erdös–Rényi law in distribution, for coin tossing and sequence matching. — *Ann. Statist.*, **18**(2), 539–570.

[12] Bakstys G. and Paulauskas V.J. (1985) Approximation of distributions of sums of Banach-valued random elements by infinitely divisible laws I. — *Litovsk. Mat. Sb.*, **26**(3) 403–414 (Russian). Transl.: *Lith. Math. J.*, **26**(3), 207-215.

[13] Bakstys G. and Paulauskas V.J. (1987) Approximation of distributions of sums of Banach-valued random elements by infinitely divisible laws II. — *Litovsk. Mat. Sb.*, **27**(2) 224–235 (Russian). Transl.: *Lith. Math. J.*, **27**(2), 106–113.

[14] Bakstys G. (1989) Approximation with accompanying laws. — *Litovsk. Mat. Sb.*, **29**(3), 423–428 (Russian). Transl.: *Lith. Math. J.*, **29**(3), 211–215.

[15] Balakrishnan N., Koutras M.V. (2001) *Runs and scans with applications.* — New York: Wiley.

[16] Banys R. (1985) A Poisson limit theorem for rare events of a discrete random field.(Russian) — *Litovsk. Mat. Sb.*, **25**(1), 3–8.

[17] Barbour A.D. and Eagleson G.K. (1983) Poisson approximation for some statistics based on exchangeable trials. — *Adv. Appl. Probab.*, **15**(3), 585–600.

[18] Barbour A.D. and Hall P. (1984) On the rate of Poisson convergence. — *Math. Proceedings Cambridge Philos. Soc.*, **95**, 473–480.

[19] Barbour A.D. (1987) Asymptotic expansions in the Poisson limit theorem. — *Ann. Probab.*, **15**(2), 748–766.

[20] Barbour A.D. and Jensen J.L. (1989) Local and tail approximations near the Poisson limit. — *Scand. J. Statist.*, **16**(1), 75–87.

[21] Barbour A.D., Holst L. and Janson S. (1992) — *Poisson Approximation.* Oxford: Clarendon Press.

[22] Barbour A.D., Chen L.H.Y. and Loh W.-L. (1992) Compound Poisson approximation for non-negative random variables via Stein's method. — *Ann. Probab.*, **20**(4), 1843–1866.

[23] Barbour A.D., Chen L.H.Y., Choi K.P. (1995) Poisson approximation for unbounded functions: independent summands. — *Statistica Sinica*, **5**(2), 749–766.
[24] Barbour A.D. and Utev S.A. (1999) Compound Poisson approximation in total variation. — *Stochastic Process. Appl.*, **82**(1), 89–125.
[25] Barbour A.D. and Xia A. (1999) Poisson perturbations. — *ESAIM Probab. Statist.*, **3**, 131–150.
[26] Barbour A.D. and Xia A. (2000) Estimating Stein's constants for compound Poisson approximation. — *Bernoulli*, **6**(4), 581–590.
[27] Barbour A.D., Chryssaphinou O. (2001) Compound Poisson approximation: a user's guide. — *Appl. Probab.*, **11**(3), 964–1002.
[28] Barbour, A. D., Chryssaphinou, O. and Vaggelatou, E. (2001) Applications of compound Poisson approximation. — In: Charalambides, C. A., Koutras, M. V. and Balakrishnan, N. (Eds.) *Probability and Statistical Models with Applications*, 41–62. Chapman & Hall.
[29] Barbour A.D. and Čekanavičius V. (2002) Total variation asymptotics for sums of independent integer random variables. — *Ann. Probab.*, **30**(2), 509–545.
[30] Barbour A.D., Novak S.Y. and Xia A. (2002) Compound Poisson approximation for the distribution of extremes. — *Adv. Appl. Probab.*, **34**(1), 223–240.
[31] Barbour A.D. and Choi K.P. (2004) A non-uniform bound for translated Poisson approximation. — *Electron. J. Probab.*, **9**, 18–36.
[32] Barbour A.D. and Lindvall T. (2005) Poisson approximation for Markov chains. — *J. Theor. Probab.*, **19**(3), 609–630.
[33] Barbour A.D. and Xia A. (2006) On Stein's factors for Poisson approximation in Wasserstein distance. — *Bernoulli*, **12**(6), 943–954.
[34] Barbour A.D., Čekanavičius V. and Xia A. (2007) On Stein's method and perturbations. — *ALEA, Lat. Am. J. Probab. Math. Stat.*, **3**, 31–53.
[35] Barbour A.D. (2009) Univariate approximations in the infinite occupancy scheme. — *ALEA, Lat. Am. J. Probab. Math. Stat.*, **6**, 415–433.
[36] Barbour A.D., Gan H.L. and Xia A. (2015) Stein factors for negative Binomial approximation in Wasserstein distance. — *Bernoulli*, **21**(2), 1002–1013.
[37] Barbour A.D., Röllin A. and Ross N. (2019) Error bounds in local limit theorems using Stein's method. — *Bernoulli*, **25**(2), 1076–1104.
[38] Barbour A. D., Johnson O., Kontoyiannis I. and Madiman M. (2010) Compound Poisson approximation via information functionals. — *Electron. J. Probab.*, **15**, 1344–1369.
[39] Bentkus V., Götze F. and Zaitsev A.Yu. (1997) Approximation of quadratic forms of independent random vectors by accompanying laws. — *Teor. Veroyatn. Primen.*, **42**(2), 308–335. Trans.: *Theory Probab. Appl.*, **42**(2), 189-212, 1998.
[40] Berbee H.C.P. (1979) *Random walks with stationary increments and renewal theory.* — Amsterdam: Mathematisch Centrum Tract 112.
[41] Bergström H. (1951) On asymptotic expansion of probability functions. — *Skand. Aktuar.*, **1**, 1–34.
[42] Bernstein S.N. (1926) Sur l'extensiori du theoreme limite du calcul des probabilites aux sommes de quantites dependantes. — *Math. Annalen*, **97**, 1–59.
[43] Bernstein S.N. (1946) *Probability Theory.* — Moscow: Nauka. (in Russian).
[44] Bobkov, S.G., Danshina, M.A., Ulyanov, V.V. (2021) Rate of convergence to the Poisson law of the numbers of cycles in the generalized random graphs. — In: Karapetyants, A.N., Pavlov, I.V., Shiryaev, A.N. (eds) *Operator Theory and Harmonic Analysis. OTHA 2020.* Springer Proceedings in Mathematics & Statistics, v. 358.
[45] Bobkov S.G., Chistyakov G.P., Gotze F. (2019) Non-uniform bounds in the Poisson approximation with applications to informational distances I. — IEEE Transactions on Information Theory, **65**(9), 5283–5293.
[46] Borisov I.S. (1996) Poisson approximation of the partial sum process in Banach spaces. — *Sibirsk. Mat. Zh.*, **37**(4), 723–731 (Russian). Transl.: *Sib. Math. J.*, **37**(4), 627–634.
[47] Borisov I.S. and Ruzankin P.S. (2002) Poisson approximation for expectations of unbounded functions of independent random variables. — *Ann. Probab.*, **30**(4), 1657–1680.
[48] Borisov I.S. (2003) A note on Dobrushin's theorem and couplings in Poisson approximation in Abelian groups. — *Teor. Veroyatn. Primen.*, **48**(3), 576–583 (Russian). Transl.: *Theory Probab. Appl.*, **48**(3), 521–528, 2004.
[49] Borisov I.S. and Vorozheikin I.S. (2008) Accuracy of approximation in the Poisson theorem in terms of χ^2 distance. — *Sibirsk. Mat. Zh.*, **49**(1), 8–22 (Russian). Transl.: *Sib. Math. J.*, **49**(1), 5–17.

[50] Borowkow A.A. (1976) *Wahrscheinlichkeitstheorie: Eine Einführung.* — Springer: Basel AG.

[51] Borovkov K.A. and Novak S.Y. (2010) On limiting cluster size distributions for processes of exceedances for stationary sequences. — *Statist. Probab. Lett.*, **80**, 1814–1818.

[52] Boutsikas M.V. and Koutras M.V. (2002) On the number of overflown urns and excess balls in an allocation model with limited urn capacity. — *J. Statist. Plan. Inf.*, **104**(2), 259–286.

[53] Boutsikas M.V. and Vaggelatou E. (2016) A new method for obtaining sharp compound Poisson approximation error estimates for sums of locally dependent random variables. — *Bernoulli*, **16**(2), 301–330.

[54] Brown T.C. and Phillips M.J. (1999) Negative binomial approximation with Stein's method. — *Method. Comput. Appl. Probab.*, **1**, 407–421.

[55] Brown T.C. and Xia A. (2001) Stein's method and birth-death processes. — *Ann. Probab.*, **29**(3), 1373–1403.

[56] Chen L.H.Y. (1975) Poisson approximation for dependent trials. — *Ann. Probab.*, **3**(3), 534–545.

[57] Chen L.H.Y. and Roos M. (1995) Compound Poisson approximation for unbounded functions on a group with application to large deviations. — *Probab. Theory Rel. Fields*, **103**, 515–528.

[58] Chen L.H.Y. and Röllin A. (2013) Approximating dependent rare events. — *Bernoulli*, **19**(4), 1243–1267.

[59] Chryssaphinou O. and Vaggelatou E. (2001) Compound Poisson approximation for long increasing sequences. — *J. Appl. Probab.*, **38**(2), 449–463.

[60] Chryssaphinou O. and Vaggelatou E. (2002) Compound Poisson approximation for a multiple runs in a Markov chain. — *Ann. Inst. Stat. Math.*, **54**(2), 411–424.

[61] Čekanavičius V. (1989) Approximation with accompanying distributions and asymptotic expansions I. — *Litovsk. Mat. Sb.*, **29**(1), 171–178 (Russian). Transl.: *Lith. Math. J.*, **29**(1), 75–80.

[62] Čekanavičius V. (1990) An approximation of integer-valued measures by generalized Poisson measures. — *In: Probabability Theory and Mathematical Statistics 2. Proceedings of the Fifth Vilnius Conference (1989)*, B.Grigelionis et al. (Eds.), 228–237.

[63] Čekanavičius V. (1991) An approximation of mixtures of distributions. — *Litovsk. Mat. Sb.*, **31**(2), 351–368 (Russian). Transl.: *Lith. Math. J.*, **31**(2), 243–257.

[64] Čekanavičius V. (1993) Non-uniform theorems for discrete measures. *Lith. Math. J.*, **33**(2), 114–126.

[65] Čekanavičius V. (1995) On asymptotic expansions in the first uniform Kolmogorov theorem. — *Statist. Probab. Lett.*, **25**(2), 145–151.

[66] Čekanavičius V. (1995) On smoothing properties of compound Poisson distributions. — *Lith. Math. J.*, **35**(2), 121–135.

[67] Čekanavičius V. (1996) On multivariate Le Cam theorem and compound Poisson measures. — *Statist. Probab. Lett.*, **28**(1), 33–39.

[68] Čekanavičius V. (1997) Asymptotic expansions in the exponent: a compound Poisson approach. — *Adv. Appl. Probab.*, **29**(2), 374–387.

[69] Čekanavičius V. (1997) Approximation of the generalized Poisson Binomial distribution: asymptotic expansions. — *Lith. Math. J.*, **37**(1), 1–12.

[70] Čekanavičius V. (1997) Asymptotic expansions for compound Poisson measures. — *Lith. Math. J.*, **37**(4), 426–447.

[71] Čekanavičius V. and Vaitkus P. (1997) Large deviations for signed compound Poisson approximations. — *Statistics and Decisions*, **15**(4), 379–396.

[72] Čekanavičius V. (1998) Poisson approximations for sequences of random variables. — *Statist. Probab. Lett.*, **39**(2), 101–107.

[73] Čekanavičius V. (1998) On signed normal-Poisson approximations. — *Probab. Theory Rel. Fields*, **111**, 563–583.

[74] Čekanavičius V. (1998) Estimates in total variation for convolutions of compound distributions. — *J. Lond. Math. Soc.*, **58**(3), 748–760.

[75] Čekanavičius V. (1999) Remarks on infinitely divisible approximations to the Binomial law. — *In: Probability Theory and Mathematical Statistics. Proceedings of the 7th Vilnius Conference*, B.Grigelionis et al. (Eds.), VSP/TEV, 135–146.

[76] Čekanavičius V. and Mikalauskas M. (1999) Signed Poisson approximations for Markov chains. — *Stochastic Process. Appl.*, **82**(2), 205–227.

[77] Čekanavičius V. (1999) On compound Poisson approximations under moment restrictions. — *Teor. Veroyatn. Primen.*, **44**(1), 74–86 (Russian). Transl. *Theory Probab. Appl.* **44**(1), 18–28, 2000.

[78] Čekanavičius V. and Kruopis J. (2000) Signed Poisson approximation: a possible alternative to normal and Poisson laws. — *Bernoulli*, **6**(4), 591–606.
[79] Čekanavičius V. and Mikalauskas M. (2001) Local theorems for the Markov Binomial distribution. — *Lith. Math. J.*, **41**(3), 219–231.
[80] Čekanavičius V. and Vaitkus P. (2001) A centered Poisson approximation via Stein's method. — *Lith. Math. J.*, **41**(4), 319–329.
[81] Čekanavičius V. (2003) Infinitely divisible approximations for discrete nonlattice variables. — *Adv. Appl. Probab.*, **35**(4), 982–1006.
[82] Čekanavičius V. and Wang Y.H. (2003) Compound Poisson approximation for sums of discrete nonlattice random variables. — *Adv. Appl. Probab.*, **35**(1), 228–250.
[83] Čekanavičius V. and Roos B. (2006) An expansion in the exponent for compound Binomial approximations. — *Lith. Math. J.*, **46**(1), 54–91.
[84] Čekanavičius V. and Roos B. (2006) Compound binomial approximations. — *Ann. Inst. Stat. Math.*, **58**, 187–210.
[85] Čekanavičius V. and Roos B. (2009) Poisson type approximations for the Markov Binomial distribution. — *Stochastic Process. Appl.*, **119**(1), 190–207.
[86] Čekanavičius V. and Vellaisamy P. (2010) Compound Poisson and signed compound Poisson approximations to the Markov Binomial law. — *Bernoulli*, **16**(4), 1114–1136.
[87] Čekanavičius V. and Elijio A. (2014) Smoothing effect of compound Poisson approximations to the distributions of weighted sums. — *Lith. Math. J.*, **54**(1), 35–47.
[88] Čekanavičius V. and Vellaisamy P. (2015) Discrete approximations for sums of m-dependent random variables. — *ALEA, Lat. Am. J. Probab. Math. Stat.*, **12**, 765–792.
[89] Čekanavičius V. (2016) *Approximation methods in Probability Theory.* — Universitext: Springer.
[90] Čekanavičius V. and Vellaisamy P. (2019) On large deviations for sums of discrete m-dependent random variables. — *Stochastics*, **91**(8), 1092–1108.
[91] Čekanavičius V. and Vellaisamy P. (2020) Lower bounds for discrete approximations to sums of m-dependent random variables. — *Probab. Math. Stat.* **40**(1), 23–35.
[92] Čekanavičius V. and Vellaisamy P. (2021) Compound Poisson approximations in ℓ_p-norm for sums of weakly dependent vectors. — *J. Theor. Probab.* **34**(4), 2241–2264.
[93] Chryssaphinou O., Papastavridis S., Vaggelatou E. (2001) Poisson approximation for the nonoverlapping appearances of several words in Markov chains. — *Combin. Probab. Comput.*, **10**(4), 293–308.
[94] Csaki E., Földes A. (1996) On the length of the longest monotone block. — *Studia Scientiarum Mathematicarum Hungarica*, **31**(1-3), 35–46.
[95] Dabrowski A., Ivanoff G., Kulik R. (2009) Some notes on Poisson limits for empirical point processes. — *Canadian J. Statist.*, **37**(3), 347–360.
[96] Daley D.J. and Vere-Jones D. (2008) *An introduction to the theory of point processes, v. II. General Theory and Structure.* — New York: Springer.
[97] Daly F. (2011) On Stein's method, smoothing estimates in total variation distance and mixture distributions. — *J. Statist. Plann. Inference*, **141**, 2228–2237.
[98] Daly F., Lefévre C. and Utev S. (2012) Stein's method and stochastic orderings. — *Adv. Appl. Prob.*, **44**(2), 343–372.
[99] Daly F. (2013) Compound Poisson approximation with association or negative association via Stein's method. — *Electron. Commun. Probab.*, **18**(30), 1–12.
[100] Daly F. (2017) On magic factors in Stein's method for compound Poisson approximation. — *Electron. Commun. Probab.*, **22**(67), 1–10.
[101] Dall' Aglio G. (1956) Sugli estremi dei momenti delle funzioni di ripartizione dopia. — *Ann. Scuola Normale Superiore Di Pisa, ser. Cl. Sci.*, **3**(1), 33–74.
[102] Dasgupta R. (1992) Nonuniform rates of convergence to the Poisson distribution. — *Sankhya: The Indian Journal of Statistics, Series A (1961-2002)*, **54**(3), 460–463.
[103] Deheuvels P. and Pfeifer D. (1986) A semigroup approach to Poisson approximation. — *Ann. Probab.*, **14**(2), 663–676.
[104] Deheuvels P. and Pfeifer D. (1988) Poisson approximation of distributions and point processes. — *J. Multivariate Anal.*, **25**(1), 65–89.
[105] Deheuvels P. and Pfeifer D. (1988) On a relationship between Uspensky's theorem and Poisson approximation. — *Ann. Inst. Statist. Math.*, **40**, 671–681.
[106] Deheuvels P., Pfeifer D., Puri M.L. (1989) A new semigroup technique in Poisson approximation. — *Semigroup Forum*, **38**, 189–201.
[107] Denzel G.E. and O'Brien G.L. (1975) Limit theorems for extreme values of chain–dependent processes. — *Ann. Probab.*, **3**(5), 773–779.

[108] Dobrushin R.L. (1953) Limit theorems for a Markov chain of two states. (Russian) — *Izv. Akad. Nauk. USSR Ser. mat.*, **17**, 291–330. Transl.: *Select. Transl. Math. Stat. and Probab.*, **1**, 97–134, 1961. Providence, R.I.: Inst. Math. Stat. Amer. Math. Soc.

[109] Dobrushin R.L. (1970) Prescribing a system of random variables by conditional distributions. — *Teor. Veroyatn. Primen.*, **15**(3), 469–497 (Russian). Transl.: *Theory Probab. Appl.*, **15**(3), 458–486.

[110] Eichelsbacher P., Roos M. (1999) Compound Poisson approximation for dissociated random variables via Stein's method. — *Combinatorics, Probability & Computing*, **8**, 335–346.

[111] Elijio A. and Čekanavičius V. (2015) Compound Poisson approximation to weighted sums of symmetric discrete variables. — *Ann. Inst. Stat. Math.*, **67**(1), 195–210.

[112] Erhardsson T. (2000) Compound Poisson approximation for counts of rare patterns in Markov chains and extreme sojourns in birth-death chains. — *Ann. Appl. Probab.*, **10**, 573–591.

[113] Erhardsson T. (2005) Stein's method for Poisson and compound Poisson approximation. — *In: An introduction to Stein's method. Lect. Notes Ser. Inst. Math. Sci. Natl. Univ. Singap., v. 4, 61–113.* Singapore: Singapore Univ. Press.

[114] Erlang A.K. (1909) The theory of probabilities and telephone conversations. — *Nyt. Tidsskr. Mat. B* **20**, 33–39 (Danish). Transl.: *Transactions of the Danish Academy of Technical Sciences*, **2**, Akademiet for de Tekniske Videnskaber, 131-137, 1948.

[115] Esseen C.–G. (1945) Fourier analysis of distribution functions. A mathematical study of the Laplace–Gaussian law. — Acta Math., v. 77, 1–125.

[116] Esseen C.–G. (1956) A moment inequality with an application to the central limit theorem. — *Skand. Aktuar.*, **39**, 160–170.

[117] Esseen C.–G. (1968) On the concentration function of a sum of independent random variables. — *Z. Wahrscheinlichkeitstheorie verw. Geb.*, **9**(4), 290–308.

[118] Eliseeva Yu.S. (2013) Multivariate estimates for the concentration function of weighted sums of independent identically distributed random variables.— *Zap. Nauchn. Sem. POMI*, **412**, 121–137. Transl.: *J. Math. Sci.*, **204**(1), 78–89, 2015.

[119] Eliseeva Yu.S. and Zaitsev A.Yu. (2012) Estimates of the concentration functions of weighted sums of independent random variables. — *Teor. Veroyatn. Primen.*, **57**(4), 768–777 (Russian). Transl.: *Theory Probab. Appl.*, **57**(4), 670–678, 2013.

[120] Franken P. (1964) Approximation der verteilungen von summen unabhängiger nichtnegativer ganzzahliger zufallsgrössen durch Poissonsche verteilunged. — *Math. Nachr.*, **27**(5-6), 303–340.

[121] Gamkrelidze N.G. (1988) On the application of smoothness function in proving a local limit theorem. — *Teor. Veroyatn. Primen.*, **33**(2), 373–376 (Russian). Transl.: *Theory Probab. Appl.*, **33**(2), 352–355, 1989.

[122] Gamkrelidze N.G. (2022) Study of lattice probability distributions. — *Itogi Nauki Teh. Ser. Mat. Pril.*, **218**, 3–66 (Russian).

[123] Gan H.L. and Xia A. (2015) Stein's method for conditional compound Poisson approximation. — *Statist. Probab. Lett.*, **100**, 19–26.

[124] Gan H.L. and Kolaczyk E.D. (2018) Approximation of the difference of two Poisson-like counts by Skellam. — *J. Appl. Probab.*, **55**(2), 416–430.

[125] Gani J. (1982) On the probability generating function of the sum of Markov-Bernoulli random variables. — *J. Appl. Probab.*, **19A**, 321–326.

[126] Genest C., Marceau E. and Mesfioui M. (2003) Compound Poisson approximations for individual models with dependent risks. — *Insurance Math. Econom.*, **32**(1), 73–91.

[127] Gerber H.U. (1979) *An introduction to mathematical risk theory.* — Philadelphia: Huebner Foundation.

[128] Gerber H.U. (1984) Error bounds for the compound Poisson approximation. *Insurance Math. Econom.*, **3**(3), 191–194.

[129] Geske M.X., Godbole A.P., Schaffner A.A., Skolnick A.M., Wallstrom G.L. (1995) Compound Poisson approximations for word patterns under Markovian hypotheses. — *J. Appl. Probab.*, **32**(4), 877–892.

[130] Gibbs A.L. and Su F. (2002) On choosing and bounding probability metrics. — *Intern. Statist. Rev.*, **70**(3), 419–435.

[131] Gini C. (1914) Di una misura delle relazioni tra le graduatorie di due caratteri. — *In: Appendix to Hancini A. Le Elezioni Generali Politiche del 1913 nel Comune di Roma.* Rome: Ludovico Cecehini.

[132] Goldstein L. (2010) Bounds on the constant in the mean central limit theorem. - *Ann. Probab.* **38**(4), 1672–1689.

[133] Gnedenko B.V. (1938) Uber die konvergenz der verteilungsgesetze von summen voneinander unabhangiger summanden. — *C.R. Acad. Sci. URSS*, **18**, 231–234.

[134] Gnedenko B.V. (1944) Limit theorems for sums of independent random variables. — Uspehi Matem. Nauk **10**, 115–165. (Russian)

[135] Gnedenko B.V., Kolmogorov A.N. (1954) *Limit distributions for sums of independent random variables.* — Addison-Wesley: Cambridge, MA.

[136] Golikova Ya.S. (2021) On calculation of constants in the Arak inequalities for concentration functions of convolutions of probability distributions. — *Zap. Nauchn. Sem. POMI*, **486**, 86–97. Transl.: *J. Math. Sci.*, **258**(6), 793–801.

[137] Götze F. and Zaitsev A.Yu. (2004) Approximation of convolutions by accompanying laws without centering. — *Zap. Nauchn. Sem. POMI*, **320**, 44–53. Transl.: *J. Math. Sci.*, **137**(1), 4510–4515, 2006.

[138] Götze F., Eliseeva Yu.S. and Zaitsev A.Yu. (2017) Arak inequalities for concentration functions and the Littlewood-Offord problem. — *Teor. Veroyatn. Primen.*, **62**(2), 241–266 (In Russian) . Transl.: *Theory Probab. Appl.*, **62**(2), 196–215, 2018.

[139] Götze F. and Zaitsev A.Yu. (2017) Rare events and Poisson point processes. — *Zap. Nauchn. Sem. POMI*, **466**, 109–119 (Russian). Transl.: *J. Math. Sci.*, **244**(1), 771–778, 2020.

[140] Götze F., Zaitsev A.Yu and Zaporozhets D. (2019) An improved multivariate version of Kolmogorov's second uniform limit theorem. — *Zap. Nauchn. Sem. POMI*, **486**, 71–85 (Russian). Transl.: *J. Math. Sci.*, **258**, 782–792, 2021.

[141] Götze F. and Zaitsev A.Yu. (2021) Convergence to infinite-dimensional compound Poisson distributions on convex polyhedra. — *Zap. Nauchn. Sem. POMI*, **501**, 118–125 (Russian).

[142] Götze F. and Zaitsev A.Yu. (2022) On alternative approximating distributions in the multivariate version of Kolmogorov's second uniform limit theorem.— *Teor. Veroyatn. Primen.*, **67**(1), 3–22 (Russian) . Transl.: *Theory Probab. Appl.*, **67**(1), 1–16.

[143] Grigelionis B. (1962) On the asymptotic expansion of the remainder term in case of convergence to the Poisson law. (Russian) — *Litovsk. Mat. Sb.*, **2**(1), 35–48.

[144] Grigelionis B. (1999) Asymptotic expansions in the compound Poisson limit theorem. — *Acta Applicand. Math.*, **58**, 125–134.

[145] Gudynas P. (1995) On convergence to Poisson law for distributions of the sum of random variables connected into chain. — *Lith. Math. J.*, **35**(3), 234–242.

[146] Haight F.A. (1967) *Handbook of the Poisson distribution.* — New York: Wiley.

[147] van Harn K. (1978) Classifying infinitely divisible distributions by functional equations. — Mathematical Centre Tracts, **103** (Mathematical Centre, Amsterdam).

[148] Harremoës P. and Ruzankin P.S. (2004) Rate of convergence to Poisson law in terms of information divergence. — *IEEE Trans. Inform Theory*, **50**(9), 2145–2149.

[149] Heinrich L. (1982) A method for the derivation of Limit theorems for sums of m-dependent random variables. — *Z. Wahrscheinlichkeitstheorie verw. Gebiete*, **60**, 501–515.

[150] Herrmann H. (1965) Variationsabstand zwischen der Verteilung einer Summe unabhängiger nichtnegativer ganzzahliger Zufallsgrössen und Poissonschen Verteilungen. — *Math. Nachr.*, **29**(5), 265–289.

[151] Hipp C. (1985) Approximation of aggregate claims distributions by compound Poisson distributions. — *Insurance Math. Econom.*, **4**(4), 227–232. Correction: **6**, p. 165, 1987.

[152] Hipp C. (1986) Improved approximations for the aggregate claims distribution in the individual model. — *ASTIN Bull.*, **16**(2), 89–100.

[153] Hsing T., Hüsler J. and Leadbetter M.R. (1988) On the exceedance point process for stationary sequence. — *Probab. Theory Rel. Fields*, **78**, 97–112.

[154] Hsing T. (1988) On the extreme order statistics for a stationary sequence. — *Stochastic Processes Appl.* **29**, 155–169.

[155] Hsing T. (1991) Estimating the parameters of rare events. — *Stochastic Process. Appl.* **37**, 117–139.

[156] Hsing T. (1993) On some estimates based on sample behavior near high level excursions. — *Probab. Theory Rel. Fields*, **95**, 331–356.

[157] Hudson W.N., Tucker H.G. and Veeh J.A. (1989) Limit distributions of sums of m-dependent Bernoulli random variables. — *Probab. Theory Rel. Fields*, **82**, 9–17.

[158] Hsiau S.R. (1997) Compound Poisson limit theorems for Markov chains. — *J. Appl. Probab.*, **34**(1), 24–34.

[159] Hwang H.-K. (1999) Asymptotics of Poisson approximation to random discrete distributions: an analytic approach. — *Adv. in Appl. Probab.*, **31**(2), 448–491.

[160] Hwang H.-K. and Zacharovas V. (2010) Uniform asymptotics of Poisson approximation to the Poisson-Binomial distribution. — *Teor. Veroyatn. Primen.*, **55**(2), 305–334. Transl.: *Theory Probab. Appl.*, **55**(2), 198–224, 2011.
[161] Huang W.T. and Tsai C.S. (1991) On a modified binomial distribution of order k. — *Stat. Prob. Lett.*, **11**, 125–131.
[162] Ibragimov I.A. and Linnik Yu.V. (1965) *Independent and stationary dependent variables.* — Groningen: Wolters-Noordhoff Publishing.
[163] Ibragimov I.A. and Presman E.L. (1973) On the rate of approach of the distributions of sums of idependent random variables to accompanying distributions. — *Teor. Veroyatn. Primen.*, **18**(4), 753–766 (Russian). Transl.: *Theory Probab. Appl.*, **18**(4), 713–727.
[164] Ibragimov R. and Sharakhmetov S. (2001) The exact constant in the Rosenthal inequality for random variables with mean zero. — *Teor. Veroyatn. Primen.*, **46**(1), 134–138 (Russian). Transl.: *Theory Probab. Appl.*, **46**(1), 127–132, 2002.
[165] Ivchenko G.I. (1974) On comparison of Binomial and Poisson distributions. — *Teor. Veroyat. Primen.*, **19**(3), 612–625 (Russian). Transl.: *Theory Probab. Appl.*, **19**(3), 584–587.
[166] Jensen J.L. (2013) On a saddlepoint approximation to the Markov Binomial distribution. — *Braz. J. Probab. Stat.*, **27**(2), 150–161.
[167] Johnson N.L., Kemp A.W. and Kotz S. (2005) *Univariate discrete distributions.* — Wiley.
[168] Kallenberg O. (2002) *Foundations of modern probability.* New York: Springer.
[169] Kanter M. (1976) Probability inequalities for convex sets and multidimensional concentration functions. — *J. Multivariate Anal.*, **6**, 222–236.
[170] Kantorovich L.V. (1942) On the translocation of masses. — *Doklady USSR Acad. Sci.*, **37**(7-8), 227–229. (Russian) Transl.: *Management Sci.*, **5**(1), 1–4, 1958, *J. Math. Sci.*, **133**(4), 1381–1382.
[171] Karymov D.N. (2004) On the accuracy of approximation in the Poisson limit theorem. — *Discrete Math. Appl.*, **16**(2), 317–327.
[172] Kendall D.G. (1967) On finite and infinite sequences of exchangable events. — *Stud. Sci. Math. Hung.*, **2**, 319–327.
[173] Kennedy J.E. and Quine M.P. (1989) The total variation distance between the binomial and Poisson distributions. — *Ann. Probab.*, **17**, 396–400.
[174] Kerstan J. (1964) Verallgemeinerung eines satzes von Prochorow und Le Cam. — *Z. Wahrsch. Verw. Gebiete*, **2**, 173–179.
[175] Khintchine A.Y. (1933) *Asymptotische Gesetze der Wahrscheinlichkeitsrechnung. Ergebnisse der Mathematik und ihrer Grenzgebiete.* — Berlin: Springer.
[176] Khintchine A.Y. (1937) Zur Theorie der unbeschränkt teilbaren Verteilungsgesetze. — *Mat. Sbornik*, **2(44)**(1), 79–119.
[177] Khintchine A.Y. (1938) *Limit laws for sums of random variables.* — Moscow: ONTI. (In Russian)
[178] Kobus M. (1995) Generalized Poisson distributions as limits of sums for arrays of dependent random vectors. — *J. Multivariate Anal.*, **52**(2), 199–244.
[179] Kolmogorov A.N. (1956) Two uniform limit theorems for sums of independent random variables. — *Teor. Veroyatn. Primen.*, **1**(4), 426–436 (Russian). Transl.: *Theory Probab. Appl.*, **1**(4), 384–394.
[180] Kolmogorov A.N. (1963) On the approximation of distributions of sums of independent terms by infinitely divisible distributions. — *Trudy Moskov. Mat. Obshch*, **12**, 437–451 (Russian). Transl.: *Selected works of A.N. Kolmogorov*, vol. 2, 484–504, 1992.
[181] Kontoyiannis I., Harremoes P. and Johnson O.T. (2005) Entropy and the law of small numbers. — *IEEE Trans. Inform. Theory*, **51**(2), 466–472.
[182] Koopman B.O. (1950) A generalization of Poisson's distribution for Markoff chains. *Proc. Nat. Acad. Sci. USA*, **36**, 202–207.
[183] Kozulyaev P.A. (1939) Asymptotic analysis of a fundamental formula of probability theory. — *U. Zap. Moskov. Gos. Univ.* **15**, 179–182. (Russian)
[184] Kruglov V.M. (2004) On ineqalities for large deviations in the Bernoulli scheme.— *Teor. Veroyatn. Primen.*, **49**(4), 785–790, (Russian). Transl.: — *Theory Probab. Appl.*, **49**(4), 695–700, 2005.
[185] Kruopis J. (1986) Precision of approximations of the generalized Binomial distribution by convolutions of Poisson measures. — *Litovsk. Mat. Sb.*, **26**(1), 53–69 (Russian). Transl.: *Lith. Math. J.*, **26**(1), 37–49.
[186] Kruopis J. (1986) Approximations for distributions of sums of lattice random variables I. — *Litovsk. Mat. Sb.*, **26**(3), 455–467 (Russian). Transl.: *Lith. Math. J.*, **26**(3), 234–244.

[187] Kruopis J. and Čekanavičius V. (2014) Compound Poisson approximations for symmetric vectors. — *J. Multivariate Anal.*, **123**, 30–42.
[188] Last G., Penrose M. (2017) *Lectures on the Poisson process.* Cambridge University Press.
[189] Leadbetter M.R., Lindgren G. and Rootzen H. (1983) *Extremes and related properties of random sequences and processes.* — New York: Springer.
[190] Le Cam L. (1960) An approximation theorem for the Poison Binomial distribution. — *Pacif. J. Math.*, **10**(4), 1181–1197.
[191] Le Cam L. (1965) On the distribution of sums of independent random variables. — In: *Bernoulli, Bayes, Laplace (Anniversary volume).* Springer: Berlin, 179–202.
[192] Liaudanskaitė G. and Čekanavičius V. (2020) Approximations for sums of three-valued 1-dependent symmetric random variables. — *Nonlinear Analysis: Modelling and Control*, **25**(4), 675–691.
[193] Liao Z., Ma Y., Xia A. (2020) On Stein's factors for Poisson approximation in Wasserstein distance with non-linear transportation costs. — *J. Theor. Probab.*, **35**(4), 2383–2412. arXiv:2003.13976v1
[194] Liapounoff A.M. (1900) Sur une proposition de la théorie des probabilités. — Bull. de l'Acad. Imp. Sci. St.-Petersbourg, v. 13, No 4, 359–386.
[195] Liapounoff A.M. (1901) Nouvelle forme du théorème sur la limite des probabilités. — Mem. Acad. Imp. Sci. St. Peterburg, v. 12, 1–24.
[196] Lindvall T. (1992) *Lectures on the coupling method.* — New York: Wiley.
[197] Liu Q. and Xia A. (2020) On moderate deviations in Poisson approximation. — *J. Appl. Probab.*, **57**(3), 1005–1027.
[198] Logunov P.L. (1990) Estimates for the convergence rate to the Poisson distribution for random sums of independent indicators. — *Teor. Veroyatn. Primen.*, **35**(3), 580–582 (Russian). Transl.: *Theory Prob. Appl.*, **35**(3), 587–590, 1991.
[199] Loynes R.M. (1965) Extreme values in uniformly mixing stationary stochastic processes. — *Ann. Math. Statist.*, **36**(3), 993–999.
[200] Lundberg F. (1903) Approximerad Framställning av Sannolikhetsfunktionen. Aterförsäkring av Kollektivrisker. — Akad. Afhandling. Uppsala: Almqvist & Wiksell.
[201] Makabe H. and Morimura H. (1956) On the approximation to some limiting distributions. — *Kodai Math. Sem. Rep.*, **8**(1), 31–40.
[202] Major P. (1990) A note on the approximation of the uniform empirical process. — *Ann. Probab.* **18**(1), 129–139.
[203] Marcinkiewicz J. (1938) Sur les fonctions independantes II. — *Func. Mat.*, **30**(1), 349–364.
[204] Mattner L. and Roos B. (2007) A shorter proof of Kanter's Bessel function concentration bound. — *Probab. Theory Rel. Fields*, **139**, 191–205.
[205] Matthes K., Kerstan J. and Mecke J. (1978) *Infinitely divisible point processes.* — New York: Wiley.
[206] Meshalkin L.D. (1960) On the approximation of polynomial distributions by infinitely-divisible laws. — *Teor. Veroyatn. Primen.*, **5**(1), 114–124 (Russian). Transl.: *Theory Probab. Appl.*, **5**(1), 106–114.
[207] Michel R. (1987) An improved error bound for the compound Poisson approximation of a nearly homogeneous portfolio. — *ASTIN Bulletin*, **17**(2), 165–169.
[208] Mikhailov V.G. (1997) On a Poisson approximation for the distribution of the number of empty cells in a nonhomogeneous allocation scheme. — *Teor. Veroyatn. Primen.*, **42**(1), 184–189 (Russian). Transl.: *Theory Probab. Appl.*, **42**(1), 174–179, 1998.
[209] Mikhailov V.G. (2001) Estimate of the accuracy of the compound Poisson approximation for the distribution of the number of matching patterns. — *Teor. Veroyatn. Primen.*, **46**(4), 713–723 (Russian). Transl.: *Theory Probab. Appl.*, **46**(4), 667–675, 2002.
[210] Mikhailov V.G. (2008) A Poisson-type limit theorem for the number of pairs of matching sequences. — *Teor. Veroyatn. Primen.*, **53**(1), 59–71 (Russian). Transl.: *Theory Probab. Appl.*, **53**(1), 106–116, 2009.
[211] Minakov A.A. (2015) Compound Poisson approximation of the number distribution for monotone strings of fixed length in a random sequence. — *Prikl. Diskr. Mat.*, **28**(2), 21–29 (Russian).
[212] Minkova L. and Omey E. (2014) A new Markov Binomial distribution. — *Comm. Statist. Theory Methods*, **43**(13), 2674–2688.
[213] de Moivre A. (1712) De Mensura Sortis. – *Philosophical Transactions*, **27**, 213–264. Transl: Hald A. (1984) A. de Moivre: 'De Mensura Sortis' or 'On the measurement of chance'. — *International Statistical Review*, **52**(3), 229–262.

[214] Mori T.F. (1990) More on the waiting time till each of some given patterns occurs as a run. — *Canadian J. Math.*, **42**(5), 915–932.
[215] Nagaev S.V. and Khodzhabagyan S.S. (1996) On an estimate for the concentration function of sums of independent random variables. — *Teor. Veroyatn. Primen.*, **41**(3), 655–665 (Russian). Transl.: *Theory Probab. Appl.*, **41**(3), 560–569, 1997.
[216] Nagaev S.V. (1998) Concentration functions and the accuracy of approximation by infinitely divisible laws in a Hilbert space. — *Dokl. Akad. Nauk*, **359**(4), 461–463 (Russian).
[217] Nagaev S.V. (2021) On the accuracy of Poisson approximation to the binomial distribution. — *Mat. Trudy*, **24**(2), 122–149 (Russian).
[218] Nagaev S.V. and Chebotarev V. (2021) On Approximation of the tails of the Binomial distribution with these of the Poisson law. — *Mathematics*, **9**, 845–865.
[219] Neuhauser C. (1994) A Poisson approximation for sequence comparisons with insertions and deletions. — *Ann. Statist.*, **22**(3), 1603–1629.
[220] Novak S.Y. (1998) On the limiting distribution of extremes. — *Siberian Adv. Math.*, **8**(2), 70–95.
[221] Novak S.Y. (2011) *Extreme value methods with applications to finance.* — London: Chapman & Hall/CRC Press. ISBN 9781439835746
[222] Novak S.Y. (2017) On the length of the longest head run. — *Statist. Probab. Lett.*, **130**, 111–114.
[223] Novak S.Y. (2019) On the accuracy of Poisson approximation. — *Extremes*, **22**(4), 729–748.
[224] Novak S.Y. (2019) Poisson approximation. — *Probability Surveys*, **16**, 228–276; (2021) **18**, 272–275; v. 18, 272-275.
[225] Novak S.Y. (2021) Poisson approximation in terms of the Gini-Kantorovich distance. — *Extremes*, **24**(1), 67–84.
[226] Novak S.Y. (2024) On Poisson approximation. — *J. Theor. Probability.* https://doi.org/10.1007/s10959-023-01310-4
[227] Okamoto M. (1959) Some inequalities relating to the partial sum of binomial probabilities. — *Ann. Inst. Statist. Math.*, **10**(1), 29–35.
[228] Omey E., Santos J. and van Gulck S. (2008) A Markov-binomial distribution. — *Appl. Anal. Discrete Math.*, **2**(1), 38–50.
[229] Petrauskienė J. and Čekanavičius V. (2011) Compound Poisson approximations for sums of one-dependent random variables I. — *Lith. Math. J.*, **50**(3), 323–336; **51**(1), 51–65.
[230] Petrov V.V. (1970) On an estimate of the concentration function of a sum of independent random variables.— *Teor. Veroyatn. Primen.*, **15**(4), 718–721 (Russian). Transl.: *Theory Probab. Appl.*, **15**(4), 701–703, 1970.
[231] Petrov V.V. (1995) *Limit theorems of Probability Theory. Sequences of independent random variables.* — Clarendon Press, Oxford University Press: New York.
[232] Pianoforte F. and Schulte M. (2021) Poisson approximation with applications to stochastic geometry. — *Electron. J. Probab.* **26**, 1–36.
[233] Pinelis I. (2020) Monotonicity properties of the Poisson approximation to the binomial distribution, with applications to statistical testing. — *Statist. Probab. Lett.*, **167**.
[234] Pittel B.G. (1981) Limiting behavior of a process of runs. — *Ann. Probab.*, **9**(1), 119–129.
[235] Pósfai A. (2009) An extension of Mineka's coupling inequality. — *Electron. Commun. Probab.*,**14**, 464–473.
[236] Presman E.L. (1973) On a multidimensional version of the Kolmogorov uniform limit theorem. — *Teor. Veroyatn. Primen.*, **18**(2), 396–402 (Russian). Transl.: *Theory Probab. Appl.*, **18**(2), 378–384, 1974.
[237] Presman E.L. (1983) Approximation of Binomial distributions by infinitely divisible ones. — *Teor. Veroyatn. Primen.*, **28**(2), 372–382 (Russian). Transl.: *Theory Probab. Appl.*, **28**(2), 393–403, 1984.
[238] Presman E.L. (1985) Approximation in variation of the distribution of a sum of independent Bernoulli variables with a Poisson law. — *Teor. Veroyatn. Primen.*, **30**(2), 391–396 (Russian). Transl.: *Theory Probab. Appl.*, **30**(2), 417–422, 1986.
[239] Prokhorov Yu.V. (1952) *Limit theorems for sums of independent random variables.* (Russian) PhD/DSc Thesis. — Moscow: Steklov Institute of Mathematics, 42 pp.
[240] Prokhorov Yu.V. (1953) Asymptotic behavior of the Binomial distribution. — *Uspehi Matem. Nauk*, **8**(55), 135–142 (Russian). Transl.: *Select. Transl. Math. Statist. and Probability*, **1**, 87–95, 1961. Providence, R.I.: Inst. Math. Statist. & Amer. Math. Soc.
[241] Prokhorov Yu.V. (1955) On sums of identically distributed random variables. — *Dokl. Akad. Nauk SSSR (N.S.)*, **105**(4), 645–647 (Russian).

[242] Prokhorov Yu.V. (1956) Convergence of random processes and limit theorems in probability theory. — *Teor. Veroyatn. Primen.*, **1**(2), 177–238 (Russian). Transl.: *Theory Probab. Appl.*, **1**(2), 157–214.
[243] Prokhorov Yu.V. (1959) An extremal problem in Probability theory. — *Theory Probab. Appl.*, **4**(2), 2, 211–214. Transl.: *Theory Probab. Appl.*, **4**(2), 201–203.
[244] Prokhorov Yu.V. (1960) On a uniform limit theorem of A.N. Kolmogorov. — *Teor. Veroyatn. Primen.*, **5**(1), 103–113 (Russian). Transl.: *Theory Probab. Appl.*, **5**(1), 98–106.
[245] Prokhorov Yu.V. (2012) *Selected works*, v. 1. — Moscow: Torus Press, 775 pp. (Russian).
[246] Quine M.P. (1994) Probability approximations for divisible discrete distributions. — *Austral. J. Statist.*, **36**(3), 339–349.
[247] Rachev S.T. (1984) The Monge-Kantorovich mass transference problem and its stochastic applications. *Teor. Veroyatn. Primen.*, **29**(4), 625–653 (Russian). Transl.: *Theory Probab. Appl.*, **29**(4), 647–676, 1985.
[248] Rachev S.T., Stoyanov S.V. and Fabozzi F.J. (2011) *A Probability metric approach to Financial Risk measures.* — Wiley-Blackwell.
[249] Raikov D.A. (1938) On the decomposition of Gauss and Poisson laws. —*Izv. Akad. Nauk Armyan. SSR Ser. Mat.*, **2**, 91–124 (Russian).
[250] Robert C.Y. (2009) Inference for the limiting cluster size distribution of extreme values. — *Ann. Statist.* **37**(1), 271–310.
[251] Rogozin B.A. (1961) On evaluating the concentration functions. — *Teor. Veroyatn. Primen.*, **6**(1), 103–105. Transl.: Theory Probab. Appl., **6**(1), 94–97.
[252] Röllin A. (2005) Approximation of sums of conditionally independent variables by the translated Poisson distribution. — *Bernoulli*, **11**(6), 1115–1128.
[253] Röllin A. (2007) Translated Poisson approximation using exchangeable pair coupling. — *Ann. Appl. Probab.*, **17**(5-6), 1596–1614.
[254] Romanowska M. (1977) A note on the upper bound for the distribution in total variation between the Binomial and the Poisson distribution. — *Stat. Neerl.*, **31**(3), 127–130.
[255] Roos M. (1994) Stein's method for compound Poisson approximation: the local approach. — *Ann. Appl. Probab.*, **4**(4), 1177–1187.
[256] Roos B. (1999) Asymptotic and sharp bounds in the Poisson approximation to the Poisson-Binomial distribution. — *Bernoulli*, **5**(6), 1021–1034.
[257] Roos B. (1999) On the rate of multivariate Poisson convergence. — *J. Multivar. Analysis*, **69**, 120–134.
[258] Roos B. (2000) Binomial approximation to the Poisson binomial distribution: the Krawtchouk expansion— *Teor. Veroyatn. Primen.*, **45**(2), 328–344. Transl.: *Theory Probab. Appl.*, **45**(2), 258–272, 2001.
[259] Roos B. (2001) Multinomial and Krawtchouk approximations to the generalized multinomial distribution. —*Teor. Veroyatn. Primen.*, **46**(1), 117–133. Transl.: *Theory Probab. Appl.* **46**(1), 103–117, 2002.
[260] Roos B. (2001) Sharp constants in the Poisson approximation. — *Statist. Probab. Lett.*, **52**, 155–168.
[261] Roos B. (2002) Kerstan's method in the multivariate Poisson approximation: an expansion in the exponent. — *Teor. Veroyatn. Primen.*, **43**(2), 397–402. Transl.: *Theory Probab. Appl.*, **47**(2), 358–363, 2003.
[262] Roos B. (2003) Improvements in the Poisson approximation of mixed Poisson distributions. — *J. Statist. Plann. Inference*, **113**(2), 467–483.
[263] Roos B. (2003) Poisson approximation via the convolution with Kornya-Presman signed measures. — *Teor. Veroyatn. Primen.*, **48**(3), 628–632. Transl.: *Theor. Probab. Appl.*, **48**(3), 555–560, 2004.
[264] Roos B. (2005) On Hipp's compound Poisson approximations via concentration functions. — *Bernoulli*, **11**(3), 533–557.
[265] Roos B. (2007) On variational bounds in the compound Poisson approximation of the individual risk model. — *Insurance Math. Econom.*, **40**(3), 403–414.
[266] Roos B. (2017) Refined total variation bounds in the multivariate and compound Poisson approximation. — *ALEA, Lat. Am. J. Probab. Math. Stat.*, **14**, 337–360.
[267] Rootzén H. (1988) Maxima and exceedances of stationary Markov processes. — *Adv. Appl. Probab.*, **20**(2), 371–390.
[268] Ruzankin P.S. (2001) On the Poisson approximation of the Binomial distribution. — *Sibirsk. Mat. Zh.*, **42**(2), 414–424 (Russian). Transl.: *Sib. Math. J.*, **42**(2), 353–363.
[269] Ruzankin P.S. (2004) On the rate of Poisson process approximation to a Bernoulli process. — *J. Appl. Probab.*, **41**(1), 271–276.

[270] Ruzankin P.S. (2010) Approximation for expectations of unbounded functions of dependent integer-valued random variables. — *J. Appl. Probab.*, **47**(2), 594–600.

[271] Ruzankin P.S., Borisov I.S. (2020) On the rate of Poisson approximation to partial sum processes. — Statist. Probab. Letters, v. 162. 108754. https://doi.org/10.1016/j.spl.2020.108754

[272] Salvemini T. (1943) Sul calcolo degli indici di concordanza tra due caratteri quantitativi. — *Atti della VI Riunione della Soc. Ital. di Statistica.*

[273] Sason I. (2013) Improved lower bounds on the total variation distance for the Poisson approximation. — *Statist. Probab. Lett.*, **83**, 2422–2431.

[274] Schbath S. (2000) An overview on the distribution of word counts in Markov chains. — *J. Comput. Biology*, **7**(1-2), 193–201.

[275] Serfling R.J. (1975) A general Poisson approximation theorem. — *Ann. Probab.*, **3**(4), 726–731.

[276] Serfling R.J. (1978) Some elementary results on Poisson approximation in a sequence of Bernoulli trials. — *SIAM Rev.*, **20**(3), 567–579.

[277] Sevastyanov B.A. (1973) Poisson limit law for a scheme of sums of dependent random variables. — *Teor. Veroyatn. Primen.*, **17**(4), 733–738 (Russian). Transl.: *Theory Probab. Appl.*, **17**(4), 695–699.

[278] Shevtsova I.G. (2013) On absolute constants in the Berry-Esseen-type inequalities for identically distributed summands. (Russian) — *Inform. Primen.*, **7**(1), 124–125. Transl.: *arXiv:1111.6554v1.*

[279] Shorgin S.Y. (1977) Approximation of a generalized Binomial distribution. — *Teor. Veroyatn. Primen.*, **22**(4), 867–871 (Russian). Transl.: *Theory Probab. Appl.*, **22**(4), 846–850, 1978.

[280] Shiryaev A.N. (2016) *Probability-1.*(third. ed.) — Springer-Verlag New York.

[281] Šiaulys J. and Čekanavičius V. (1988) Approximation of distributions of integer-valued additive functions by discrete charges I. — *Litovsk. Mat. Sb.*, **28**(4), 795–810 (Russian). Transl.: *Lith. Math. J.*, **28**(4), 392–401.

[282] Simons G. and Johnson N.L. (1971) On the convergence of Binomial to Poisson distributions. — *Ann. Math. Statist.*, **42**(5), 1735–1736.

[283] Skellam J.G. (1946) The frequency distribution of the difference between two Poisson variates belonging to different populations. — *J. R. Statist. Soc. (N.S.)*, **109**, 296.

[284] Šliogerė J. and Čekanavičius V. (2015) Two limit theorems for Markov binomial distribution. — *Lith. Math. J.*, **55**(3), 451–463.

[285] Šliogerė J. and Čekanavičius V. (2016) Approximation of symmetric three-state Markov chain by compound Poisson law. — *Lith. Math. J.*, **56**(3), 417–438.

[286] Smith R.L. (1988) Extreme value theory for dependent sequences via the Stein–Chen method of Poisson approximation. — *Stochastic Process. Appl.*, **30**(2), 317–327.

[287] Statulevičius V. (1965) Limit theorems for densities and asymptotic expansions for distributions of sums of independent random variables. — *Teor. Veroyatn. Primen.*, **10**(4), 546–659 (Russian). Transl.: *Theory Probab. Appl.*, **10**(4), 582–595.

[288] Stein C. (1992) A way of using auxiliary randomization. — In: *Probability Theory.* Proc. Singapore Probab. Conf., 159–180. Berlin: de Gruyter.

[289] Studnev Yu.P. (1960) An approximation to the distribution of sums by infinitely divisible laws. — *Teor. Veroyatn. Primen.*, **5**(4), 465–469 (Russian). Transl.: *Theory Probab. Appl.*, **5**(4), 421–426.

[290] Takeuchi K. and Takemura A. (1987) On sum of 0-1 random variables I. Univariate case. — *Ann. Inst. Statist. Math.*, **39**(1), 85–102.

[291] Tsaregradskii I.P. (1958) On uniform approximation of the Binomial distribution by infinitely divisible laws. — *Teor. Veroyatn. Primen.*, **3**(4), 470–474 (Russian). Transl.: *Theory Probab. Appl.*, **3**(4), 434–438.

[292] Tsybakov A.B. (2009) *Introduction to nonparametric estimation.* — Springer.

[293] Tyurin I.S. (2011) A refinement of the remainder in the Lyapunov theorem. —— *Teor. Veroyatn. Primen.*, **56**(4), 808–811 (Russian). Transl.: *Theory Probab. Appl.*, **56**(4), 693–696, 2012.

[294] Tyurin I.S. (2012) Some optimal bounds in the central limit theorem using zero biasing. — *Statist. Probab. Lett.*, **82**, 514–518.

[295] Upadhye N.S (2009) Compound negative binomial approximation to sums of random variables. — *Phd. Thesis, Indian Institute of Technology Bombay.*

[296] Upadhye N.S. and Kumar A.N. (2018) Pseudo-Binomial approximation to (k_1, k_2)-runs. — *Statis. Probab. Lett.* **141**, 19–30.

[297] Upadhye N.S., Vellaisamy P. and Čekanavičius, V. (2017) On Stein operators for discrete approximations. *Bernoulli*, **23**(4A), 2828–2859.

[298] Uspensky J.V. (1931) On Ch.Jordan's series for probability. — *Ann. Math.*, **32**(2), 306–312.
[299] Uspensky I.V. (1937) *Introduction to Mathematical Probability.* — New York: McGraw-Hill.
[300] Utev S.A. (1985) Extremal problems in moment inequalities. — *Trudy Inst. Mat. Sib. Otd. AN SSSR*, **5**, 56–75 (Russian).
[301] Utev S.A. (1992) *Extremal problems, characterisation, and limit theorems of Probability Theory.* — DSc Thesis. Novosibirsk: Novosibirsk Inst. Math., 280 pp. (Russian).
[302] Vaitkus P. and Čekanavičius (1998) On centered Poisson approximation. — *Lith. Math. J.*, **38**(4), 391–404.
[303] Vallender S.S. (1973) Calculation of the Wasserstein distance between probability distributions on the line. — *Teorya Veroyatn. Primen.*, **18**(4), 824–827 (Russian). Transl.: *Theory Probab. Appl.*, **18**(4), 784–786, 824–827, 1974.
[304] Vatutin V.A. and Mikhailov V.G. (1982) Limit theorems for the number of empty cells in an equiprobable scheme for group allocation of particles. — *Teorya Veroyatn. Primen.*, **27**(4), 684–692 (Russian). Transl.: *Theory Probab. Appl.*, **27**(4), 734–743, 1984.
[305] Vellaisamy P. (2004) Poisson approximation for $(k_1; k_2)$ events via the Stein-Chen method. — *J. Appl. Probab.*. **41**(4), 1081–1092.
[306] Vellaisamy P. and Čekanavičius V. (2014) Nonuniform approximations for sums of discrete m-dependent random variables. — In: *Contemporary Developments in Statistical Theory.* A Festschrift for Hira Lal Koul, 375–393.
[307] Vellaisamy P., Upadhye N.S. and Čekanavičius V. (2012) On negative Binomial approximation. — *Teorya Veroyatn. Primen.*, **57**(1), 141–154. Reprinted in *Theory Probab. Appl.*, **57**(1), 97–109, 2013.
[308] Vellaisamy P. and Čekanavičius V. (2018) Infinitely divisible approximations for sums of m-dependent random variables. — *J. Theor. Probab.*, **31**(4), 2432–2445.
[309] Vervaat W. (1969) Upper bounds for the distance in total variation between the Binomial or negative Binomial and the Poisson distribution. — *Stat. Neerl.*, **23**(1), 79–86.
[310] Volkonskiy V.A. and Rozanov Yu.A. (1959) Some limit theorems for random functions I. — *Teorya Veroyatn. Primen.*, **4**(2), 186–207 (Russian). Transl.: *Theory Probab. Appl.*, **4**(2), 178–197.
[311] Wang Y.H. (1981) On the limit of the Markov Binomial distribution. — *J. Appl. Probab.*, **18**(4), 937–942.
[312] Wang Y.H. (1991) A compound Poisson convergence theorem. — *Ann. Probab.*, **19**(1), 452–455.
[313] Wang Y.H., Chang H.-F. and Chen S.-Y. (2003) Convergence theorems for the lengths of consecutive successes of Markov Bernoulli sequences. — *J. Appl. Probab.*, **14**(3), 741–749.
[314] Wang Y.H. and Tang L. (2003) Poisson style convergence theorems for additive processes defined on Markov chains. — *Statistica Sinica*, **13**, 227–242.
[315] Wang X. and Xia A. (2008) On Negative Binomial Approximation to k-Runs. — *J. Appl. Probab.*, **45**(2), 456–471.
[316] Witte H.-J. (1990) A unification of some approaches to Poisson approximation. — *J. Appl. Probab.*, **27**(3), 611–621.
[317] Xia A. (1997) On using the first difference in the Stein–Chen method. — *Ann. Appl. Probab.*, **7**(4), 899–916.
[318] Xia A. and Zhang M. (2009) On approximation of Markov Binomial distributions. — *Bernoulli*, **15**(4), 1335–1350.
[319] Yakshyavichus Sh. (1997) On a method of expansion of the probabilities of lattice random variables. — *Teor. Veroyatn. Primen.*, **42**(2), 294–307 (Russian). Transl.: *Theory Probab. Appl.*, **42**(2), 271–282.
[320] Yannaros N. (1991) Poisson approximation for random sums of Bernoulli random variables. — *Statist. Probab. Lett.*, **11**(2), 161–165.
[321] Zacharovas V. and Hwang H.-K. (2010) A Charlier-Parseval approach to Poisson approximation and its applications. — *Lith. Math. J.*, **50**(1), 88–119.
[322] Zaitsev A.Yu. (1981) Some properties of n-fold convolutions of distributions. — *Teor. Veroyatn. Primen.*, **26**(1), 152–156 (Russian). Transl.: *Theory Probab. Appl.*, **26**(1), 148–152.
[323] Zaitsev A.Yu. and Arak T. (1983) On the rate of convergence in the second Kolmogorov's uniform limit theorem. — *Teor. Veroyatn. Primen.*, **28**(2), 333–353. Transl.: *Theory Probab. Appl.*, **28**(2), 351–374, 1984.
[324] Zaitsev A.Yu. (1983) On the accuracy of approximation of distributions of sums of independent random variables, which are non-zero with a small probability, by accompanying laws. — *Teor. Veroyatn. Primen.*, **28**(4), 625–636 (Russian). Transl.: *Theory Probab. Appl.*, **28**(4), 657–669, 1984.

[325] Zaitsev A.Yu. (1987) On the uniform approximation of distributions of sums of independent random variables. — *Teor. Veroyatn. Primen.*, **32**(1), 45–52 (Russian). Transl.: *Theory Probab. Appl.*, **32**(1), 40–47, 1988.
[326] Zaitsev A.Yu. (1988) Estimates for the closeness of successive convolutions of multidimensional symmetric distributions. — *Probab. Theory Rel. Fields*, **79**, 175–200.
[327] Zaitsev A.Yu. (1989) Multidimensional version of the second uniform limit theorem of Kolmogorov. — *Teor. Veroyatn. Primen.*, **34**(1), 128–151 (Russian). Transl.: *Theory Prob. Appl.*, **34**(1), 108–128, 1990.
[328] Zaitsev A.Yu.(1989) Approximation of convolutions of multi-dimensional symmetric distributions by accompanying laws. — *Zap. Nauchn. Sem. LOMI AN SSSR*, **177**, 55–72 (Russian). Transl.: *J. Soviet Mathematics*, **61**(1), 1859–1872, 1992.
[329] Zaitsev A.Y. (1990) Certain class of nonuniform estimates in multidimensional limit theorems. — *Zap. Nauchn. Sem. POMI*, **184**, 92–100 (Russian). Transl.: (1994), *J. Math. Sci.*, **68**, 459–468, 1994.
[330] Zaitsev A.Yu. (1991) An example of a distribution whose set of n-fold convolutions is uniformly separated from the set of infinitely divisible laws in distance in variation. — *Teor. Veroyatn. Primen.*, **36**(2), 356–361 (Russian). Transl.: *Theory Probab. Appl.*, **36**(2), 419–425, 1992.
[331] Zaitsev A.Yu. (1996) Approximation of convolutions by accompanying laws under the existence of moments of low orders. — *Zap. Nauchn. Sem. POMI*, **228**, 135–141 (Russian). Transl.: *J. Math. Sci.*, **93**(3), 336–340, 1999.
[332] Zaitsev A.Yu. (2003) Approximation of a sample by a Poisson point process. — *Zap. Nauchn. Sem. POMI*, **298**, 111–125 (Russian). Transl.: *J. Math. Sci.*, **128**(1), 2556–2563, 2005.
[333] Zhang M. (2011) Approximation for counts of head runs. — *Sci. China Math.*, **54**(2), 311–324.
[334] Zubkov A.M. and Mikhailov V.G. (1979) Repetitions of s-tuples in a sequence of independent trials. — *Teor. Veroyatn. Primen.*, **24**(2), 267–279 (Russian). Transl.: *Theory Probab. Appl.*, **24**(2), 269–282, 1980.
[335] Zubkov A.M. and Serov A.A. (2012) A complete proof of universal inequalities for the distribution function of the Binomial law. — *Teor. Veroyatn. Primen.*, **57**(3), 597–602 (Russian). Transl.: *Theory Probab. Appl.*, **57**(3), 539–544, 2013.

Index